Developmental Genetics and Plant Evolution

The Systematics Association Special Volume Series

Series Editor

Alan Warren
Department of Zoology, The Natural History Museum,
Cromwell Road, London, SW7 5BD, UK.

The Systematics Association provides a forum for discussing systematic problems and integrating new information from genetics, ecology and other specific fields into taxonomic concepts and activities. It has achieved great success since the Association was founded in 1937 by promoting major meetings covering all areas of biology and palaeontology, supporting systematic research and training courses through the award of grants, production of a membership newsletter and publication of review volumes by its publishers Taylor & Francis. Its membership is open to both amateur and professional scientists in all branches of biology who are entitled to purchase its volumes at a discounted price.

The first of the Systematics Association's publications, *The New Systematics*, edited by its then president Sir Julian Huxley, was a classic work. Over 50 volumes have now been published in the Association's 'Special Volume' series often in rapidly expanding areas of science where a modern synthesis is required. Its *modus operandi* is to encourage leading exponents to organise a symposium with a view to publishing a multi-authored volume in its series based upon the meeting. The Association also publishes volumes that are not linked to meetings in its 'Volume' series.

Anyone wishing to know more about the Systematics Association and its publications are invited to contact the series editor.

Forthcoming titles in the series:

Telling the Evolutionary Time: Molecular Clocks and the Fossil Record
P. C. J. Donoghue and M. P. Smith

Milestones in Systematics
D. M. Williams and P. L. Forey

Other Systematics Association publications are listed after the index for this volume.

The Systematics Association Special Volume Series 65

Developmental Genetics and Plant Evolution

Edited by Quentin C. B. Cronk
Richard M. Bateman and
Julie A. Hawkins

With the assistance of Lynn M. Sanders

CRC Press
Taylor & Francis Group
Boca Raton London New York

CRC Press is an imprint of the
Taylor & Francis Group, an **informa** business

First published 2002 by Taylor & Francis
11 New Fetter Lane, London EC4P 4EE

Simultaneously published in the USA and Canada
by Taylor & Francis Inc,
29 West 35th Street, New York, NY 10001

Taylor & Francis is an imprint of the Taylor & Francis Group

Every effort has been made to ensure that the advice and information in this book is true and accurate at the
time of going to press. However, neither the publisher nor the authors can accept any legal responsibility or lia-
bility for any errors or omissions that may be made. In the case of drug administration, any medical procedure
or the use of technical equipment mentioned within this book, you are strongly advised to consult the manu-
facturer's guidelines.

British Library Cataloguing in Publication Data
A catalogue record for this book is available from the British Library

Library of Congress Cataloging in Publication Data
A catalog record has been requested.

ISBN 0-415-25790-5 (hbk)
ISBN 0-415-25791-3 (pbk)

Contents

Contributors

Abbott, R. J., Division of Environmental and Evolutionary Biology, Harold Mitchell Building, School of Biology, University of St Andrews, St Andrews, Fife, KY16 9TH, UK.

Ainsworth, C., Department of Biology, Imperial College at Wye, Wye, Ashford, Kent, TN25 5AH, UK.

Albert, V. A., Natural History Museums and Botanical Gardens, University of Oslo, Sars' Gate 1, N-0562, Oslo, Norway.

Bateman, R. M., Department of Botany, Natural History Museum, Cromwell Road, London, SW7 5BD, UK.

Baum, D. A., Department of Organismic and Evolutionary Biology, Harvard University, 22 Divinity Avenue, Cambridge, MA 02138, USA. (Current address: Department of Botany, University of Wisconsin, 430 Lincoln Drive, Madison, WI 53706, USA.)

Becker, A., Max-Planck-Institut für Züchtungsforschung, Abteilung Molekulare Pflanzengenetik, Carl-von-Linné-Weg 10, D-50829 Köln, Germany. (Present address: Friedrich-Schiller-Universität, Lehrstuhl für Genetik, Philosophenweg 12, D-07743 Jena, Germany.)

Buzgo, M., Institute of Systematic Botany, University of Zürich, Zollikerstrasse 107, CH-8008 Zürich, Switzerland.

Coen, E. S., Department of Genetics, John Innes Centre, Colney Lane, Norwich, NR4 7UH, UK.

Corley, S. B., Department of Plant Sciences, University of Oxford, South Parks Road, Oxford, OX1 3RB, UK.

Cox, E. J., Department of Botany, The Natural History Museum, Cromwell Road, London, SW7 5BD, UK.

Cranfill, R., University Herbarium, University of California, Berkeley, CA 94720-2465, USA.

Cronk, Q. C. B., University of Edinburgh, Institute of Cell and Molecular Biology, King's Buildings, Mayfield Road, Edinburgh, EH9 3JH, *and* Royal Botanic Garden, 20A Inverleith Row, Edinburgh, EH3 5LR, UK.

Cubas, P., Centro Nacional de Biotecnología, Departamento de Genetica Molecular de Plantas, Centro Universidad Autonoma de Madrid, Cantoblanco, 28049, Madrid, Spain.

DiMichele, W. A., Department of Paleobiology, National Museum of Natural History, Smithsonian Institution, Washington, DC 20560, USA.

Donoghue, M. J., Department of Ecology and Evolutionary Biology, Yale University, P.O. Box 208106, New Haven, CT 06520, USA.

Doust, A. N., Department of Biology, University of Missouri-St Louis, 8001 Natural Bridge Road, St Louis, MO 63130, USA.

Elomaa, P., Department of Applied Biology, P.O. Box 27, FIN-00014 University of Helsinki, Finland.

Frederiksen, S., Botanical Institute, Gothersgade 140, DK-1123, Copenhagen K, Denmark.

Frohlich, M. W., Herbarium, University of Michigan, NUBS Bldg, 1205 North University Avenue, Ann Arbor, MI 48109-1057, USA. (Present address: Department of Botany, The Natural History Museum, Cromwell Road, London, SW7 5BD, UK.)

Gillies, A. C. M., Division of Environmental and Evolutionary Biology, Harold Mitchell Building, School of Biology, University of St Andrews, St Andrews, Fife, KY16 9TH, UK.

Gleissberg, S., Institut für Spezielle Botanik, Universität Mainz, Bentzelweg 9a, D-55099 Mainz, Germany.

Glover, B. J., Department of Plant Sciences, University of Cambridge, Downing Street, Cambridge, CB2 3EA, UK.

Hake, S., Department of Plant and Microbial Biology, 111 Koshland Hall, University of California, Berkeley, USA.

Hämäläinen, J., Department of Applied Biology, P.O. Box 27, FIN-00014 University of Helsinki, Finland.

Harrison, C. J., Institute of Cell and Molecular Biology, University of Edinburgh, King's Buildings, Mayfield Road, Edinburgh, EH9 3JH, UK and Royal Botanic Garden, 20A Inverleith Row, Edinburgh, EH3 5LR, UK.

Hawkins, J. A., Centre for Plant Diversity and Systematics, School of Plant Sciences, University of Reading, Reading, RG6 6AS, UK.

Hay, A., Department of Plant and Microbial Biology, 111 Koshland Hall, University of California, Berkeley, CA, USA.

Henderson, I., John Innes Centre, Norwich Research Park, Colney Lane, Norwich, NR4 7UH, UK.

Holtan, H., Department of Plant and Microbial Biology, 111 Koshland Hall, University of California, Berkeley, CA, USA.

Hudson, A., Institute of Cell and Molecular Biology, University of Edinburgh, King's Buildings, Mayfield Road, Edinburgh, EH9 3JH, UK.

Johansen, B., Botanical Institute, Gotersgade 140, DK-1123, Copenhagen K, Denmark.

Kaur, H., Department of Plant Sciences, University of Oxford, South Parks Road, Oxford, OX1 3RB, UK.

Kellogg, E. A., Department of Biology, University of Missouri-St Louis, 8001 Natural Bridge Road, St Louis, MO 63121, USA.

Kenrick, P., Department of Palaeontology, The Natural History Museum, Cromwell Road, London, SW7 5BD, UK.

Kirchner, C., Maiwald Patentanwalts. GmbH Elisenhof, Elisenstr. 3, D-80335 München, Germany.

Knapp, S., Department of Botany, The Natural History Museum, Cromwell Road, London, SW7 5BD, UK.

Kotilainen, M., Department of Applied Biology, P.O. Box 27, FIN-00014 University of Helsinki, Finland.

Langdale, J. A., Department of Plant Sciences, University of Oxford, South Parks Road, Oxford, OX1 3RB, UK.

Martin, C., Department of Genetics, John Innes Centre, Norwich Research Park, Colney Lane, Norwich, NR4 7UH, UK.

McCormick, S., Department of Plant and Microbial Biology, 111 Koshland Hall, University of California, Berkeley, USA.

McLellan, T., School of Molecular and Cell Biology, University of the Witwatersrand, Private Bag 4, WITS 2050, South Africa.

Münster, T., Max-Planck-Institut für Züchtungsforschung, Abteilung Molekulare Pflanzengenetik, Carl-von-Linné-Weg 10, D-50829 Köln, Germany.

Ori, N., Department of Plant and Microbial Biology, 111 Koshland Hall, University of California, Berkeley, USA.

Pöllänen, E., Department of Applied Biology, P.O. Box 27, FIN-00014 University of Helsinki, Finland.

Pryer, K. M., Department of Biology, Duke University, Durham, NC 27708-0338, USA.

Rudall, P. J., Jodrell Laboratory, Royal Botanic Gardens Kew, Richmond, Surrey, TW9 3DS, UK.

Saedler, H., Max-Planck-Institut für Züchtungsforschung, Abteilung Molekulare Pflanzengenetik, Carl-von-Linné-Weg 10, D-50829 Köln, Germany.

Schneider, H., Department of Biology, Duke University, Durham, NC 27708, USA.

Scotland, R. W., Department of Plant Sciences, University of Oxford, South Parks Road, Oxford, OX1 3RB, UK.

Shephard, H. L., Department of Biology, Imperial College at Wye, Wye, Ashford, Kent, TN25 5AH, UK.

Smith, A. R., University Herbarium, University of California, Berkeley, CA 94720-2465, USA.

Teeri, T. H., Department of Applied Biology, P.O. Box 27, FIN-00014 University of Helsinki, Finland.

Theißen, G., Max-Planck-Institut für Züchtungsforschung, Abteilung Molekulare Pflanzengenetik, Carl-von-Linné-Weg 10, D-50829 Köln, Germany. (Present address: Friedrich-Schiller-Universität, Lehrstuhl für Genetik, Philosophenweg 12, D-07743 Jena, Germany.)

Tsiantis, M., Department of Plant Sciences, University of Oxford, South Parks Road, Oxford, OX1 3RB, UK.

Uimari, A., Department of Applied Biology, P.O. Box 27, FIN-00014 University of Helsinki, Finland.

Walbot, V., Department of Biological Sciences MC5020, 385 Serra Mall, Stanford University, Stanford, CA 94305-5020, USA.

Winter, K.-U., Max-Planck-Institut für Züchtungsforschung, Abteilung Molekulare Pflanzengenetik, Carl-von-Linné-Weg 10, D-50829 Köln, Germany. (Present address: GE CompuNet, Geschäftsstelle Köln, Industriestrasse 161e, D-50999 Köln, Germany.)

Wolf, P. G., Department of Biology, Utah State University, Logan, UT 84322-5305, USA.

Preface

Context

Biology is currently the most rapidly developing of all the sciences, reflecting the fact that disciplines with a long and rich history, such as organismal systematics and the often mathematically-based allelic approaches to evolutionary theory, have been joined by increasingly diverse and vibrant molecular sciences. Within this sphere of biology, evolutionary–developmental genetics (colloquially, 'evo-devo') is arguably among the most rapidly growing. Its status as an increasingly influential discipline is due in part to the range of biologists participating, since the discipline seeks common ground and synergy between molecular biologists, developmental geneticists and process morphologists. Yet, despite all the recent hyperbole, most biologists of all persuasions have only a hazy idea of the nature of evo-devo, how it relates to other biological disciplines, what potential it really offers for understanding evolution, and how close it is to realising that potential.

There has been a long tradition of conceptual trends in both evolutionary theory and developmental genetics being set by zoologists, typically those specialising in a very narrow range of model animals such as humans, mice, fruit-flies and nematode worms. However, in recent years biological coverage in the mass media has been dominated not by animals but by plants; specifically, by genetically modified crop plants, the most tangible products of the recent revolution in molecular genetics. The phenomenal economic potential of such crops (and, more recently, the environmental and ethical issues that they raise) has promoted an exponential increase in plant- (mostly angiosperm-) related molecular genetics and genomics research, and that has begun to revitalise studies of plant growth and development. When this synergy is in turn linked to an increasing recognition of the value of comparative approaches, driven primarily by the phylogenetic revolution of the last two decades, the stage is set for a thorough revision of plant evolutionary biology.

This volume is the proceedings of an international conference *Developmental Genetics and Plant Evolution* (DGPE), organised by the editors and sponsored by the UK Systematics Association and the Linnean Society of London, with the support of the Natural History Museum, London. The conference took place on 20–22 September 2000 in the Flett Theatre of the Natural History Museum, and was attended by 167 delegates from 16 countries. Our primary aim was to build on the excitement of the now famous plant evo-devo conference held at Taos, New Mexico in 1992 (R. Chasan, 1993, *Plant Cell*, 5: 363) while demonstrating (ultimately in print) the wealth of data that have accumulated, and the many conceptual advances that have been made, during the subsequent eight years of this exponentially growing field.

Indeed, the DGPE conference was held in the final year of the millennium, which was notable for the unveiling of the first draft (a draft of remarkably high quality) of the complete genome of a plant (*Arabidopsis* Genome Initiative, 2000, *Nature*, 408: 796). Such advances are highly significant, as they promise to propel the research programme outlined in this volume forward at a dizzying speed. Given the difficulty of determining how a clock works by studying a small percentage of its cogs and springs in isolation, gaining access to the whole mechanism is undoubtedly a major advance, even if it is not immediately interpretable.

It is always exciting in science when disparate phenomena, previously studied in isolation, are cross-linked into a more unified discipline. Systematics (the study of the hierarchical nesting of phenotypic pattern), molecular phylogenetics (the study of the evolutionary history of genomic components, often to infer organismal history), morphology (the study of developmental pattern), developmental genetics (the study of developmental process), population genetics and ecological genetics (the study of the fate of alleles in the environment), are all disciplines that have at their core some form of heritable variation. However, technical difficulties have until now prevented correlation of allelic variation with specific nucleotide sequences and phenotypes, limiting the ranges of both analysable species and analysable genes.

Structure

Fortunately, recent technical advances mean that, at least in theory, the same homologous string of nucleotides can be studied in a systematic, developmental and ecological context, thus permitting the 'eco-molecular evolutionary synthesis' predicted by Cronk (Chapter 1). The subsequent six chapters (Chapters 2–7) emphasise theoretical and general issues, beginning with the generation of genetic diversity by transposable element activity (Walbot). Homology is the basis of comparative evolutionary science, and two contributions (Hawkins; Baum and Donoghue) examine from different perspectives how the square peg of morphological homology can be driven into the round hole of molecular homology. Although it is widely recognised that evolution occurs by the origin and fixation of new alleles, the nature of these alleles and the processes leading to fixation are more contentious. Kellogg considers whether micro- and macro-evolution are, in fact, the same phenomenon viewed through opposite ends of the phylogenetic telescope, or whether they may derive from distinct and different processes. Bateman and DiMichele explore macro-evolution and adaptive versus non-adaptive evolutionary change in the context of the generation and ecological filtration of saltational phenotypes. Frohlich considers the question of whether testable hypotheses can be formulated at the molecular level to explain major morphological transitions in plant evolution. His own 'mostly male theory' arguably comes closest so far to achieving this laudable goal.

The middle block of contributions (Chapters 8–15) addresses more specific issues concerning the evolution of flowers and inflorescences. Theißen *et al.* present a lucid overview of the MADS-box genes that have been fundamental to the evolution of reproductive structures in seed plants. Johansen and Fredericksen and Teeri *et al.*

examine the role of MADS-box genes in flowers with specialised morphology (orchids and composites, respectively). This is essential if we are to assess the applicability of the *Arabidopsis/Antirrhinum*-derived ABC model and determine what is general and what is specific in that model. Characterisation of *SEPALLATA* genes has confirmed suspicions that floral development has further genetic complexity, even in *Arabidopsis*. Apart from the meristic and cyclic aspects of floral development, dorsiventral patterning (zygomorphy) is immensely important, having a major role in modulating plant–pollinator interactions; some of the largest flowering plant families are predominantly zygomorphic (Orchidaceae, Asteraceae, Scrophulariaceae, Fabaceae = Leguminosae).

Gillies *et al.* and Cubas review current knowledge of dorsiventral patterning in the Asteraceae and the Lamiales respectively. Nowhere is the richness of plant diversity reflected better than in the variety of floral morphology that exists even within closely related groups. Knapp and Doust and Kellogg guide the reader through the flowers of Solanaceae and Poaceae, respectively, and question the nature of the underlying molecular basis. The contribution of Glover and Martin is notable, as their experimental system allows direct estimation of the adaptive significance of a particular gene as it affects pollinator choice via the micromorphology of the petals, thus linking developmental genetics with ecological genetics.

The subsequent eight contributions (Chapters 16–23) explore the vegetative structures of land plants. McLellan *et al.* discuss the application and interpretation of QTL approaches, notably in their study of the diversity of leaf size and shape. Schneider *et al.* consider innovations important in the evolution of land plants, and along with Kenrick offer neobotanical and palaeobotanical morphological perspectives on leaf–shoot divergence. Langdale *et al.* and Harrison *et al.* summarise our knowledge of the developmental genetics of leaves in an evolutionary context of the origin of microphyllous and megaphyllous leaves. Similarly, Gleissberg and Tsiantis *et al.* examine angiosperm leaf diversity, and show how developmental genetic approaches will contribute to our understanding of the evolution of the compound leaf. Rudall and Buzgo conclude the vegetative section by illustrating and reinterpreting the morphological and evolutionary complexity involved in the diversification of monocot leaves.

Completing the empirical contributions, Cox (Chapter 24) reviews the remarkable range of developmental diversity that can be present even in unicellular organisms such as diatoms, raising methodological questions about how the new synthesis can be extended to groups about which we have little genomic information. The final contribution, by Baum (Chapter 25), draws together several methodological threads and suggests a framework of approaches for the immediate future.

Although plant evolutionary–developmental genetics is maturing rapidly, it is nonetheless a discipline with two contrasting terminological domains. Therefore, for the convenience of readers from either (or neither) side of the fence, a reasonably comprehensive glossary and a brief account of standard notations for genes and gene products are provided immediately prior to the taxon and subject indexes.

Cross-cutting themes

Survey of the 25 chapters (the work of 58 authors) reveals several cross-cutting themes that merit brief mention.

Most of the chapters take as their starting point phylogenies, using the trees as explicit comparative frameworks for discussing molecular and morphological (including physiological) transitions, and recognising that one of the key challenges is identifying direct links between genotypic and phenotypic transitions. Various kinds of phylogenetic constraint are widely inferred. Several contributions use these phylogenies to attempt to identify phenotypic key innovations (e.g. Kellogg, Frohlich, Bateman and DiMichele, Theißen *et al.*), some authors also relating the putative innovations to co-evolution (Glover and Martin, Knapp).

The generation of allelic novelty is considered especially carefully by Walbot, but subsequent emphasis may either treat the genetic novelty and its bearer as immediately synonymous (Bateman and DiMichele) or more conventionally as requiring population-level fixation to achieve evolutionary significance via adaptation (Glover and Martin, Knapp). Although under-explored in the volume, the importance of ecological filters in winnowing phenotypic variation is noted in both experimental (Glover and Martin) and natural (Cronk, Bateman and DiMichele, Knapp) contexts.

Many authors interpret key evolutionary transitions in the context of pattern-based terminology, most then attempting to bridge the gap between pattern and process. Most frequent among these concepts are heterochrony and heterotopy (including homeosis and transfer of function), usefully redefined by Baum and Donoghue and placed in a teratological context by Bateman and DiMichele, who echo Kellogg in suggesting that heterotopy generally causes more profound morphogenetic shifts than heterochrony.

The use of evo-devo data for hypothesis testing is emphasised, both for assessing homology (e.g. Hawkins, Rudall and Buzgo, Cox) and for the assessment of putative sequences of acquisition of sets of phenotypic characters (e.g. Frohlich), leading for example to contrasting perspectives on the palaeobotanically-inspired telome theory (cf. Langdale *et al.*, Kenrick, Schneider *et al.*, Harrison *et al.*). Most of these authors invoke at least some fossil data, but there is considerable variation in the degree to which these authors view plants as holistic versus modular entities.

The distinction made by McLellan *et al.* between primary genes that are the direct cause of a phenotypic shift and secondary genes that are expressed downstream has been readily adopted by several authors. Many authors refer to the putative primary genes under investigation as candidate genes, and different approaches to identifying and exploring the actions of these genes are contrasted. Quantitative trait locus (QTL) surveys (McLellan *et al.*, Doust and Kellogg), *in situ* transcript localisation approaches (Johansen and Frederiksen), and the so far under-explored array-based genomic comparisons (Cronk, Baum) are contrasted (cf. Soltis *et al.* and Baum *et al.*, 2002, *Tr. Plant Sci.*, 7: 22).

As yet, the number of primary genes sampled for study that are understood in any detail is remarkably small. The well-known ABC genes of floral whorl control (e.g. Theißen *et al.*, Johansen and Frederiksen) belong to the well-studied class of MADS-box genes. Among the best-understood non-MADS-box genes are *KNOX* (knotted-like homeobox) genes of vegetative expression (Gleissberg, Tsiantis *et al.*), and the multi-purpose *CYCLOIDEA* gene family (Gillies *et al.*, Cubas). Similarly, the range of taxa that have legitimate claims to being viewed as 'models' remains relatively small (e.g. *Arabidopsis, Antirrhinum, Zea*), though this volume reports progress in the use of several relative newcomers such as Asteraceae, Orchidaceae, Solanaceae, the grass *Setaria* and diatoms. This should help to negate the previously phylogenetically clustered and morphologically unrepresentative taxon sampling.

Summary

In commissioning and compiling these 25 chapters from leading practitioners of the discipline we have encouraged uniformity of presentation and terminology, and have where possible enhanced connectedness. Nonetheless, this work remains a tapestry of many disparate threads; much diversity of methodological approach and interpretative opinion persists. We view this diversity as more of a strength than a weakness, as we believe that the discipline of evo-devo has reached a stage of maturity where it is increasingly subject to a range of selection pressures and continued diversity will be needed if further evolution is to proceed with the requisite speed. We hope that the study of evolution and the genome will not merely be 'ships that pass in the night' (Kreitman and Gaasterland, 2001, *Curr. Opin. Genet. Dev.*, 11, 609), but rather that this volume will promote further productive dialogue and exchange.

Acknowledgements

We conclude this Preface by warmly thanking the conference co-ordinator, Lynn Sanders (NHM), upon whose shoulders fell a large part of the burden of the organisation of the meeting and the subsequent assembly of materials for this volume. Thanks are also due to Paula Rudall for much additional editorial work, John Galey for compiling the indexes, and Dilys Alam, Grant Soanes and their team at Taylor & Francis, together with Claire Dunstan and Carl Gillingham at Wearset Pre-press Book Services, for the speed with which they expedited publication once the copy-edited manuscripts had been delivered to them. Finally, we thank our institutional supporters, the Systematics Association (UK), the Linnean Society of London and the Natural History Museum, London, all organisations possessing a distinguished history of supporting scholarly and innovative discourse.

Quentin Cronk[1], Richard Bateman[2], Julie Hawkins[3] (September 2001)

[1] ICMB, Edinburgh University/Royal Botanic Garden, Edinburgh, UK
[2] Department of Botany, Natural History Museum, London, UK
[3] School of Plant Sciences, University of Reading, UK

Plates

Plate 1

(a–b) Comparison of (a) a typical flower of the moth-pollinated orchid *Platanthera chloran-tha* showing its long, nectiferous spur and well-differentiated sepals, lateral petals and labellum, with (b) the flower of an adjacent teratological individual showing pseudopeloria: the petals of the normal flower, including the labellum and spur, have been replaced by three additional sepal-like structures (Bateman and DiMichele, **Chapter 7**).

(c–d) Comparison of zygomorphic, wild-type (left) and actinomorphic, peloric (right) flowers of *Linaria vulgaris* (c) and *Saintpaulia* cultivar (d) (Cubas, **Chapter 13**).

Plate 2

(a) Down-regulation of *GRCD1* causes a homeotic transformation in the marginal female flowers of the *Gerbera* capitulum; in the control plant (left) development of stamens initiates on whorl 3 but subsequently arrests, whereas in the *GRCD1* antisense transformant line (right) narrow petals (arrowed) replace withered staminodes (Teeri et al., **Chapter 11**).

(b) *Streptocarpus wendlandii* has escaped stem–leaf developmental canalisation to generate this 'unifoliate' morphology; the giant leaf is one of the two cotyledons, continuing to grow throughout the life of the plant, and generating inforescences in the absence of a shoot apical meristem (Cronk, **Chapter 1**: Photo, Debbie White).

(c–d) *Lathyrus sylvestris* (c), a papilionoid legume showing the typical zygomorphic flower with strong differentiation among the petals, contrasted with *Cadia purpurea* (d), a legume from Yemen phylogenetically nested within the papilionoids but showing a rare reversion to fully actinomorphic flowers (Cronk, **Chapter 1**: Photo, Debbie White).

Plate 3

(a) Three exceptional *Ginkgo* leaves bearing several ectopic ovules in distal sinuses of leaves; those at left and middle have each formed one seed (Frohlich, **Chapter 6**).

(b–c) Strongly zygomorphic perianth of *Schizanthus gilliesii* (b) contrasted with the more subtly zygomorphic androecium of *Hyoscyamus niger* (c); both are members of the more typically actinomorphic family Solanaceae (Knapp, **Chapter 14**).

(d) Leaf–stem junction in a monocot (the ginger *Zingiber officinale*) showing a prominent ligule at the junction of sheath and blade (Rudall and Buzgo, **Chapter 23**).

(e–f) *Arabidopsis* mutants with altered margin (*am*) phenotypes display inappropriate *KNAT1* expression in the tips of leaf serrations identified by the blue precipitate; wild type (e) is compared with *aml* mutant (f) (Tsiantis et al., **Chapter 22**).

Plate I

Plate 2

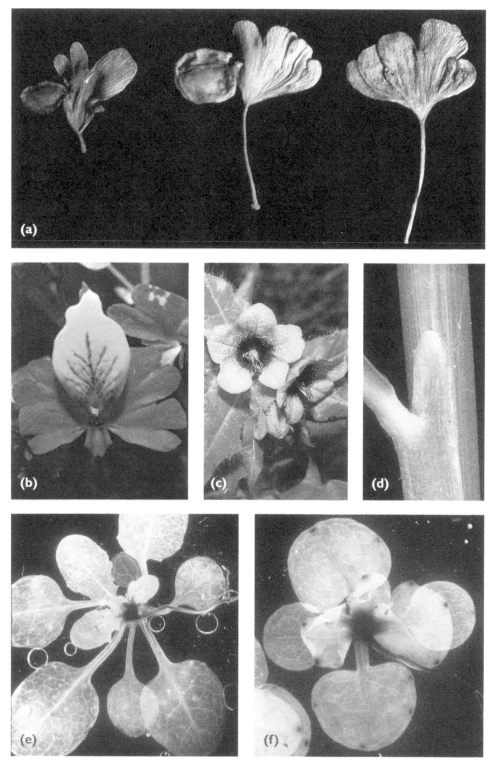

Plate 3

Chapter 1

Perspectives and paradigms in plant evo-devo

Quentin C. B. Cronk

ABSTRACT

Morphological innovations of land plants (such as leaves, roots and flowers) have profoundly changed the terrestrial biosphere. Plant innovations have affected global geochemistry, hydrology, atmospheric composition and the evolution of other biota. The study of plant evolution can therefore be viewed as a study of the way in which nucleotide changes have affected the functioning of the planet. These innovations result from alterations in the original developmental pathways responsible for the relatively simple construction of early land plants, about which we know little, as most of our current knowledge concerns the highly derived development of *Arabidopsis*. However, even in angiosperms many of the most important phenotypic characters are of unknown molecular developmental origin. With several completely sequenced plant genomes imminent, there is a real prospect of obtaining a new evolutionary synthesis that combines insights from molecular biology, developmental biology and ecology.

1.1 Darwin, Fisher and evo-devo: changing perspectives

> No doubt it is a very surprising fact that characters should reappear after having been lost for many, probably for hundreds of generations. But when a breed has been crossed only once by some other breed, the offspring occasionally show for many generations a tendency to revert in character to the foreign breed, – some say, for a dozen or even a score of generations. After twelve generations, the proportion of blood, to use a common expression, from one ancestor, is only 1 in 2048; and yet, as we see, it is generally believed that a tendency to reversion is retained by this remnant of foreign blood.
>
> Charles Darwin, *On the Origin of Species* (1859)

> If two plants which differ constantly in one or several characters be crossed, numerous experiments have demonstrated that the common characters are transmitted unchanged to the hybrids and their progeny; but each pair of differentiating characters, on the other hand, unite in the hybrid to form a new character, which in the progeny of the hybrid is usually variable. The object of the experiment was to observe these variations in the case of each pair of

In *Developmental Genetics and Plant Evolution* (2002) (eds Q. C. B. Cronk, R. M. Bateman and J. A. Hawkins), Taylor & Francis, London, pp. 1–14.

differentiating characters, and to deduce the law according to which they appear in successive generations.

Gregor Mendel, *Experiments in Plant Hybridization [Versuche huber Pflanzenhybriden]* (1865, translated 1965)

Darwin's theory of natural selection, introduced in 1859, proposed a mechanism of evolution at the organismal level. As Darwin had no acceptable theory of genetics, natural selection foundered and was bitterly opposed in the early twentieth century. The metaphor of 'blood' for the material of inheritance in English (see the extract above), with its connotations of liquid blending, was too great an intellectual hurdle for Darwin's thought to cross to a particulate theory of inheritance. Mendel, on the other hand, while having a clear view of particulate inheritance, was not sympathetic to Darwinism (Bishop, 1996).

However, the rediscovery of Mendelian (i.e. particulate) rules of inheritance by De Vries and others finally allowed natural selection to shift to the gene level. R. A. Fisher (1890–1962) brilliantly re-established natural selection as the dominant evolutionary mechanism by showing mathematically its potency in acting on allele frequencies. The paradigm of natural selection acting at the gene level led to the 'modern synthesis' and has lasted to the present day.

Now, however, another major paradigm shift is in prospect as the study of natural selection shifts to the nucleotide level, emphasising in particular the selective effect of nucleotide changes in the cis-regulatory elements of transcription factors and other regulatory genes. Such changes, leading to altered specificity of DNA–protein binding, may in turn lead to heterotopic and heterochronic changes in the expression domains of key regulators. It is likely that, over the next few years, evidence will emerge about which are the dominant genomic channels for driving evolutionary (including adaptive) changes in the phenotype.

As the coding sequences of developmental genes are highly conserved, they are unlikely to be sources of organismal diversity. Instead, regions which control the expression of such genes (e.g. cis-regulatory elements) are possible candidates. Transcription factors are important controllers of gene expression in developmental pathways, so mechanisms which control the activity of these 'controllers' are potent sources of evolutionary variation. The idea that polymorphism in the cis-regulatory regions of transcription factors is a major source for morphological evolution derives (in a plant context at least) from the work of John Doebley and co-workers (Doebley and Lukens, 1998; reviewed by Cronk, 2001). This 'Doebley hypothesis' is testable and should be a stimulus for future work. In addition to cis-regulation by promoter elements, transcription factors can also be regulated by phosphorylation/methylation, ligand binding, heterodimerisation between transcription factors and by accessory factors mediating DNA binding or transcription (Calkhoven, 1996).

The advances of the last 150 years have bridged the gap between Darwin and Mendel and gone much further. Mendel's original particles of inheritance in pea are already known at the nucleotide level (Bhattacharya *et al.*, 1990; Fincham, 1990; Lester *et al.*, 1997). The future holds a prospect of a much more complete understanding of evolution, from nucleotide to organism. Complex connections between genotype and phenotype will be elucidated by molecular developmentalists.

Nucleotide sequences of genes can be used to design probes for mRNA expression (Johansen and Frederiksen, 2002); we can therefore connect genes to development in a direct way by examining where particular genes are expressed. We can also construct phylogenies of the same genes (Theißen *et al.*, 2002) to give a comprehensive picture of development and evolution.

1.2 Epigenetics

A challenge to this simplistic approach, however, comes from the strong epigenetic effects that are known to influence the link between gene and phenotype (Waddington, 1957). It is obvious from alternation of generations in plants that it is difficult to predict the phenotype from the genotype. In bryophytes and pteridophytes, for instance, a very different organism results from development of a spore and a zygote, yet apart from the haploid/diploid difference the genotypes are identical; what differs is their epigenotype. Epigenetics, in Waddington's sense, refers to the expression pattern of genes, which will depend on the genetic background, or epigenotype (epigenetics has more recently come to be used more specifically for the control of gene expression by methylation, an unfortunate terminological constraint). Microarray technology will provide a powerful tool for examining the epigenotype but it is possible that genetic changes may lead to minimal, poorly predictable or even chaotic changes in the phenotype because of the non-linearity of epigenetic effects.

Waddington conceived a metaphor for this link between genotype and phenotype, suggesting that development was controlled by an 'epigenetic landscape', with the phenotype represented by two-dimensional spatial co-ordinates, and the vertical dimension indicating the probability of transition from one phenotype to another during development. In his metaphor the organism was perceived as a ball rolling around on this surface; obviously it would be unlikely to occupy all of the landscape but would tend to run down the valleys, thereby canalising development. Waddington imagined the shape of the developmental landscape resulting from an underlying gene network, holding up the undulating phenotypic surface as a series of interlacing struts. Evolution changes the contours of this landscape and so results in organisms progressing down different developmental trajectories. This thought-experiment is somewhat laboured, but it makes certain testable predictions. The most important of these, from an evolutionary point of view, is that some developmental trajectories will be strongly canalised (steep-sided valleys formed by the underlying gene network) and will not be able to evolve readily. Changes in individual genes will be ineffective in lowering the sides of the 'valleys' so that evolutionary transitions may be made from one developmental trajectory to the next.

Is there any evidence for evolutionary inertia caused by the epigenotype? Certainly there are some features that are very stable in evolution, such as the papilionoid flower of most legumes, whose basic ground-plan has been maintained during the diversification of the group into many thousands of species. However, there are exceptions among the legumes, such as the flower of the genistoid *Cadia* which is a pendent bell. Tucker (2000) considers the 'chaotic' and variable floral organs in *Ceratonia* to result from a breakdown of developmental canalisation. Another example is the conservatism of the leaf–stem dichotomy in plants, with stems being indeterminate organs and leaves being determinate. However, the genus *Streptocarpus* (Plate 2b) has evolved many species with indeterminate leaves: leaves which produce further leaves and flowers without a vegetative stem being formed (Möller and Cronk, 2001; Harrison *et al.*, 2002). What is not clear is whether these examples of conservatism (i.e. conserved ground-plan) are due to epigenetic canalisation (developmental constraint) or adaptive canalisation (fitness constraint). It is also possible to think of an adaptive landscape, where certain phenotypes occupy adaptive peaks; consequently, any deviations will be eliminated by selection. As the science of genomics matures into epigenomics, and we understand more about the selective pressures on development, at least some answers to the epigenetic canalisation versus adaptive canalisation issue should be forthcoming.

1.3 Synthesis

1.3.1 Integration of paradigms

Developmental biology and selection at the nucleotide level are the two process-based phenomena that complement the two pattern-based disciplines of systematics and morphology. Any future biodiversity science will be incomplete without an integration of all four disciplines.

In later editions of the *Origin*, Darwin added a glossary prepared by W. S. Dallas, which gave a central role to developmental biology in distinguishing homology (in the broad sense) from analogy: 'Homology: That relation between parts which results from their development from corresponding embryonic parts.' The study of development is now needed more than ever to help elucidate the homology of morphological characters (for detailed discussion see Hawkins, 2002). However, only by studying such homologies in the context of phylogenetic trees can strict homology (taxic homology or synapomorphy) be distinguished from convergent evolution or homoplasy. It is the recognition of taxic homology versus homoplasy that allows evolutionary and ecological hypotheses (scenarios) of character evolution to be explicitly formulated. This deeper understanding feeds into the study of development and the cycle repeats. However, investigating the relationship between evolution and development requires the recognition of some key integrative concepts (homeosis, heterochrony, heterotopy, phylogenetic prepatterning, gain of function, loss of function), as set out below.

1.3.2 Homeosis

The conversion of one organ into another is potentially an important evolutionary phenomenon (Bateman and DiMichele, 2002; Baum and Donoghue, 2002). The simplest model for this at the molecular level is the loss of expression of genes responsible for particular organ identity, so that these organs take on the identity of adjacent primordia. Loss-of-function mutations of MADS-box genes controlling flower morphology are of this type (Theißen *et al.*, 2002). Another example comes from peloric flowers (Cubas, 2002). *Cyc/dich* double mutants of *Antirrhinum* lack the adaxial action of these genes so all primordia are abaxialised (Luo *et al.*, 1999). What is needed now are good examples of homeosis in driving evolutionary (as opposed to teratologic) morphological innovation. *Cadia* (Fabaceae) is a relatively derived legume which does not have the usual strongly zygomorphic pea-flower but instead has a derived radial flower (Plate 2c, d). This may be the result of conversion of wing and keel petals homeotically into standard petals, as in the legume mutant *Clitoria ternatea* var. *pleniflora*, which has five standards and, consequently, radial symmetry. However, it may alternatively be the result of heterochrony (S. Tucker, pers. comm. 2001), as early stages of flower development are radially symmetrical in Fabaceae.

1.3.3 Heterochrony

Heterochronic evolution (the change in timing of developmental processes) has repeatedly been demonstrated (e.g. Kellogg, 2002) to be an important (perhaps the *most* important) feature of plant evolution. Bateman (1994) gives a useful detailed review of the concepts as applied to plants. At a molecular level it is easy to see how selection on the cis-regulatory elements of transcription factors could alter the timing of expression during development and hence of the developmental processes driven by the relevant transcription factor. The early expression of otherwise late-expressed genes may lead to paedomorphosis: the expression of late developmental features in an otherwise juvenile state. For example, Friedman and Carmichael (1998) found that the reduced form of the female gametophyte in *Gnetum* resulted from paedomorphosis. Because spatial relations change with time, some heterochronic changes will lead to spatial changes within an organ, thus relating heterochrony and heterotopy (Baum and Donoghue, 2002), but relatively minor spatial changes resulting from heterochrony can and should be distinguished from changes in topological relations resulting from heterotopy.

1.3.4 Heterotopy

Although heterotopy, a topological change in the expression of a character, was coined by Haeckel (1866) as a counterpart of heterochrony, it appears to have proved less fruitful in evolutionary thought. This may be because heterotopic change is rarer than heterochronic change: if so there may be a molecular explanation. Heterotopic change can readily be induced artificially: a dramatic and well-known example is the formation of ectopic eyes by expression of the vertebrate gene *PAX6* in *Drosophila*, or by targeted mis-expression of its *Drosophila* homologue, *EYELESS* (Halder *et al.*, 1995). At a molecular level, a substantial change in the promoter of a gene that controls the induction of a complex trait, which then becomes regulated by an alternative transcription factor with a topologically different expression pattern, could cause heterotopy. Such major change in a promoter, putting a developmental cascade under the control of a different transcriptional regulator, is likely to be much rarer than a relatively minor change in a cis-regulatory element required to alter the timing of gene expression and thus bring about heterochronic change. Whatever the cause, topological relations tend to be conservative, and for this reason the topological criterion is useful for assessing homology. This indicates that heterotopic change happens rarely (or at least is rarely established: Bateman and DiMichele, 2002). However, although rare, heterotopic change in evolution may nonetheless be significant, as it allows evolutionary novelty. Schneider *et al.* (2002) give a range of examples in their survey of vascular plant evolution. Frohlich (2002) postulates a heterotopic expression of ovules on male cones as a key event in the evolution of the angiosperm flower. In another example, Tsiantis *et al.* (2002) showed that *KNOX* gene expression, which in *Arabidopsis* is restricted to the apical meristem, is expressed in the compound leaves of tomato. This is a true heterotopic change, with the character conferred by the gene expression (indeterminacy) appearing in a new organ. However, changes in the spatial distribution of KNOX gene expression within leaves, which varies among species, may also result from heterochronic changes.

1.3.5 Phylogenetic prepatterning

In the vegetable kingdom we have a case of analogous variation, in the enlarged stems, or as commonly called roots, of the Swedish turnip and Ruta-baga, plants which several botanists rank as varieties produced by cultivation from a common parent: if this be not so, the case will then be one of analogous variation in two so-called distinct species; and to these a third may be added, namely, the common turnip. According to the ordinary view of each species having been independently created, we should have to attribute this similarity in the enlarged stems of these three plants, not to the *vera causa* of community of descent, and a consequent tendency to vary in a like manner, but to three separate yet closely related acts of creation. Many similar cases of analogous variation have been observed by Naudin in the great gourd family, and by various authors in our cereals.

Charles Darwin, *On the Origin of Species* (1859) (last sentence added in later editions)

In plant evolution, perhaps more so than in animal evolution, homoplasy is rife, and when such characters evolve in a particular group they can do so repeatedly, coming 'not as single spies but in battalions'. An interesting question is whether there might be underlying homologies which permit the repeated origin of a character, thus raising the possibility that there is a cryptic molecular basis for particular homoplastic traits: a 'latent homology' or 'underlying synapomorphy'.

Nodulation, the formation of nitrogen-fixing tumour-like growths on roots, is a very important aspect of the agricultural and natural nitrogen cycles. It is also very complex developmentally, so our intuition would tell us that it is unlikely to have evolved very often. However, even within the Fabaceae, nodulation, following the same complex developmental paths, has evolved three times: in faboids, mimosoids and in the caesalpinoid *Chamaecrista* (Doyle, 1994, 1998). How can it be that such a complex trait is homoplasious?

Nodulation is associated with the formation of nodulins (proteins involved in nodulation) in the roots. Novel nodulins may be derived by recombination or duplication events and therefore they may be found to be synapomorphies for the evolution of nodulation in a particular group. Existing genes may be 'pirated' to new functions, so although there may be no homology at the level of nodulation, there may be homology of control genes defining them as 'nodulins'. Of particular note are the leghaemoglobins (Lb) which are 'late' nodulins (genes switched on after nodules have formed). They are responsible for the red colour of cut nodules exposed to air. Lb facilitates oxygen diffusion to plant tissues, while its strong affinity for oxygen keeps oxygen concentration sufficiently low for the oxygen-sensitive bacterial nitrogenases to function in the highly reducing nitrogen-fixation process. Nodulation is an extremely complex phenomenon and therefore difficult to dissect in terms of separate and sequential character evolution. Both plant and bacterial characters are involved. The plant characters include flavonoid signalling molecules which attract the bacteria, early nodulins such as peribacterioid membrane proteins and late nodulins such as leghaemoglobins. The bacterial characters include various exopolysacchride nod-factors. We can only assume that the Fabaceae have various pre-adaptations (developmental 'prepatterns') which facilitate the formation of symbioses with free-living rhizobia. This would explain the high levels of convergence among the independent events. It would also suggest that these homoplastic events are homologous at the level of the prepattern.

Furthermore, other families of the eurosid 1 clade, to which Fabaceae belongs, have evolved nodulation independently many times (Swenson, 1996). Therefore, although nodulation is homoplastic, there may be a synapomorphy of the rosids which predisposes this group to evolve nodulation in parallel. The taxonomic distribution of nodulation in eurosid 1 is as follows.

Rhizobial nodules: Fabaceae (various), Ulmaceae (*Parasponia*); Actinomycete (*Frankia*) nodules: Betulaceae (*Alnus*), Casuarinaceae (*Allocasuarina, Casuarina, Ceuthostoma, Gymnostoma*), Coriariaceae (*Coriaria*), Datiscaceae (*Datisca*), Eleagnaceae (*Eleagnus, Hippophae, Shepherdia*), Myricaceae (*Comptonia, Myrica*), Rhamnaceae (*Ceanothus, Colletia, Discaria, Kentrothamnus, Retanilla, Talguenea, Trevoa*), Rosaceae (*Cercocarpus, Chamaebatia, Cowania, Dryas, Purshia*).

The restriction of nodulation to the eurosid 1 clade strongly suggests that there is a prepattern at the level of this clade as well as a legume (rhizobial) prepattern. Unfortunately, the different nature of the actinorhizal and the rhizobial nodule has been an impediment to detecting common patterns, particularly as rhizobial nodules have a stem-like vasculature and *Frankia* nodules have a root-like vasculature (Sprent, 2001). However, Gualtieri and Bisseling (2000) have pointed out that the first stage of nodulation in both rhizobial and actinorhizal forms is the formation of a pre-nodule. It is the second stage, the infection vascularisation phase, that differs between the two groups. Thus, it may be capacity for pre-nodule formation that is the prepattern characteristic of the eurosid 1 clade.

Another case of prepatterning is found in floral zygomorphy, which has been independently evolved in (for instance) Fabales and in various asterid groups, including the Lamiales (Donoghue *et al.*, 1998). However, the gene *CYCLOIDEA*, which plays a large part in the establishment of zygomorphy in *Antirrhinum*, has orthologues in most dicots so far examined (Cubas, 2002). Furthermore, even the non-zygomorphic plant *Arabidopsis* has a florally expressed orthologue (*TCP1*) which is asymmetrically expressed in floral meristems. This asymmetrical expression is therefore a prepattern which is likely to facilitate the multiple origin of zygomorphy (Cubas *et al.*, 2001).

The evolution of heterospory was one of the most important events in the evolutionary history of land plants (Cronk, 2001), leading ultimately to the seed (see also Schneider *et al.*, 2002). However, it evolved not just once but many times, prompting

Bateman and DiMichele (1994) to call it 'the most iterative key innovation in the history of the plant kingdom'. Given the perspective of prepatterning, we would expect a genetic background at the level of the homosporous vascular plants which is permissive of the evolution of sexual differentiation and differential spore growth (possibly the two are developmentally linked) under suitable ecological conditions. The nature of this putative heterospory prepattern is completely unknown. We might seek it in homosporous *Equisetum*, which shows functional heterospory in some species (Duckett and Duckett, 1980), and whose fossil relatives *Archaeocalamites* (Bateman, 1991) and *Calamocarpon* were heterosporous (*Calamocarpon* very strongly so).

1.3.6 Gain of function

Evolutionary novelty is caused by genetic changes which may be recessive, following the loss of a previous character, or dominant, resulting from a gain of function. Gain-of-function mutations may be produced by (1) spatial or temporal increase in the expression domain of a gene, (2) the involvement of a gene in an extra pathway, or (3) the evolution of alternative splicing of the same gene. Many such changes probably result from changes in cis-regulatory elements of genes. Accumulated gains of function are one reason why humans (an estimated 35,000 genes) can possess scarcely more genes than a nematode (*c.* 20,000 genes) yet be vastly more complex at the morphological level.

Gene duplication is another powerful means for gains of function in evolution. Cronk (2001) argued that apparent 'gene redundancy' is likely to be illusory. If there is no negative selective consequence to losing a gene, it is likely to be removed quickly by the mutation background and become a pseudogene. If, however, the first mutations in an identical gene (initially masked from strong selection by the other locus) result in an advantageous gain of function, then both loci will be maintained in the population by selection. It should be noted that this gain of function may be very subtle, and if it is only revealed under stress conditions, for instance, its existence will not be obvious from greenhouse experiments. Evolution by duplicational gain may have been very important in plant evolution, accentuated by the high frequency of polyploidy in plants, caused by regular segmental duplication events. As such mutations are dominant they are exposed to selection immediately. The duplication of *KNOX* genes at the base of the land plants may have permitted the evolution of complex organisation. Plants, including mosses but none of the algae so far examined, have two major groups of *KNOX* genes, resulting from an ancient duplication. One of these groups is of central importance for the maintenance of apical meristems.

1.3.7 Loss of function

In contrast to gain-of-function mutations, loss-of-function mutations may be important in plant evolution allowing the deletion of developmental pathways and hence 'reversionary' evolution. The only problem with such loss-of-function mutations is that they are recessive and will not be exposed to selection until segregation produces homozygotes. A good example of a loss-of-function mutation of evolutionary significance in the wild is the change of *Arabidopsis thaliana* (a winter annual requiring vernalisation) to summer annual behaviour in the warmer climate of the Cape Verde Islands. The Cape Verde population has acquired a deletion at the 3-prime end of the gene which inactivates the vernalisation process and hence eliminates the physiological requirement for vernalisation (Johanson *et al.*, 2000) that is deleterious in an aseasonal equatorial environment.

1.4 Ecological integration

The link between genotype and phenotype is, in plants particularly, mediated by the environment. Organismic development in modular iterative organisms is a continual process of feedback between genotype, phenotype and environment, as environmental variables are transduced by signalling cascades which trigger activation of developmental and growth control genes. Phylogenetic, rather than ontogenetic, changes in phenotype are also controlled by ecological context but as feedback, driven by natural selection on nucleotide polymorphisms in evolutionary time. So far, molecular genetics has been slow to explore the ecological dimension and this needs to be addressed in future evo-devo studies.

The intimate association between evolutionary innovations and ecology can be illustrated by the leaf, which originated multiple times in land plants (Cronk, 2001). Although the leaf evolved as a result of nucleotide changes driven by ecological factors, the appearance of the leaf had profound ecological consequences of its own.

It is difficult now to imagine a world without leaves, but the first vegetative fossils of land plants (from the Silurian) are of leafless and rootless dichotomising axes: the Rhyniophytes (Kenrick, 2002; Langdale *et al.*, 2002). Given ample light and low competition leaves were not necessary and the developmental mechanisms for creating them were not present. Later, in the more competitive environment of the Devonian, a wide variety of leaf and leaf-like structures evolved, increasing the efficiency of light capture. Through geological history the evolutionary and phenotypic plasticity of leaves have enabled plants to adapt to global change, particularly changes of temperature and precipitation, and leaf fossils have therefore been used as indicators of past climate (Wilf, 1997; Wilf *et al.*, 1998). The shape and size of leaves influence important physiological features such as transpiration and, through convection and radiation, the heat balance of the leaf. Leaf characteristics such as marginal toothing have therefore changed over time. Other features such as the change in stomatal density with increasing carbon dioxide concentrations in the atmosphere are purely phenotypic responses (Beerling *et al.*, 1998), but changing carbon dioxide concentration may have had lasting effects on leaf form in evolution (Beerling *et al.*, 2001). Internal anatomy of leaves has co-evolved with biochemical pathways as an adaptive response, as in the appearance of Kranz anatomy with C4 pathways (Ehleringer *et al.*, 1997).

Maximum relative growth rates are achieved when the photosynthetic area is high compared with the mass of the plant, and the leaf can be seen as a way of maximising area for given resources. Leaves are rich in the main assimilatory protein, RUBISCO (the most abundant protein on earth), and the availability of this protein in leaves can therefore sustain a large global population of herbivorous animals. Leaves also have global effects on evapotranspiration, carbon dioxide flux, albedo and ultimately on climate. Furthermore, leaves have evolved senescence–abscission mechanisms and are consequently 'throwaway' organs in a way that most stems are not. This maintains high photosynthetic efficiency and rapid nutrient cycling but also allows adaptation to drought and cold. Without leaf abscission, high productivity could not be maintained in boreal regions. Finally, leaves characteristically produce a canopy which alters the wavelengths of light passing through it. The sensing of light that has passed through, or has been scattered from, leaves by detection of the red/far red ratio is a major control of plant growth (Ballaré *et al.*, 1990).

The evolution of the leaf is thus one of the great transitions of evolution, one which changed the nature and functioning of the terrestrial biosphere, and one whose evo-devo origin is now being elucidated (Harrison *et al.*, 2002; Langdale *et al.*, 2002). The nucleotide changes which gave rise to the leaf took place in an ecological context and in turn had profound ecological consequences. Any complete biodiversity science will therefore need to understand molecular innovation at an ecological level by generating an eco-molecular synthesis.

1.5 The future

1.5.1 New tasks

As noted elsewhere (Cronk, 2001) there is a significant difference in focus between genome sequencing efforts in animals and plants. Large-scale gene-sequencing projects in animals have sampled taxonomic groups widely, with worm, fly and mammal already fully sequenced. In contrast, almost all the genomic information thus far obtained from plants comes from one group: the angiosperms. As a result, progress has been slow in unravelling some of the most important evolutionary transitions in plants, such as the innovations (e.g. cuticle, stomata, sporopollenin, vascular tissue: Edwards *et al.*, 1996; Bateman *et al.*, 1998) associated with the colonisation of the land. The complete genome sequence of a bryophyte would be an important tool for the study of the whole of land plant evolution. On the benefit side, progress in unravelling purely angiosperm features such as floral morphology has been rapid (e.g. Teeri *et al.*, 2002; Theißen *et al.*, 2002). As genomic information grows, various questions, such as those listed below, will become easier to answer.

1   Where in the genome is the nucleotide variation responsible for phenotypic diversity, and where is the nucleotide variation on which selection acts to produce adaptive phenotypic traits? Information is needed about the nucleotide basis of interspecies variation, specifically to test the 'Doebley hypothesis' that variation in the cis-regulatory elements of transcription factors is the major source of morphological differences among species.
2   How do differences in nucleotides between alleles give rise to different phenotypes? Related to the first question, we also need to know how nucleotide changes are transduced, by interactions with other genes in developmental pathways, into altered phenotypes. Studies of single-locus morphological polymorphisms in the natural ecosystems will be essential, using emerging complex analytical techniques to correlate genetic and phenotypic polymorphism.
3   How are different developmental paths filtered by the environment? Where phe-

notypic traits are adaptive, studies of the selective coefficients on allelic variation of developmental significance will be particularly illuminating in integrating the environment with nucleotide change.

4   How do we study organisms not amenable to classical genetics? To work on organisms as diverse as orchids and oak trees, there is a need to develop new plant-friendly tools for reverse genetics and gene expression studies. These new tools must be adaptable for high-throughput use and they must be robust if they are to be used with diverse species. Reliance on sensitive techniques, which require a great deal of species-specific optimisation and cannot readily be used with large numbers of samples, will only delay progress.

1.5.2 Replacing the 'modern synthesis' with an eco-molecular synthesis

We have seen how the 'modern synthesis' (e.g. Huxley, 1942; Stebbins, 1950, 1983), which triumphantly unified Darwinian natural selection and Mendelian genetics, needs to be replaced by a new paradigm of selection acting at the nucleotide and genome level (rather than the gene/allele level). In doing so we might find there is a place for mutations of large effect (Bateman and DiMichele, 2002), which have been notably excluded from the modern synthesis. The eco-molecular synthesis must incorporate the findings of molecular genetics in revealing how gene action leads to the phenotype. It will also incorporate genomics and knowledge of how complex gene networks evolve. The comparative biology of the phenotype–environment interaction will be illuminated using the tools of phylogenetics, molecular developmental biology, ecology and genomics together. How does the natural environment interact with (and produce changes in) a complex regulatory network of some 20,000 interacting genes, through the medium of a structurally complex phenotype? The chapters in this book are an early salvo in what will be a long but highly productive endeavour.

REFERENCES

Ballaré, C. L., Scopel, A. L. and Sánchez, R. A. (1990) Far-red radiation reflected from adjacent leaves: an early signal of competition in plant canopies. *Science*, 247, 329–332.

Bateman, R. M. (1991) Palaeobiological and phylogenetic implications of anatomically-preserved *Archaeocalamites* from the Dinantian of Oxroad Bay and Loch Humphrey Burn, southern Scotland. *Palaeontographica Abt. B*, 223, 1–59.

Bateman, R. M. (1994) Evolutionary–developmental change in the growth architecture of fossil rhizomorphic lycopsids: scenarios constructed on cladistic foundations. *Biological Reviews of the Cambridge Philosophical Society*, 69, 527–597.

Bateman, R. M., Crane, P. R., DiMichele, W. A., Kenrick, P. R., Rowe, N. P., Speck, T. and Stein, W. E. (1998) Early evolution of land plants: phylogeny, physiology, and ecology of the primary terrestrial radiation. *Annual Reviews of Ecology and Systematics*, 29, 263–292.

Bateman, R. M. and DiMichele, W. A. (1994) Heterospory: the most iterative key innovation in the evolutionary history of the plant kingdom. *Biological Reviews*, 69, 345–417.

Bateman, R. M. and DiMichele, W. A. (2002) Generating and filtering major phenotypic novelties: neoGoldschmidtian saltation revisited, in *Developmental Genetics and Plant*

Evolution (eds Q. C. B. Cronk, R. M. Bateman and J. A. Hawkins), Taylor & Francis, London, pp. 109–159.

Baum, D. A. and Donoghue, M. J. (2002) Transference of function, heterotopy and the evolution of plant development, in *Developmental Genetics and Plant Evolution* (eds Q. C. B. Cronk, R. M. Bateman and J. A. Hawkins), Taylor & Francis, London, pp. 52–69.

Bhattacharya, M. K., Smith, A. K., Ellis, T. H. N., Hedley, C. and Martin, C. (1990) The wrinkled-seed character of pea described by Mendel is caused by a transposon-like insertion in a gene encoding starch-branching enzyme. *Cell*, 60, 115–122.

Beerling, D. J., McElwain, J. C. and Osborne, C. P. (1998) Stomatal responses of the 'living fossil' *Ginkgo biloba* L. to changes in atmospheric CO_2 concentrations. *Journal of Experimental Botany*, 49, 1603–1607.

Beerling, D. J., Osborne, C. P. and Chaloner, W. G. (2001) Evolution of leaf-form in land plants linked to atmospheric CO_2 decline in the Late Palaeozoic era. *Nature*, 410, 352–354.

Bishop, B. E. (1996) Mendel's opposition to evolution and to Darwin. *Journal of Heredity*, 87, 205–213.

Calkhoven, C. F. (1996) Multiple steps in the regulation of transcription-factor level and activity. *Biochemical Journal*, 317, 329–342.

Cronk, Q. C. B. (2001) Plant evolution and development in a post-genomic context. *Nature Reviews Genetics*, 2, 607–619.

Cubas, P. (2002) Role of TCP genes in the evolution of morphological characters in angiosperms, in *Developmental Genetics and Plant Evolution* (eds Q. C. B. Cronk, R. M. Bateman and J. A. Hawkins), Taylor & Francis, London, pp. 247–266.

Cubas, P., Coen, E. and Martinez-Zapater, J. M. (2001) Ancient asymmetries in the evolution of flowers. *Current Biology*, 11, 1050–1052.

Darwin, C. (1859) *On the Origin of Species*. Murray, London.

Doebley, J. and Lukens, L. (1998) Transcriptional regulators and the evolution of plant form. *Plant Cell*, 10, 1075–1082.

Donoghue, M. J., Ree, R. H. and Baum, D. A. (1998) Phylogeny and the evolution of flower symmetry in the Asteridae. *Trends in Plant Science*, 3, 311–317.

Doyle, J. J. (1994) Phylogeny of the legume family: an approach to understanding the origins of nodulation. *Annual Reviews of Ecology and Systematics*, 25, 325–349.

Doyle, J. J. (1998) Phylogenetic perspectives on nodulation: evolving views of plants and symbiotic bacteria. *Trends in Plant Science*, 3, 473–478.

Duckett, J. G. and Duckett, A. R. (1980) Reproductive biology and population dynamics of wild gametophytes of *Equisetum*. *Botanical Journal of the Linnean Society*, 80, 1–40.

Edwards, D., Abbott, G. D. and Raven, J. A. (1996) Cuticles in early land plants: a palaeo-ecophysiological evaluation, in *Plant Cuticles* (ed. G. Kierstens), Bios Scientific Publishers, Oxford, pp. 1–31.

Ehleringer, J. R., Cerling, T. E. and Helliker, B. R. (1997) C4 photosynthesis, atmospheric CO_2, and climate. *Oecologia*, 112, 285–299.

Fincham, J. R. S. (1990) Mendel – now down to the molecular level. *Nature*, 343, 208–209.

Friedman, W. E. and Carmichael, J. S. (1998) Heterochrony and developmental innovation: evolution of female gametophyte ontogeny in *Gnetum*, a highly apomorphic seed plant. *Evolution*, 52, 1016–1030.

Frohlich, M. W. (2002) The Mostly Male theory of flower origins: summary and update regarding the Jurassic pteridosperm *Pteroma*, in *Developmental Genetics and Plant Evolution* (eds Q. C. B. Cronk, R. M. Bateman and J. A. Hawkins), Taylor & Francis, London, pp. 85–108.

Gualtieri, G. and Bisseling, T. (2000) The evolution of nodulation. *Plant Molecular Biology*, 42, 181–194.

Haeckel, E. (1866) *Generelle Morphologie der Organismen*, 2 vols. G. Reimer, Berlin.

Halder, G., Callaerts, P. and Gehring, W. J. (1995) Induction of ectopic eyes by targeted expression of eyeless gene in *Drosophila*. *Science*, 267, 1788–1792.

Harrison, C. J., Cronk, Q. C. B. and Hudson, A. (2002) An overview of leaf evolution, in *Developmental Genetics and Plant Evolution* (eds Q. C. B. Cronk, R. M. Bateman and J. A. Hawkins), Taylor & Francis, London, pp. 395–403.

Hawkins, J. A. (2002) Evolutionary developmental biology: impact on systematic theory and practice, and the contribution of systematics, in *Developmental Genetics and Plant Evolution* (eds Q. C. B. Cronk, R. M. Bateman and J. A. Hawkins), Taylor & Francis, London, pp. 32–51.

Huxley, J. S. (1942) *Evolution, The Modern Synthesis*. Allen and Unwin, London.

Johansen, B. and Frederiksen, S. (2002) Orchid flowers: evolution and molecular development, in *Developmental Genetics and Plant Evolution* (eds Q. C. B. Cronk, R. M. Bateman and J. A. Hawkins), Taylor & Francis, London, pp. 206–219.

Johanson, U., West, J., Lister, C., Michaels, S., Amasino, R. and Dean, C. (2000) Molecular analysis of *FRIGIDA*, a major determinant of natural variation in *Arabidopsis* flowering time. *Science*, 290, 344–347.

Kellogg, E. A. (2002) Are macroevolution and microevolution qualitatively different? Evidence from Poaceae and other families, in *Developmental Genetics and Plant Evolution* (eds Q. C. B. Cronk, R. M. Bateman and J. A. Hawkins), Taylor & Francis, London, pp. 70–84.

Kenrick, P. (2002) The Telome Theory, in *Developmental Genetics and Plant Evolution* (eds Q. C. B. Cronk, R. M. Bateman and J. A. Hawkins), Taylor & Francis, London, pp. 365–387.

Langdale, J. A., Scotland, R. W. and Corley, S. B. (2002) A developmental perspective on the evolution of leaves, in *Developmental Genetics and Plant Evolution* (eds Q. C. B. Cronk, R. M. Bateman and J. A. Hawkins), Taylor & Francis, London, pp. 388–394.

Lester, D. R., Ross, J. J., Davies, P. J. and Reid, J. B. (1997) Mendel's stem length gene (Le) encodes a gibberellin 3 beta-hydroxylase. *Plant Cell*, 9, 1435–1443.

Luo, D., Carpenter, R., Vincent, C., Copsey, L., Clark, J. and Coen, E. (1999) Control of organ asymmetry in flowers of *Antirrhinum*. *Cell*, 99, 367–376.

Mendel, G. (1965) [1865, 1869] *Experiments in Plant Hybridisation* [*Versuche huber Pflanzenhybriden*] (ed. J. H. Bennett), Oliver & Boyd, Edinburgh.

Möller, M. and Cronk, Q. C. B. (2001) Evolution of morphological novelty: a phylogenetic analysis of growth patterns in *Streptocarpus* (Gesneriaceae). *Evolution*, 55, 918–929.

Schneider, H., Pryer, K. M., Cranfill, R., Smith A. R. and Wolf, P. G. (2002) Evolution of vascular plant body plans: a phylogenetic perspective, in *Developmental Genetics and Plant Evolution* (eds Q. C. B. Cronk, R. M. Bateman and J. A. Hawkins), Taylor & Francis, London, pp. 330–364.

Sprent, J. I. (2001) *Nodulation in Legumes*. Royal Botanic Gardens, Kew.

Stebbins, G. L. (1950) *Variation and Evolution in Plants*. Columbia University Press, New York.

Stebbins, G. L. (1983) Mosaic evolution: an integrating principle for the modern synthesis. *Experientia*, 39, 823–834.

Swensen, S. M. (1996) The evolution of actinorhizal symbioses: evidence for multiple origins of the symbiotic association. *American Journal of Botany*, 83, 1503–1512.

Teeri, T. H., Albert, V. A., Elomaa, P., Hämäläinen, J., Kotilainen, M., Pöllänen, E. and Uimari, A. (2002) Involvement of non-ABC MADS-box genes in determining stamen and carpel identity in *Gerbera hybrida* (Asteraceae), in *Developmental Genetics and Plant Evolution* (eds Q. C. B. Cronk, R. M. Bateman and J. A. Hawkins), Taylor & Francis, London, pp. 220–232.

Theißen, G., Becker, A., Winter, K.-U., Münster, T., Kirchner, C. and Saedler, H. (2002) How the land plants learned their floral ABCs: the role of MADS-box genes in the evolutionary origin of flowers, in *Developmental Genetics and Plant Evolution* (eds Q. C. B. Cronk, R. M. Bateman and J. A. Hawkins), Taylor & Francis, London, pp. 173–205.

Tsiantis, M., Hay, A., Ori, N., Kaur, H., Holtan, H., McCormick, S. and Hake, S. (2002) Developmental signals regulating leaf form, in *Developmental Genetics and Plant Evolution* (eds Q. C. B. Cronk, R. M. Bateman and J. A. Hawkins), Taylor & Francis, London, pp. 418–430.

Tucker, S. C. (2000) Organ loss in detarioid and other leguminous flowers, and the possibility of saltatory evolution, in *Advances in Legume Systematics* 9 (eds P. Herendeen and A. Bruneau), Royal Botanic Gardens, Kew, pp. 107–120.

Waddington, C. H. (1957) *The Strategy of the Genes*. Allen and Unwin, London.

Wilf, P. (1997) When are leaves good thermometers? A new case for leaf margin analysis. *Paleobiology*, 23, 373–390.

Wilf, P., Wing, S. L., Greenwood, D. R. and Greenwood, C. L. (1998) Using fossil leaves as paleoprecipitation indicators: an Eocene example. *Geology*, 26, 203–206.

Chapter 2

Impact of transposons on plant genomes

Virginia Walbot

ABSTRACT

Transposons constitute the major fraction of the maize genome; their abundance distinguishes small grass genomes from the much larger maize genome. In other taxa, transposon copy number is similarly important as a contributor to genome size. Retrotransposons are abundant in intergenic regions, and Miniature Inverted Repeat Transposons (MITEs) and other DNA-based transposons are often associated with genes. Retrotransposons are likely to contribute to overall genome structure, including the ability of chromosomes to pair; recent evidence suggests that retrotransposon number can be modulated rapidly in response to environmental change. DNA transposons inserted in or near transcription units can alter gene transcription both quantitatively and qualitatively; transposon excisions create tremendous allelic diversity by introducing deletions and insertions into genes. Epigenetic regulation of transposons can create epi-alleles of host genes exhibiting novel quantitative regulation. Maize and other plants are thought to tolerate transposon activity because the host can epigenetically silence transposons and because haploid gametic selection reduces the impact of deleterious mutations. The utility of transposons in creating allelic diversity within the plant soma is proposed as an explanation for their retention in the genome.

2.1 Introduction

Transposable DNA elements were discovered as the source of variegated maize phenotypes by Barbara McClintock, an illustrious pioneer of twentieth-century genetics (Fedoroff, 1989). Earlier geneticists had studied variegation, which is obvious in maize seeds (Figure 2.1) and many other plant parts; they had classified alleles as demonstrating such somatic variation in contrast to the stable colored and stable colorless alleles. More detailed analysis by Marcus Rhoades (1938) demonstrated the first case of *trans*-activation of gene expression; specifically, the response of the *a1-dt* allele to a dominant *Dotted* factor elsewhere in the genome. In the absence of *Dotted*, the *a1-dt* allele gives a pure white kernel. In contrast, the presence of *Dotted* results in a white field with numerous purple sectors. We now know that the *Dotted* gene makes a transposase protein that excises the *dt* element from the *a1-dt* allele. McClintock's work established that mobile DNA elements could be responsible for the *in cis* component of mutability, a discovery that surprised everyone.

In *Developmental Genetics and Plant Evolution* (2002) (eds Q. C. B. Cronk, R. M. Bateman and J. A. Hawkins), Taylor & Francis, London, pp. 15–31.

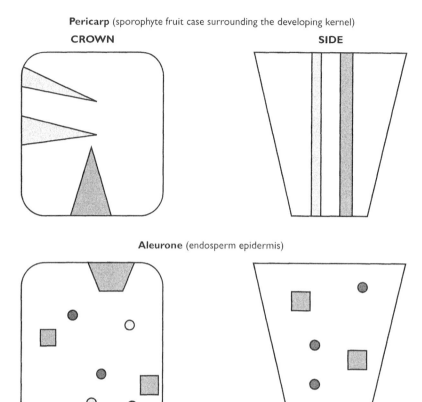

Pericarp (sporophyte fruit case surrounding the developing kernel)

CROWN   SIDE

Aleurone (endosperm epidermis)

Figure 2.1 Comparison of sector shape from somatic excision of transposons in the pericarp and aleurone. DNA transposon excisions result in revertant tissue sectors that vary in shape and size in the pericarp and endosperm of the maize kernel. Size reflects the time during development when excision occurred, because continued mitosis of a revertant cell produces a clonal sector of revertant tissue. Sector shape reflects the pattern of cell division and expansion in that tissue.

    Top panel. On the left is a representation of the crown of a maize kernel, and on the right is a side view. Somatic variegation in the *P-vv* allele, a regulatory gene required for phlobaphene pigment accumulation in the pericarp, is mediated by *Ac*. The pericarp is part of the fruit case, and differentiates during the ontogeny of the ear (inflorescence) of the vegetative plant. In the absence of phlobaphene pigment the pericarp is clear. In plants that have an active *Ac* element and the *P-vv* allele, colouration indicative of somatic excision is observed as stripes running from the base to the top of the kernel (to the position where the silk emerges from the pericarp). The pericarp forms from the fusion of three leaf-like primordia, and sectors follow the elongated shape of these pericarp components. The *P* locus is compound, and there are mutationally separable components contributing to pericarp and cob pigment (Coe *et al.*, 1988).

    Bottom panel. Somatic variation resulting from restoration of anthocyanin pigmentation in the aleurone is depicted. The representations of maize kernels are as listed for the top panel. The aleurone is the epidermal layer of the triploid endosperm; the founding cell results from the double fertilisation typical of flowering plants in which a diploid primary endosperm nucleus in the megagametophyte fuses with a haploid sperm nucleus. The

Many examples of somatic color variegation had been recovered historically from corn strains developed by Native Americans, who must have prized the colorful patterns of stripes and dots on kernels to have selected and maintained maize lines with active transposable elements. That these maize lines were also a major food source demonstrates that an active transposable element can be tolerated. Today, many horticultural variants exhibit signs of transposable element behavior, and natural species also exhibit similar heritable variegation (Griffiths and Ganders, 1983).

Although mobility was the most surprising property of the transposons mapped and analyzed by McClintock, she coined the term "controlling elements" to refer to these newly described genetic factors. She worked with two autonomous transposons, *Ac* and *Spm*, which we now know each encode a transposase protein. She showed that active forms of these elements were required *in trans* to mobilize derivative elements *Ds* and *dSpm*, respectively. More important to her, she demonstrated that alleles that acquired a transposon insertion fell under the control of the mobile DNA system and for that reason she named the transposons controlling elements. McClintock speculated that the controlling elements might be important in molding host gene expression patterns, an insight for which there is growing molecular evidence. It is also equally true that diverse host processes control transposons. Transposons and their hosts have co-evolved tolerance: the transposon is not completely silenced in the host, and the host is not killed by the transposon.

In the fifty years since McClintock's original reports, multiple types of transposons have been discovered. The two basic categories are retrotransposons and DNA transposons. Retroelements insert through the production of an RNA copy of an existing transposon; this RNA transcript is copied by reverse transcriptase into DNA, which inserts into a new chromosomal location. Retrotransposons can proliferate and insert, but they do not excise; as a consequence, retrotransposon-induced mutations have a stable, usually null phenotype. In contrast, the "classic" DNA elements discovered by McClintock such as *Ac/Ds* and *Spm/dSpm* move as pieces of DNA. These elements are organized into families of autonomous and non-autonomous elements. For example, *Activator* (*Ac*) is autonomous because it encodes a transposase that recognizes motifs in the termini of *Ac* and *Ds* elements; *Ds* elements are defective, non-autonomous *Ac* derivatives that do not encode transposase but can still respond to it. The transposase produces nicks at or near the termini of *Ac/Ds* elements, excising the element from its original chromosomal location (Figure 2.2). Newly excised elements can then insert elsewhere. The exposed 3'OH at the terminus of the transposon DNA and the transposase protein cooperate to produce a double-stranded break in the target DNA. The transposon 3' end is

resulting triploid nucleus undergoes multiple rounds of division without cytokinesis, forming a syncitium of 256–512 nuclei. Ultimately, these are arrayed along the plasma membrane of the very large endosperm cell; subsequent cellularisation and specialisation of internal and external cells results in formation of a single cell layer of epidermal cells. The epidermal cells undergo primarily anticlinal cell divisions and hence remain a single layer. Somatic excision of a transposon from an anthocyanin biosynthetic gene early in endosperm ontogeny will result in a very large sector, but the majority of events occur after cellularisation, resulting in patches and dots of purple pigment.

Step 1

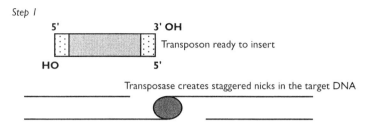

Step 2

A phosphodiester bond forms between the exposed 3' OH of the transposon and the 5' base at the staggered nick at both ends of the transposon. There are single-stranded gaps reflecting the staggered nicks created by the transposase at the beginning of the reaction.

Step 3

Host DNA repair synthesis fills in the bases missing at each end of the element and ligates the 5' ends of the transposon into the repaired DNA duplex; this step creates a direct host sequence duplication (➤) next to the transposon termini.

Figure 2.2 DNA transposon activities. Insertion of the transposon (box with stippled ends representing the termini of the element) requires a double-stranded break in the host chromosome. The 3' OH of each transposon end, in concert with the transposase protein, creates the breaks in each strand and attaches the reactive end of the transposon to the host DNA. Fill-in repair followed by ligation is required to attach the 5' end of each transposon strand to the host DNA to restore chromosomal integrity. The initial breaks mediated by transposases are deduced to be staggered (offset on the upper and lower strand), because each element family makes a characteristic host sequence duplication. These duplications arise during chromosomal repair, which involves using the single-stranded overhangs at the site of breakage as templates for restoring a double-stranded sequence. For example, *Ac/Ds* creates 8-bp host sequence duplication, *Spm/dSpm* elements create 3-bp host sequence duplication, and *MuDR/Mu* elements create a 9-bp host sequence duplication.

then attached into the new chromosomal location at the site of attack (single-stranded attachment to the chromosome), followed by host DNA repair synthesis to recreate an intact, double-stranded chromosome.

There is considerable diversity in the outcome of DNA transposon excision events (for chapters illustrating the diversity of transposition mechanisms and host repair pathways see Craig *et al.*, 2002). Some elements, such as *P* of *Drosophila melanogaster*, excise and move to a new location; however, the "old" location is often repaired by copying a *P* element sequence from a sister chromatid or an ectopic genome location. With this manner of repair, both the "old" and "new" locations are occupied by *P* elements, which therefore increase in copy number in the genome. *Ac/Ds* elements typically move after local DNA replication into a nearby region that has not yet replicated; as a consequence, there is one *Ac/Ds* left at the old location (on one of the two newly replicated chromosomes) but the new location is represented on both newly replicated chromosome regions. As a consequence, *Ac/Ds* elements can increase in copy number as well.

Unforeseen from the genetic analysis of DNA transposons and the mechanisms of their step-wise increases in copy number was the magnitude of the contribution of mobile elements to the composition of the maize genome, and indeed to the genomes of higher eukaryotes in general. There are dozens of transposon families in the sequenced *Arabidopsis thaliana* genome (Le *et al.*, 2000; McCombie *et al.*, 2000). Individual transposon types can be present from 1 to 100,000 copies in the maize genome, and collectively they comprise more than two-thirds of the genome (Table 2.1; SanMiguel *et al.*, 1996). The much smaller rice (Song *et al.*, 1998; Mao *et al.*, 2000) and sorghum (Tikhonov *et al.*, 1999) genomes also contain numerous transposons; they comprise a smaller fraction of the genome than in maize. Clearly, nearly all transposable elements are quiescent or defective nearly all of the time, otherwise stable chromosomes would not exist. Periodically, however, transposons can be activated, and while active, they can alter genome structure, gene expression patterns, and allelic diversity.

The primary goal of this chapter is to illustrate these phenomena. The curious reader further sampling the literature will be rewarded with color photographs of variegated phenotypes and insights into the diversity of mechanisms transposons employ to persist and multiply in nature.

2.2 Transposon types and their impact on the maize genome

2.2.1 Genome structure

Genes in maize and other higher plants are compact, with short introns and short untranslated regions. It seems reasonable to assume that approximately the same number of genes is required in all flowering plants, within a factor of two for the estimate in *Arabidopsis* of ~26,000 (Walbot, 2000). Why then do genome sizes

Table 2.1 Composition of the maize genome

Sequence component	Estimated contribution (%)
Ji retrotransposon	16
Opie retrotransposon	15
Tekay retrotransposon	10
Huck retrotransposon	10
Cinful retrotransposon	7
Grande retrotransposon	4
Low and middle copy number elements	16
Genes and all other components	22

Source: Data supplied by Jeff Bennetzen.

range over many orders of magnitude, from ~100 MB to >10,000 MB? This "C-value paradox" was unresolved until recently when it became clear that two genome features contribute to large size: polyploidy and transposon amplification. Although there are thousands of examples of mutations in single loci that yield a visible phenotype in maize, the maize genome is the result of relatively recent tetraploidization (Gaut *et al.*, 2000). In many cases the "duplicate" loci have now diverged sufficiently in function so that mutation of one of the copies yields a scorable phenotype. Complete duplication of the genome and numerous rearrangements distinguish the 2500 MB maize genome organized in ten chromosomes from the 450 MB rice genome organized in twelve chromosomes (Wilson *et al.*, 1999).

Highly repetitive (>10,000 copies per genome) and mid-repetitive (100–1000 copies) DNA was discovered in maize using DNA reassociation kinetics; these genome components constituted most of the genome (Hake and Walbot, 1980). Furthermore, the reassociation kinetic experiments indicated that single-copy DNA (the presumed genes) was finely interspersed with much more repetitive sequences. Mapping and DNA sequencing has demonstrated that retrotransposons (Kumar and Bennetzen, 1999) constitute the majority of the mid- and highly repetitive components of the maize genome (SanMiguel *et al.*, 1996). Large genomes such as maize presumably accumulate retrotransposons through repeated episodes of element amplification, whereas smaller genomes either restrict retrotransposon amplification or have a mechanism (such as recombination) to eliminate such elements periodically (Figure 2.3).

Support for the dynamic nature of transposon gain and loss is found when comparing closely related taxa. Retrotransposon types can be nearly species-specific (Kumar and Bennetzen, 1999). Although the surveys have not been sufficiently comprehensive to ask whether a genome contains just a few copies of a particular retrotransposon that is abundant in a sister taxon, it is still impressive that some highly abundant element families appear to be restricted to just a single grass species. In contrast, most DNA transposons are ubiquitous in the flowering plants, although the elements differ in copy number and biological activity. In most species or in most individuals of a species, the DNA transposons are epigenetically silenced or have suffered mutations that render them non-functional.

Changes in transposon copy number can be correlated with other aspects of species biology. In a recent example, Kanazawa *et al.* (2000) concluded that the presence or absence of MITEs was correlated with speciation in rice and its relatives. Their observation was that there was low variability within a species or ecotype but numerous differences among species or ecotypes. For example, specific MITE distributions were found in *Oryza rufipogon* ecotypes that exhibit an annual or perennial life cycle. Kanazawa *et al.* (2000: 327) conclude:

> either that gene flow has been highly restricted between different species, as well as between different ecotypes of *O. rufipogon* after they were differentiated, or that loci with or without MITEs have been selected in nature together with the linked genes that are responsible for adaptation to environments. In addition, a very low polymorphism with regard to the presence or absence of MITEs within each species or each ecotype suggests that the frequency of transposition of MITEs is very low, assuming that the loci that contain MITEs are free from selection pressure.

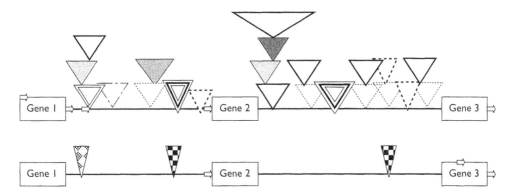

Figure 2.3 Accumulation of transposons in intergenic regions of maize (top) compared with grasses with smaller genomes (bottom). The physical distance between maize genes is typically much greater than found in grasses with smaller genomes such as sorghum (Bennetzen, 2000b). This difference appears to be accounted for, in the limited number of homologous regions examined, by the presence of many retrotransposons in the intergenic regions of maize. In this diagram, the triangles represent retrotransposons, with each triangle type representing a different retroelement type. Triangle size suggests the length heterogeneity within and between retrotransposon families; in real examples, the density of retrotransposons is much higher than illustrated here (SanMiguel *et al.*, 1996). Retroelements insert into each other; occasionally they are found near a gene. The density of the line on the transposons indicates the age of the element, estimated by the degree of divergence within that family type. It is assumed that elements entered the region and then acquired point mutations over time. As a consequence, more recently arrived retroelements typically contain fewer mutations compared to other family members. MITEs, represented by short arrows, are clustered near genes; as discussed in the text, the diverse types of MITEs are very numerous in grass genomes. Recent evidence suggests that MITEs are DNA transposons, but they are illustrated here because they are so numerous. The predominant retrotransposon types are species-specific when grasses as divergent as >10MYA or more are compared, as illustrated in the lower panel. This diagrammatic small genome grass has different, and far fewer, retrotransposons in the intergenic regions. The total distance of the region from gene 1 to gene 3 is much shorter than in maize.

Rapid changes in plant genome size have been documented historically without understanding a mechanism that could achieve rapid change (reviewed by Walbot and Cullis, 1985). With discovery of appropriate molecular probes to follow transposon copy numbers, it is possible to ask whether rapid morphological or physiological adaptation is accompanied by genomic changes involving transposons. There is evidence for transposon changes in barley varieties where retrotransposon contribution is different at an examined chromosome location (Shirasu *et al.*, 2000) or in the genome as a whole (Vicient *et al.*, 1999). A recent study has correlated the copy number and distribution of the *Bare1* elements of barley with specific environmental conditions, suggesting that dynamic changes in copy number can occur quickly and be stabilized in contiguous populations (Kalendar *et al.*, 2000). Furthermore, brief low doses of UV-B radiation are sufficient to activate quiescent DNA transposons such as *Mu* elements of maize (Walbot, 1999). Although still fragmentary, such evidence allows the fascinating speculation that external conditions activate amplification or removal of retrotransposons as part of the response of the organism to "stress." This scenario was predicted by Barbara McClintock in her Nobel address (McClintock, 1984).

Because retrotransposons comprise such a large fraction of many genomes, it is possible that they can contribute to chromosome pairing (Moore, 2000). Alterations in the fidelity of, or even capacity for, pairing often contribute to speciation events. It is thus possible that selection for proliferation of rare retrotransposon types to high copy number underlies some speciation events. As more is known about the retrotransposon composition of the complete rice genome, instructive comparisons can be made with segments of sorghum, maize, barley, wheat and other grass genomes that are selectively sequenced. For these species, there are several genome projects sequencing bacterial artificial chromosomes (BACs) containing ~100–200 kB fragments of a grass genome. The BACs are chosen by hybridization with specific gene probes so that those regions with similar genes can be compared between species. (More information about such projects can be obtained at http://barleygenomics.wsu.edu and http://pgir.rutgers.edu.)

Our present understanding is that most retrotransposons are quiescent; indeed, given their high copy numbers surprisingly few retrotransposon transcripts are found in plant cDNA collections in GenBank. Because an RNA intermediate is required for increases in copy number, the lack of transcripts is a good indicator that the retrotransposons are not actively replicating. Occasionally, very low copy number retrotransposons of maize are responsible for a new mutation; for example, the *Bs1* element of maize (Kumar and Bennetzen, 1999). My conclusion is that massive amplification of high-copy retrotransposons is very infrequent because each new copy could mutate a gene by insertion and each new insertion is a new site for homologous recombination during meiosis and somatic mitotic divisions (for further discussion of this point, see Bennetzen, 2000a). I think that the current "restriction" of retrotransposons to intergenic regions could reflect gametic failure for cells that experienced insertions in essential genes; it is also plausible that retroelements preferentially insert into a type of chromatin found outside the transcriptional units. In either case, until the chromatin is suitably arranged to suppress recombination in newly amplified retroelements of identical sequence, their proliferation presents an incipient disaster of rearrangements and recombination events. These events could result in gene losses and duplication of chromosome segments.

DNA transposons, on the other hand, often explain restriction fragment length polymorphisms in or near maize genes (Figure 2.4). Insertion of DNA transposons appears to be preferential for transcription units or the neighboring DNA. The presence of these insertions must be tolerated for the allele to survive. The fact that maize exhibits very high allelic polymorphism is accounted for mainly by the action of transposable elements; in many other species, allelic diversity is low and there are fewer transposons associated with the genes. MITEs are particularly abundant in or near maize genes and genes in other grasses (Bureau *et al.*, 1996; Bureau and Wessler, 1992, 1994; Hu *et al.*, 2000; Wessler *et al.*, 1995; Zhang *et al.*, 2000). These elements are short (just a few hundred base pairs long) but contain motifs that can alter gene expression. The presence of DNA transposons can confer new regulatory properties to genes, alter the protein product from the gene, and generate allelic diversity. The changes to alleles produced by transposon excision are discussed in the next section; short insertion and deletion mutations distinguish most maize

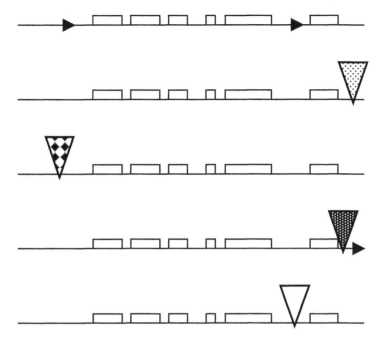

Figure 2.4 Comparisons among functional maize alleles typically uncover large polymorphisms attributable to historic transposon insertions (triangles) and MITEs (smaller arrowheads). As a consequence, restriction enzyme fragment length polymorphisms are very common in maize genes.

alleles from other alleles of the same locus. Although the transposon is no longer present, these insertion and deletion events can be best explained by transposon activities.

2.2.2 Allelic diversity

DNA transposons can both insert and excise when an autonomous, transposase-encoding element of their family is present. It is this group of elements that generates the somatic mutability generally regarded as characteristic of transposon activities (Figure 2.1). During insertion, each transposon family creates a characteristic host sequence duplication, reflecting the number of bases between the staggered nicks introduced in the target site (Figure 2.2). For example, *MuDR/Mu* elements have a 9-bp duplication, *Ac/Ds* an 8-bp duplication, and *Spm/dSpm* a 3-bp duplication. As a consequence, the host allele is altered not only by the transposon insertion but also by the duplication generated during insertion. Transposon excision from this altered allele is the starting point for generating allelic diversity.

As illustrated in Figure 2.5, an allele harboring an active transposon is a hotspot for secondary mutations. In the case of *Ac/Ds* the frequency of an insertion into any location is only $\sim 10^{-6}$, but once they are present excisions that restore gene expression occur in 10^{-1} to 10^{-2} of the progeny (Walbot, 1992). When DNA transposons excise, all or part of the host sequence duplication can be retained. In the case of

Ac/Ds insertion into an exon, the 8-bp host sequence duplication would create a frameshift mutation. Deletion of 2 bp, however, creates a two- amino acid insertion (6 bp) relative to the progenitor allele and restores the reading frame. Such short insertions and deletions are very common in maize alleles (Alfenito *et al.*, 1998). Ectopic base additions also occur; that is, bases that were not originally present in the gene are added during chromosome repair. One proposed mechanism for generating ectopic base additions involves formation of a sealed loop at the site of a staggered chromosome break; the single-stranded overhang can be ligated to the opposite strand. During restoration of the chromosome, the loop is nicked but at a different site than the original end of the single-stranded overhang, resulting in base changes. Chromosome repair in maize can also be accomplished by additions of several hundred bp of "filler DNA" that is copied from nearby or unknown chromosome locations; "filler DNA" was first characterized in spontaneous deletions (Wessler *et al.*, 1990). Similar events are also recovered after *Mu* element excision (Raizada *et al.*, 2001).

In the case of *MuDR/Mu* elements large deletions and ectopic additions result in changes of dozens of amino acids. In one study in which deletions of up to 35 bp were found, as well as ectopic base additions of 1–5 bp (Britt and Walbot, 1991), the range of possible in-frame excision alleles was >10^5 (Nordborg and Walbot, 1995). All of this diversity could occur from a *Mu* element inserted at just one location in a gene. Other transposons such as *Ac/Ds* and *Spm/dSpm* typically result in deletions and additions of only a few bp; as a consequence, they are predicted to generate only dozens of allelic types at each point of insertion. Even this lower output of new alleles is impressive, however, considering that *in situ* mutagenesis can occur at every base pair of a gene over a very long time period. Laboratory strategies for site-directed mutagenesis that alter one amino acid at a time are much more conservative than the impact made by transposons on maize alleles *in vivo*.

Transposons can also stimulate recombination that results in larger rearrangements of chromosomes, notably large deletions, inversions and duplications. When transposase is present the nicks created as a prelude to transposition can result in host-mediated chromosomal changes in which the transposon is retained at the site. The double-stranded breaks that occur during successful excision of a DNA element can be the substrates for such large-scale chromosomal events; in these events, the inciting transposon is missing from the final product (Figure 2.5).

Maize DNA transposons exhibit very different frequencies of new allele generation. The *MuDR/Mu* family generates a high frequency of very diverse somatic alleles, but very few (~10^{-5}) germinal events. For this family, new insertion events are actually more common than germinal excision events (Walbot, 1992). In contrast, both *Ac/Ds* (Baran *et al.*, 1992) and *Spm/dSpm* (reviewed by Nordborg and Walbot, 1995) generate many germinal revertants, albeit of a more limited diversity than *MuDR/Mu*. Accumulation of altered alleles over evolutionary time will consequently depend on several factors, including:

1   the frequency of insertion of transposons (which varies among the known element types and depends on their copy number and activity);
2   the frequency of novel allele generation (which varies at least one thousand-fold among studied types), and;

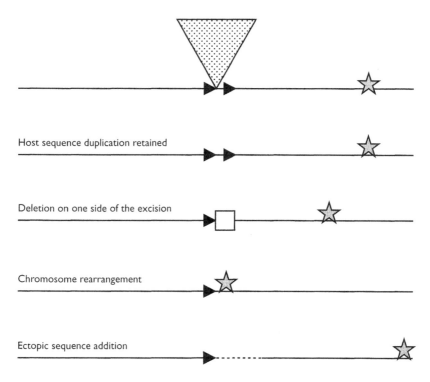

Figure 2.5 DNA transposon excision can generate allelic diversity. All or part of the host sequence duplication can be retained and be accompanied by further deletions, ectopic base additions, or rearrangements involving the gene. See the text for details. The star indicates a nearby site that remains in the same relative position to the insertion site in most cases, but during inversions or large deletions that position is found at or closer to the insertion site.

3   the likelihood of novel allele transmission (which requires that excision occurs in pre-germinal or germinal cells and does not result in haploid lethality).

2.2.3 Gene expression

Resident transposons can alter gene expression merely by their presence. Historically, transposon insertions may have been a major contributor of new promoter information, new introns and new untranslated regions. Figure 2.6 illustrates some of the common alterations in gene transcript production that occur after transposon insertion. In these examples, we are assuming that excision of the transposon is not occurring; rather, it is the presence of the transposon that alters gene expression. It is clear that transposons can affect the type of product and the quantity of product(s) by diverse alterations in RNA transcription and processing (Luehrsen *et al.*, 1994). For many alleles with a transposon insertion, modest production of the normal product persists and no mutant phenotype is visible in the host. For some alleles, an altered product may confer a new phenotype. For other alleles, expression of product at higher levels or in new locations can confer a

(1) Novel transcription start site from within the transposon confers new quantitative and qualitative regulation of the same protein product

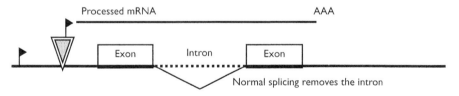

(2) Splice sites in the transposon create a chimeric transcript encoding novel amino acids from part of the transposon and deletion of exon I sequences after the transposon insertion site

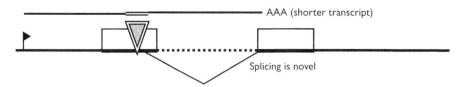

(3) Transposon insertion decreases the efficiency of splicing, resulting in low production of the normal transcript and the unspliced transcripts are destroyed

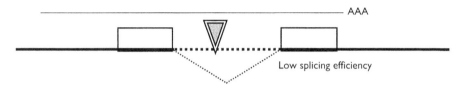

(4) Transcription into the transposon results in a truncated protein product with the carboxy terminus encoded by the transposon

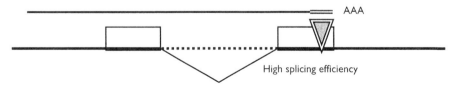

(5) Unspliced transcript results in a transcript encoding a truncated protein product with the carboxy terminus encoded by intron sequences

Figure 2.6 How the presence of a DNA transposon can alter host gene expression, transcript splicing outcome and efficiency, and the protein product encoded by the allele.

phenotype. Because of the diversity of possibilities, each transposon insertion requires characterization to understand the changes in gene expression programmed by an affected allele.

There are several documented cases of human selection for phenotypes, including starch composition (waxy or non-waxy) in maize and rice, in which transposon mutations conferred the selected phenotype. The lectinless soybean is also the result of transposon insertion. In observing somatic variegation (Figure 2.1) you can often find a graded series of phenotypes in color or composition. Considering the allelic diversity created by transposon excision it is not surprising that alleles varying in phenotype are produced dynamically during plant development. Similarly, distinctive quantitative variation may be an important facet of stable alleles containing a transposon (Figure 2.6).

Because transposons are subject to frequent epigenetic silencing, typically involving host methylation of the transposon ends (Chandler and Walbot, 1986), the presence of a transposon in the promoter may result in novel "epi-alleles" that differ in expression levels. For example, the standard *R* gene of maize, a regulator of anthocyanin production, contains a *doppia* transposon in the promoter. This transposon is subject to methylation, and this epigenetic modification parallels reduced expression of *R* (Walker, 1998).

McClintock named *Spm Suppressor-mutator* to indicate that even the inactive transposon had an impact on gene expression (Fedoroff, 1989). When the transposon is active, kernels of *a1-Spm* are white with purple dots, indicative of excision; the white background indicates that as long as *Spm* is present there is little or no gene activity from the allele until there is an excision event. When *Spm* epigenetically loses activity, however, the kernels have a pale purple, uniform pigmentation. In this epigenetic state there are no excisions but nonetheless partial gene expression is restored. Similarly, *Mu* insertions in promoter and 5′ untranslated leader regions typically block gene expression when the transposase is present. After epigenetic silencing, however, a "read out" promoter in the *Mu* termini is activated, thereby restoring gene expression to the insertion allele (Barkan and Martienssen, 1991). Not surprisingly, epigenetics is currently a "hot topic" (Wolffe and Matzke, 1999). It is a means of altering gene expression without DNA mutations, and consequently can greatly increase the range of allelic variants from a single DNA sequence. It seems likely that transposons will be major contributors to the creation and modulation of epi-alleles in maize and other species.

There are numerous examples of epigenetic states involving transposons. This extra level of complexity in their regulation provides a mechanism for generating many quantitative variants in host gene expression as a result of events occurring in a nearby transposon.

2.3 Why are plants so tolerant of transposon insertions?

Given the substantial impact that transposons can have on genome structure, allelic diversity, and gene expression, it is important to consider how the host tolerates these mobile elements. I will therefore now summarize recent evidence on the interaction between the host and its transposons. Extended discussions of the implications of transposon activity in plant evolution are available in recent reviews (e.g. Bennetzen, 2000b; Fedoroff, 2000a, b).

For many DNA transposons diversity is generated mainly in the somatic tissues; this also means that most of the potentially deleterious new mutations and chromosome breaks occur in the somatic cells. Stringent regulation of events preceding meiosis and in the resulting haploid cells could readily be selected based on survivorship in the progeny. It is interesting to consider that somatic diversity may itself be a selected feature. Phenotypic variation in the soma can be very beneficial to plants in avoiding herbivory and pathogen attack; one branch that expresses novel chemicals may survive intact while other branches are defoliated (Walbot, 1996). Consequently, maintenance of a low level of transposon activity in somatic tissues may be selected for based on somatic survival and the fact that flowers arise from somatic tissues. The flowers on the hypothetical branch of novel phenotype may set more seed because photosynthesis locally can support better fruit development. As a consequence, progeny of these fruit could inherit a novel allele created by transposon activity as well as the active transposable element system. Somatic selection within plants requires entertaining the idea that the various apical meristems likely to produce flowers are initially identical siblings that diverge as a result of mutation and epigenetic events.

Plants may also be more tolerant of transposons than animals as a result of the alternation of generations (Walbot, 1985). In this life strategy, there is haploid genetic activity in the gametophytes prior to gametogenesis and in the gametes prior to fertilization. For those loci expressed in the haploid, deleterious alleles are likely purged from the population by gametophytic arrest or fertilization failure. This purifying selection may provide plants with more latitude in allele generation and hence greater tolerance for active transposons.

It is also clear that, in general, the host is dominant to the transposons, because nearly all transposons are transcriptionally and transpositionally inactive. Drastic selection for control over newly activated and/or horizontally transmitted transposons may have occurred episodically throughout plant evolution. If true, then hosts are likely to have multiple, effective mechanisms for curbing transposons. Indeed, one reason to explain the frequent silencing of transgenes (Fagard and Vaucheret, 2000) is that these newly added DNA segments trigger well-honed host responses to transposon insertions. Periodic transposon activation involves only a tiny fraction of the transposons in the genome and occurs in only a small fraction of the population. Despite the potential beneficial impact of somatic diversification and even transposon-mediated gene mutation, host suppression of all transposon activities is the major outcome of the relationship between maize transposons and their host plant.

2.4 Future studies of maize transposons

The analysis of grass genomes will receive a tremendous boost from publication of the full sequence of the rice genome and selective sequencing of the maize genome. Appreciation of the range in the types of transposons and their copy numbers will be far easier to grasp by observing much larger data sets. Activation of transposons by environmental perturbations appears to be a fruitful experimental route to determining which transposon types are still functional, despite sequence divergence. For maize, transposon tagging is the primary route to gene discovery and gene mutations

(Walbot, 1992; http://zmdb.iastate.edu). Consequently, analysis of gene expression and structure will also provide information on transposon properties. These biotechnological applications also demonstrate how a fundamental discovery can result in powerful new methodology for applied science.

ACKNOWLEDGMENTS

I thank the organizers of the recent meeting on developmental genetics and plant evolution for an invitation to participate. I thank Jeff Bennetzen for supplying an unpublished summary of maize genome composition for Table 2.1. My work on *MuDR/Mu* transposons of maize is supported by a grant from the US National Institutes of Health (GM49681).

REFERENCES

Alfenito, M. R., Souer, E., Buell, R., Koes, R., Mol, J. and Walbot, V. (1998) Functional complementation of anthocyanin sequestration in the vacuole by widely divergent glutathione S-transferases. *Plant Cell*, 10, 1135–1149.

Baran, G., Echt, C., Bureau, T. and Wessler, S. (1992) Molecular analysis of the maize *wx*-B3 allele indicates that precise excision of the transposable element Ac is rare. *Genetics*, 130, 377–384.

Barkan, A. and Martienssen, R. A. (1991) Inactivation of maize transposon *Mu* suppresses a mutant phenotype by activating an outward-reading promoter near the end of *Mu1*. *Proceedings of the National Academy of Sciences USA*, 88, 3502–3506.

Bennetzen, J. L. (2000a) Comparative sequence analysis of plant nuclear genomes: microcolinearity and its many exceptions. *Plant Cell*, 12, 1021–1029.

Bennetzen, J. L. (2000b) Transposable element contributions to plant gene and genome evolution. *Plant Molecular Biology*, 42, 251–269.

Britt, A. B. and Walbot, V. (1991) Germinal and somatic products of excision of *Mu1* from the *Bronze-1* gene of *Zea mays*. *Molecular General Genetics*, 227, 267–276.

Bureau, T. E. and Wessler, S. R. (1992) *Tourist*: a large family of small inverted repeat elements frequently associated with maize genes. *Plant Cell*, 4, 1283–1294.

Bureau, T. E. and Wessler, S. R. (1994) Mobile inverted-repeat elements of the *Tourist* family are associated with the genes of many cereal grasses. *Proceedings of the National Academy of Sciences USA*, 91, 1411–1415.

Bureau, T. E., Ronald, P. C. and Wessler, S. R. (1996) A computer-based systematic survey reveals the predominance of small inverted-repeat elements in wild-type rice genes. *Proceedings of the National Academy of Sciences USA*, 93, 8524–8529.

Chandler, V. L. and Walbot, V. (1986) DNA modification of a transposable element of maize correlates with loss of activity. *Proceedings of the National Academy of Sciences USA*, 83, 1767–1771.

Coe, E. H., Neuffer, M. G. and Hoisington, D. A. (1988) The genetics of corn, in *Corn and Corn Improvement* (eds G. F. Sprague and J. W. Dudley), American Society of Agronomy, Madison, WI, pp. 81–258.

Craig, N. L., Craigie, R., Gellert, M. and Lambowitz, A. (eds) (2002) *Mobile DNA II*. American Society of Microbiology, Washington, DC.

Fagard, M. and Vaucheret, H. (2000) (Trans)gene silencing in plants: how many mechanisms? *Annual Review of Plant Physiology and Plant Molecular Biology*, 51, 167–194.

Fedoroff, N. V. (1989) Maize transposable elements, in *Mobile DNA* (eds D. Berg and M. M. Howe), American Society of Microbiology, Washington, DC, pp. 375–411.

Fedoroff, N. V. (2000a) Transposable elements as a molecular evolutionary force. *Annals of the New York Academy of Sciences*, 870, 251–264.

Fedoroff, N. V. (2000b) Transposons and genome evolution in plants. *Proceedings of the National Academy of Sciences USA*, 97, 7002–7007.

Gaut, B. S., d'Ennequin, M. L., Peek, A. S. and Sawkins, M. C. (2000) Maize as a model for the evolution of plant nuclear genomes. *Proceedings of the National Academy of Sciences USA*, 97, 7008–7015.

Griffiths, A. J. F. and Ganders, F. R. (1983) *Wildflower Genetics*. Flight Press, Vancouver.

Hake, S. and Walbot, V. (1980) The genome of *Zea mays*, its organization and homology to related grasses. *Chromosoma*, 79, 251–270.

Hu, J. P., Reddy, V. S. and Wessler, S. R. (2000) The rice *R* gene family: two distinct subfamilies containing several miniature inverted-repeat transposable elements. *Plant Molecular Biology*, 42, 667–678.

Kalendar, R., Tanskanen, J., Immonen, S., Nevo, E. and Schulman, A. H. (2000) Genome evolution of wild barley (*Hordeum spontaneum*) by BARE-1 retrotransposon dynamics in response to sharp microclimatic divergence. *Proceedings of the National Academy of Sciences USA*, 97, 6603–6607.

Kanazawa, A., Akimoto, M., Morishima, H. and Shimamoto, Y. (2000) Inter- and intra-specific distribution of *Stowaway* transposable elements in AA-genome species of wild rice. *Theoretical and Applied Genetics*, 101, 327–335.

Kumar, A. and Bennetzen, J. L. (1999) Plant retrotransposons. *Annual Review of Genetics*, 33, 479–532.

Le, Q.-H., Wright, S., Yu, Z.-H. and Bureau, T. (2000) Transposon diversity in *Arabidopsis thaliana*. *Proceedings of the National Academy of Sciences USA*, 97, 7376–7381.

Luehrsen, K. R., Taha, S. and Walbot, V. (1994) Nuclear pre-mRNA processing in higher plants. *Progress in Nucleic Acid Biochemistry and Molecular Biology*, 47, 149–193.

McClintock, B. (1984) The significance of responses of the genome to challenge. *Science*, 226, 792–801.

McCombie, W. R., delaBastide, M., Habermann, K., Parnell, L., Dedhia, N., Gnoj, L. *et al.* (2000) The complete sequence of a heterochromatic island from a higher eukaryote. *Cell*, 100, 377–386.

Mao, L., Wood ,T. C., Yu, Y.-S., Budiman, M. A., Tomkins, J., Woo, S.-S. *et al.* (2000) Rice transposable elements: a survey of 73,000 sequence-tagged-connectors. *Genome Research*, 10, 982–990.

Moore, G. (2000) Cereal chromosome structure, evolution, and pairing. *Annual Review of Plant Physiology and Plant Molecular Biology*, 51, 195–222.

Nordborg, M. and Walbot, V. (1995) Estimating allelic diversity generated by excision of different transposons types. *Theoretical and Applied Genetics*, 90, 771–775.

Raizada, M. N., Nan, G. L. and Walbot, V. (2001) Somatic and germinal mobility of the *RescueMu* transposon in transgenic maize. *Plant Cell*, 13, 1587–1608.

Rhoades, M. M. (1938) Effect of the *Dt* gene on the mutability of the *a₁* allele in maize. *Genetics*, 23, 377–397.

SanMiguel, P., Tikhonov, A., Jin, Y.-K., Motchoulskaia, N., Zakharov, D., Melake-Berhan, A. *et al.* (1996) Nested retrotransposons in the intergenic regions of the maize genome. *Science*, 274, 765–768.

Shirasu, K., Schulman, A. H., Lahaye, T. and SchulzeLefert, P. (2000) A contiguous 66-kb barley DNA sequence provides evidence for reversible genome expansion. *Genome Research*, 10, 908–915.

Song, W. Y., Pi, L. Y., Bureau, T. E. and Ronald, P. C. (1998) Identification and characterization of 14 transposon-like elements in the noncoding regions of members of the Xa21 family of disease resistance genes in rice. *Molecular and General Genetics*, 258, 449–456.

Tikhonov, A. P., SanMiguel, P. J., Nakajima, Y., Gorenstein, N. M., Bennetzen, J. L. and Avramova, Z. (1999) Colinearity and its exceptions in orthologous *adh* regions of maize and sorghum. *Proceedings of the National Academy of Sciences USA*, 96, 7409–7414.

Vicient, C. M., Suoniemi, A., Anamthamat-Jónsson, K., Tanskanen, J., Beharav, A., Nevo, E. and Schulman, A. H. (1999) Retrotransposon BARE-1 and its role in genome evolution in the genus *Hordeum*. *Plant Cell*, 11, 1769–1784.

Walbot, V. (1985) On the life strategies of plants and animals. *Trends in Genetics*, 1, 165–169.

Walbot, V. (1992) Strategies for mutagenesis and gene cloning using transposon tagging and T-DNA insertional mutagenesis. *Annual Review of Plant Physiology and Plant Molecular Biology*, 43, 49–82.

Walbot, V. (1996) Sources and consequences of phenotypic and genotypic plasticity in flowering plants. *Trends in Plant Science*, 1, 27–32.

Walbot, V. (1999) UV-B damage amplified by transposons in maize. *Nature*, 397, 398–399.

Walbot, V. (2000) Green chapter in the book of life. *Nature*, 408, 794–795.

Walbot, V. and Cullis, C. A. (1985) Rapid genomic change in plants. *Annual Review of Plant Physiology*, 36, 367–396.

Walker, E. L. (1998) Paramutation of the *r1* locus of maize is associated with increased cytosine methylation. *Genetics*, 148, 1973–1981.

Wessler, S., Tarpley, A., Purugganan, M., Spell, M. and Okagaki, R. (1990) Filler DNA is associated with spontaneous deletions in maize. *Proceedings of the National Academy of Sciences USA*, 87, 8731–8735.

Wessler, S. R., Bureau, T. E. and White, S. E. (1995) LTR-retrotransposons and MITEs: important players in the evolution of plant genomes. *Current Opinion in Genetics and Development*, 5, 814–821.

Wilson, W. A., Harrington, S. E., Woodman, W. L., Lee, M., Sorrells, M. E. and McCouch, S. R. (1999) Inferences on the genome structure of progenitor maize through comparative analysis of rice, maize and the domesticated panicoids. *Genetics*, 153, 453–473.

Wolffe, A. P. and Matzke, M. A. (1999) Epigenetics: regulation through repression. *Science*, 286, 481–486.

Zhang, Q., Arbuckle, J. and Wessler, S. R. (2000) Recent, extensive, and preferential insertion of members of the miniature inverted-repeat transposable element family *Heartbreaker* into genic regions of maize. *Proceedings of the National Academy of Sciences USA*, 97, 1160–1165.

Chapter 3

Evolutionary developmental biology: impact on systematic theory and practice, and the contribution of systematics

Julie A. Hawkins

ABSTRACT

Evolutionary developmental genetics brings together systematists, morphologists and developmental geneticists; it will therefore impact considerably on each of these component disciplines. The goals and methods of phylogenetic analysis are reviewed here, and the contribution of evolutionary developmental genetics to morphological systematics, in terms of character conceptualisation and primary homology assessment, is discussed. Evolutionary developmental genetics, like its component disciplines, phylogenetic systematics and comparative morphology, is concerned with homology concepts. Phylogenetic concepts of homology and their limitations are considered here, and the need for independent homology statements at different levels of biological organisation is evaluated. The role of systematics in evolutionary developmental genetics is outlined. Phylogenetic systematics and comparative morphology will suggest effective sampling strategies to developmental geneticists. Phylogenetic systematics provides hypotheses of character evolution (including parallel evolution and convergence), stimulating investigations into the evolutionary gains and losses of morphologies. Comparative morphology identifies those structures that are not easily amenable to typological categorisation, and that may be of particular interest in terms of developmental genetics. The concepts of latent homology and genetic recall may also prove useful in the evolutionary interpretation of developmental genetic data.

3.1 Introduction

Plant evolutionary developmental genetics is a cross-disciplinary science whose component disciplines have until recently been more or less separate. In botany, as Weston (2000) noted, the research streams of morphology and systematics are independent to a surprising degree. Collaborations between plant developmental geneticists and morphologists/systematists have been even fewer. Although the independent activities of plant scientists might suggest otherwise, phylogeny, development, gross morphology and genetics are causally linked as different aspects of the same phenomena (Roth, 1988). Figure 3.1 shows the interrelationships of these spheres of interest. Five levels of investigation are indicated: genetics (level 1), development (level 2), morphology (levels 3 and 4) and phylogeny (level 5). Research in these component disciplines has been limited in terms of the number of levels

In *Developmental Genetics and Plant Evolution* (2002) (eds Q. C. B. Cronk, R. M. Bateman and J. A. Hawkins), Taylor & Francis, London, pp. 32–51.

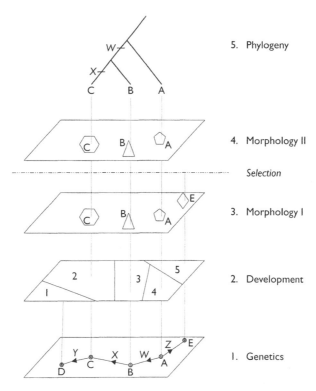

Figure 3.1 The relationship between genetics, development, morphology and phylogenetic systematics. Modified from Oster and Alberch (1982). Genetic changes W, X, Y and Z have given rise to organisms B, C, D and E from their progenitor A. These genetic changes are accompanied by changes in development (pathways 1 to 5) and morphology. Not all novel morphs are viable; organisms D and E are screened from the population of morphologies, D because of the need to maintain ontogenetic integration, and E because the novel morphology is lost due to selection or drift. The final morphologies are informative of phylogenetic relationship. Although the 'linearity' of relationship between genome, ontogeny and morphology is arguably an over-simplification (Minelli, 1998), it is accepted that molecular genetic changes direct changes in ontogeny, and that evolutionary diversification is limited by morphogenetic and morphological viability.

addressed or the number of organisms studied. Developmental geneticists build on the developmental biologists' descriptions of developmental processes to discover the developmental genetic basis of these processes and therefore the genetic basis of gross morphologies. Their interest is at levels 1, 2 and 3 of Figure 3.1, but only in terms of the genetics, development and morphology of selected model organisms and their mutants. The study of the development of an individual (level 2) has been the morphologists' endeavour. Comparative morphologists also seek to identify morphological correspondences – the relationship between two or more morphological structures. These are revealed through comparisons of development as well as gross morphology; comparative morphologists are therefore interested in within-species and between-species comparisons at levels 2 and 4. Population biologists are able to describe the evolutionary genetic processes (between levels 3 and 4) that permit a subset of morphological innovations to persist. They may have interests at or below the species level. Phylogenetic systematics is concerned with the reconstruction of relationships among species or higher taxa. Historically, comparative morphology (level 4) has provided the data used for phylogeny reconstruction, but during the last decade DNA sequences have increasingly become the markers of choice for the reconstruction of phylogenetic relationship. The genetic differences underlying morphological divergence (level 1) have yet to be thoroughly investigated as phylogenetic markers.

The advent of the 'new synthesis' between genetics, development, morphology and phylogeny can be attributed to two important new developments. First, the reconstruction of phylogeny from sequence data has provided more robust phylogenies which stimulate more rigorous interpretation of morphology (Endress *et al.*, 2000; Pennington and Gemeinholzer, 2000; Rudall, 2000). We now know much more about evolutionary relationships and the evolution of morphologies. Second, the arrival of developmental genetic data has made possible real understanding of the molecular genetic basis of development and morphology (Howell, 1998). Prior to these advances the relationship between evolution and development merited more theoretical discussion (e.g. Nelson, 1978; Patterson, 1983; Humphries, 1988; Hall, 1992), but practically the fields had not been unified into a discipline with its own research strategy.

Plant evolutionary developmental genetics is becoming established as a discipline, but in the foreseeable future it will of necessity rely on contributions from biologists trained in different research areas. Few attempts have been made to codify the role of phylogenetic systematics in evolutionary developmental genetics (Donoghue *et al.*, 1998; Baum, 2002), and even fewer to examine the implications of evolutionary developmental genetics for phylogenetic systematics. This chapter provides an overview of evolutionary developmental genetics from a systematics perspective, examines the role of phylogenetic systematics, and suggests some ways in which deeper understanding of evolutionary mechanism might influence the work of systematists.

3.2 The goals of phylogenetic systematics

Phylogenetic systematics (or cladistics) is an explicit method that now dominates comparative biology. Its goals are the reconstruction of relationships and homology assessment. Relationships are represented in the form of a phylogeny (tree or cladogram). Morphological or molecular data may be used for the reconstruction of relationship. Figure 3.2 shows the steps typical of a morphological cladistic analysis. Morphological cladistic analysis begins with an assessment of morphological variation for a group of organisms and the representation of that variation as characters and character states. For example, some plant species might possess determinate spinescent axillary shoots whereas the axillary shoots of other related species are indeterminate branches. This variation might be coded in the following way:

axillary shoots: spinescent, determinate (0), indeterminate (1)
[character: state (0), state (1)]

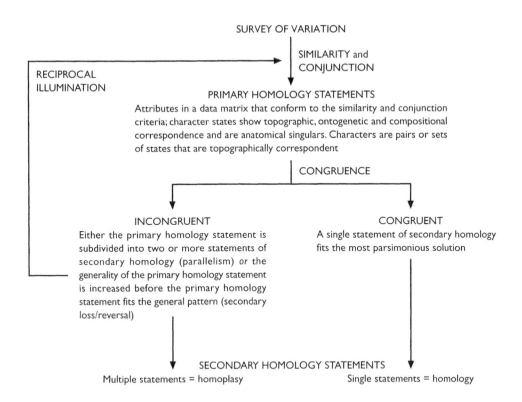

SURVEY OF VARIATION

RECIPROCAL
ILLUMINATION

SIMILARITY and
CONJUNCTION

PRIMARY HOMOLOGY STATEMENTS

Attributes in a data matrix that conform to the similarity and conjunction criteria; character states show topographic, ontogenetic and compositional correspondence and are anatomical singulars. Characters are pairs or sets of states that are topographically correspondent

CONGRUENCE

INCONGRUENT

Either the primary homology statement is subdivided into two or more statements of secondary homology (parallelism) or the generality of the primary homology statement is increased before the primary homology statement fits the general pattern (secondary loss/reversal)

CONGRUENT

A single statement of secondary homology fits the most parsimonious solution

SECONDARY HOMOLOGY STATEMENTS

Multiple statements = homoplasy

Single statements = homology

Figure 3.2 Morphological cladistic analysis. According to this framework, drawn from the work of Patterson (1982), presented by Rieppel (1988) and discussed by De Pinna (1991) and Hawkins (2000), variation is examined and primary homology statements are conceptualised following similarity and conjunction criteria. Following the application of the similarity and conjunction criteria, the congruence test is applied and primary homology statements are identified as single statements of secondary homology (congruence = homology) or multiple statements of secondary homology (incongruence = homoplasy). Where there are multiple statements of secondary homology, primary homology statements are re-examined through the process of reciprocal illumination.

Patterson (1982) formally described the similarity test, used to formulate characters and states. Character states, or putative homologies, show topographic, ontogenetic and compositional similarity. The spinescent shoots show topographic similarity in that the spines are in the axis of the leaf; ontogenetic similarity in that the developing axillary shoots have terminal buds that become senescent; and compositional similarity in that the spines share decreased histological complexity. The character state describing the spinescent forms of the axillary shoot passes all three tests, so does the character state describing the indeterminate axillary shoot. The different character states only show topographic correspondence; ontogenetically and compositionally they clearly differ. Description of a character in standard cladistic analysis therefore requires recognition of both state identity (same) and transformation (same but different) (Hawkins, 2000).

The data matrix is a set of primary homology statements. In morphological studies, parsimony analysis is used to discover the scheme of relationships that minimises the number of character state changes (other optimality criteria may be used for molecular data). When the data are traced most parsimoniously onto a phylogeny we consider them a set of secondary homology statements. Looking at the tree, we can view character state changes as transformations that occur once or more than once. In the case of homoplasy, either the primary homology statement is subdivided into two or more statements of secondary homology (parallelism) or the generality of the primary homology statement is increased before the primary homology statement fits the general pattern (secondary loss/reversal). A primary homology statement that is resolved as a synapomorphy (with single evolutionary origin and no reversals) passes the congruence test and is considered homology. The process of reconstructing the sequence of character state transformations on the tree is referred to as character optimisation (Figure 3.3).

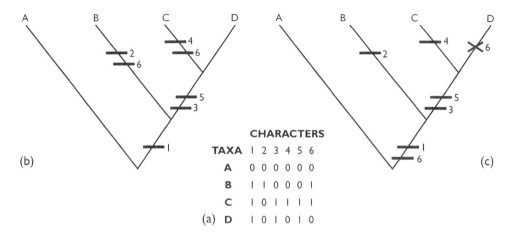

CHARACTERS

TAXA	1	2	3	4	5	6
A	0	0	0	0	0	0
B	1	1	0	0	0	1
C	1	0	1	1	1	1
(a) D	1	0	1	0	1	0

Figure 3.3 Synapomorphy and homoplasy. This simplified example shows the data matrix (3.3a) for four taxa (A, B, C and D) scored for six characters. The character states are represented as 0's and 1's. We wish to know the relationships of B, C and D relative to taxon A. The tree topology, with taxon B the sister to a group including taxon C and D, is the most parsimonious for this data matrix. Character 1 state 1 is a synapomorphy for (B, C, D). For characters 3 and 5, character state 1 is a synapomorphy for (C, D). Character state 2 of character 1 is an autapomorphy for taxon B. Character state 1 of character 6 can be optimised in two equally most parsimonious ways on the tree. Either the primary homology statement represented by character 6 is subdivided into two or more statements of secondary homology (parallelism; 3.3b) *or* the generality of the primary homology statement is increased before the primary homology statement fits the general pattern (there is secondary loss/reversal represented by the cross; 3.3c). Either way, there is more than one character transformation mapped, and we have identified homoplasy. The scheme of relationships represents a set of hypothesis of transformation. For example, there is a transformation between state 0 and state 1 of character 3 before the ancestor of C and D diverged from taxon B.

3.3 Primary homology assessment: typological thinking and evidence for transformation

Primary homology assessment, the representation of variation as characters and character states, is a fundamental and influential stage of cladistic analysis (Scotland and Pennington, 2000). The concept of a character is sometimes considered a vestige of typological thought in that characters are ideal structures with an 'eternal essence' (Mayr, 1982); their essence is still identifiable though the structures may be substantially modified by evolution (transformation). The cladistic 'way of seeing' is therefore both typological and transformational. Typological and transformational approaches to primary homology assessment are both controversial and difficult to apply (Carine and Scotland, 1999; Hawkins, 2000), and merit re-evaluation in the light of developmental genetics (Bang et al., 2000).

3.3.1 Some problems with typological thinking

Cladistic methods rely on structural categorisation: fragmentation of organisms into parts, characters and character states. The structural categorisation and typological thinking implicit in this fragmentation may be problematic. Sattler is probably the most influential critic of these approaches, rejecting structural categorisation and preferring to consider whole organisms as a continuum of process combinations rather than as comparable structures with static, alternate states (Sattler, 1984, 1994). Sattler's ideas and his critique of the cladistic method were reviewed by Weston (2000). It is pertinent here to ask why typological thinking fails and whether there are likely to be developmental genetic explanations for these failures.

Typological thinking fails when two organisms have structures that defy 1:1 categorisation. Actinomorphic and zygomorphic flowers are categorised as flowers, as they pass the similarity test (all flowers show topographic correspondence; actinomorphic flowers show ontogenetic and structural correspondence, and so do zygomorphic flowers). Although we may wish to investigate the developmental genetic basis of the evolution of the forms, and provide a process-based explanation for their relationship (see Section 3.3.2), their recognition as flowers is not necessarily problematic. There are other plant morphologies that seem to be more difficult to categorise. A few years ago I struggled to homologise pinnate and bipinnate leaves as part of a study of relationships for a group of legumes (Parkinsonia; Hawkins, 1996). Should I take a leaflet-down approach and equate rachises with pinnular rachises or a primary axis-up approach so that leaflets in the pinnate leaf are homologised with pinnular rachises in the bipinnate leaf (Figure 3.4)? Leaflets in pinnate leaves show structural and ontogenetic correspondence, whereas the primary rachises of the bipinnate leaf and the rachis of the pinnate leaf subtend the axillary bud or shoot and are topographically correspondent. It seems that topographic and structural criteria are in conflict here.

As Patterson (1982) argued, and I have shown in Section 3.2, in standard cladistic analysis characters are defined on the basis of topographic correspondence alone, whereas states also share ontogenetic and compositional correspondence. In my opinion, there need not be a coding problem for characters such as pinnate and bipinnate leaves if one accepts the primacy of topographic correspondence in character

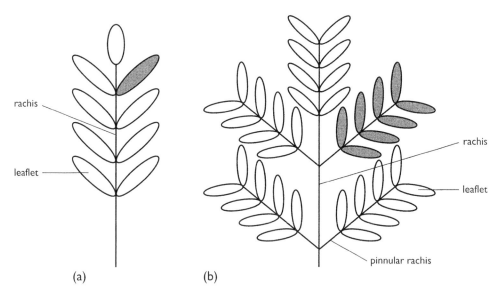

Figure 3.4 Homology assessment, pinnate and bipinnate leaves. A leaflet-down approach equates the rachis of the pinnate leaf (a) with the pinnular rachis of the bipinnate leaf (b). A primary axis-up approach equates the leaflet of the pinnate leaf with the pinnular rachis of the bipinnate leaf. If we consider the bipinnate leaf as the result of homeosis, with the leaflet of the pinnate leaf replaced by another pinnate leaf, then the leaflet of the pinnate leaf (a) is homologous to the pinnular rachis and leaflets of the bipinnate leaf (b).

definition. The primary rachis of the bipinnate leaf and the rachis of the pinnate leaf are equivalent because both subtend the axillary bud. The secondary rachis and leaflets of the bipinnate leaf correspond with (i.e. are topographically correspondent to) the leaflets of the pinnate leaf. One interpretation is that a bipinnate leaf is equivalent to a pinnate leaf that has its leaflets replaced by another set of pinnate leaves (Figure 3.4). This 'replacement' is homeosis, the complete or partial replacement of one part or structure by another part or structure of the same organism (Sattler, 1994).

It is a measure of the lack of intellectual interaction between phylogenetic systematists and morphologists noted by Weston (2000) that none of the systematists with whom I discussed the pinnate–bipinnate leaf problem with were aware of Sattler and Rutishauser's (1992) discussion of partial homology between leaves and shoots. Sattler and Rutishauser (1992) developed Arber's (1950) partial shoot theory, and showed that the early developmental stages of some pinnate leaves resemble those of a shoot and later stages become more leaf-like. Sattler and Rutishauser (1992) and Sattler (1994) argued for a continuum between simple leaves, pinnate leaves and shoots, suggesting that these structures are not amenable to structural categorisation. Sattler and Rutishauser (1992) did not refer to homeosis as an explanation for the pinnate and bipinnate leaf. Is there experimental justification for such a stance, and does it conflict with the partial shoot theory? If bipinnate leaves are to be considered homeotic mutants, with leaflets replaced by pinnate leaves, we might expect pinnate leaves and the pinnular rachises of bipinnate leaves to be equivalent in developmental genetic terms. There is molecular genetic evidence supporting the partial shoot theory: genes found in shoot meristems but absent from simple leaves are expressed in some pinnate leaves (Sinha, 1999). This expression of 'shootness' where a leaf is expected topographically is analogous to expression of 'shootness' where one would expect a leaflet. As yet, there are no data to support or refute homeosis as an explanation for bipinnate leaves.

Standard cladistic analysis uses topographic correspondence as the decisive criterion of sameness which relates structures that are different in different organisms. I have suggested that primacy of the topographic criterion is preferable in the case of pinnate and bipinnate leaves; if we consider bipinnate leaves as homeotic mutants, with leaflets replaced by pinnate leaves, we have a solution to the character coding problem. Sattler (1994) argued that primacy of topographic correspondence amounts to an *ad hoc* decision, and instead proposed that the terms 'homotopy' and 'heterotopy' be used. Homotopy should be used to describe structures that show the same or similar relative position; homomorphy would describe structures that have the same or similar special quality of structures (Sattler, 1994). Increasingly, plant biologists are arguing that heterotopic gene expression accounts for significant evolutionary innovation (e.g. Bateman and DiMichele, 2002; Baum and Donoghue, 2002; Frohlich, 2002; Schneider *et al.*, 2002). I would expect therefore that character coding which recognises homomorphy will be employed more widely in morphological cladistic analyses. In my survey of character coding (Hawkins, 2000), I referred to positional coding. For example, we might observe a contrasting distribution of glandular hairs:

glandular hairs: on pinnular rachis (0) on peduncle (1)

Of more than 1400 phylogenetic characters examined, only seven used positional coding (Hawkins, 2000). The paucity of positional characters might reflect the rarity of heterotopy in the groups studied, but is more likely due to reluctance on the part of cladists to overturn convention and recognise characters with shared structural similarity uniting states that describe contrasting topological position.

Much of the time there is general agreement between morphological and molecular phylogenies (Patterson *et al.*, 1993). This observation suggests that many morphological structures, such as actinomorphic and zygomorphic flowers, can be conceptualised as characters and states in a way that is informative of evolutionary history. A paucity of morphological characters in cladistic analysis is more often attributable to difficulties in handling continuous variables (Stevens, 1991) than to difficulties in identifying 1:1 correspondences; in practice, one cannot always distinguish actinomorphic and zygomorphic symmetry (Sattler, 1994). Molecular genetic characterisation of continuously varying traits will certainly contribute to evolutionary knowledge, but should also inform character conceptualisation.

3.3.2 Can developmental genetics provide process-based explanations for transformations?

The concept of transformation has been understood in many different ways by evolutionary biologists (Brady, 1994). Transformation may refer to a process inferred to occur between any of the following:

a.   a single structure in an extant organism and a single structure in its hypothetical ancestor;
b.   a single structure in an extant organism and a single structure in another extant organism;
c.   a single structure in an extant organism and a single structure in a fossil;
d.   a single structure in a fossil and a single structure in another fossil;
e.   a replicated structure and a single structure in another organism;
f.   a single structure to another single structure within an individual;
g.   a single structure to a series of structures within an individual.

Only the transformation between a single structure in an extant organism or fossil and a single structure in another extant organism or fossil is of interest to systematists framing primary homology assessments. In the hypothetical example outlined in Section 3.2, the spinescent axillary shoot of taxon A has a transformational relationship with the indeterminate axillary shoot of taxon B. This is the 'same but different' relation of Hawkins (2000), whereby two forms of a structure occur in mutually exclusive groups of organisms. Although the recognition of transformational relationship of this kind is central to standard cladistic analysis, transformation has become almost disreputable for some systematists, as an untested hypothesis supported only weakly by topographic correspondence alone, and with nothing concrete to say about relationship (Patterson, 1982). Three item statement analysis has been proposed as an alternative to standard cladistic analysis that avoids untestable statements of transformation (e.g. Carine and Scotland, 1999). Setting aside the intense controversy among cladists, all of these transformational relationships are of potential interest to evolutionary developmental geneticists, as all have been described as homologies (see Section 3.4).

What evidence is there for historical transformation between single structures in organisms? Hay and Mabberley (1994) argued that characters and their states are hypotheses and therefore cannot be considered to have participated in the real process of phylogenetic transformation. Transformation is necessarily an abstraction at one level: we cannot, for example, examine two extant plants and ask whether the zygomorphic flower of one is a transformation of the actinomorphic flower of another. One flower does not give rise directly to another flower (Sattler, 1984). However, it is possible to identify the genetic differences between organisms with contrasting flower structures (Baum, 2002; McLellan et al., 2002). A correlation between patterns of gene expression and morphology is suggestive of the process underlying the historical transformation: we assume that the change in form is mediated by a change in gene expression pattern. Additional evidence would be provided by a transformation experiment that produced, by introducing or down-regulating target genes, asymmetric flowers in a species that otherwise showed symmetric flowers. This experimental genetic approach could be termed a primary homology experiment if the structures being compared are 'the same but different' in a pair of extant organisms. Primary homology experiments will comprise one of the key endeavours of evolutionary developmental genetics, but could inform systematists seeking to reconstruct phylogeny.

3.4 Homology

The products of phylogenetic systematics are first and foremost a scheme of relationships, and secondarily a set of homology assessments. Patterson (1982), in explicitly equating homology with synapomorphy, provided the most prescriptive and arguably the most useful definition of homology for systematists. According to Patterson, homologous structures pass both the similarity test (they are structurally, topographically and ontogenetically correspondent) and the congruence test (i.e. they have a single evolutionary origin). A third test, the conjunction test, is not widely used; it states that homologous structures must be anatomical singulars. Homology as synapomorphy is taxic homology, since, unlike transformational homology, its recognition depends on reconstruction of the relationships among taxa.

Patterson's definition of homology is not well known in the wider biological community, where the term suffers much broader usage. As comparative developmental genetic data accumulate and are used to speak to the problem of homology assessment, the use of the term becomes broader still. The current view of homology promoted by developmental geneticists redefines it as the relationship of corresponding structures that have shared patterns of gene expression during development (Bolker and Raff, 1996). These correspondences may be between structures in one organism, for example:

> Although it was generally thought that the shoot and root meristems share only superficial similarities, recent findings indicate that the transcription factor networks that control epidermal and radial patterning in the root perform a similar function in the shoot. This argues for evolutionary homology, suggesting that one meristem was derived from another or that an ancient patterning process pre-dated the origin of either meristem and that root and shoot meristems incorporated these primitive regulatory systems.
>
> (Benfey and Weigel, 2001)

Alternatively, the patterns of gene expression may be shared by structures found in different organisms. In this case, comparison tends to be between structures whose identity is controversial. For example, Yu et al. (1999) used a transgenic approach, as well as expression analysis, to show that Gerbera pappus bristles are highly modified true sepals, and thus that pappus bristles and sepals are homologous. Mooney and Freeling (1997) have proposed that pea stipules might be homologous to the maize ligule, and argued that shared gene expression patterns would provide evidence of homology.

The equation of homology with synapomorphy excludes homologies such as shoots as roots. Only putative homologies between structures such as the ligule and stipule are transformational in the cladistic sense, with the two forms of a structure found in mutually exclusive groups of organisms (the 'same but different' relation of Hawkins, 2000). Only the relationship between, for example, maize ligules and the ligules of the most closely related grasses ('same' *sensu* Hawkins, 2000) would be considered taxic homology *sensu* Patterson (1982), since they are a synapomorphy for those grasses. These substantially different concepts have led not only to the definition of taxic and transformational homology (Eldredge, 1979; Patterson, 1982), but also biological and taxic homology (Sluys, 1996) and historical, biological and generative homology (Butler and Saidel, 2000). These authors recognised homology as synapomorphy (taxic or historical homology), homology following evolutionary transformation (transformational or biological homology), and homology as shared development (biological or generative homology). Although several authors have retreated from Patterson's 'austere definition' (Bang *et al.*, 2000: 24) and proposed alternative and complementary definitions of homology, more inclusive definitions of homology that speak to all biological levels of organisation and all relationships (taxic and transformational) are notably absent.

The trend towards primacy of expression analysis over traditional criteria in homology assessment has been criticised by McLellan *et al.* (2002), who note that a gene may be expressed in a wide variety of non-homologous structures. The dissociation between homology as evidenced by traditional criteria and as evidenced by expression analysis has been documented by several zoologists (e.g. Bolker and Raff, 1996; Wray and Abouheif, 1998), and was attributed by Wagner *et al.* (2000) to the continuing evolution of developmental mechanisms after a character has originated. Dissociation between morphological and developmental genetic homology was also noted by Albert *et al.* (1998) and Gustaffsson and Albert (1999). Gustaffsson and Albert (ibid.: 404) considered that organs and tissues are homologous in the genetic sense, 'morphology and ontogeny aside', if they express the same genes during development. Albert *et al.* (1998) suggested a new term, 'process orthology', to describe these organs and tissues with shared gene expression patterns. Butler and Saidel (2000) also looked for new terms to describe dissociations between morphological and developmental genetic homology. They suggested 'syngeny' to refer to morphological homologies with the same developmental genetic basis, and 'allogeny' to refer to morphological homologies which differ at the genetic level. Zoologists have long made a case for the uncoupling of morphological (topographic and structural/histogenetic) concepts and ontogenetic concepts (reviewed by Hall, 1992), and more recent calls for a multi-tiered system of homology recognition have also come from zoologists (e.g. Butler and Saidel, 2000). The case for an independent homology statement for *every* level of biological organisation, with separate statements for morphology, ontogeny and for developmental genes and development genetic networks, may be reiterated by botanists if data accumulate that clearly show dissociations between levels. Aside from arguments about homology, if evolution at the different levels of the organisational hierarchy is uncoupled, then these differences should be explored. How often are plant morphologies held within narrow bounds despite the diversification of the developmental genetic processes that give rise to the morphologies? What explanations are there for morphological stasis accompanied by underlying ontogenetic or genetic plasticity?

There are few examples in the botanical literature of homology assessments which conflict at different levels of biological organisation. Kaplan (2001) notes uncoupling of anatomy/histology and gross morphology: histological organisation cannot serve as a marker for the basic morphological regions of a leaf, and many of the plant genes characterised to date control histogenesis rather than morphogenesis. Although the utility, for botanists, of multi-tiered homology concepts is yet to be seen, what is more certain is that the application of a gene expression criterion in homology assessment will be problematic, both conceptually and practically. Homology may be partial, rather than absolute, and gene expression may show quantitative, positional and temporal variation (Baum and Donoghue, 2002). Should protein or transcript localisation take priority, given that the location of the transcript may not be equivalent to the site of protein activity (Lucas *et al.*, 1995)? Which components of a genetic network should one consider as indictors of homology? Are deep pattern-formation genes less informative of homology than upstream or downstream genes, or must primary and secondary genes *sensu* McLellan (2002) be distinguished in order to adequately determine morphological homologies?

Genetic comparisons are made at the genome level (where the gene itself is the unit of comparison) and at the gene level (where comparison is of nucleotide or amino acid differences). Comparisons are amenable to components of the similarity test: topographic (relative gene order, relative base pair order) and ontogenetic (intron processing) comparisons can be made. Primary homology assessments based on similarity can be evaluated in a cladistic framework, and a statement of homology at all these levels can become a statement of taxic homology. Comparison of developmental genetic networks or pathways has also been made; where networks are similar and have single origins they show taxic homology (Abouhief, 1999).

Since the early days of molecular biology, the term 'homology' has been used to refer to genetic relationships other than taxic homology as evidenced by similarity and congruence. At the gene level Patterson (1988) noted widespread use of the term 'homology' to describe sequence similarity alone; genes are often considered 95 per cent homologous if they share 95 per cent base pair identity. Such a measure is based on site identity, but site identity is not equivalent to homology since sequence data invariably contain homoplasy, such that two sequences have the same base at a position as a result of reversal or parallelism rather than shared ancestry. Most notably and most widely used are the terms 'orthology' and 'paralogy' (Fitch, 1970). Orthology (ortho = exact) refers to taxic homology *sensu* Patterson; orthologues are homologues which reflect the descent of a species. Paralagous sequences (para = parallel) are sometimes referred to as homologous, though they are duplicate genes and their phylogeny reflects gene history but not organismal history. Although Fitch's terminology is used by geneticists, a new emphasis placed on process, or function, is found in the literature with reference to gene homology. For example, '*TFL1* in *Arabidopsis* has been cloned by virtue of its homology (similarity in function) with a cloned gene in *Antirrhinum* called *CENTRORADIALIS*' (Howell, 1998: 181). Conservation of function is useful when we seek to discover orthologous genes in new species, but it is less useful as a homology concept since function may diversify as genes are recruited to secondary roles. For example, *FLO/LFY* has a role in the establishment of floral meristem identity (Coen *et al.*, 1990), and has since been shown to have been recruited to a secondary role in the specification of leaf structure in pea (Hofer *et al.*, 1997). Observations such as these tell us more about the evolution of developmental genetic networks (and gene network homology) than morphological or single-gene homology.

More recently, ways of comparing gene networks have been proposed (Abouheif, 1999). Abouheif recognised genetic networks as a distinct level of biological organisation separate from genes, morphogenesis and morphology, and provided similarity and phylogenetic criteria for assessing gene network homologies. As yet, the existing data are too partial for meaningful analysis of plant developmental networks in a phylogenetic context; nonetheless, the assessment of homologies between gene regulatory networks should soon become one of the most exciting challenges facing plant biologists. Systematists will provide the conceptual framework which will enable us to meet the challenge.

3.5 Species relationships and evolutionary developmental genetic hypotheses

The most important contribution of phylogenetic systematics for evolutionary developmental genetics is the identification of taxa and morphologies for study. The scheme of relationships has embedded within it biological hypotheses of character state gains and losses (Figure 3.3). Phylogenetic systematics allows us to pose 'How did it arise?' questions, which seek to explain morphological innovations (the origin of structures). The developmental genetic processes of the morphological structure in question are elucidated, candidate genes are identified (usually through studies of mutants), and these candidate genes are characterised in sister species lacking the innovation. Phylogenetic systematics determines the hierarchical level at which the structure arises and identifies the closest relatives lacking the structure. Exemplars for higher-level studies can be identified: these should be taxa that are not highly derived and which have few if any autapomorphies. The scheme of relationships also allows us to identify morphologies that have arisen repeatedly. In this case we might ask whether structures that have arisen more than once could be homologous at the developmental genetic level. When investigating homology or homoplasy, phylogenetics provides developmental geneticists with viable sampling strategies; without a phylogenetic framework, we do not have hypothesis-driven comparative science. For systematists, evolutionary developmental genetics may suggest new ways of thinking about parallelism and convergence (Section 3.5.1), and provide tools for the investigation of latent homology and genetic recall (Section 3.5.2).

3.5.1 Convergent and parallel evolution

Patterson (1982) provided an often-cited scheme for recognising parallelism and convergence (Table 3.1). Homology passes both the congruence and similarity tests, parallelism passes the similarity test and fails the congruence test, and convergence fails both tests. The similarity test *sensu* Patterson (1982) traditionally comprises topographic, structural/histogenetic and ontogenetic correspondence. However, alternative interpretations may be more inclusive. A maximally inclusive concept of the similarity test demands developmental genetic correspondence at gene and network levels, as well as at the topographic, structural/histogenetic and ontogenetic levels. Following a maximally inclusive concept, if we can demonstrate that a mor-

Table 3.1 Distinguishing parallelism and convergence

Morphological relation	Similarity	Congruence
homology	pass	pass
parallelism	pass	fail
convergence	fail	fail

phological attribute that has evolved many times does not show similarity at all levels, then we have convergence rather than parallelism. Attempts to explain the distribution of homoplastic characters often ask whether the same transformations (e.g. actinomorphic to zygomorphic flowers: Donoghue *et al.*, 1998) have the same developmental genetic basis. If they do, then they are parallelisms, if they do not then, according to the maximally inclusive similarity concept, they are convergences.

3.5.2 Latent homology and genetic recall

Although phylogenetic studies may show that particular innovations have occurred multiple times in different lineages, the complexity of the innovation might discredit a hypothesis of repeated *de novo* elaboration (Figure 3.5). Evolutionary biologists have long been aware that, in some cases, patterns of homoplasy may point to

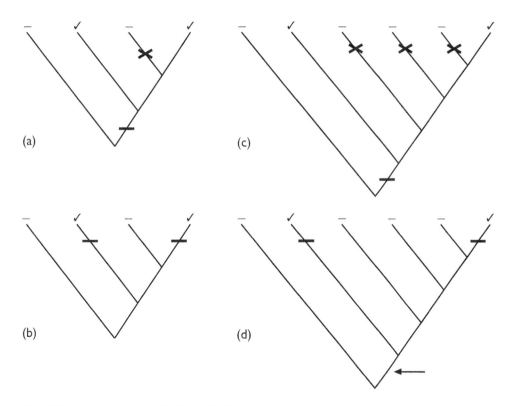

Figure 3.5 Latent homology. De Pinna (1992) argued that since absence cannot conform to any of the similarity criteria, a single gain with secondary loss is more consistent with the initial assumption of primary homology. For example, considering the presence or absence of a structure such as glandular hairs (presence = ✓; absence = −), we should prefer to optimise characters so that the structure is lost from some taxa (3.5a), rather than evolving independently more than once (3.5b). In some cases a scheme invoking secondary loss is less parsimonious (tree 3.5c has more steps than tree 3.5d). If the ability to develop the structure is a property of a clade we could think of it as a synapomorphy (arrowed), even though not all the members of that monophyletic group express the feature.

underlying similarities that reflect the shared heritage of a particular clade. De Beer (1971) discussed this problem thirty years ago, though not in these terms. Rather, he described latent homology: 'latent homology conveys the impression that beneath the homology of phenotypes there is a genetically-based homology which provides some evidence of affinity between the groups that show it' (de Beer, 1971: 10). Explicitly phylogenetic discussions of latent homology have since been provided by Brundin (1976), Cantino (1982, 1985), Saether (1983) and Sluys (1989), who argued that the occurrence of repeatedly evolved characters in several but not all of the study taxa should be used to support the inference that the group is monophyletic.

Latent homology might be explained by reiterated recruitment of retained developmental genetic pathways. Is it likely that we will find molecular genetic evidence of latent homology, or is 'rampant parallelism' really convergence? When species diverged recently latent homology seems plausible, but at higher taxonomic levels we might expect convergence (Cantino, 1985). There has been some discussion of the likelihood of re-expression of long-silent genes. Marshall et al. (1994) modelled the rate at which unexpressed genes might be expected to lose their function through mutation. They estimated that successful reactivation might be expected to occur up to 6 million years after a trait was lost. Fryer (1999) argued that observations in nature suggest that re-activation may occur over a much greater time-scale, and suggested that DNA repair mechanisms might account for the longer life-span of silent genes. Recent work on the fate of duplicated genes in vertebrates suggests a more compelling explanation for the restoration of function (Wagner, 1998). Wagner showed that approximately 50 per cent of all gene duplications will lead to functional divergence, with the remainder of duplicated genes becoming pseudogenes. Restoration of primary function to a gene which has a secondary role would seem more likely than resurrection of a pseudogene, since the nonsense mutations that characterise pseudogenes would be selected against in the functional duplicate.

There are several possible examples of latent homology in the botanical literature. Burtt (1994) reviewed recurrent forms and changes of form in angiosperms, and suggested that the secondary losses of rays in composites was due to suppression of their development and 'a mutation very little different in value from the sort that puts a block in the biosynthetic pathway of anthocyanin and results in white flowers' (Burtt, 1994: 144). Cronk (2002) considers the evolution of root nodules in Leguminosae a classic case since the morphology is complex and the phylogenetic evidence for multiple origins of the different nodulation types is strong (Sprent, 2000). The best-substantiated example genetically is the evolution of petals in the angiosperms. The traditional view is that petals have evolved several times. However, recent floral organ identity gene studies suggest that the common ancestor of monocots and eudicots possessed B-group genes and that the presence of petals in 'derived' groups is therefore an expression of latent homology (Baum and Whitlock, 1999).

The hypothetical example in Figure 3.5, and the petal and root nodule examples discussed here, consider multiple origins of structures, distinguishing presence from absence. Harlan (1982) noted the numerous multiple reversals to the plesiomorphic character states which are found in the angiosperms. He argued that evolutionary innovation could be the result of suppression, and referred to the loss of suppression as 'genetic recall', giving as an example the restoration of fertility in the lateral spikelets of barley. Genetic recall, like latent homology, is clearly worthy of developmental genetic study.

3.6 Closing comments

Sequence data have a valuable role in generating robust and rigorous phylogeny reconstructions. Evolutionary developmental genetics relies on knowledge of relationship among species; the majority of trees used to frame hypotheses for evolutionary developmental genetic investigation will be reconstructed using DNA sequence data. Yet the primary goal of evolutionary developmental genetics is deeper understanding of the relationship between morphological structures: thus evolutionary developmental genetics returns morphology to the centre stage of systematics. Now almost discredited as a source of data for the reconstruction of relationship, the role of morphological data in the reconstruction of evolutionary history has been defended by a dwindling (if vocal) minority (e.g. Bateman, 1999; Endress *et al.*, 2000). Some systematists may argue that by reconstructing relationships based on sequence data, and by mapping morphologies onto trees *a posteriori*, we gain adequate understanding of morphological evolution. There are some excellent examples of careful morphological work interpreted using sequence-based trees (e.g. Gleissberg and Kadereit, 1999). However, it is morphological cladistic analysis that has given us the framework for discussing gains and losses, convergence and parallelism, and taxic and transformational homology. Although other ways of viewing morphologies, such as Sattler's process morphology (Sattler, 1994), are extremely valuable, both the framework for conceptualisation of morphological cladistic characters and the difficulties that may be encountered in character conceptualisation are extremely illuminating. As a result of the rapidly increasing interest in evolutionary developmental genetics, morphological cladistic analysis should now experience a genuine renaissance.

ACKNOWLEDGEMENTS

I thank Quentin and Richard, co-editors, and Lynn Sanders for their support; Richard and Mark Carine for reviewing the manuscript; Elspeth Haston for discussion of leaf morphologies and for drawing Figure 3.4, Sue Mott for editorial assistance, and Sue Mitchell for help with the remaining figures.

REFERENCES

Abouheif, E. (1999) Establishing homology criteria for regulatory gene networks: prospects and challenges, in *Homology* (ed. J. A. Churchill), Novartis Foundation Symposium 222, Wiley, Chichester, pp. 207–225.

Albert, V. A., Gustafsson, M. H. G. and DiLaurenzio, L. (1998) Ontogenetic systematics, molecular developmental genetics, and the angiosperm petal, in *Molecular Systematics of*

Plants II: DNA Sequencing (eds D. E. Soltis, P. S. Soltis and J. J. Doyle), Chapman & Hall, London, pp. 349–374.

Arber, A. (1950) *The Natural Philosophy of Plant Form*. Cambridge University Press, Cambridge.

Bang, R., DeSalle, R. and Wheeler W. (2000) Transformationalism, taxism, and developmental biology in systematics. *Systematic Biology*, 49, 19–27.

Bateman, R. M. (1999) Integrating molecular and morphological evidence of evolutionary radiations, in *Molecular Systematics and Plant Evolution* (eds P. M. Hollingsworth, R. M. Batemen and R. J. Gornall), Taylor & Francis, London, pp. 432–471.

Bateman, R. M. and DiMichele, W. A. (2002) Generating and filtering major phenotypic novelties: neoGoldschmidtian saltation revisited, in *Developmental Genetics and Plant Evolution* (eds Q. C. B. Cronk, R. M. Bateman and J. A. Hawkins), Taylor & Francis, London, pp. 109–159.

Baum, D. A. (2002) Identifying the genetic causes of phenotypic evolution: a review of experimental strategies, in *Developmental Genetics and Plant Evolution* (eds Q. C. B. Cronk, R. M. Bateman and J. A. Hawkins), Taylor & Francis, London, pp. 493–507.

Baum, D. A. and Donoghue, M. J. (2002) Transference of function, heterotopy, and the evolution of plant development, in *Developmental Genetics and Plant Evolution* (eds Q. C. B. Cronk, R. M. Bateman and J. A. Hawkins), Taylor & Francis, London, pp. 52–69.

Baum, D. A. and Whitlock, B. A. (1999) Genetic clues to petal evolution. *Current Biology*, 9, R525–R527.

Benfey, P. N. and Weigel, D. (2001) Transcriptional networks controlling plant development. *Plant Physiology*, 125, 109–111.

Bolker, J. A. and Raff, R. A. (1996) Developmental genetics and traditional homology. *Bioessays*, 18, 489–494.

Brady, R. (1994) Explanation, description and the meaning of transformation in taxonomic evidence, in *Models in Phylogeny Reconstruction* (eds R. W. Scotland, D. J. Siebert and D. M. Williams), Clarendon Press, Oxford, pp. 11–29.

Brundin, L. (1976) A Neocomian chironomid and Podominae–Aphroteniinae (Diptera) in the light of phylogenetics and biogeography. *Zoologica Scripta*, 5, 139–160.

Burtt, B. L. (1994) A commentary on some recurrent forms and changes of form in angiosperms, in *Shape and Form in Plants and Fungi* (eds D. S. Ingram and A. Hudson), Academic Press, London, pp. 143–152.

Butler, A. B. and Saidel, W. M. (2000) Defining sameness: historical, biological, and generative homology. *Bioessays*, 22, 846–853.

Cantino, P. D. (1982) Affinities of the Lamiales: a cladistic analysis. *Systematic Botany*, 7, 237–248.

Cantino, P. D. (1985) Phylogenetic inference from non-universal derived character states. *Systematic Botany*, 10, 119–122.

Carine, M. A. and Scotland, R. W. (1999) Taxic and transformational homology: different ways of seeing. *Cladistics*, 15, 121–129.

Coen, E. S., Romero, J. M., Doyle, S., Elliot, R., Murphy, G. and Carpenter, R. (1990) *Floricaula*: a homeotic gene required for flower development in *Antirrhinum majus*. *Cell*, 63, 1311–1322.

Cronk, Q. C. B. (2002) Perspectives and paradigms in plant evo-devo, in *Developmental Genetics and Plant Evolution* (eds Q. C. B. Cronk, R. M. Bateman and J. A. Hawkins), Taylor & Francis, London, pp. 1–14.

de Beer, G. R. (1971) *Homology, an Unsolved Problem*. Oxford University Press, London.

De Pinna, M. C. C. (1992) Concepts and tests of homology in the cladistic paradigm. *Cladistics*, 7, 367–394.

Donoghue, M. J., Ree, R. H. and Baum, D. A. (1998) Phylogeny and the evolution of flower symmetry in the Asteridae. *Trends in Plant Science*, 3, 311–317.

Eldredge, N. (1979) Alternative approaches to evolutionary theory. *Bulletin of the Carnegie Museum of Natural History*, 13, 7–19.

Endress, P. K., Baas, P. and Gregory, M. (2000) Systematic plant morphology and anatomy – 50 years of progress. *Taxon*, 49, 401–434.

Fitch, W. M. (1970) Distinguishing homologous from analogous proteins. *Systematic Zoology*, 19, 99–113.

Frohlich, M. W. (2002) The Mostly Male theory of flower origins: summary and update regarding the Jurassic pteridosperm *Pteroma*, in *Developmental Genetics and Plant Evolution* (eds Q. C. B. Cronk, R. M. Bateman and J. A. Hawkins), Taylor & Francis, London, pp. 85–108.

Fryer, G. (1999) The case of the one-eyed brine shrimp: are ancient atavisms possible? *Journal of Natural History*, 33, 791–798.

Gleissberg, S. and Kadereit, J. W. (1999) Evolution of leaf morphogenesis: evidence from developmental and phylogenetic data in Papaveraceae. *International Journal of Plant Sciences*, 160, 787–794.

Gustaffson, M. H. G. and Albert, V. A. (1999) Inferior ovaries and angiosperm diversification, in *Molecular Systematics and Plant Evolution* (eds P. M. Hollingsworth, R. M. Bateman and R. J. Gornall), Taylor & Francis, London, pp. 403–431.

Hall, B. K. (1992) *Evolutionary Developmental Biology*. Chapman & Hall, London.

Harlan, J. R. (1982) Human interference with grass systematics, in *Grasses and Grasslands* (eds J. R. Estes, R. J. Tyrl and J. N. Brunker), University of Oklahoma Press, Norman, pp. 42–47.

Hawkins, J. A. (1996) *Systematics of* Parkinsonia L. *and* Cercidium Tul. *(Leguminosae: Caesalpinioideae)*. D.Phil. thesis, University of Oxford, UK.

Hawkins, J. A. (2000) A survey of primary homology assessment: different botanists perceive and define characters in different ways, in *Homology and Systematics: Coding Characters for Phylogenetic Analysis* (eds R. W. Scotland and R. T. Pennington), Taylor & Francis, London, pp. 22–53.

Hay, A. and Mabberley, D. J. (1994) On perception of plant morphology: some implications for phylogeny, in *Shape and Form in Plants and Fungi* (eds D. S. Ingram, and A. Hudson), Academic Press, London, pp. 101–118.

Hofer, J., Turner, L., Hellens, R., Ambrose, M., Matthews, P., Michael, A. and Ellis, N. (1997) UNIFOLIATA regulates leaf and flower morphogenesis in pea. *Current Biology*, 7, 581–587.

Howell, S. H. (1998) *Molecular Genetics of Plant Development*. Cambridge University Press, Cambridge.

Humphries, C. J. (ed.) (1988) *Ontogeny and Systematics*. British Museum (Natural History), London.

Kaplan, D. R. (2001) Fundamental concepts of leaf morphology and morphogenesis: a contribution to the interpretation of molecular genetic mutants. *International Journal of Plant Sciences*, 162, 465–474.

Lucas, W. J., Bouchepillon, S., Jackson, D. P., Nguyen, L., Baker, L., Ding, B. and Hake, S. (1995) Selective trafficking of KNOTTED1 homeodomain protein and its messenger-RNA through plasmodesmata. *Science*, 270, 1980–1983.

McLellan, T., Shephard, H. L. and Ainsworth, C. (2002) Identification of genes involved in evolutionary diversification of leaf morphology, in *Developmental Genetics and Plant Evolution* (eds Q. C. B. Cronk, R. M. Bateman and J. A. Hawkins), Taylor & Francis, London, pp. 315–329.

Marshall, C. R., Raff, E. C. and Raff, R. A. (1994) Dollo's law and the death and resurrection of genes. *Proceedings of the National Academy of Sciences*, 91, 12283–12287.

Mayr, E. (1982) *The Growth of Biological Thought. Diversity, Evolution, and Inheritance.* Belknap Press, Cambridge, MA.

Minelli, A. (1998) Molecules, developmental modules, and phenotypes: a combinatorial approach to homology. *Molecular Phylogenetics and Evolution*, 9, 340–347.

Mooney, M. and Freeling, M. (1997) Using regulatory genes to investigate the evolution of leaf form. *Maydica*, 42, 173–184.

Nelson, G. (1978) Ontogeny, phylogeny, paleontology, and the biogenetic law. *Systematic Zoology*, 27, 324–345.

Oster, G. and Alberch, P. (1982) Evolution and bifurcation of developmental systems. *Evolution*, 36, 444–459.

Patterson, C. (1982) Morphological characters and homology, in *Problems of Phylogeny Reconstruction* (eds K. A. Joysey and A. E. Friday), Systematics Association Special Volume 21, Academic Press, London, pp. 21–74.

Patterson, C. (1983) How does phylogeny differ from ontogeny?, in *Development and Evolution* (eds B. C. Goodwin, N. Holder and C. C. Wylie), British Society for Developmental Biology Symposium 6, Cambridge University Press, Cambridge, pp. 1–31.

Patterson, C. (1988) Homology in classical and molecular biology. *Molecular Biology and Evolution*, 5, 603–625.

Patterson, C., Williams, D. M. and Humphries, C. J. (1993) Congruence between molecular and morphological phylogenies. *Annual Review of Ecology and Systematics*, 24, 153–188.

Pennington, R. T. and Gemeinholzer, B. (2000) Cryptic clades, fruit wall morphology and biology of *Andira* (Leguminosae: Papilionoideae). *Botanical Journal of the Linnean Society*, 134, 267–286.

Roth, V. L. (1988) The biological basis of homology, in *Ontogeny and Systematics* (ed. C. J. Humphries), British Museum (Natural History), London, pp. 1–27.

Rudall, P. J. (2000) 'Cryptic' characters in monocotyledons: homology and coding, in *Homology and Systematics: Coding Characters for Phylogenetic Analysis* (eds R. W. Scotland and R. T. Pennington), Taylor & Francis, London, pp. 114–123.

Saether, O. A. (1983) The canalised evolutionary potential: inconsistencies in phylogenetic reasoning. *Systematic Zoology*, 32, 343–359.

Sattler, R. (1984) Homology – a continuing challenge. *Systematic Botany*, 9, 382–394.

Sattler, R. (1994) Homology, homeosis and process morphology in plants, in *Homology: The Hierarchical Basis of Comparative Biology* (ed. B. K. Hall), Academic Press, San Diego, CA, pp. 424–475.

Sattler, R. and Rutishauser, R. (1992) Partial homology of pinnate leaves and shoots. Orientation of leaflet inception. *Botanische Jahrbücher für Systematik*, 114, 61–79.

Schneider, H., Pryer, K. M., Cranfill, R., Smith, A. R. and Wolf, P. G. (2002) Evolution of vascular plant body plans – a phylogenetic perspective, in *Developmental Genetics and Plant Evolution* (eds Q. C. B. Cronk, R. M. Bateman and J. A. Hawkins), Taylor & Francis, London, pp. 330–364.

Scotland, R. and Pennington, R. T. (eds) (2000) *Homology and Systematics: Coding Characters for Phylogenetic Analysis.* Taylor & Francis, London.

Sinha, N. (1999) Leaf development in angiosperms. *Annual Review of Plant Physiology and Plant Molecular Biology*, 50, 419–446.

Sluys, R. (1989) Rampant parallelism: an appraisal of the use of nonuniversal derived character states in phylogenetic reconstruction. *Systematic Zoology*, 38, 350–370.

Sluys, R. (1996) The notion of homology in current comparative biology. *Journal of Zoological Systematics and Evolutionary Research*, 34, 145–152.

Sprent, J. I. (2000) Nodulation as a taxonomic tool, in *Advances in Legume Systematics, Part 9* (eds P. S. Herendeen and A. Bruneau), Royal Botanic Gardens, Kew, pp. 21–39.

Stevens, P. F. (1991) Character states, morphological variation, and phylogenetic analysis: a review. *Systematic Botany*, 16, 553–583.

Wagner, A. (1998) The fate of duplicated genes: loss or new function? *Bioessays*, 20, 785–788.

Wagner, G. P., Chiu, C. H. and Laubichler, M. (2000) Developmental evolution as a mechanistic science: the inference from developmental mechanisms to evolutionary processes. *American Zoologist*, 40, 819–831.

Weston, P. H. (2000) Process morphology from a cladistic perspective, in *Homology and Systematics: Coding Characters for Phylogenetic Analysis* (eds R. W. Scotland and R. T. Pennington), Taylor & Francis, London, pp. 124–144.

Wray, G. A. and Abouheif, E. (1998) When is homology not homology? *Current Opinion in Genetics and Development*, 8, 675–680.

Yu, D. Y., Kotilainen, M., Pollanen, E., Mehto, M., Elomaa, P., Helariutta, Y., Albert, V. A. and Teeri, T. H. (1999) Organ identity genes and modified patterns of flower development in *Gerbera hybrida* (Asteraceae). *Plant Journal*, 17, 51–62.

Chapter 4

Transference of function, heterotopy and the evolution of plant development

David A. Baum and Michael J. Donoghue

ABSTRACT

The concept of transference of function, developed by E. J. H. Corner, refers to situations in which a particular ecological function carried out by one part of an ancestor is transferred to another spatial location in a descendant species. Under this view it is necessary that both the ancestral and derived structures fulfill the same biological role and that there be phylogenetic continuity between the use of the ancestral and derived structures. Transference of function can entail heterotopy, where a genetic program formerly expressed in the ancestral location comes to be expressed in the derived location. We refer to transfer of function via heterotopy as homologous transference of function, because the expression of genes typical of one structure in another structure constitutes the sharing of genetic identity between those structures. We distinguish "homeoheterotopy," the transfer of genetic identity among pre-existing structural modules, which can explain transference of function, from "neoheterotopy," the production of a module in a novel location, which cannot. We discuss the hypothesis that transference of function generally involves structures in close physical proximity, briefly explore the consequences of homeoheterotopy for concepts of homology, and consider the possibility that certain groups of organisms, such as angiosperms, are especially prone to homeoheteropy and transference of function.

4.1 Introduction

The great tropical botanist E. J. H. Corner was a keen observer of plant structure and diversity. In the course of his work he observed a pattern that he considered noteworthy because of its implications for the mode of morphological evolution. He noticed that among closely related plants one often finds that different species fulfill the same ecological function through the elaboration of different organs. For example, fleshy structures associated with seeds can be derived from quite different organs yet fulfill the same role of attracting vertebrate dispersers (Corner, 1949a, b). In a paper presented to the Linnean Society to commemorate the 100th anniversary of the Darwin and Wallace presentations on natural selection, Corner (1958) enumerated many putative examples of transference of function (Table 4.1) and discussed developmental mechanisms that might underlie this phenomenon. At that time, however, too little was know about phylogenetic theory and mechanisms of

In *Developmental Genetics and Plant Evolution* (2002) (eds Q. C. B. Cronk, R. M. Bateman and J. A. Hawkins), Taylor & Francis, London, pp. 52–69.

Table 4.1 Examples of transference of function proposed by Corner (1958)

Function	Alternative locations	Taxa involved
Anthocyanin production	Leaf underside; petiole; node; internode; young leaves; old leaves	Angiosperms
Petal function	Petals; calyx[1]	*Saraca* vs. *Pahudia* (Fabaceae)
Initiation of flowers	Leaf axil; leaf primordium itself	Angiosperms vs. Brassicaceae (*Capsella*)[2]
Flower	Flower; inflorescence	*Euphorbia*, Asteraceae etc.
Compound fruit	Infructescence; simple fruit	*Parkia* vs. *Archidendron* (Fabaceae)[3]
Production of ovules	Carpels; receptacular tissue or sepal internodes	Apocarpous vs. syncarpous angiosperms
Fleshy fruit-associated structures	Peduncle; pericarp	*Anacardium* vs. *Mangifera* (Anacardiaceae)
	Pericarp; receptacle	*Rubus* vs. *Fragaria* (Rosaceae)
	Pericarp; sepals; inflorescence axis	Unspecified Moraceae vs. *Morus* vs. *Artocarpus*
	Pericarp; inflorescence axis	*Carludovica* (Cyclanthaceae) or *Monstera* (Araceae) vs. Arecaceae
	Pericarp; sepals; receptacle	*Pernettya* vs. *Gaultheria* vs. *Vaccinium* (Ericaceae)[7]
Wings	Seeds; fruit	Species of *Ricotia* (Brassicaceae)
	Seeds; fruit	*Sterculia alata* vs. *Tarrietia* (Malvaceae, Sterculioideae)
	Carpel ridges; persistent sepals	*Rheum* vs. *Rumex* (Chenopodiaceae)
Viscin layer	External vascular bundles	Loranthoideae vs. Viscoideae (Loranthaceae)
Accumulation of fibers, crystals and specialized cells	Outer integument of seed; "middle" integument	Various Annonaceae[4]
Malpighian cells (macrosclereids)	Outer seed epidermis; hypodermis; outer integument; inner integument	Fabaceae vs. Cucurbitaceae vs. Myristicaceae vs. Malvaceae *s.l.*, Euphorbiaceae, etc.
Fleshy seed-associated structures	Sarcotesta; funicular aril; micropylar aril; chalazal aril; aril borne on the seed opposite the chalaza	Various angiosperms[5]
Leaf development	Leaf; shoot	Phyllantheae (Euphorbiaceae)
Primary stem development	Primary SAM; cotyledon axillary SAM	Most angiosperms vs. some Fabaceae
Bearing of inflorescences	Shoot; root	Most angiosperms vs. *Taeniophyllum* (Orchidaceae)
Photosynthesis	Leaves; roots	Most angiosperms vs. some Orchidaceae
Bearing of flowers and roots	Stem; cotyledon	Most Gesneriaceae vs. some *Streptocarpus*[6]
Mesophyll production	Leaf lamina; petiole	Most dicots vs. e.g. *Plantago*
Light interception	Leaflets; petiole; stipules	Different species of *Lathyrus*
	Leaves; stems	Most dicots vs. Cactaceae

Notes:
1   Specifically Corner emphasizes the transfer of vascular supply.
2   Corner's interpretation is contradicted by recent studies suggesting that the flower meristem remains axillary but the subtending leaf is suppressed (McConnell and Barton, 1998).
3   *Archidendron* has several free carpels per flower and, in fruit, resembles the infructescence of *Parkia*.
4   Discussed in more detail by Corner (1949a).
5   Discussed in more detail by Corner (1949a, b).
6   There have been several independent origins of the phyllomorph condition (Jong and Burtt, 1975; Möller and Cronk, 2001).
7   See the phylogenetic analysis of Powell and Kron (2001).

plant development for transference of function to be framed as a testable evolutionary–developmental hypothesis. Our aim in this chapter is to revisit Corner's idea with the help of a phylogenetic perspective and an improved understanding of plant developmental genetics. We clarify the definition of transference of function and related concepts, and then speculate on the importance of these phenomena in plant evolution.

4.2 Definition of transference of function

Corner (1958: 33) provided the following description of transference of function:

> a property which occurs in an organ, tissue or cell-layer in one case may occur in other parts of the body in other cases. The property is the same, but its site of development has shifted ... the transference of function, as I have called the process, is a method of evolution.

In trying to provide a modern redefinition of transference of function we faced a difficult decision. One option would be to follow Corner's likely intent and define the concept so as to require the transfer of some developmental program between structures. However, to evolutionary biologists the term "function" refers to a biological role fulfilled by a structure, often with the further implication that the structure evolved by natural selection to fulfill that biological role (Gould and Vrba, 1982). Bearing this in mind, one could instead use the term "transference of function" to refer to cases in which biological roles are transferred among structures regardless of whether any developmental programs have been transferred.

To clarify these two alternative definitional schemes, consider a hypothetical case in which there is a transition from wind-dispersed fruit endowed with a pappus of feathery (calyx-derived) hairs to fruit with (receptacle-derived) wings. Assuming that this transfer did not involve the transfer of any genetic programs from hairs to wings (which is likely because the properties that make a hair an efficient aid to dispersal are not the same properties as are relevant in the case of a wing), should this event be considered "transference of function?" Given the literal reading of the phrase, the answer is "yes," since an ecological function, namely dispersal, has been transferred. Given a genetic view, the answer would be "no," since no heritable properties have been transferred from the calyx to the receptacle.

We propose framing the definition of transference of function in terms of the biological roles fulfilled by structures:

> Transference of function from structure A to structure B means that an ecologically meaningful biological role is carried out by structure A in species *a* and structure B in species *b*, and there was a direct transition from using A to using B in an ancestor of *b*.

Direct transition could (and probably will often) entail a transitional stage during which the function is fulfilled by both structure A and B. To accommodate those cases in which transfer of function is accomplished by the transfer of genetic identity we recognize a subcategory, homologous transfer of function (see Section 4.5.1).

4.3 Testing hypotheses of transference of function

The concept of transference of function is a useful tool when interpreting plant diversity. Corner expected little more than this from the concept. Nonetheless, much more will be gained if transference of function is framed in terms of testable hypotheses. Given our definition, a hypothesis of transference of function can be subjected to two phylogenetic tests. These are comparable to tests used in the study of adaptation, in that one test deals with historical genesis and the other with current utility (Gould and Vrba, 1982; Baum and Larson, 1991).

4.3.1 Phylogenetic continuity

The definition of transference of function requires that there be a direct transition from the use of the donor to recipient structure. This requires, first, that all intervening ancestors should have carried out the ecological function with one or the other or both structures (Figure 4.1). Additionally, it is necessary that the common ancestor of the focal species should have used one or the other structure, *but not both*. This follows because otherwise there was not transference of function but differential contraction of function (Figure 4.1d). These two criteria can be evaluated using phylogenetic information about the taxa in question combined with methods of ancestral-state reconstruction (e.g. Schluter *et al.*, 1997; Ree and Donoghue, 1998). Below, we use two examples to illustrate the two ways in which phylogenetic data can refute a hypothesis of transference of function.

Fragaria (strawberries) and *Rubus* (including blackberries and raspberries) both have colorful, palatable tissues associated with the fruit that facilitate endozoochory. In *Rubus* the walls of the free carpels fulfill this function, whereas in *Fragaria* it is the swollen receptacle. Corner (1958) hypothesized that this represented transference of function. Current knowledge of the phylogeny of Rosoideae (Eriksson *et al.*, 1998), however, implies that *Rubus* and *Fragaria* are only distantly related: *Fragaria* is nested within *Potentilla* (a genus whose traditional circumscription makes it massively paraphyletic), whereas *Rubus* is affiliated with the genera *Geum*, *Fallugia* and *Waldsteinia*. Given that *Potentilla* (with one possible exception) and all remaining Rosoideae have dry fruits, the reconstruction of ancestral states using simple parsimony suggests that the common ancestor of *Rubus* and *Fragaria* had neither a fleshy pericarp nor a swollen receptacle. It is necessary to weight gains over losses more than 4:1 in generalized parsimony reconstruction before the ancestral state is switched to fleshiness. If we conclude on this basis that fleshiness did in fact evolve independently in blackberries and strawberries, this would serve to refute Corner's (1958) hypothesis of transference of function.

Another example can be used to illustrate the idea that the ancestral function should be restricted to one or the other structure in order to imply transference of function. Corner (1958) hypothesized that the swollen, fleshy, brightly-colored peduncle (hypocarp) of the cashews (*Anacardium* spp.) arose by transference of function from the fleshy fruit wall (pericarp) that attracts dispersal agents in mangoes (*Mangifera*). Based on molecular data (Pell, pers. comm. 2000), *Mangifera* and *Anacardium* fall in a well-supported clade with three other genera of Anacardiaceae: *Semecarpus*, *Gluta* and *Sorindeia*. The topology of this clade is resolved convincingly as (*Semecarpus*(*Anacardium*(*Gluta*(*Mangifera*,*Sorindeia*)))).

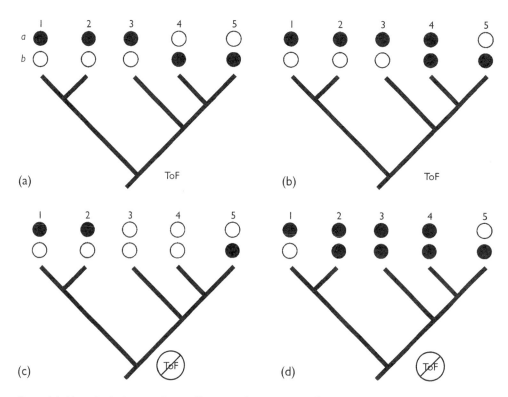

Figure 4.1 Hypothetical examples to illustrate the criterion of phylogenetic continuity in tests of transference of function (ToF). Five species (1–5) are related as shown. Taxa 1 and 5 differ in that taxon 1 uses structure *a* whereas taxon 5 uses structure *b*. The scenario depicted in Figure 4.1a implies (using parsimony) transference of function directly from using *a* to *b* and thus satisfies the criterion of phylogenetic continuity. The scenario depicted in Figure 4.1b implies progressive transference of function from using *a* to using *a* + *b* to using just *b* and thus also satisfies the criterion of phylogenetic continuity. The scenario depicted in Figure 4.1c implies (using parsimony) the existence of ancestors that used neither *a* nor *b* to fulfill the function and thus contradicts the criterion of phylogenetic continuity. The scenario depicted in Figure 4.1d implies (using parsimony) an ancestor that used both *a* and *b* to fulfil the function with independent contraction to using either *a* or *b* and, thus, fails the criterion of phylogenetic continuity.

All species in this clade, except those in *Anacardium*, have a fleshy pericarp, suggesting that a fleshy pericarp is plesiomorphic. However, *Semecarpus* also has a swollen hypocarp resembling that seen in *Anacardium* (but less juicy and colorful). As a result, it is plausible that initially fleshiness was expressed in both the pericarp *and* hypocarp and that later cashews and mangoes *lost* this property from the pericarp and hypocarp respectively. This result, if confirmed, would contradict Corner's hypothesis of transference of function.

As shown above, one does have the ability to reject a hypothesis of transference of function based on phylogenetic data. Nonetheless, it should be noted that in some cases one might be tempted to rescue a hypothesis by proposing the existence of a further structure that has acted as a repository of the function. For example, one might look at the strawberry/raspberry case and suggest that some other part of the plant, perhaps the stem, was the source of the fleshiness that was transferred independently to the receptacle and pericarp. However, this would not so much rescue the original hypothesis as replace it with a new hypothesis, subject also to a test of phylogenetic continuity.

4.3.2 Functional equivalence

The second, testable implication of a hypothesis of transference of function is that the structures should be fulfilling the same ecological function. This could be explored by standard ecological/functional studies. To clarify the content of such tests we will use an example that may well pass the test of phylogenetic continuity. Most banisterioid Malpighiaceae have wings that develop from the carpel wall during fruit maturation and are assumed to aid in fruit dispersal. The genus *Dicella*, however, lacks carpel wings but its five calyx lobes are expanded and adopt a wing-like form in fruit (Chase, 1981). The hypothesis one might posit is that this represents transference of fruit dispersal function from the gynoecium to the calyx. Based on a recent phylogenetic analysis (Davis *et al.*, 2001), *Dicella* is embedded in a large clade of wing-fruited taxa. The sister to *Dicella*, *Tricomaria*, has membranous outgrowth of the pericarp, but instead of forming wings these have been modified into bristles, which are unlikely to play a role in wind dispersal due to their small size compared to the mass of the fruit. While the Davis *et al.* tree is consistent with phylogenetic continuity between wind dispersal via carpel and calyx wings, one cannot entirely rule out a loss of wind dispersal in the common ancestor of *Dicella* and *Tricomaria*, followed by its re-evolution (via calyx wings) in *Dicella*.

If we assume that *Dicella* does indeed pass the test of phylogenetic continuity, it would be appropriate to consider tests of the ecological function of these traits. Tests of ecological equivalence would involve showing that both the carpel and sepal wings function as aids to dispersal. For example, in both *Dicella* and species with carpel wings one could determine whether the dispersal shadow obtained when wings are removed is skewed towards short dispersal distances relative to those of intact fruit. Such experiments have not, to our knowledge, been carried out in Malpighiaceae but have been used to study other taxa with winged diaspores (Augspurger, 1986; Augspurger and Franson, 1987; Matlack, 1987; Sipe and Linnerooth, 1995). If one did not find evidence for a role of wings in seed dispersal alternative hypotheses, such as function in water dispersal (Chase, 1981), might then be considered.

One might suppose that there would be an evolutionary shift in the structure performing a function only if the derived structure fulfilled the ecological role better than the original structure. Thus, in the *Dicella* example, one might predict that sepal wings would be more effective at wind dispersal than carpel wings. However, derived characters are not always selectively better because natural selection is not always involved. And, even if selection was responsible for the transition, the functional transfer could have been driven by other components of fitness (Baum and Larson, 1991). For example, wings could shift location due to impacts on predator evasion, seed germination, or desiccation avoidance.

4.4 Genetic identity of organismic structures

Transference of function implies the movement of properties between the structural elements that make up an organism. Therefore, before initiating a discussion of the mechanisms underlying transference of function, it is important to clarify the notion of the genetic identity of organismic structures. This is problematic because of uncertainty as to whether the parts of organisms ("characters") can be viewed as individuals and, if so, what underlying phenomena serve to individuate them (e.g. Wagner, 1989). We will skirt these issues and assume that characters are somehow individuated and that they may be grouped into classes that exist in multiple organisms in a population and/or many times in a single organism (e.g. epidermal cells, petals, cotyledons, root hairs).

Considering a particular class of characters (say, those organs we call petals), there will be a particular combination of genes that is characteristic of that kind of structure in a species under study. If we consider this combination of genes to give the structure its current genetic identity, then expansion or movement of the expression of those genes to a different location would result in a transfer of genetic identity. Note that genetic identity is here understood to be a local property of a particular species at a particular time, rather than a global property of a "type" (in the classic sense) of structure. Because the genes that characterize categories such as "petal" or "leaf" will constantly evolve, it is very unlikely that there exists a set of genes that characterize every organ conventionally assigned to these typological classes. Thus, gene expression data may be of little help in assigning structures to such categories.

Under our view of genetic identity, when a structure expresses new genes, its genetic identity changes. If those newly expressed genes were formerly part of the genetic identity of another structure, then genetic identity has, in a sense, been transferred from the donor to the recipient character. For example, if all the genes expressed in the petal of an ancestor came to be expressed in a position that was previously occupied by a sepal (and sepal-specific genes were turned off), the resulting structure would share genetic identity with a petal while showing positional homology to a sepal. Alternatively, some but not all of the genes could show modified expression such that a subset of the petal-specific genes came to be expressed in the calyx whorl and a subset of the sepal genes were turned off (or were not turned on) in that whorl. In this case one can interpret the resultant structure as having a mixture of petal and sepal properties. The implications of this view of genetic identity are discussed further in Section 4.6.2. Here, the key point is that the identity of a structure is in some way linked to the developmental regulatory genes that are expressed in that structure relative to those that were expressed in ancestral organisms.

4.5 Developmental causes of transference of function

4.5.1 Homologous versus non-homologous transfer of function

Given our modern understanding of developmental and evolutionary mechanisms it is useful to distinguish two broad explanations for the pattern of transference of function. The first explanation is independent evolution or non-homology. This was illustrated in Section 4.2 with a hypothetical example of the transfer of dispersal function between hairs and wings. In this case none of the properties that permit the new structure to fulfill the function evolved by activating the expression of those genes that allowed the old structure to complete that function. Alternatively transference of function could involve the transfer of genetic identity between structures. Given that we believe there is a close link between homology and genetic identity, we think it is appropriate to refer to the latter pattern as "homologous transference of function." However, even in cases of homologous transference of function, there could be some properties that facilitate completion of the ecological function but that evolved *de novo* in the derived structure rather than being transferred from the ancestral structure. We would consider these to be valid cases of homologous transference of function so long as *some* genetic properties arose by transfer from the donor structure.

Homologous transference of function was the basis of Corner's (1949b, 1958) claim that superficially similar structures that are traditionally interpreted as being non-homologous may, in terms of genetic identity, show some degree of "hidden" homology (see also Sattler, 1988). For example, consider the superficial similarity of pinnate leaves and branch systems, despite the supposed homology of pinnate leaves and simple leaves (as indicated traditionally by determinate growth and the lack of axillary meristems). Corner's (1958: 37) suggestion that this similarity could reflect the sharing of genetic mechanisms has since been supported by morphological developmental work (Sattler and Rutishauser, 1992). Additionally, molecular studies have found that genes that act in shoot meristems and are absent from simple leaves may be expressed in compound leaves (e.g. Sinha *et al.*, 1993; Hareven *et al.*, 1996; Chen *et al.*, 1997). If this interpretation is correct, the transfer of shoot properties and associated ecological functions to leaves would be an example of homologous transference of function.

4.5.2 A definitional scheme for heterotopy

Homologous transference of function implies that developmental events come to take place in different parts of the organism. Hence, it must entail heterotopy: evolutionary change in the spatial location of a developmental program (Bateman, 1994; Zelditch and Fink, 1996). Heterotopy is a concept that has achieved little attention relative to the reams of literature on heterochrony: evolutionary change in the relative timing of developmental processes (e.g. Alberch *et al.*, 1979; Guerrant, 1988; Raff and Wray, 1989; Bateman, 1994). Nonetheless, it would seem to be an important mechanism in plant developmental evolution (Kellogg, 2000). It has been claimed that plant evolution may often entail changes to the promoter/enhancer elements that cause changes in the expression patterns of regulatory genes (e.g. Doebley and Lukens, 1998). Such changes in the expression of developmental regulatory genes will usually result in changes in the spatial location of downstream developmental phenomena and, thus, would frequently result in heterotopy.

We think it is valuable to distinguish two kinds of heterotopy. *Neoheterotopy* refers to cases in which a structure is generated in a novel location. For example, *Phyllonoma* is usually interpreted as having an inflorescence that has been transferred to the surface of a leaf (Dickinson and Sattler, 1974). *Homeoheterotopy*, in contrast, refers to cases in which genetic identity is transferred from a donor structure to another, pre-existing recipient structure. For example, the transfer of wing-like properties from the carpels to the sepals in the ancestor of *Dicella* could have entailed homeoheterotopy. *Homeosis* is here defined as a special case of homeoheterotopy in which all aspects of genetic identity are conferred upon the recipient structure.

Our definitional scheme is summarized in Figure 4.2. It is important to note that, notwithstanding our admiration for his contributions to developmental botany, we have chosen not to use the definitions proposed by Sattler (1988). His broader use of the term "homeosis" is virtually synonymous with our homeoheterotopy. We prefer to maintain a narrow definition of homeosis as complete homeoheterotopy because this terminology accords with current usage in molecular developmental biology. Additionally, Sattler (1988) took a narrow view of "heterotopy" that closely resembles our neoheterotopy. We prefer a broader notion because it frees up "homeosis" for the more restricted usage and because under Sattler's (1988) scheme there is no term for the combination of homeoheterotopy and neoheterotopy.

The definitional scheme proposed here suggests that homologous transference of function must entail homeoheterotopy (including homeosis). This follows because transference of function involves the transfer of properties among pre-existing structures. However, homeoheterotopy and homologous transference of function are not synonymous because heterotopy can occur without any ecological function being translocated.

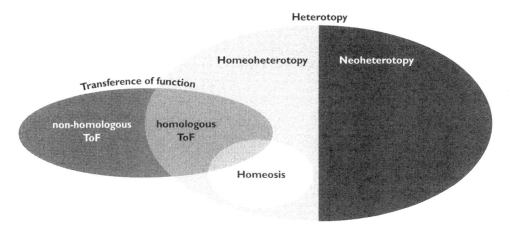

Figure 4.2 Schematic showing the relationship between some key developmental concepts. The large oval represents heterotopy: evolution via a change in the spatial location of developmental phenomena. This is divided into two halves based on whether (a) the developmental program is shifted to a pre-existing structural module (homeoheterotopy) or (b) a new structural module arises (neoheterotopy). Homeosis is here understood as complete homeoheterotopy, wherein all elements of the genetic identity of one structure come to replace those in another structure. Transference of function (ToF) refers to cases in which an ecological function performed by one structure comes to be carried out by another structure. This can involve the transfer of genetic properties from the old to the new structure (homologous ToF: a subset of homeoheterotopy) or it might not involve any genetic transfer (non-homologous ToF).

4.5.3 Fuzzy boundaries between categories

Definitional schemes for biological concepts may have heuristic value even when the boundaries between concepts are not always absolute or objective. We hope this is the case here because many of the boundaries depicted with solid lines in Figure 4.2 are, in fact, fuzzy.

The distinction between homeo- and neoheterotopy is a case in point. Consider the hypothetical transfer of the property of redness from an aril to the testa (as suggested within *Pithecellobium* by Corner, 1958), which could be considered as homeoheterotopy via the expression of pigment biosynthetic genes in the testa. If instead the property transferred was the production of glandular trichomes, then one might lean towards an interpretation of neoheterotopy in which trichome production came to be ectopically activated in a new location. This may seem curious since indumentum and pigmentation are both properties of epidermal tissues. The distinction between homeoheterotopy and neoheterotopy thus depends on whether one considers the property transferred to be a property of a character or an individuated character in its own right. Pigmentation is hard to see as anything but a property of cells and tissues, whereas multicellular trichomes may be viewed as individuated structures. However, as intimated earlier, the notion of the individuation of structural modules is challenging. Therefore, the distinction between neo- and homeoheterotopy is somewhat subjective, depending on how one delimits characters and how one defines homology (see also Sattler, 1988).

The issue of whether one considers a structure to be an individuated character is often influenced by how one interprets evolutionary history and, thus, the decision to treat a case as homeo- versus neoheterotopy can turn on historical inferences. For example, the case used in Section 4.5.2 to illustrate neoheterotopy – epiphyllous inflorescences – is subject to an alternative homeoheterotopic interpretation in which a leaf genetic program comes to be expressed in an inflorescence peduncle. This latter interpretation seems less plausible, however, because although a stem segment (peduncle) and a leaf are both developmental modules, they are generally considered to be different kinds of modules. As a result, it seems unlikely that the peduncle should be converted so perfectly to a leaf while at the same time maintaining the property of bearing flowers that was "inherited" from the peduncle. It seems, instead, much easier to imagine the ectopic formation of shoot meristems on leaves, as can occur in tissue culture or when shoot meristem identity genes are overexpressed (e.g. Chuck *et al.*, 1996). Thus, the claim that epiphyllous flowering represents neoheterotopy implies that an inflorescence is an individuated module and that it evolved by the ectopic initiation of shoot meristems on the surface of a leaf.

Just as the foregoing paragraphs have highlighted some blurriness in the distinction between homeoheterotopy and neoheterotopy, a similar fuzziness can be recognized between heterotopy and heterochrony. Since different structural modules develop at different times in ontogeny, heterotopy can perhaps always be reinterpreted as heterochrony, and vice versa (Sattler, 1988). For example, imagine the transition from a simple to a compound umbel in which the function of positioning flowers for pollinator visits has been transferred from the primary peduncle to the secondary peduncles by converting the floral meristems into inflorescence meristems. This can be interpreted as heterochrony, in that the onset of floral meristem identity has been delayed. However, it can also be viewed as heterotopy because the part of the plant producing the floral organs has changed. In this particular case it seems equally accurate to invoke heterochrony or heterotopy. Other cases may be shifted to one end of the spectrum or another. For example, if one envisages transfer of the petal genetic program to a sepal, then heterotopy is a more natural explanation, despite the fact that sepals usually develop earlier than petals. Conversely, the cases in which floral form is modified by changes in the timing of anthesis relative to floral development (Guerrant, 1982, 1988) are easier to interpret in terms of hetero-chrony.

4.6 Evolutionary implications of transference of function and homeoheterotopy

We have suggested that the concept of transference of function provides a useful prism for interpreting certain patterns of plant evolution and for suggesting fruitful avenues for further research. To justify this claim we now explore an assortment of topics that emerge from a consideration of this phenomenon.

4.6.1 Spatial propinquity

Corner (1958) suggested that most instances of transference of function occur between structures in close spatial proximity, for example, between petals and sepals. There are, doubtless, exceptions to this "rule." For example, in *Salvia*, petals or bracts may be showy but never, so far as we know, sepals. Nonetheless, based on our subjective assessment, the hypothesis that spatial propinquity influences the like-lihood of transference of function is plausible.

We can suggest two classes of explanation that would account for an increased tendency for transference of function between closely-spaced or adjacent structures. First, this could reflect underlying developmental mechanisms. For example, the commonest cause of transference of function could be via expansion and then con-traction of the expression of regulatory genes. The second possible explanation is that structures in close proximity are better situated to take up a similar function. For example, it is plausible that only floral organs or bracts closely associated with flowers could serve the role of attracting pollinators or dispersers. On the other hand, although ecological function may favor physical proximity between donor and recipient organs, it may sometimes prevent transfer between adjacent organs. For example, in *Dicella* we hypothesized transference of the function of seed disper-sal from the carpels to the calyx. While it is likely that a structure associated with dispersal would have to develop within the flower/inflorescence, it is unlikely to involve the adjacent structures to the carpels, namely the stamens, because those structures fulfill other important ecological functions related to sexual reproduction.

It would take a broad-scale comparative phylogenetic study to rigorously evaluate whether there is some truth to Corner's intuition and, if so, to evaluate whether the explanation for the pattern is developmental or functional. To determine whether spatial propinquity dominates, one might use a randomization procedure to ask whether transfers more commonly involve adjacent structures than one would expect by chance. In order to distinguish the developmental and functional explanations one could exploit the fact that the developmental explanation could apply only in cases of homologous transfers of function, whereas the functional explanation would apply to both homologous and non-homologous cases. Consequently, finding that the proportion of adjacent transfers was higher in homologous than non-homologous transfers would serve to support the developmental explanation. That being said, we are a long way from possessing the kind of empirical data needed to conduct such a comparative meta-analysis.

4.6.2 Mixed homology

Homologous transference of function involves the movement of genetic identity from one spatial location to another. In cases of complete homeosis it might be said that there has simply been a movement in the spatial location of a particular structure or, if the structure persists in the original location, a module duplication. In other cases of homeoheterotopy some, but not all, of the genetic properties are transferred, resulting in a structure with mixed homology. This notion can be illustrated by "inflorescence-flowers."

Inflorescence-flowers are compact inflorescences where the whole structure functions as a single blossom (see Faegri and van der Pijl, 196: 21–23). In many cases these inflorescences are also organized much like a flower, with enlarged petal-like bracts or sterile florets around the periphery and, sometimes, a subdivision into concentric zones of differing sexual identity. The most obvious example is the capitulum of Asteraceae, but flower-inflorescences with sterile marginal structures are abundant in other angiosperm groups, including *Cornus*, *Euphorbia*, *Eryngium*, *Hydrangea*, *Parkia*, *Protea*, *Psychotria* and *Viburnum*.

Corner (1958) hypothesized that flower-inflorescences arise through partial transfer of a floral genetic program to the whole inflorescence (see also Albert *et al.*, 1998). This hypothesis would be explicable, given current developmental knowledge, if compact inflorescences showed broad expression of floral meristem identity genes such as *LEAFY* (*LFY*). This in turn could activate floral organ identity genes in a manner that is partially floral. For example, one could hypothesize that enhanced expression of C-class floral organ identity genes in florets situated towards the center of the inflorescence is the explanation for fertile flowers being produced centrally and sterile flowers marginally. Likewise, the increased activation of B-class genes peripherally could serve to enhance the petaloid quality of the marginal florets. This hypothesis has not yet been tested (but see Albert *et al.*, 1998).

Coincidentally, an experiment conducted in one of our labs (DAB) using wildtype *Arabidopsis* plants containing a *35S::LFY:GR* transgene (see Wagner *et al.*, 1999) bears on this issue. After infloresence meristems were initiated, *LFY* was activated by application of dexamethosone. This should result in the production of physiologically active *LFY* in all parts of the plant, including the inflorescence meristems, which normally lack *LFY* activity (Weigel *et al.*, 1992). Interestingly, in almost all treated plants (but no control plants) inflorescence apices became converted into condensed multiflowered structures that lack internode elongation and, thus, resemble capitula (H.-S. Yoon and D. A. Baum, unpubl. obs.). One interpretation of this result is that the flower meristem identity gene *LEAFY* is conferring some floral properties, specifically the lack of internode elongation, upon the inflorescence meristem. This evidence is circumstantial, but it provides support for the notion that the identity of flowers and inflorescences can become mixed.

Another case of mixed homology which is gaining experimental support relates to the claim that compound leaves (e.g. Corner, 1958; Sattler, 1988; Sattler and Rutishauser, 1992; Rutishauser, 1995; Goliber *et al.*, 1999) and gesneriad phylo-morphs (Corner, 1958; Jong and Burtt, 1975) have mixed shoot–leaf identity. This inference has gained support from evidence that shoot meristem identity genes may be expressed in developing compound leaves, but not in simple leaves, and that these genes influence the degree of compounding in tomatoes (Sinha *et al.*, 1993; Hareven *et al.*, 1996; Chen *et al.*, 1997).

The notion of mixed homology bears obvious similarities to the concept of partial homology, promoted especially by Rolf Sattler (e.g. Sattler, 1984, 1988, 1991; see also Roth, 1991; Minelli, 1998). Both are based upon the idea that developmental processes can be transferred piecemeal from one structure to another such that characters in different species need not have 1:1 relationships with each other. Thus, Sattler suggested that a character could show a defined degree of partial homology to each of several structures in its ancestor. Thus, a petal could be imagined to be 70 percent homologous to a stamen and 30 percent to a sepal.

"Partial" and "mixed" homology may differ subtly, however, in orientation. In Sattler's approach the challenge seems to be to devise a semi-quantitative index of the proportion of whole structures that have different identities. In contrast, we wish to focus our efforts on decomposing structures into their genetic elements. Thus, our notion of "mixed" homology is more historical in outlook and suggests that we strive to reconstruct how and why different genetic modules come to co-exist in particular structures.

Mixed and partial homology both call into question much evolutionary developmental research whose primary aim seems to be identifying a single correct homology statement for highly derived structures. For example, some have hoped to use the expression of B-group MADS-box genes to determine whether petals are derived from sepals/bracts or stamens (Albert *et al.*, 1998; Kramer and Irish, 1999). However, if, as we suspect, petals represent a mixture of sepal and stamen developmental programs, B-gene expression will not provide a complete answer (Baum and Whitlock, 1999; Kramer and Irish, 2000).

Not only does mixed homology influence evolutionary developmental genetics, it can also complicate the phylogenetic analysis of morphological data. The first step in morphological phylogenetics is to construct a data matrix in which the columns represent inferred homologies at one level in the hierarchy and shared character-states within a column represent hypothesized homologies at a lower hierarchical level (Hawkins, 2002). This matrix therefore assumes hierarchically nested homology relationships between organismic characters. However, if a compound leaf is both a leaf *and* a shoot, then the homology relationships become reticulate (see also Minelli, 1998). The leaf of a closely related simple-leaved species could be simultaneously homologous with the leaflet *and* the leaf of the compound-leaved species. Thus, if one scored a "leaf" property such as petiole length one might not know whether to score the petiole of the compound leaf as a whole or the petiolules of the leaflets (Hawkins, 2002). This situation is analogous to phylogenetic analysis using a duplicated gene wherein one species has a "hybrid" gene arising by partial gene conversion between paralogs. In that case there is no meaningful way to include that species without first breaking it up into the regions that are derived from each of the two paralogs. The same is true of morphological structures with mixed homology except that, deprived of the linearity of gene sequences, it is much more difficult to identify the source of developmental modules.

4.6.3 Variation in the propensity to show homologous transference of function

As discussed in Section 4.5.2, homologous transference of function requires homeo-heterotopy. In order to manifest homeoheterotopy a developmental program needs to be composed of distinct modules and it needs to be possible, in the sense of not being massively deleterious, for genetic subroutines that are "adapted" for one module to be activated elsewhere. (A genetic subroutine is an integrated piece of a developmental program analogous to a subroutine in a computer program.) In this section we consider the possibility that angiosperms may be especially prone to transference of function because their developmental systems are highly modular and have evolved to tolerate shifts of genetic subroutines from module to module.

There is little hard data comparing the number of structural modules in different eukaryotic lineages. However, while there is a risk that we are being biased by human perception, we are fairly confident that, among land plants, angiosperms show, on average, the greatest degree of modular proliferation (a possible exception being the rhizomorphic lycopsids: Bateman, 1994). This can be seen by considering the numerous determinate, leaf-like modules that are found in the typical angiosperm, including cotyledons, juvenile leaves, vegetative leaves, reproductive leaves, bracts, sepals, petals, stamens and carpels. The diversity of leaf modules may be further enhanced by specializations (e.g. anisophylly, involucral bracts), additional developmental phases (spring versus summer leaves), or variants exposed to different environmental conditions and therefore expressing different suites of genes (e.g. sun versus shade; under insect attack versus intact). In contrast, even a relatively complex moss is unlikely to have more than a handful of structural modules.

Although angiosperms might tend to be more highly modular than other land plants, one would have a hard time arguing that they are more modular than animals. After all, animals such as mammals have innumerable distinct anatomical modules (e.g. eyes, femurs, finger nails, hearts and kidneys). The presence of multiple modules is, however, insufficient to predict extensive homeoheterotopy. It is also necessary that the genetic program be composed of genetic subroutines that are specialized for particular modules but that can become activated in a new module without having massively detrimental effects.

It has frequently been claimed, based on anecdotal evidence, that plants can tolerate more massive genetic alterations than animals (Van Steenis, 1969, 1977, 1978; Gottlieb, 1984). Although this claim is in need of empirical verification, a theoretical consideration of the structure of plant and animal developmental programs provides some justification for the idea that plants are more tolerant than animals of the misexpression of genetic subroutines.

It is generally agreed that animals and plants differ in their developmental architecture. Animal (especially vertebrate) development is "closed" in the sense that it is largely buffered against environmental variability such that final form is for the most part explicable in terms of the underlying genotype. In contrast, plants have more open developmental systems consisting of a limited number of modular units whose deployment is highly responsive to local environmental cues. Why should these two developmental styles differ in the expected degree of disruption caused by a homeo-heterotopic mutation?

The essence of an open developmental system is that structural modules, such as leaves, are deployed repeatedly at times and in positions governed by the environment. Consequently, a structural module will find itself in different settings (e.g. leaves in shade versus full sun), making it necessary that structural modules show some plasticity, turning on or off developmental subroutines as needed. Furthermore, since the environmental cues to which a plant responds are varied (e.g. light, temperature, pathogen attack), it seems necessary that a subroutine activated in response to one cue should function in all modules, regardless of which combination of other subroutines are active. As a result, one might expect the evolution of multiple subroutines, each of which is an integrated unit that can function in different modules that vary somewhat in their genetic identity. The properties of a given structural module would then be determined in a combinatorial manner by the set of subroutines that are activated. As a consequence, generalized subroutines may be easily activated in a novel structural module without severe detrimental consequences (see Gottlieb, 1984).

Vertebrates produce many structural modules at various hierarchical levels (e.g. heart, kidney, glomeruli, femurs). However, most modules are reiterated only once or twice. For example, a vertebrate produces only one heart, and that heart experiences basically the same "environment" from organism to organism. As a result, there is less need for plasticity when a module develops, and greater potential for functional specialization. Thus, we expect that genetic subroutines expressed in the heart will tend to be specialized for the heart and will depend upon that specific "environment" for proper deployment.

Based on the reasoning presented above, we suspect that plant developmental subroutines, unlike in most animals, are robust to the cellular environment in which they are activated. Consequently, it is likely that homeoheterotopy may be more important in angiosperm evolution than in animal evolution. On the other hand, it is important to note that there are many animals that have a more open, plant-like developmental system (e.g. bryozoa, colonial cnidarians), there are some aspects of animals that more closely resemble plant modular development (e.g. vertebrae, feathers, scales, arthropod appendages), and there are some aspects of plant development that are more animal-like (e.g. embryogenesis). Nonetheless, as a working hypothesis aimed at promoting further work, we suggest that plants (especially angiosperms) are likely to show more homeoheterotopy than animals.

4.6.4 Future prospects

Corner built his notion of transference of function upon a foundation of careful observations of plant morphology. This bedrock of botanical knowledge is perhaps the main reason why his idea still has relevance fifty years later, despite massive changes in evolutionary theory, systematics and developmental biology. His intuitions resonate remarkably well with modern perspectives, such as recent claims for the importance of changes in the enhancer/promoter domains of developmental regulatory genes (Doebley and Lukens, 1998). Furthermore, Corner (1958) is rich in novel interpretations and creative hypotheses that warrant rigorous exploration. Our hope is that this chapter will bring the concept of transference of function to the attention of a wider audience and will promote additional studies of its mechanisms, distribution and consequences.

ACKNOWLEDGMENTS

We would like to thank the organizers of the DGPE symposium and the numerous people who helped by way of discussion or critical comments on the ideas herein: Richard Bateman, Quentin Cronk, Julie Hawkins, Lena Hileman, Toby Kellogg, Elena Kramer, Tracy McLellan, Peter Stevens, Richard Thomas, Günter Wagner, Alan Yen and Ho-Sung Yoon. Several other colleagues were kind enough to provide expertise on specific examples that we included or considered including: Philip Cantino (*Salvia*), Charles Davis (Malpighiaceae), Torsten Eriksson (Rosaceae), Raymond Harley (*Salvia*), Katherine Kron (*Gaultheria/Pernettya*), Alison Miller (Anacardiaceae), Susan Pell (Anacardiaceae), Lena Struwe (Anacardiaceae) and Kenneth Sytsma (Moraceae). Research described in this chapter was funded by the National Science Foundation (DEB-9876070; IBN-0078161).

REFERENCES

Alberch, P., Gould, S. J., Oster, G. F. and Wake, D. B. (1979) Size and shape in ontogeny and phylogeny. *Paleobiology*, 5, 296–317.

Albert, V. A., Gustafsson, M. H. G. and Di Laurenzio, L. (1998) Ontogenetic systematics, molecular developmental genetics, and the angiosperm flower, in *Molecular Systematics of Plants II* (eds P. Soltis, D. Soltis and J. J. Doyle), Chapman and Hall, New York, pp. 349–374.

Augspurger, C. K. (1986) Morphology and dispersal potential of wind-dispersed diaspores of neotropical trees. *American Journal of Botany*, 73, 353–363.

Augspurger, C. K. and Franson, S. E. (1987) Wind dispersal of artificial fruits varying in mass, area, and morphology. *Ecology*, 68, 27–42.

Bateman, R. M. (1994) Evolutionary–developmental change in the growth architecture of fossil rhizomorphic lycopsids: scenarios constructed on cladistic foundations. *Biological Review*, 69, 527–597.

Baum, D. A. and Larson, A. (1991) Adaptation reviewed: a phylogenetic methodology for studying character macroevolution. *Systematic Zoology*, 40, 1–18.

Baum, D. A. and Whitlock. B. A. (1999) Genetic clues to petal evolution. *Current Biology*, 9, R525–R527.

Chase, M. W. (1981) A revision of *Dicella* (Malpighiaceae). *Systematic Botany*, 6, 159–171.

Chen, J.-J., Hanssen, B.-J., Williams, A. and Sinha, N. (1997) A gene fusion at a homeobox locus: alterations in leaf shape and implications for morphological evolution. *Plant Cell*, 9, 1289–1304.

Chuck, G., Lincoln, C. and Hake, S. (1996) *KNAT1* induces lobed leaves with ectopic meristems when overexpressed in *Arabidopsis*. *Plant Cell*, 8, 1277–1289.

Corner, E. J. H. (1949a) The durian theory. *Annals of Botany*, 13, 368–414.

Corner, E. J. H. (1949b) The annonaceous seed and its four integuments. *New Phytologist*, 48, 332–364.

Corner, E. J. H. (1958) Transference of function. *Botanical Journal of the Linnean Society*, 56, 33–40.

Davis, C. C., Anderson, W. R. and Donoghue, M. J. (2001) Phylogeny of Malpighiaceae: evidence from chloroplast *ndhF* and *trnL-F* nucleotide sequences. *American Journal of Botany*, 88, 1830–1846.

Dickinson, T. A. and Sattler, R. (1974) Development of the epiphyllous inflorescence of *Phyllonoma integerrima* (Turcz.) Loes.: implications for comparative morphology. *Botanical Journal of the Linnean Society*, 69, 1–13.

Doebley, J. and Lukens, L. (1998) Transcriptional regulators and the evolution of plant form. *Plant Cell*, 10, 1075–1082.

Eriksson, T., Donoghue, M. J. and Hibbs, M. S. (1998) Phylogenetic analysis of *Potentilla* using DNA sequences of nuclear ribosomal internal transcribed spacers (ITS), and its implications for the classification of Rosoideae (Rosaceae). *Plant Systematics and Evolution*, 211, 155–179.

Faegri, K. and van der Pijl, L. (1966) *The Principles of Pollination Ecology*. Pergamon Press, Toronto.

Goliber, T., Kessler, S., Chen, J. J., Bharathan, G. and Sinha, N. (1999) Genetic, molecular, and morphological analysis of compound leaf development. *Current Topics in Developmental Biology*, 43, 259–290.

Gould, S. J. and Vrba, E. S. (1982) Exaptation – a missing term in the science of form. *Paleobiology* 8, 4–15.

Gottlieb, L. D. (1984) Genetics and morphological evolution in plants. *American Naturalist*, 123, 681–709.

Guerrant, E. O. (1982) Neotenic evolution of *Delphinium nudicaule* (Ranunculaceae) – a hummingbird-pollinated larkspur. *Evolution*, 36, 699–712.

Guerrant, E. O. (1988) Heterochrony in plants: the intersection of evolution ecology and ontogeny, in *Heterochrony in Evolution: a Multidisciplinary Approach* (ed. M. L. McKinney), Plenum Press, New York, pp. 111–133.

Hareven, D., Gutfinger, T., Parnis, A., Eshed, Y. and Lifschitz, E. (1996) The making of a compound leaf: genetic manipulation of leaf architecture in tomato. *Cell*, 84, 735–744.

Hawkins, J. A. (2002) Evolutionary developmental biology: impact on systematic theory and practice, and the contribution of systematics, in *Developmental Genetics and Plant Evolution* (eds Q. C. B. Cronk, R. M. Bateman and J. A. Hawkins), Taylor & Francis, London, pp. 32–51.

Jong, K. and Burtt, B. L. (1975) The evolution of morphological novelty exemplified in the growth patterns of some Gesneriaceae. *New Phytologist*, 75, 297–311.

Kellogg E. A. (2000) The grasses: a case study in macroevolution. *Annual Review of Ecology and Systematics*, 31, 217–238.

Kramer, E. M. and Irish, V. F. (1999) Evolution of genetic mechanisms controlling petal development. *Nature*, 399, 144–148.

Kramer, E. M. and Irish, V. F. (2000) Evolution of the petal and stamen developmental programs: evidence from comparative studies of the lower eudicots and basal angiosperms. *International Journal of Plant Science*, 161(6 suppl.), S29–S40.

Matlack, G. R. (1987) Diaspore size, shape, and fall behavior in wind-dispersed plant species. *American Journal of Botany*, 74, 1150–1160.

McConnell, J. R. and Barton, M. K. (1998) Leaf polarity and meristem formation in *Arabidopsis*. *Development*, 125, 2935–2942.

Möller, M. and Cronk, Q. C. B. (2001) Evolution of morphological novelty: a phylogenetic analysis of growth patterns in *Streptocarpus* (Gesneriaceae). *Evolution*, 55, 918–929.

Minelli, A. (1998) Molecules, developmental modules, and phenotypes: a combinatorial approach to homology. *Molecular Phylogenetics and Evolution*, 9, 340–347.

Powell, E. A. and Kron, K. A. (2001) An analysis of the phylogenetic relationships in the wintergreen group (*Diphylcosia, Gaultheria, Pernettya, Tepuia*; Ericaceae). *Systematic Botany*, 26, 808–817.

Raff, R. A. and Wray, G. A. (1989) Heterochrony: developmental mechanisms and evolutionary results. *Journal of Evolutionary Biology*, 2, 409–434.

Ree, R. H. and Donoghue, M. J. (1998) Step matrices and the interpretation of homoplasy. *Systematic Biology*, 47, 582–588.

Roth, V. L. (1991) Homology and hierarchies: problems solved and unresolved. *Journal of Evolutionary Biology*, 4, 167–194.

Rutishauser, R. (1995) Developmental patterns of leaves in Podostemaceae compared with more typical flowering plants: saltational evolution and fuzzy morphology. *Canadian Journal of Botany*, 73, 1305–1317.

Sattler, R. (1984) Homology – a continuing challenge. *Systematic Botany*, 9, 382–394.

Sattler, R. (1988) Homeosis in plants. *American Journal of Botany*, 75, 1606–1617.

Sattler, R. (1991) Process homology: structural dynamics in development and evolution. *Canadian Journal of Botany*, 70, 708–714.

Sattler, R. and Rutishauser, R. (1992) Partial homology of pinnate leaves and shoots – orientation of leaflet inception. *Botanische Jahrbücher für Systematik, Pflanzengeschichte und Pflanzengeographie*, 114, 61–79.

Schluter, D., Price, T., Mooers, A. Ø. and Ludwig, D. (1997) Likelihood of ancestor states in adaptive radiation. *Evolution*, 51, 1699–1711.

Sinha, N., Williams, R. and Hake, S. (1993) Overexpression of the maize homeobox gene, *KNOTTED-1* causes a switch from determinate to indeterminate cell fates. *Genes and Development*, 7, 787–795.

Sipe, T. W. and Linnerooth, A. R. (1995) Intraspecific variation in samara morphology and flight behavior in *Acer saccharinum* (Aceraceae). *American Journal of Botany*, 82, 1412–1419.

Takhtajan, A. L. (1991) *Evolutionary Trends in Flowering Plants*. Columbia University Press, New York.

Van Steenis, C. G. G. J. (1969) Plant speciation in Malesia, with special reference to the theory of non-adaptive saltatory evolution. *Biological Journal of the Linnean Society*, 1, 97–133.

Van Steenis, C. G. G. J. (1977) Autonomous evolution in plants. *Garden's Bulletin, Singapore*, 29, 103–126.

Van Steenis, C. G. G. J. (1978) *Patio ludens* and extinction of plants. *Notes from the Royal Botanic Garden Edinburgh*, 36, 317–323.

Wagner, D., Sablowski, R. W. M. and Meyerowitz, E. M. (1999) Transcriptional activation of *APETALA1* by *LEAFY*. *Science*, 285, 582–584.

Wagner, G. P. (1989) The origin of morphological characters and the biological basis of homology. *Evolution*, 43, 1157–1171.

Weigel D., Alvarez, J., Smyth, D. R., Yanofsky, M. F. and Meyerowitz, E. M. (1992) *LEAFY* controls floral meristem identity in *Arabidopsis*. *Cell*, 69, 843–859.

Zelditch, M. L. and Fink, W. L. (1996) Heterochrony and heterotopy: stability and innovation in the evolution of form. *Paleobiology*, 22, 241–254.

Chapter 5

Are macroevolution and microevolution qualitatively different? Evidence from Poaceae and other families

Elizabeth A. Kellogg

ABSTRACT

In phylogenies, ancient developmental changes mark the origin of major clades, whereas more recent events mark the origin of closely related species. Character changes marking ancient events can often be described as heterotopic, where a genetic program is deployed in a different place. In contrast, character change marking recent events is more often quantitative and/or heterochronic. There are multiple reasons why the difference between ancient and recent events might be artifactual, but it is also possible that it is a genuine phenomenon that reflects the effects of extinction, selection, gradual divergence of gene function and/or the relative likelihood of reversal of characters.

5.1 Introduction

One fundamental result of the New Synthesis was the conclusion that all evolutionary change begins as a shift in gene frequencies in populations (Dobzhansky, 1951). From this precept, it follows that the kinds of changes that mark major radiations must be qualitatively the same as those marking closely related species or even slightly differentiated populations. Macroevolution, the origin of higher taxa, should thus resemble microevolution. Rephrasing this statement in a more phylogenetic fashion, the sorts of mutations that characterize recent nodes in a phylogeny should also characterize more ancient nodes, and should do so in roughly the same proportions. I have attempted to show this graphically in Figure 5.1a, where half of the terminal taxa are distinguished by some particular sort of mutation (e.g. a point mutation in a regulatory gene), indicated by a black dot. Approximately one third of the taxa are marked by a hollow square "mutation" and the remainder by an X "mutation." Therefore we expect that changes marking deeper nodes should be about half black dot type, a third hollow squares and the rest Xs. (This is also the basis of the idea that rank is arbitrary – a higher taxon is an arbitrarily named clade and therefore has no special properties distinct from "lower" taxa: Stevens, 1998.)

In contrast to this view, many authors have noticed apparent differences among the characters that change at different levels in the phylogeny. For example, Mishler (2000) distinguishes between "deep" phylogenetic nodes and "shallow" ones. At shallow nodes, he states, "characters, *at least at the morphological level*, may be quite subtle..." (italics mine). Similarly, Chase *et al.* (2000) argued that currently

In *Developmental Genetics and Plant Evolution* (2002) (eds Q. C. B. Cronk, R. M. Bateman and J. A. Hawkins), Taylor & Francis, London, pp. 70–84.

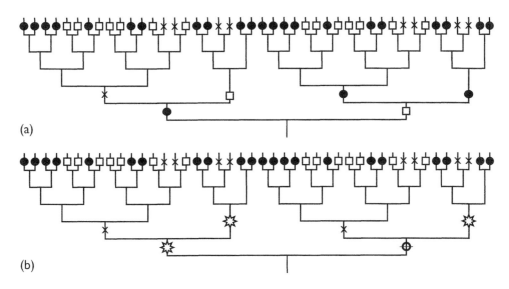

(a)

(b)

Figure 5.1 Hypothetical phylogeny. Symbols on branches indicate different types of mutations. (a) Under the modern synthesis, the types of mutations that mark recent branches should be the same kind and occur at the same frequency as those that mark deeper branches. (b) Evidence cited in this chapter suggests the possibility that changes on deeper branches appear qualitatively different from those on recent branches.

recognized angiosperm families represent "biological reality." The families are "marked to a large extent by canalized suites of functionally correlated traits; they are biologically optimal organizational units." Both authors thus point to differences in the morphological characters at deep and shallow nodes. This would lead to a predicted pattern such as in Figure 5.1b, in which there are different sorts of characters at deep nodes rather than shallow ones, or alternatively that the characters at deep nodes were a subset of the ones at shallow nodes.

Recent advances in phylogeny reconstruction, developmental morphology and molecular genetics allow us to begin to test whether there are in fact differences in deep versus shallow nodes. A good group for such a study is the grass family, comprising 8000–10,000 species, and thus comparable in size to Aves (birds) and about twice as large as Mammalia, but only half the size of Orchidaceae or Asteraceae. The family includes the cereals, as well as many ecological dominants covering approximately 20 percent of the Earth's land surface (Shantz, 1954). A well-supported phylogeny of the family is now available, based on combined data from seven genes (6972 aligned sites) plus fifty-three morphological characters (GPWG, 2000) (Figure 5.2). Fossil pollen grains suggest that the family originated 55–70 million years ago (Linder, 1987; Jacobs *et al.*, 1999). Although grains found in the Palaeocene are indubitably grasses, the older Cretaceous grains may represent non-grass sister taxa. With the phylogenetic framework thus established, efforts have begun to characterize the morphological changes marking each node in the phylogeny.

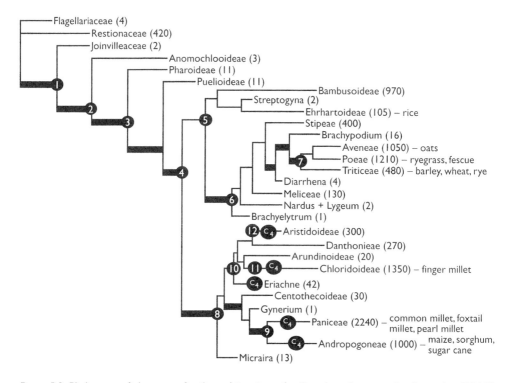

Flagellariaceae (4)
Restionaceae (420)
Joinvilleaceae (2)
Anomochlooideae (3)
Pharoideae (11)
Puelioideae (11)
Bambusoideae (970)
Streptogyna (2)
Ehrhartoideae (105) – rice
Stipeae (400)
Brachypodium (16)
Aveneae (1050) – oats
Poeae (1210) – ryegrass, fescue
Triticeae (480) – barley, wheat, rye
Diarrhena (4)
Meliceae (130)
Nardus + Lygeum (2)
Brachyelytrum (1)
Aristidoideae (300)
Danthonieae (270)
Arundinoideae (20)
Chloridoideae (1350) – finger millet
Eriachne (42)
Centothecoideae (30)
Gynerium (1)
Paniceae (2240) – common millet, foxtail millet, pearl millet
Andropogoneae (1000) – maize, sorghum, sugar cane
Micraira (13)

Figure 5.2 Phylogeny of the grass family and its sister families, based on results from the GPWG (2000, 2001). Numbers in parentheses indicate approximate species richness of each clade. Branch lengths are approximately proportional to numbers of mutations, as indicated by parsimony analyses. Heavy branches indicate groups that are found in all molecular phylogenies, and thus appear in the semi-strict consensus and/or are supported by 90–100 percent bootstrap values in analyses that include one or more molecular data sets. Numbered nodes refer to discussion in text. Lineages marked with C_4 consist wholly or partly of species with the C_4 photosynthetic pathway. Redrawn from Kellogg (2000).

5.2 Quantitative change is common, particularly among closely related species

Among species of most taxa, the diagnostic characters often concern size of parts. Stevens (1991) lists a large set of studies in which quantitative characters were used to infer phylogenies, generally among species within a genus. Few of these studies have been followed directly by comprehensive species-level molecular phylogenies, so it is difficult to determine exactly where on the trees the quantitative characters map. Because most of the studies include only congeneric species, however, the characters all appear to mark shallow nodes.

There are few if any species-level molecular phylogenies across which all the key morphological characters have been mapped. Available data, however, suggest that many speciation events led to modest changes in sizes of parts, often in many different combinations. Here I cite a small sample of studies extracted from recent issues of the *American Journal of Botany* and *Systematic Botany*. These cover a range of families, and suggest that the pattern is a general one.

Kelly (2001) reported on relationships among species in *Asarum* sect. *Asarum* (Aristolochiaceae). In this group, *A. debile* and *A. caudigerellum* are sister species, as demonstrated by a phylogenetic analysis of morphology combined with ITS sequences. They are distinguished from other species in the section by connation of sepals beyond their attachment to the ovary. The two species are distinguished from each other by the number of stamens, the lengths of the flowers and whether the sepal lobes are revolute. The other twelve species are distinguished from each other by such characters as whether the sepal lobes are erect, spreading or reflexed; whether plants are densely or sparsely covered with hairs; whether the foliage leaves are solitary or paired; whether foliage leaves are marcescent or deciduous; whether stamens and styles are exserted from the calyx or not; and whether annual growth increments are less than 1 cm or more than 2 cm long. These characters are homoplasious when mapped on the phylogeny.

Schultz and Soltis (2001) investigated subspecies of *Leptodactylon californicum* (Polemoniaceae). They demonstrated the distinctiveness of three subspecies, based on hair length and density, leaf lobe number and length, corolla lobe width and intensity of flower color. Baldwin *et al.* (2001) used molecular data to demonstrate the close relationship between, but sharp discontinuities separating, two species of *Blepharizonia* (Asteraceae). The two differed morphologically in pubescence, inflorescence branching pattern and pappus length.

Wolfe and Randle (2001) investigated relationships among species in *Hyobanche* (Orobanchaceae) using sequences of ITS, *rbcL* and data from inter-simple sequence repeats. The four species investigated fell into two pairs, with the pairs distinguished by corolla pubescence, corolla shape, and width and length of the corolla mouth. The species pair *H. atropurpurea* and *H. sanguinea* were clearly distinct on the basis of molecular data, and also distinguishable by the color of their corollas. *Hyobanche glabrata* and *H. rubra* were much less distinct on the molecular level, but could be distinguished morphologically by corolla color and shape, corolla mouth length and width, anther and stigma exsertion, and calyx pubescence.

Fjellheim *et al.* (2001) identified four species of *Festuca* distinguished by RAPD multilocus phenotypes, isozymes and chromosome numbers. The species were also distinguished by a set of quantitative morphological characters, including amount of anthocyanin in the panicle, hairs on the stem, length of lemmas and glumes, and width of leaves.

In a study of inflorescence development among species of *Setaria*, we have found that inflorescence form varies quantitatively (Doust and Kellogg, 2002). These characters are heritable and thus do not reflect simple phenotypic plasticity. The species differ in the number of orders of branching, the number of primordia produced in each branch order, and the extent of elongation of inflorescence internodes, all quantitative differences.

The genus *Poa* provides a similar example. Species of *Poa* differ in the arrangement of hairs on the floral bracts and in the developmental timing of anther maturation (Kellogg, 1990). The sample of species studied for timing of anther maturation is too small to permit phylogenetic inferences, but the hair characters show extensive homoplasy when mapped across Soreng's (1990) restriction site phylogeny for the genus (Figure 5.3). Different hair patterns are caused by truncation or extension of a single developmental pathway. Hairs may be formed only on the keel of the lemma (equivalent to the midvein of the bract), or on the keel plus the marginal veins, or keel plus marginal plus midveins, or all over the lemma. This corresponds to a developmental sequence. On an individual lemma, hairs form first on the keel, then on the marginal veins, then the midveins, and finally in the intercostal regions. The genetic basis of the changes remains unknown.

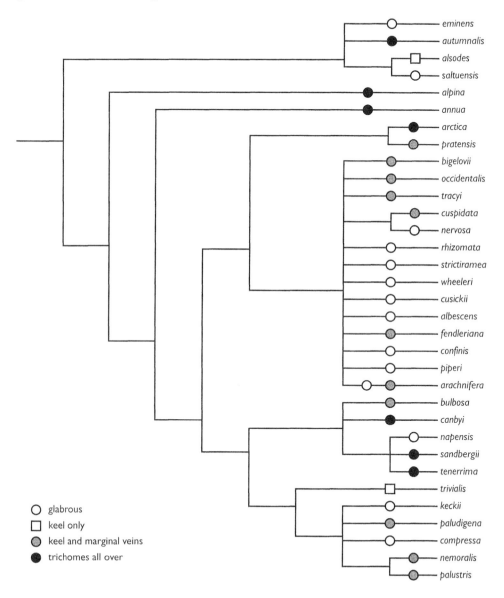

Figure 5.3 Phylogeny of the genus *Poa*, based on chloroplast restriction site data; pruned and redrawn from Soreng (1990). Differently shaded circles indicate different patterns of tri-chome development on the lemma.

In the foregoing studies, genetically distinct species (as indicated by molecular data) are often distinguished morphologically by combinations of character states, and only rarely by apomorphies. Multiple combinations of character states appear on phylogenies as homoplasy, which is apparently high among closely related taxa. Reference to Doust and Kellogg (2002) or to Figure 5.3 shows that this is certainly true for the examples from *Setaria* and *Poa*, cited above.

This phenomenon is not a pattern unique to plants, so is not correlated with modular growth or sessile growth habit. Wagner (2000) compiled numerous studies of fossil animals, all of which showed that available morphospace seems to become filled rapidly. In fossil animals, as in extant grasses, this pattern could reflect either developmental constraints or selection repeatedly choosing similar forms.

Work on the origin of *Zea mays* subsp. *mays* is the only instance in which change has been characterized at the level of genes. In this case, expression of the gene *teosinte branched1* is elevated in the cultigen relative to its wild ancestor (Doebley *et al.*, 1997). Increased amounts of RNA correlate with suppression of tillers (basal branches) in *Z. mays* subsp. *mays*, whereas lower amounts of RNA correlate with the highly branched architecture of the ancestral subsp. *parviglumis*. The distinction between the two is thought to reflect a change in the regulatory region upstream of the gene (Wang *et al.*, 1999). This evolutionary change is thus a modest shift in the amount of RNA produced, which then leads to an apparently major change in plant architecture.

Among closely related species, few phenotypic changes seem to be more complex than just modifications in the sizes of parts. We have, however, found a few examples. The genera *Pennisetum* and *Cenchrus*, for example, are similar to *Setaria* in early development, but the primary branch axis ultimately enlarges and becomes wider than long (Doust and Kellogg, 2002). In *Cenchrus*, the bristles also become wide and flat during development. It is possible that broadening of the bristles reflects the same sorts of cellular changes that control broadening of the primary branches, and thus may be caused by a change in the position of expression of a gene or developmental program.

A similar example comes from C_4 photosynthesis, a highly efficient photosynthetic pathway that has arisen multiple times in grass evolution (Kellogg, 1999). Giussani *et al.* (2001) identified six to eight origins in the panicoid grasses alone (all taxa above node 9 in Figure 5.2), including some that appear to be phylogenetically quite recent. The perceived number of origins depends on optimization of the character on their cladogram, and it is only slightly less parsimonious to postulate a single origin, multiple losses, and then several re-gains of the character. Either optimization requires multiple origins; in the first scenario the origins are independent, whereas in the second they are reversals following an origin and a loss. If the first scenario is correct, then the changes that lead to C_4 photosynthesis might appear frequently in evolutionary time, but be difficult or impossible to reverse once they have occurred. Under the second scenario, C_4 originates rarely, but then is evolutionarily labile once it has appeared in a clade. As with most of the other changes described here, we do not know the genetic basis of C_4 photosynthesis, except that it requires multiple regulatory changes in multiple genes. It also requires vein development in sets of cells where this does not normally occur.

5.3 Quantitative change appears less common at deep nodes

Elsewhere (Kellogg, 2000) I have shown that changes at the deep nodes of the phylogeny – at least where they can be characterized developmentally – involve predominantly qualitative changes, many of them apparently created by a change in position of expression of a developmental program. Curiously, the characters at these deep nodes rarely involve changes in the sizes of parts. Thus, node 1 in Figure 5.2, which may represent a Late Cretaceous or Early Tertiary divergence, is marked by a change in the location of asymmetrical cell division. Node 2 is marked by an acceleration of embryo development relative to seed maturation. Node 7 is marked by a striking and apparently abrupt increase in genome size, presumably caused by amplification of retroelements (Bennetzen and Kellogg, 1997); correlated with this expansion is a modest rearrangement of the nuclear genome (Kellogg, 1998). Node 9, dated by various molecular clock estimates to be 25 to 32 million years old (Gaut and Doebley, 1997), coincides with a shift in position of a cell-death program from the tapetal cells in the anther to the gynoecium (LeRoux and Kellogg, 1999; Zaitchik et al., 2000). Nodes 9, 10, 11 and 12 are all instances of the origin of C_4 photosynthesis, which correlates with a change in position of vascular development (leading to what is sometimes called Kranz anatomy: Hattersley and Watson, 1975; Dengler and Dengler, 1990), as well as a change in tissue specific expression of metabolic and photosynthetic enzymes (Langdale et al., 1988; Langdale and Nelson, 1991; Nelson and Langdale, 1992; Sinha and Kellogg, 1996; Kanai and Edwards, 1999; Leegood and Walker, 1999).

This result is probably more general, but requires more detailed study of morphological character change than is currently available. For example, Kenrick and Crane (1997) cite numerous characters that must have changed during the early evolution of the land plants, yet we know almost nothing about how such changes might have occurred genetically or developmentally. The origin of secondary growth might perhaps be interpreted as a heterotopic change, in which meristematic activity has been moved from apical cells to cells in the cortex. But it is hard to know how to view such characters as sporangium symmetry or sporophyll shape (to choose just a couple at random). Perhaps these are developmentally and genetically no different from the quantitative characters that mark closely related extant species.

Similarly, Soltis et al. (2000), Doyle and Endress (2000), and Endress and Igersheim (2000), among many others, cite multiple characters that mark lineages among the early angiosperm radiation, but few if any of these are understood developmentally and/or genetically. A change from crassinucellar to tenuinucellar ovules could perhaps be interpreted as a heterotopic change in position of the megaspore mother cell, whereas changes between secretory and plasmodial tapetum might be described as heterochronic changes in the timing of cell-wall breakdown (Furness and Rudall, 1998).

5.4 Heterochrony versus heterotopy, quantitative versus qualitative change

Many of the changes described above could be classified as heterochrony – a change in the rate or duration of development of one part of an organism relative to other parts. Heterochrony has been proposed by some authors to be an important mode of evolution of development (e.g. Gould, 1977; Alberch *et al.*, 1979; Raff and Wray, 1989). In these papers, development was described as a linear set of transformations which ended with sexual maturity. By changing the relative time of sexual maturity, the developmental progam could be extended or truncated, producing differences in adult form. This model was later extended to include change in timing of any developmental event relative to any other (Raff and Wray, 1989). In this later formulation, however, the term heterochrony risks explaining everything and therefore explaining nothing (Rice, 1997).

Possibly the most valuable contribution of the concept of heterochrony is the idea that aspects of development can be decoupled. Whether a particular pattern fits with a definition of heterochrony or not may be less important than the observation that one aspect of a plant can develop quite independently of other aspects.

The qualitative changes appearing preferentially at deeper nodes could be described as heterotopic, in that a developmental program is expressed in a different place. For example, leaf-borne flowers or "ectopic" meristems may be described as heterotopic change. As observed by Baum and Donoghue (2002), a change in position of a developmental program generally also requires a change in timing, so heterotopic change is probably also heterochronic (but not conversely). However, quantitative change (size, shape) may well be caused by heterochrony, whereas heterotopy may easily lead to qualitative changes.

Whether to classify particular changes as heterochronic or heterotopic is often a matter of opinion, and also a question of how one defines a "developmental program." If the program is defined as, for example, "make hairs on lemma," then the different hair patterns described above for *Poa* are produced by the amount of time the program is active (i.e. heterochrony). If the program is defined as "trichome development," then the program is activated sequentially in different sets of cells, resulting in a change in position of hairs (i.e. heterotopy).

5.5 Macromutation is a meaningless term

Neither heterochrony nor heterotopy really defines a *bona fide* mechanism – they are simply convenient phenotypic descriptors that say nothing about the underlying molecular processes. All evolutionary change *could* in principle be caused by point mutations. Doebley and Lukens (1998) argued convincingly that evolutionary change is likely to come via regulatory genes, and several groups have now shown that promoters function as analog devices that integrate inputs from multiple (regulatory) sources (Yuh *et al.*, 1998; Flores *et al.*, 2000; Ghazi and VijayRaghavan, 2000; Halfon *et al.*, 2000; Xu *et al.*, 2000). If this is generally true then even modest changes in strength of binding of enhancers to promoters could lead to large phenotypic effects. This observation also suggests that the term "macromutation" is meaningless, since it is not clear whether it refers to genotype or phenotype.

Bateman and DiMichele (1994: 66) pointed out exactly this problem in their definition of "saltation," which is related to the question of macromutation. Their definition of saltation is "a genetic modification that is expressed as a profound phenotypic change across a single generation and results in a potentially independent evolutionary lineage (for the present, we prefer to evade our responsibility to quantify 'profound')."

The central difficulty in defining terms such as macromutation and saltation is quantification of morphological change. The reluctance of Bateman and DiMichele to define "profound" is quite understandable since profundity is in the eye of the beholder. It is a familiar experience among taxonomists, when speaking to someone about his/her study group, that he/she will describe it as enormously diverse, while to the uninitiated the taxa all look remarkably alike. In my experience, many people have felt this way about what to me looks like the exceptional diversity of the grasses – all of them look alike to most people, even taxonomists. So what to me would seem a remarkable "macro" mutation, might seem to an orchidologist as a mere evolutionary hiccup. In the absence of a standard measure of phenotypic divergence, the definition of the amount of change will depend on one's own point of reference.

If we define macromutation as a large genetic change, the same problem appears. A point mutation may be considered small ("micro") and a chromosomal translocation large ("macro"), but what about a deletion of five base pairs? Or 500 base pairs?

Compounding the basic definitional problem is the lack of correlation between particular sorts of genetic versus phenotypic change. For example, a point mutation ("micro") may cause the elevated RNA levels of *teosinte branched 1* in maize (Wang *et al.*, 1999), which causes a difference in plant architecture ("macro"?), whereas large differences in chromosome size or number ("macro"?) often have no detectable phenotypic effect (sub-"micro"?) (e.g. *Claytonia virginiana*: Rothwell, 1959). In many cases, what appears to be a large amount of morphological variation occurs simultaneously with only a few mutations in neutrally evolving genes. Well-documented cases include *Aquilegia* (Ranunculaceae: Hodges and Arnold, 1994), Hawaiian silverswords (Asteraceae: Barrier *et al.*, 1999), and *Gaertnera* (Rubiaceae: Malcomber, 2002). One possible explanation is that the morphology in question is easy to change and may in fact be controlled by just a few genes (Hilu, 1983; Gottlieb, 1984). When we know exactly what genes control the structures we see, this supposition can be tested.

5.6 How might deep nodes appear to differ from shallow ones?

What are the possible reasons that characters at deep nodes appear to be qualitatively different from those at shallow nodes? One reason, of course, is that it is simply a problem of human perception – there isn't a real difference. The numbers of events studied are still very small, and in many cases the morphological changes involved are not yet characterized developmentally, much less genetically. A case in point is the origin of the grass floret, which may have arisen at node 2 (Figure 5.2) and then been lost in Anomochlooideae, or may have occurred at node 3. Although lemmas and paleas are conventionally described as bract and prophyll respectively, some evidence suggests they are derived from sepals. This observation, along with studies of B and C class genes, suggest that the leaf-like morphology of glumes, lemmas and paleas may have been overlaid on a conventional floral gene expression pattern. Crucially, we have no idea what changes occurred at node 3.

If future studies support any of the distinctions made here, however, how might the apparent qualitative difference between old and recent speciation events come about? It seems likely that millions of years of evolution are themselves the explanation. Processes occur over tens of millions of years that simply do not occur in the few hundred thousand to few million years that distinguish many species. I outline some of these in the following sections. These possibilities are not mutually exclusive, but may all conspire to produce a genuine difference between deep and shallow nodes.

Extinction. Chase *et al.* (2000) suggest that the difference between deep and shallow nodes reflects selective superiority; all species that lacked the winning combination of traits have gone extinct, thus eliminating much of the apparently intermediate variation. Mishler (2000) likewise suggests that extinction creates its own particular patterns, although he does not actually specify what they might be. There is no clear way to test the effect of extinction. Certainly, the effect of extinction will depend on whether it is the result of selection against particular clades and phenotypes, or if it is random with respect to the phylogeny. If the latter is true, then the characters at deep nodes should be a subset of the characters at shallow ones. If the extinction is selective, however, then the phenotypes that persist could be a highly biased subset of those that existed before the extinction event.

Probability of reversal. A character that is unlikely to reverse will, purely by chance, come to characterize major clades. For example, heterotopic change may occur relatively infrequently, but may be less likely to reverse and thus may come to characterize major clades. The close vein spacing described above for nodes 9, 10, 11 and 12 actually originates at least seven times above node 9, but is apparently never lost once it is acquired. This one character alone accounts for multiple heterotopic changes on the phylogeny presented here.

Teotónio and Rose (2001) note that reverse evolution has received relatively little attention. Although it is clear that evolution can reverse itself, it is not clear how often that occurs and how distinct the reversed character is from the original one. When size characters are mapped onto a phylogeny, they appear to reverse frequently. I know of few characters mapping to deep nodes that are purely size characters. How many families are distinguished, for example, by the size of their leaves or by the length of their stipules?

Selection, epistasis and pleiotropy all might prevent or reduce reversibility (Teotónio and Rose, 2001). In addition, duplications of genes, chromosomes or whole genomes seem to be largely (although not inevitably) unidirectional changes. At the moment, we observe reversibility *a posteriori*, by examining how a character behaves on a cladogram. Ultimately, however, an understanding of the genetic basis of characters may permit assessment of reversibility *a priori*, and may permit assessment of whether certain sorts of characters will be recognizable over long periods of evolutionary time.

Gradual change in gene expression patterns. Quantitative differences might come to look more heterotopic over time. For example, there are two C-class genes in the grasses, one expressed predominantly in stamens (*zmm2*) and the other predominantly in carpels (*zag1*). Ancestrally, these must have been expressed in both organ types, and their function has gradually diverged quantitatively over time (Schmidt *et al.*, 1993; Mena *et al.*, 1996). Continued specialization could lead to the appearance of a shift to an anther-specific expression pattern for *zmm2*.

The duplication and differentiation of C-class genes is one possible example of the duplication–degeneration–complementation (DDC) model proposed by Force *et al.* (1999). They suggested that gene duplications might lead to differentiation of function of the duplicate copies via mutations in the promoters. The DDC model predicts that the ancestral C-class gene had separate regulatory sequences controlling stamen and carpel expression. One of the descendant genes would have lost the enhancers that elevate stamen expression and thus be expressed predominantly in carpels, whereas the other descendant gene would have lost the enhancers for carpel expression. There are, of course, other models that one could imagine. For example, mutations in regulatory sequences could instead increase the strength of binding of repressors, actively shutting down gene expression in one tissue or the other. These predictions are clear and testable.

The observation of frequent heterotopy does not require a different class of mutational mechanisms, but simply that some mutations lead to different tissue specificity, rather than generating more or less of a particular gene product in the same cells or tissues. The mutations changing tissue specificity could be as small as point mutations or as dramatic as genomic rearrangements. Thus, I am not making any claim about the nature of the mutation *per se*, but rather about its phenotypic consequences. The gene-level changes could be completely uniform through evolutionary time, but the perceived variation in the phenotype may be far more episodic.

Current studies on the evolution of development make this an empirical problem. When Goldschmidt (1940) proposed the idea of "hopeful monsters" there was no meaningful method of testing his ideas, because there was no way to determine the developmental or genetic basis of most phenotypes (Bateman and DiMichele, 2002). With the current enormous growth of genomic data (especially in functional genomics), we have a hope of understanding the phenotype of all plants, not just a few model systems. The key challenge will be to combine genetic approaches with appropriate tools of molecular biology to determine the genetic basis of morphological characters – characters that themselves have been the subject of investigation by taxonomists for centuries.

ACKNOWLEDGMENTS

I thank R. Bateman, A. Doust, S. Malcomber and an anonymous reviewer for comments that greatly improved the manuscript. Research underlying portions of this chapter is supported by NSF grant DEB 9815392, and by the E. Desmond Lee and Family Endowment to the University of Missouri-St Louis.

REFERENCES

Alberch, P., Gould, S. J., Oster, G. F. and Wake, D. B. (1979) Size and shape in ontogeny and phylogeny. *Paleobiology*, 5, 296–317.

Baldwin, B. G., Preston, R. E., Wessa, B. L. and Wetherwax, M. (2001) A biosystematic and phylogenetic assessment of sympatric taxa in *Blepharizonia* (Compositae–Madiinae). *Systematic Botany*, 26, 184–194.

Barrier, M., Baldwin, B. G., Robichaux, R. H. and Purugganan, M. D. (1999) Interspecific hybrid ancestry of a plant adaptive radiation: allopolyploidy of the Hawaiian silversword alliance (Asteraceae) inferred from floral homeotic gene duplications. *Molecular Biology and Evolution*, 16, 1105–1113.

Bateman, R. M. and DiMichele, W. A. (1994) Saltational evolution of form in vascular plants: a neoGoldschmidtian synthesis, in *Shape and Form in Plants and Fungi* (eds D. S. Ingram and A. Hudson), Academic Press, London, pp. 61–100.

Bateman, R. M. and DiMichele, W. A. (2002) Generating and filtering major phenotypic novelties: neoGoldschmidtian saltation revisited, in *Developmental Genetics and Plant Evolution* (eds Q. C. B. Cronk, R. M. Bateman and J. A. Hawkins), Taylor & Francis, London, pp. 109–159.

Baum, D. A. and Donoghue, M. J. (2002) Transference of function, heteropy, and the evolution of plant development, in *Developmental Genetics and Plant Evolution* (eds Q. C. B. Cronk, R. M. Bateman and J. A. Hawkins), Taylor & Francis, London, pp. 52–69.

Bennetzen, J. L. and Kellogg, E. A. (1997) Do plants have a one-way ticket to genomic obesity? *Plant Cell*, 9, 1509–1514.

Chase, M. W., Fay, M. F. and Savolainen, V. (2000) Higher-level classification in the angiosperms: new insights from the perspective of DNA sequence data. *Taxon*, 49, 685–704.

Dengler, R. E. and Dengler, N. G. (1990) Leaf vascular architecture in the atypical C_4 NADP-malic enzyme grass *Arundinella hirta*. *Canadian Journal of Botany*, 68, 1208–1221.

Dobzhansky, T. (1951) *Genetics and the Origin of Species*. Columbia University Press, New York.

Doebley, J. and Lukens, L. (1998) Transcriptional regulators and the evolution of plant form. *Plant Cell*, 10, 1075–1082.

Doebley, J., Stec, A. and Hubbard, L. (1997) The evolution of apical dominance in maize. *Nature*, 386, 485–488.

Doust, A. N. and Kellogg, E. A. (2002) Integrating phylogeny, developmental morphology and genetics: a case study of inflorescence evolution in the "bristle grass" clade (Panicoideae: Poaceae), in *Developmental Genetics and Plant Evolution* (eds Q. C. B. Cronk, R. M. Bateman and J. A. Hawkins), Taylor & Francis, London, pp. 298–314.

Doyle, J. A. and Endress, P. K. (2000) Morphological phylogenetic analysis of basal angiosperms: comparison and combination with molecular data. *International Journal of Plant Sciences*, 161 (6 Suppl.), S121–S153.

Endress, P. K. and Igersheim, A. (2000) Gynoecium structure and evolution in basal angiosperms. *International Journal of Plant Sciences*, 161 (6 Suppl.), S211–S223.

Fjellheim, S., Elven, R. and Brochmann, C. (2001) Molecules and morphology in concert. II. The *Festuca brachyphylla* complex (Poaceae) in Svalbard. *American Journal of Botany*, 88, 869–882.

Flores, G. V., Duan, H., Yan, H. J., Nagaraj, R., Fu, W. M., Zou, Y., Noll, M. and Banerjee, U. (2000) Combinatorial signaling in the specification of unique cell fates. *Cell*, 103, 75–85.

Force, A., Lynch, M., Pickett, F. B., Amores, A., Yan, Y. and Postlethwaite, J. (1999) Preservation of duplicate genes by complementary, degenerate mutations. *Genetics*, 151, 1531–1545.

Furness, C. A. and Rudall, P. J. (1998) The tapetum and systematics in monocotyledons. *Botanical Review*, 64, 201–239.

Gaut, B. S. and Doebley, J. F. (1997) DNA sequence evidence for the segmental allotetraploid origin of maize. *Proceedings of the National Academy of Sciences of the USA*, 94, 68090–68094.

Ghazi, A. and VijayRaghavan, K. (2000) Control by combinatorial codes. *Nature*, 408, 419–420.

Giussani, L. M., Cota-Sánchez, J. H., Zuloaga, F. O. and Kellogg, E. A. (2001) A molecular phylogeny of the grass subfamily Panicoideae (Poaceae) shows multiple origins of C_4 photosynthesis. *American Journal of Botany*, 88, 1993–2012.

Goldschmidt, R. (1940) *The Material Basis of Evolution*. Yale University Press, New Haven, CT.

Gottlieb, L. D. (1984) Genetics and morphological evolution in plants. *American Naturalist*, 123, 681–709.

Gould, S. J. (1977) *Ontogeny and Phylogeny*. Harvard University Press, Cambridge, MA.

Grass Phylogeny Working Group (GPWG) (2000) A phylogeny of the grass family (Poaceae), as inferred from eight character sets, in *Grasses: Systematics and Evolution* (eds S. W. L. Jacobs and J. E. Everett), CSIRO, Collingwood, Victoria, pp. 3–7.

Grass Phylogeny Working Group (GPWG) (2001) Phylogeny and subfamilial classification of the Poaceae. *Annals of the Missouri Botanical Gardens*, 88, 373–457.

Halfon, M. S., Carmena, A., Gisselbrecht, S., Sackerson, C. M., Jimenez, F., Baylies, M. K. and Michelson, A. M. (2000) Ras pathway specificity is determined by the integration of multiple signal-activated and tissue-restricted transcription factors. *Cell*, 103, 63–74.

Hattersley, P. W. and Watson, L. (1975) Anatomical parameters for predicting photosynthetic pathways of grass leaves: the "maximum lateral cell count" and the "maximum cells distant count." *Phytomorphology*, 25, 325–333.

Hilu, K. W. (1983) The role of single-gene mutations in the evolution of flowering plants, in *Evolutionary Biology* (eds M. K. Hecht, B. Wallace and G. T. Prance), Plenum Publishing Corporation, New York, pp. 97–128.

Hodges, S. A. and Arnold, M. L. (1994) Columbines: a geographically widespread species flock. *Proceedings of the National Academy of Sciences of the USA*, 91, 5129–5132.

Jacobs, B. F., Kingston, J. D. and Jacobs, L. L. (1999) The origin of grass-dominated ecosystems. *Annals of the Missouri Botanical Garden*, 86, 590–643.

Kanai, R. and Edwards, G. E. (1999) The biochemistry of C_4 photosynthesis, in C_4 *Plant Biology* (eds R. F. Sage and R. K. Monson), Academic Press, San Diego, pp. 49–87.

Kellogg, E. A. (1990) Ontogenetic studies of florets in *Poa* (Gramineae): allometry and heterochrony. *Evolution*, 44, 1978–1989.

Kellogg, E. A. (1998) Relationships of cereal crops and other grasses. *Proceedings of the National Academy of Sciences of the USA*, 95, 2005–2010.

Kellogg, E. A. (1999) Phylogenetic aspects of the evolution of C_4 photosynthesis, in C_4 *Plant Biology* (eds R. F. Sage and R. K. Monson), Academic Press, San Diego, pp. 411–444.

Kellogg, E. A. (2000) The grasses: a case study in macroevolution. *Annual Review of Ecology and Systematics*, 31, 217–238.

Kelly, L. M. (2001) Taxonomy of *Asarum* sect. *Asarum* (Aristolochiaeae). *Systematic Botany*, 26, 17–53.

Kenrick, P. and Crane, P. R. (1997) *The Origin and Early Diversification of Land Plants*. Smithsonian Institution Press, Washington, DC.

Langdale, J. A. and Nelson, T. (1991) Spatial regulation of photosynthetic development in C_4 plants. *Trends in Genetics*, 7, 191–196.

Langdale, J. A., Rothermel, B. A. and Nelson, T. (1988) Cellular pattern of photosynthetic gene expression in developing maize leaves. *Genes and Development*, 2, 106–115.

Leegood, R. C. and Walker, R. P. (1999) Regulation of the C_4 pathway, in C_4 *Plant Biology* (eds R. F. Sage and R. K. Monson), Academic Press, San Diego, pp. 89–131.

LeRoux, L. G. and Kellogg, E. A. (1999) Floral development and the formation of unisexual spikelets in the Andropogoneae (Poaceae). *American Journal of Botany*, 86, 354–366.

Linder, H. P. (1987) The evolutionary history of the Poales/Restionales – a hypothesis. *Kew Bulletin*, 42, 297–318.

Malcomber, S. T. (2002) Phylogeny of *Gaertnera* Lam. (Rubiaceae) based on multiple DNA markers: evidence of a rapid radiation in a widespread, morphologically diverse genus. *Evolution*, 56, in press.

Mena, M., Ambrose, B., Meeley, R. B., Briggs, S. P., Yanofsky, M. F. and Schmidt, R. J. (1996) Diversification of C-function activity in maize flower development. *Science*, 274, 1537–1540.

Mishler, B. D. (2000) Deep phylogenetic relationships among "plants" and their implications for classification. *Taxon*, 49, 661–683.

Nelson, T. and Langdale, J. A. (1992) Developmental genetics of C_4 photosynthesis. *Annual Review of Plant Physiology and Plant Molecular Biology*, 43, 25–47.

Raff, R. A. and Wray, G. A. (1989) Heterochrony: developmental mechanisms and evolutionary results. *Journal of Evolutionary Biology*, 2, 409–434.

Rice, S. H. (1997) The analysis of ontogenetic trajectories: when a change in size or shape is not heterochrony. *Proceedings of the National Academy of Sciences of the USA*, 94, 907–912.

Rothwell, N. V. (1959) Aneuploidy in *Claytonia virginica*. *American Journal of Botany*, 46, 353–360.

Schmidt, R. J., Veit, B., Mandel, M. A., Mena, M., Hake, S. and Yanofsky, M. F. (1993) Identification and molecular characterization of *ZAG1*, the maize homolog of the *Arabidopsis* floral homeotic gene *AGAMOUS*. *Plant Cell*, 5, 729–737.

Schultz, J. L. and Soltis, P. S. (2001) Geographic divergence in *Leptodactylon californicum* (Polemoniaceae): insights from morphology, enzyme electrophoresis, and restriction site analysis of rDNA. *Systematic Botany*, 26, 75–91.

Shantz, H. L. (1954) The place of grasslands in the Earth's cover of vegetation. *Ecology*, 35, 143–145.

Sinha, N. R. and Kellogg, E. A. (1996) Parallelism and diversity in multiple origins of C_4 photosynthesis in grasses. *American Journal of Botany*, 83, 1458–1470.

Soltis, P. S., Soltis, D. E., Zanis, M. J. and Kim, S. (2000) Basal lineages of angiosperms: relationships and implications for floral evolution. *International Journal of Plant Sciences*, 161 (6 Suppl.), S97–S107.

Soreng, R. J. (1990) Chloroplast-DNA phylogenetics and biogeography in a reticulating group: study in *Poa* (Poaceae). *American Journal of Botany*, 77, 1383–1400.

Stevens, P. F. (1991) Character states, morphological variation, and phylogenetic analysis: a review. *Systematic Botany*, 16, 553–583.

Stevens, P. F. (1998) What kind of classification should the practising taxonomist use to be saved? in *Plant Diversity in Malesia III: Proceedings of the 3rd International Flora Malesiana Symposium* (eds J. Dransfield, M. J. E. Coode and D. A. Simpson), Royal Botanic Gardens, Kew, pp. 295–319.

Teotónio, H. and Rose, M. R. (2001) Perspective: reverse evolution. *Evolution*, 55, 653–660.

Wagner, P. J. (2000) Exhaustion of morphologic character states among fossil taxa. *Evolution*, 54, 365–386.

Wang, R.-L., Stec, A., Hey, J., Lukens, L. and Doebley, J. (1999) The limits of selection during maize domestication. *Nature*, 398, 236–239.

Wolfe, A. D. and Randle, C. P. (2001) Relationships within and among species of the holoparasitic genus *Hyobanche* (Orobanchaceae) inferred from ISSR banding patterns and nucleotide sequences. *Systematic Botany*, 26, 120–130.

Xu, C. Y., Kauffmann, R. C., Zhang, J. J., Kladny, S. and Carthew, R. W. (2000) Overlapping activators and repressors delimit transcriptional response to receptor tyrosine kinase signals in the Drosophila eye. *Cell*, 103, 87–97.

Yuh, C.-H., Bolouri, H. and Davidson, E. H. (1998) Genomic cis-regulatory logic: experimental and computational analysis of a sea urchin gene. *Science*, 279, 1896–1902.

Zaitchik, B. F., LeRoux, L. G. and Kellogg, E. A. (2000) Development of male flowers in *Zizania aquatica* (North American wild-rice; Gramineae). *International Journal of Plant Sciences*, 161, 345–351.

Chapter 6

The Mostly Male theory of flower origins: summary and update regarding the Jurassic pteridosperm *Pteroma*

Michael W. Frohlich

ABSTRACT

The Mostly Male theory is the first explanation of the evolutionary origins of the flower to be based on evidence from gene phylogeny and gene function as well as from plant morphology. Evidence supporting the theory is described. Though based primarily on studies of modern plants, the fossil pteridosperm group Corystospermales fits the theory well, if one assumes that *Pteroma*, the latest-occurring male corystosperm reproductive structure, is a stem bearing simple microsporophylls, like the older, better understood fossil *Pteruchus*. A new specimen of *Pteroma* shows that the microsporangium-bearing structures were helically arranged on the axis, suggesting that they do indeed represent simple microsporophylls borne on a stem, as required for the theory. Gene changes that cause evolutionary novelty have been termed "primary genes." Observations that may allow "candidate" primary genes to be recognized are discussed, and these criteria are applied to the Mostly Male theory, to suggest whether candidate primary genes might be recognized for particular evolutionary steps. Possible gene-based tests of the theory are also proposed.

6.1 Introduction

The evolutionary origin of flowers has long been a central problem in plant science, made famous as Darwin's "abominable mystery" (Darwin and Seward, 1903, vol. 2: 20). Many theories have been proposed to explain flower origins, but nearly all have difficulty in bridging the large morphological gap between flowering plants and their potential gymnosperm relatives and ancestors (Stebbins, 1974). Even in the recently ascendant anthophyte theory this difficulty was manifest, generating very divergent evolutionary scenarios and very different inferences of homology due to only slight changes in the phylogenetic data (Doyle, 1996). Phylogenies that support the anthophyte theory are incompatible with the recent strong evidence, based on many genes, that extant gymnosperms are monophyletic (Samigullin *et al.*, 1999; Bowe *et al.*, 2000; Chaw *et al.*, 2000; Donoghue and Doyle, 2000; Frohlich and Parker, 2000), hence these theories now have no more support than the many others proposed over the years.

The Mostly Male theory includes one instance of heterotopy that greatly reduces the remaining morphological gap between flowering plants and the best-known

In *Developmental Genetics and Plant Evolution* (2002) (eds Q. C. B. Cronk, R. M. Bateman and J. A. Hawkins), Taylor & Francis, London, pp. 85–108.

members of one fossil gymnosperm group, the Corystospermales (Frohlich and Parker, 2000; Frohlich, 2001). Although this theory was originally based on evidence from living plants, the existence of a fossil group that fits the theory lends considerable support. Gaps do remain, but one of the most important is temporal: the best-known corystosperms – including the permineralized male structures that most impressively fit the theory – are Triassic, 130 million years older than the first flowers. Only two fossil corystosperms are known after the Triassic (Yao *et al.*, 1995). One plant (*Ktalenia*) of the early Aptian (Cretaceous) is contemporaneous with angiosperms, but its male structures are unknown (Taylor and Archangelsky, 1985). The other, *Pteroma*, is a Jurassic compression fossil known only from its male structures. Harris' (1964) description of *Pteroma* in the Yorkshire Jurassic Flora does not resolve the most critical morphological question related to the Mostly Male theory, that the microsporophylls be simple and borne on a specialized stem. This symposium, held at the Natural History Museum in London, has allowed me to study the *Pteroma* fossils, which are all housed there, and thus to largely resolve this issue, as described below in the third section of this chapter, following a review of the Mostly Male theory.

McLellan *et al.* (2002) highlight the distinction between the "primary" gene(s) which *cause(s)* a change in morphology, either through a change in expression pattern or through a change in protein sequence, versus the "secondary" genes that show changes in expression or that evolve sequence change as secondary effects of the primary gene(s). If one seeks the genetic change(s) responsible for evolutionary innovation then the primary gene(s) are the goal. It is worth considering what the primary genes responsible for the innovations of the Mostly Male theory might have been, as described in Section 6.4.

A theory is valuable only if it can be tested. The most directly applicable tests of the Mostly Male theory derive from its developmental–genetic predictions (Frohlich and Parker, 2000; Frohlich, 2001). These are best examined by the study of many genes, using a genomics approach. Preparation for such a study requires consideration of how such genes might be found, how data regarding them should be interpreted, and how effectively different classes of genes could be used for testing the theory (see Section 6.4).

6.2 The Mostly Male theory

6.2.1 Description of the theory

The Mostly Male theory focuses on the origin of flower bisexuality from the unisexual reproductive structures found in nearly all gymnosperms. How could male and female structures, formerly borne separately, come to be produced in the same organ complex in flowering plants? The developmental–genetic changes that gave rise to bisexuality may be portrayed as a continuum of possibilities, symbolized by the line in Figure 6.1. In one possibility, represented by the midpoint of this line, many or all of the genes formerly active in the unisexual male structure, as well as a comparable number of genes formerly active in the unisexual female structure, would all be expressed together in the protoflower. Surely this would provide sufficient developmental–genetic information to generate both male and female structures. Alternatively, the ancestral male and female structures might not have made equal

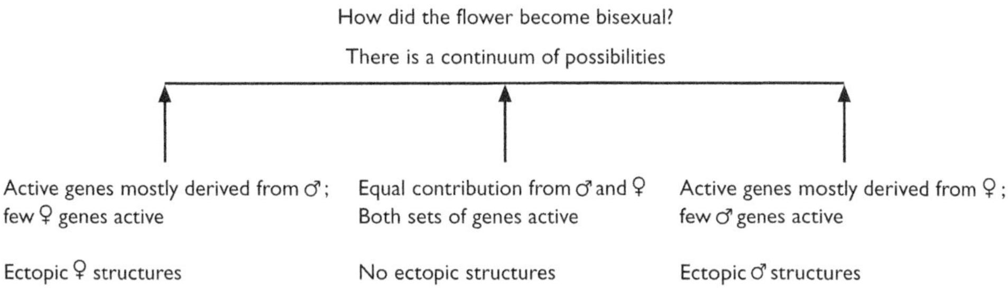

How did the flower become bisexual?

There is a continuum of possibilities

Active genes mostly derived from ♂; few ♀ genes active	Equal contribution from ♂ and ♀ Both sets of genes active	Active genes mostly derived from ♀; few ♂ genes active
Ectopic ♀ structures	No ectopic structures	Ectopic ♂ structures

Figure 6.1 The continuum of developmental–genetic mechanisms that could have generated bisexual reproductive structures (modified from Frohlich and Parker, 2000).

contributions to the set of genes expressed in the protoflower, as symbolized by the line segments extending left and right of the midpoint. To the left of the midpoint there are progressively fewer genes derived from those formerly expressed in the ancestral female structure. The endpoint represents expression of only those female-derived genes required to generate the minimal female structure, the ovule, along with most or all of the genes originally expressed in the male structure. At the left endpoint the reproductive complex is organized almost completely by male-derived genes, with ectopic ovules borne on a fundamentally male reproductive unit. To the right of the midpoint this trend is reversed; expressed genes derived from the ancestral male structure become progressively fewer, with the endpoint representing expression of the minimal set of male-derived genes, just sufficient to generate ectopic microsporangia, on a fundamentally female structure.

This continuum includes the full range of relative contributions of genes derived from the ancestral male and female units, so truth must lie somewhere along it. The Mostly Male theory grows from evidence that flowers lie to the left of the midpoint of this continuum, and may well lie at the extreme left, where a basically male reproductive unit is rendered bisexual by ectopic ovules that formed upon it.

Evidence bearing on the actual position of flowering plants along this continuum derives (1) from phylogenetic and functional studies of the *FLORICAULA/LEAFY* gene, (2) from gene overexpression studies and teratologies showing that ectopic ovules are easily produced, (3) from the relative morphological variability of stamens and carpels in flowering plants, (4) from the pollination syndrome of modern Gnetales, which resembles that suggested for the first plants with ectopic ovules, and (5) from studies of fossils that fit the theory.

In the full scenario of the Mostly Male theory, the gymnosperm ancestor of flowering plants would have had simple microsporophylls borne spirally on a stem, with the ovule antecedent borne elsewhere on the plant. The ovule antecedent would have become borne ectopically on (at least some of) the microsporophylls. This teratology would be immediately selected for, perhaps because the liquid pollination droplet produced by the ectopic ovule would attract pollinating insects, as in the modern Gnetales. Later, the ectopic ovules would become important as seed producers, so the separately borne ovule antecedents would eventually be lost. The microsporophylls bearing ectopic ovules would lose their microsporangia and eventually enclose the ovules, generating the carpel.

6.2.2 FLORICAULA/LEAFY *gene phylogeny*

FLORICAULA/LEAFY (*LFY*) is a homeotic transcription factor first cloned from *Antirrhinum*, and subsequently found in *Arabidopsis* and other plants (Coen *et al.*, 1990; Weigel *et al.*, 1992). It is expressed in a graded manner in young leaf primordia and in the shoot apex. Many pathways influence the level of *LFY* expression. In *Arabidopsis*, when this expression becomes sufficiently high, it triggers a genetic circuit that further raises expression to high levels in the apical meristem, causing the meristem to be determined as floral, so it develops into a flower rather than a vegetative shoot (Blázquez *et al.*, 1997; Hempel *et al.*, 1997; Blázquez and Weigel, 2000). In *Arabidopsis, LFY* directly stimulates expression of the MADS box genes *AP1* and *AG* (Busch *et al.*, 1999; Wagner *et al.*, 1999), and is required for initial expression of *AP3* and *PI* (Chen *et al.*, 2000), which are crucial elements of the "ABC" gene system that makes primordia develop into sepals, petals, stamens or carpels, which are the major flower organs.

LFY is single-copy in studied diploid flowering plants, including the basal angiosperm *Nymphaea* (with one possible exception in higher eudicots), but all four groups of extant gymnosperms have two copies of *LFY*, resulting from a single ancient duplication (Frohlich and Meyerowitz, 1997; Frohlich and Parker, 2000). Evidence from pine shows that one copy of *LFY* (the "Leaf" paralog *PRFLL*) is highly expressed only in the early developing male cone, starting at the very earliest stages, before any specifically male structures have yet formed, but is expressed at low levels in early developing female cones and in vegetative shoot apices (Mellerowicz *et al.*, 1998). The other copy (the "Needle" paralog, *NEEDLY*) is expressed at high levels only in developing female cones, again starting at the very earliest stages, before any specifically female structures have formed, but at low levels in early developing male cones and in vegetative shoot apices (Mouradov *et al.*, 1998). This is consistent with the Needle paralog of *LFY* specifying the female reproductive structures of gymnosperms, whereas the Leaf paralog specifies the male structures. That the two paralogs should have these functions is not surprising, as we had previously speculated (Frohlich and Meyerowitz, 1997), because it seems reasonable that structures long borne separately might well become specified by different genes. Gymnosperm male and female reproductive structures have been borne separately since the origin of the group in the late Devonian.

In the *LFY* gene tree, all angiosperm *LFY*s form a single clade that attaches to the Leaf clade of gymnosperm genes, *above* the duplication event that created the two gene copies found in gymnosperms (Frohlich and Parker, 2000). This means that the duplication event occurred *before* the lineage (of organisms) leading to angiosperms separated from the lineage leading to the extant gymnosperms. Hence, the angiosperm ancestor would have inherited both the Leaf and Needle paralogs from its gymnosperm ancestor. This is most easily seen by placing the gene tree within the tree of organisms (Figure 6.2). The gene tree is indicated by thin lines, whereas the much wider tree of organisms is marked only by an outline, so that the gene tree within is clearly visible. The gene duplication event occurs within an internode of the tree of organisms. At every speciation event all genes present in the ancestor are inherited by both descendent lineages. A diamond indicates the attachment of the angiosperm *LFY*s to the gymnosperm Leaf clade. A Needle paralog gene (shown as a dashed line) would have been inherited by the angiosperm clade, but no Needle paralog has been found in modern angiosperms. The Needle paralog is certainly absent from *Arabidopsis*, and our search in the water lily *Nymphaea*, a member of the second clade of flowering plants, yielded only a pseudogene of the known water lily gene (Frohlich and Parker, 2000). It is most reasonable to infer that the Needle paralog was lost before diversification of the modern angiosperm lineages; this loss is indicated by the black dot terminating the dashed line in Figure 6.2.

LFY stimulates expression of a diverse cascade of genes that determine manifold aspects of flower development. The Leaf and Needle paralogs in gymnosperms presumably stimulate cascades of genes in the male and female reproductive structures, respectively. Retention of the Leaf paralog in angiosperms should result in expression of numerous male-derived genes in the angiosperm flower, while loss of the Needle paralog should result in fewer female-derived genes being expressed in the flower. Hence, the flower would lie to the left of the midpoint in Figure 6.1.

6.2.3 FLORICAULA/LEAFY *mutants*

Strong loss-of-function mutants of *LFY* cause flowers to develop more or less as vegetative shoots. In *Antirrhinum* this transformation is profound, but in *Arabidopsis* the flower still shows a shortened axis with congested, spirally arranged organs that

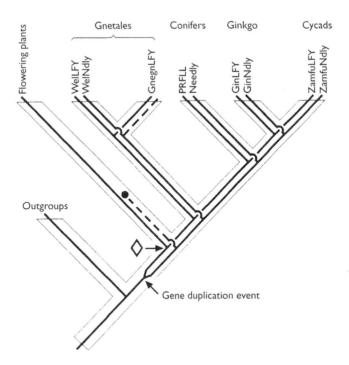

Figure 6.2 LEAFY gene tree (heavy line) placed within the tree of organisms (thin outline). Heavy dashed lines indicate genes that have not been found. Relationships among the living gymnosperm groups are uncertain, but do not affect the inference that a Needle paralog of *LFY* was lost in the flowering plant lineage (black dot) (from Frohlich and Parker, 2000).

develop as sepals surrounding a group of fused carpels (Coen *et al.*, 1990; Weigel *et al.*, 1992). The carpels are partly fertile, in spite of the numerous deformities of these flowers. A mass of carpels may also terminate the inflorescence in *Arabidopsis*, without any subtending sepals. The complete absence of stamens, while carpels are still formed, supports the closer association of the angiosperm *LFY* with the male program in the flower, whereas other genes, acting in parallel with *LFY*, can still produce carpels.

LFY shows additional functions in pea (*Pisum*), where it is required for the formation of compound leaves (Hofer *et al.*, 1997), and in grasses (*Oryza* and *Lolium*), where it is expressed at high levels in the developing inflorescence (spikelets) (Kyozuka *et al.*, 1998; Gocal *et al.*, 2001). These must represent late-evolved functions of *LFY*, because the highly branched grass inflorescence and the compound leaves of legumes are specialized features that arose high in the angiosperm tree.

6.2.4 The occurrence of ectopic ovules

Although the ovule is a highly complex structure, ectopic ovules occur in both flowering plants and in gymnosperms. Colombo *et al.* (1995) overexpressed a single *Petunia* gene, *FPB11*, in *Petunia*, with the result that ectopic ovules formed on both the sepals and the petals. These were fully formed ovules, possessing both integuments, the nucellus and a normal-appearing embryo sac. These were borne on sepals and petals that had *not* been converted into carpels; thus, it is an example of ectopic ovules, not ectopic carpels. This observation led Colombo *et al.* (1995) to suggest that the placenta, the structure that bears ovules in flowering plants, should be considered the fifth major flower organ, along with sepals, petals, stamens and carpels. Their view is consistent with the ovule having first appeared in the flower as an ectopic structure.

Several *Ginkgo* trees have been reported in China and Japan that bear ectopic ovules on a small fraction of their leaves (Fujii, 1896; Sakisaka, 1929; Ma and Li, 1991). Some of these ovules even develop into seeds, proving that the ectopic ovules can be fully functional. The cause of these naturally occurring teratologies is not known, though Sakisaka (1929) suggested that it was a result of extreme age of the individual tree. Fujii (1896) and Sakisaka (1929) also reported ectopic microsporangia on leaves of some male *Ginkgo* trees.

A *Ginkgo* tree in Cincinnati, Ohio, USA, likewise bears ectopic ovules on leaves. Through the courtesy of F. Wagner I have been able to examine dried leaves of this tree originally collected for W. H. Wagner by C. Schram and subsequently by R. Naczi. This includes dozens of leaves bearing seeds and many others bearing the remains of presumed ovules that did not develop into seeds (Figure 6.3; Plate 3a). They were collected from the ground after being shed in the Fall. The leaves bearing ectopic structures are smaller than normal leaves, ranging from half their width to very small, with hardly any blade formed. Ectopic ovules are borne in deep sinuses in the leaf blade. A prominent thickened vein extends below the ectopic ovule, gradually merging into the typical vein pattern of the leaf. Most commonly there are one or two ovules per leaf, but there can be as many as five, with each in its own sinus, more-or-less evenly distributed across the leaf blade.

Figure 6.3 *Ginkgo* leaves bearing ectopic ovules in sinuses of leaves. Leaves at left and middle have one ovule each that has formed a seed. Leaves were collected from the ground after shedding by C. Schram and R. Naczi for W. H. Wagner (courtesy of F. S. Wagner; specimens at MICH).

The leaf lobes on either side of the sinuses are sometimes contorted so the leaf blade does not lie in a single plane. Leaves that bear seeds average an even smaller size than those with non-seed producing ovules. When present, seeds are attached in the sinus, and usually are pressed onto the surface of the leaf. The middle leaf in Figure 6.3 is folded over, so the seed extends beyond it. In these collections there are no leaves that show abscission of the seed from the leaf, but many leaves have seeds still attached, suggesting that normal abscission of the seed from the leaf does not occur. Moreover, a similar *Ginkgo* tree was recently discovered outside the gardens at Kew by R. Bateman and P. Rudall (pers. comm., 2001).

The fact that ectopic ovules can be formed in *Petunia* by overexpression of a single *Petunia* gene, and that ectopic ovules occur naturally in *Ginkgo*, suggests that appearance of ectopic ovules is not too implausible an event to postulate for an evolutionary scenario.

Bateman and DiMichele (2002) discuss the Paleozoic gymnosperm *Pullaritheca longii*, in which a single teratological female reproductive unit (a cupule) bears microsporangia as well as ovules. This provides another example of ectopic reproductive structures, again suggesting that such events are not implausible.

6.2.5 Morphology of carpels and stamens

Flowering plants show little variation in the number or placement of microsporangia borne on each stamen, but carpels show dramatic variation in both number(s) and location(s) of ovules. A typical stamen has two "thecae," each of which contains two microsporangia embedded in its tissues, yielding a total of four microsporangia per stamen. Common modifications reflect loss of one theca or loss of one microsporangium in each theca, resulting in stamens bearing two microsporangia. In other seed plants the number of microsporangia borne on each microsporophyll varies greatly (for example, hundreds are borne on cycad microsporophylls), but no such increase in microsporangium number occurs in flowering plants. The only possible exceptions are in the Malvales, but these are an advanced group, so this feature represents a very late acquisition within angiosperms. Further, the special features of Malvales may represent multiple stamens that are fused together, rather than individual stamens with extra microsporangia.

By contrast, variation in ovule position is so great that this is an important character used to identify major groups of angiosperms (Cronquist, 1981). Ovules may be axile, parietal, basal, apical or free central within the ovary. These terms hide very many further variations in the details of ovule positioning, especially in plants with many ovules borne crowded together. The number of ovules per carpel varies from one (or even less than one) to hundreds of thousands in some orchids. Regardless of the original number and position of ovules in the first angiosperm, some of this variation in ovule position and number must represent new, ectopic placement of some ovules. The fact that this variation occurs at high taxonomic levels within angiosperms suggests that it reflects features of angiosperm development near the base of the angiosperm tree, which subsequently became fixed in some large clades. If ovules were originally ectopic in the first flower then it is reasonable that their position and number would not be rigidly circumscribed in subsequent evolution. The variation in ovule number and placement, contrasted with the stability of microsporangium number and placement, is consistent with ovules, but not microsporangia, having been ectopic in the first flowers.

Some individual flowers do contain many microsporangia, but this is accomplished by increasing the number of stamens, not the number of microsporangia per stamen. The numbers of major flower organs (sepals, petals, stamens and carpels) are variable, both over evolutionary time, as shown by variation in organ number within some angiosperm orders and even families, and by the frequent mutations which alter the numbers of parts in flowers, or convert one structure into another. In many cultivated plants double forms are known with increased petal number, often also generating organs that show intermediacy between petals and stamens. Even within the family Brassicaceae, which shows few natural variations in its floral formula, many different *Arabidopsis* mutants are known that increase petal, stamen and/or carpel number (Running and Hake, 2001). Yet among all these mutations none are known that produce stamens with extra microsporangia, nor are there mutations that yield ectopic microsporangia. Microsporangia occur only on stamens, or on organs that are partly converted into stamens. This suggests that formation of microsporangia is firmly dependent on the developmental milieu of the stamen, as might occur after a very long evolutionary history in which microsporangia are produced only on stamens; hence, microsporangia were unlikely to have been ectopic in the original flower, in contrast to the ovule.

6.2.6 Gnetalean pollination syndrome as an analog of the earliest flowers

For ectopic ovules to be evolutionarily significant this feature must have experienced positive selection as soon as it appeared, allowing the trait to be retained in the population and to be further refined. Complete neutrality of selection could, in principle, have allowed persistence, but such neutrality is unlikely for such a major modification. Although ectopic ovules might make additional seeds, as in the case of the *Ginkgo* ectopic ovules discussed above, a simpler effect could have generated the positive selection. The typical gymnosperm ovule exudes a pollination droplet from its micropyle; the droplet catches pollen grains and is then withdrawn, bringing the pollen into the ovule (Gifford and Foster, 1989; Tomlinson *et al.*, 1997). In all three extant genera of Gnetales there are sterile ovules in the functionally male reproductive complex. In *Welwitschia* the sterile ovule is borne on the same short axis that produces the pollen-organs. These ovules never develop into seeds, but they do exude large pollination droplets (Hufford, 1996). These droplets attract insects to the male, while the droplets of the functional ovules attract insects to the female, resulting in insect pollination (Endress, 1996). Current molecular phylogenetic evidence shows Gnetales are most closely related to conifers, rather than to flowering plants; hence, contrary to early versions of the Anthophyte theory (Doyle and Donoghue, 1986, 1987; Doyle, 1994), the presence of a sterile ovule in the male structures of *Welwitschia* (and other Gnetales) cannot reflect homology with the general arrangement of parts in the flower. Rather, this feature is analogous to the occurrence of ovules on male structures in the scenario of the Mostly Male theory. The fact that this innovation clearly did occur in Gnetales suggests that it could have happened in the ancestors of flowering plants as well.

6.2.7 Existence of fossils that fit the Mostly Male theory

The Mostly Male theory grew from evidence based on modern plants. Though I was aware of fossils that figured in older theories of flower origins (Stebbins, 1974; Doyle, 1978; Crane, 1985), the Mostly Male theory was not erected with possible fossil antecedents in mind. Because the ovules in the Mostly Male theory are ectopic, the theory places few restrictions on female structures of potential ancestors of flowering plants. It is only necessary that there be a structure that could have been the antecedent of the angiosperm ovule.

The theory does establish stringent requirements for male structures of any potential gymnosperm ancestor of flowering plants. The microsporophylls must be simple, rather than compound, and must not be strongly lobed, and they must be borne in groups along a stem specialized to bear them. These features are required so that the male unit becomes a good antecedent for the flower after ectopic ovules appear on (some) microsporophylls. Distal microsporophylls along the stem would lose their microsporangia and evolve to enclose the ovules, creating carpels. Basal microsporophylls would be modified into stamens (Frohlich and Parker, 2000; Frohlich, 2001).

Gymnosperm groups commonly cited as possible flowering plant ancestors (or relatives of ancestors) seemed to lack the crucial male reproductive features required by the Mostly Male theory. In particular, *Caytonia*, often cited because its female structures provide a reasonable ovule antecedent, almost certainly bore *Caytonanthus* microsporophylls, which were pectinate and are not known to have been borne in groups (Harris, 1964; Doyle, 1978; Crane, 1985; Stewart and Rothwell, 1993; Taylor and Taylor, 1993; Osborn, 1994). The corystosperms, which have an even more credible ovule antecedent, were generally thought to possess large compound microsporophylls (Crane, 1985; Retallack and Dilcher, 1988; Taylor and Taylor, 1993).

After conceiving the theory, I learned that the corystosperm male structure had been reinterpreted based on a newly discovered permineralized fossil of the Triassic corystosperm *Pteruchus*. This clearly showed that the microsporophylls were simple, unlobed and spirally arranged on a stem (Yao *et al.*, 1995). This precisely fits requirements of the Mostly Male theory. Though corystosperms had a reasonable ovule antecedent, it was borne on a branching stem system (Axsmith *et al.*, 2000), which provides no reasonable antecedent for the carpel. In contrast, ectopic ovules placed on the upper surface of the *Pteruchus* microsporophyll would make an excellent carpel antecedent: the microsporophyll would only need to lose its microsporangia and enclose the ovules to closely resemble a carpel. *Pteruchus* microsporophylls even possessed paired secondary veins (Yao *et al.*, 1995); these could generate the vein pattern of ovules of basal angiosperms, which commonly have a main vein with a secondary vein placed symmetrically on each side of the primary vein.

In our original discussion of the Mostly Male theory (Frohlich and Parker, 2000) we also relied on *Pteroma*, from the Yorkshire Jurassic Flora, because it is later in time than *Pteruchus*, and because *Pteroma* resembles angiosperms in having the microsporangia embedded in the tissues of the microsporophyll. In the absence of evidence indicating whether *Pteroma* had a simple or compound microsporophyll, we assumed that it resembled *Pteruchus*. The need for this assumption was a weak point in our original argument.

6.3 The *Pteroma* fossils

6.3.1 Introduction to Pteroma

Pteroma has generally been considered a corystosperm, because of its resemblance to the corystosperm male reproductive structure *Pteruchus* (Harris, 1964). They both have flat, oval, laterally stalked microsporangium-bearing structures ("sporangial heads") attached to an otherwise unbranched elongate axis. Unlike *Pteruchus*, *Pteroma* has microsporangia embedded in the tissues of the microsporophyll. The Jurassic *Pteroma* is much later in time than the Triassic *Pteruchus*, making *Pteroma* of especial interest for the Mostly Male theory.

Since Harris described *Pteroma* in 1964, based on "two small blocks and three tiny fragments," more than a dozen additional specimens have been identified and prepared, principally by Christopher R. Hill, although he did not publish descriptions of them. All of these specimens are housed at the Natural History Museum, and were kindly made available to me by Dr David Lewis and Dr Paul Davies. These new specimens confirm Harris' (1964) description of *Pteroma*, and allow valuable additional observations.

6.3.2 Pteroma *specimens often appear entangled with each other*

Most blocks (except for the very smallest) show numerous sporangial heads. Some heads are linearly arrayed, due to their attachment to the elongate axis, which is often visible; however, in most blocks the sporangial heads are too widely separated to have all been borne on a single straight axis, and/or multiple elongate axes are visible. The heads are *not* arrayed in any standardized pattern, so this is *not* due to some hidden, regular pattern of branching of the main axes. Rather, it appears that two or more separate individual axes and their sporangial heads are tangled together in most blocks. The relatively small number of blocks that contain *Pteroma* fossils indicates that this cannot be due to random association during deposition. Rather, I suspect the *Pteroma* specimens became tangled with each other close to the plant, and were transported, already entangled, to the point of deposition. No other plant parts are included in these tangles.

The presence of opened slits on all sporangia (where this can be observed) and the general absence of pollen suggests that they had already completed their function of pollen release. After this, it is reasonable that the axes bearing sporangial heads would have been shed from the plant. The absence of leaves or other organs attached to these axes is consistent with a regular mechanism of separation. Shedding over a short interval could help account for their becoming entangled, if they fell on top of each other on the ground. Although the structures involved are in no way homologous, this process might have resembled the shedding of male catkins from birch or other modern trees, which frequently become entangled after shedding.

6.3.3 *Arrangement of sporangial heads on the elongate axis*

Sporangial heads that are linearly arrayed on the rock surface are *not* borne in any simple repeating pattern that would imply a planar organization of the axis with its attached heads, hinting that the heads could have been spirally arranged. Intact specimens cannot be removed from the matrix, due to numerous tiny cracks through the specimens (Harris, 1964), so one is limited to observation of specimens in the rock. A spiral arrangement, if present, would make some sporangial heads point into the rock, so some luck is required for their positions to be observable based on one exposed surface.

One specimen (V61418) does allow the relative positions of a series of sporangial heads to be determined. This specimen is not completely flat; in particular, some stalks leading to sporangial heads clearly point slightly upwards or downwards, out of the plane of the main axis. This is apparent in Figure 6.4, which shows a stereo pair of photos of this specimen. Figure 6.5 shows the labeling of the stalks leading to sporangial heads. Stalks 3 and 4 are especially clear in their orientations; 3 points steeply downward while stalk 4 points slightly downward. Stalk 5 appears to be above the plane of the axis at its attachment point, although it then bends downward, so the sporangial head is below the main axis. The relative positions of these three stalks are consistent with a right-handed helical arrangement of appendages. If so, the next stalk, beyond number 5, should be attached to the underside of the axis.

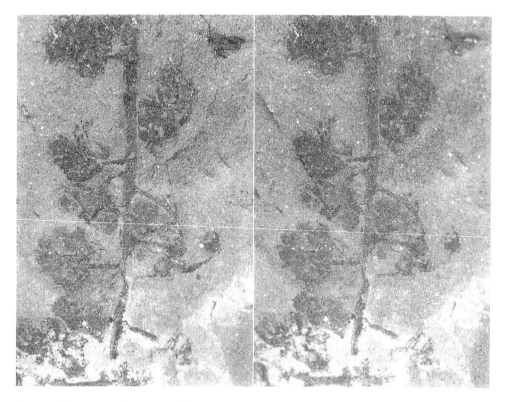

Figure 6.4 Stereo-pair illustration of *Pteroma*, showing relative insertions of microsporangiate heads above or below the plane of the long axis. Disparity is exaggerated for clarity (specimen V61418, BM).

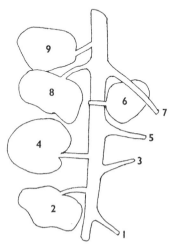

Figure 6.5 Diagram of the *Pteroma* specimen in Figure 6.4 to show numbering of sequential microsporangiate heads.

Viewed under high magnification, stalk number 6 clearly extends all the way across the top of the elongate axis; also, the head lies very close to the elongate axis. This is consistent with stalk 6 being attached to the underside of the elongate axis and having become folded over the top of the axis; which fits the right-hand helical arrangement. The positions of stalk 7 (above the axis) and stalk 8 (below) are also consistent with a continuing right-handed helix. The angle of stalk 9 is not apparent, but its position to the left of the axis would fit the helix, assuming that it had been directed steeply upward from the axis. Beyond stalk 9 there is a considerable gap before the next visible stalk.

In the other direction from stalks 3 and 4, stalk 2 appears to be directed slightly upward, though its orientation is not absolutely clear. The black remnant of stalk 2 has largely chipped away, but it has left a clear impression visible at high magnification. The orientation of stalk 1 is unclear. The distal portion appears to be lifted due to something underneath it in the rock. There are no appendages on the axis beyond stalk number 1. The spacing between these nine heads is fairly uniform, suggesting that all the heads on this segment of the axis are visible in the fossil.

The structure of this fossil is most consistent with the stalked sporangial heads being borne in a right-handed helix on the axis. The sporangial heads certainly do not appear to be borne in a plane, in two ranks on either side of the main axis, which would be the most reasonable alternative. A helical arrangement of sporangial heads resembles that of *Pteruchus*, supporting the view that *Pteroma* is a corystosperm. It also supports the interpretation of the sporangial heads as simple microsporophylls borne on a stem resembling *Pteruchus*, as was assumed by Frohlich and Parker (2000), and as the Mostly Male theory requires for putative ancestors or relatives of flowering plants.

It would be most interesting to discover other parts of the plant that bore *Pteroma* male structures. If the suggestion that these male structures were shed from the parent plant, entangled and transported to this site is correct, then it would be less likely that other parts of the plant would be deposited in close association with *Pteroma* specimens, as seeds, cupules and leaves might well have been shed at some other season(s). Further, the relatively small *Pteroma* axes might have been transported longer distances or to different resting places compared to leaves, seeds or vegetative stems. It may be difficult to find other parts of this plant and to prove their association with *Pteroma*.

6.4 Primary genes

6.4.1 What makes a gene primary?

To achieve the most profound understanding of an evolutionary innovation, one might attempt to identify the genetic changes that brought about the innovation. McLellan *et al.* (2002) highlight this goal. They term the genes that *cause* a morphological innovation – whether by changed expression or by altered protein sequence – the "primary" genes. Other genes, that show changed expression pattern as a consequence of the primary genes, are termed "secondary" genes. The secondary genes may be essential effector genes necessary to bring about the novel morphology, but they would not be said to "cause" the novelty if their changed actions are completely determined by primary genes that lie upstream in the control hierarchy. McLellan *et al.* (2002) studied morphological innovation within a species, so quantitative genetic methods are effective to locate the regions in the genetic map where allelic differences between morphotypes appear to cause the observed morphological differences.

There is no obvious experimental system operable over wide evolutionary distances that can identify primary genes. Even the definition of a primary gene becomes difficult over large evolutionary distances, as a series of smaller evolutionary steps may together be responsible for the major phenotypic difference one wishes to explain. Different classes of gene(s) may have been causal at each of these steps; does that make them all "primary" genes? Should genes responsible for the initial steps be the only ones considered "primary," especially if subsequent changes were only possible in the altered context created by the initial changes? Even if one had a satisfactory concept of evolutionary–genetic causation, one rarely has sufficiently detailed knowledge of the historical sequence of evolutionary innovations to apply such concepts, making this question moot.

Even if the series of morphological changes were known in detail, there is no guarantee that the genes responsible at each step in this historical process would be identifiable today. Developmental homeostasis is a well-known phenomenon that allows organisms to achieve their correct morphologies in spite of moderate insults from the environment, or even from mutations, that they may suffer (Fenster and Galloway, 1997; Moller, 1997). Developmental homeostasis is presumably accomplished through redundancy in the developmental–genetic program. This may involve related genes that have overlapping or equivalent functions or, at the other extreme, it may involve systems that operate in wholly different ways but push the developing organism toward a similar final form. One such developmental–genetic system may have been the initial cause of a morphological change, with other systems appearing later and helping to guarantee correct development even in extreme environments. At first the initial system would be the most important in controlling development. Over evolutionary time, the secondary systems could become increasingly important, and eventually may become dominant as determiners of the developmental path taken by the organism. An apparent example of this phenomenon has been found in insects, where the system that organizes the parasegments in grasshoppers (a near-basal short germ-band insect) has been relegated to a secondary role in the advanced (long germband) fruit fly (Akam and Dawes, 1992; Salazar-Ciudad *et al.*, 2001).

Such wholesale shifts in genetic control systems are thought to be responsible for the existence of the "phylotypic stage" in some animal groups. Within these animal groups, the earliest developmental stages may differ wildly among subgroups, yet *all* develop into an intermediate phylotypic stage that is highly stereotyped, though attained by very different developmental routes. An example occurs in echinoderms, in which some taxa develop via free-swimming larvae that later metamorphose into the phylotypic stage, and then into the adult, versus others that develop inside the egg directly into the phylotypic stage, and then emerge from the egg as tiny adults (Raff, 1996: 208). If such a shift has occurred in the evolutionary history of a group, then the primary genes may have lost their formerly crucial function, making their historically important role extremely difficult to identify.

Is there any hope of finding the primary genes responsible for a profound evolutionary event? First it *is* essential to identify the critical morphological–evolutionary event(s) (abbreviated below as "CMEE") that caused (or permitted) the acquisition of the evolutionary innovation under scrutiny. Then, in spite of the above caveats, the primary gene might be recognizable as exhibiting some of the following features; such a gene may be called a "candidate primary gene."

6.4.2.1 The gene is at the apex of a large cascade of effector genes that specify a morphological fate identified as the CMEE

If no shift in developmental control has occurred, then the primary gene may be recognized by this criterion, but the absence of shifts may be difficult to demonstrate. If several gene cascades promote the same final form, the genes at the apex of each cascade must be considered as candidate primary genes.

6.4.2.2 The gene belongs to an ancient, distinct lineage among organisms exhibiting the CMEE

This suggests that the gene was already distinct from any sister paralogs *before* the morphological evolutionary event occurred, hence it was in existence and available to participate in evolutionary events when the CMEE occurred.

6.4.2.3 Among genes with recent duplications, a paralog shows features that are plesiomorphic within the group of organisms possessing the CMEE

The gene responsible for a CMEE would likely be very important, at least for a period, until possible shifts in developmental control might reduce its significance. Such a gene may have undergone rapid change around the time of the CMEE, but subsequently it may become evolutionarily conservative. Even if subsequent duplications generate multiple paralogs, one (or more) paralogs are likely to retain functions related to the CMEE; these paralogs might be recognized by retention of similar amino acid sequences, expression patterns and upstream controls, which would be interpreted as plesiomorphic features among the recent paralogs in the gene tree. To be significant, this criterion requires support from other listed criteria. In phylogenetic studies plesiomorphic features are usually of little interest, but in the search for primary genes such features may be highly informative; hence, this is highlighted by listing as a separate criterion.

6.4.2.4 The gene shows differences in functional role between organisms with the CMEE and outgroup organisms lacking the CMEE

This feature is based on phylogenetic thinking, because in combination with criterion 2 above it would place a genetic change at the point in the organismal phylogeny where the CMEE occurred. Functional data are frequently unavailable, but the ratio of DNA base substitutions that do cause a change in the amino acid sequence versus those that do not (d_N/d_S) can identify branches in a cladogram where selection seems to have altered the protein sequence. This strongly suggests functional change of the protein on that branch (Yang and Bielawski, 2000).

Unfortunately, many secondary genes will also show changes in their roles at the same point in the tree. Although this attribute is likely to apply to primary genes, it is not as diagnostic as one might wish.

6.4.2.5 The gene could have initiated a developmental–evolutionary path favored with an immediate selective advantage

Greater understanding of the evolution of development and of ecological interactions may allow a reasonable assessment of the impact of selection on hypothetical intermediates of an evolutionary scenario. Some scenarios may include a genetic mechanism for the CMEE that allows continuous positive selection, unlike other genetic explanations. The genetic mechanism that the former posits involves the more reasonable candidate primary gene(s), because continuous positive selection is likely to be required for major evolutionary change. Obviously, this interpretation depends on one's confidence in the ability to evaluate selective regimes of the past.

6.4.2.6 The gene is able to generate a homeotic transformation analogous to the CMEE, when gene expression is altered experimentally

This may be the strongest available evidence to identify a candidate primary gene. To be most convincing, the experimental transformation should be possible in disparate lineages, implying that the only required cofactors are plesiomorphic attributes and thus may have been present at the base of the group, when the CMEE actually occurred. If the transformation is only possible within one small terminal clade, then required cofactors may be apomorphic features of that one clade. The gene and its cofactors may be part of a new developmental–genetic mechanism that represents a shift in developmental control, and hence may be a poor candidate for a primary gene. In general, such studies are typically performed in few organisms. Failure to find the homeotic transformation in one or a few lineages is only weak evidence against a candidate primary gene, as shifts in developmental control may defeat the morphological transformation in some lineages. Strong evidence against a candidate primary gene would require testing multiple distantly related lineages and finding none capable of the homeotic transformation. If homeotic transformation *is* found in distant lineages, indicating that cofactors are widespread, it is possible that one of the cofactors could actually have served as the primary gene.

6.4.2.7 A gene loss or duplication event occurred at the phylogenetic position where the CMEE occurred

Gene loss indicates that the function of the gene was no longer required. If the CMEE involves *loss* of a structure, then loss of the gene that specifies that structure could have been causal, making it the primary gene. This evidence is weakened, though, because many downstream genes specific to that structure will also be lost. The causal event could have been upstream of the organ-specifying gene, with a subsequent loss of the specifying gene serving as a guarantee that the structure would not be formed. In this case, loss of the organ-specifying gene could be viewed as achieving developmental homeostasis. Some gene families show very frequent duplication and loss events, whereas these are infrequent in other families. Correlation with a CMEE is more suggestive of causality for families where such changes are infrequent.

Nearly equivalent to complete loss of a gene is the loss of function of a gene in controlling a large genetic hierarchy, even if the gene retains minor functions later in development. It is a commonplace that crucial, early-acting genes also may have minor, later roles as well. For example, *Ultrabithorax*, a critical early gene in *Drosophila* development, also causes the difference between related species in the abundance of sensory hairs on the leg (Stern, 1998). Evolution recruits whatever genes are available for whatever roles they can serve. Because of such late functions, a gene may be retained in the genome when it is no longer important in determining early development. Such a gene could have had a primary role in the loss of a structure that it formerly specified (e.g. if a change in its promoter kept it off early in development), though of course the causal change could have been upstream of the gene, or even downstream, if a single downstream genetic modification can prevent formation of the structure.

A gene duplication event *per se* rarely results in a strongly altered phenotype, unless a concomitant change in the promoter region alters expression. Subsequent change in expression or in amino acid sequence of the new paralog is a standard scenario for evolutionary change, and may well be common. Though the duplication event is likely to antedate the change in gene function, extinction of closely related lineages that lack the CMEE will probably make the duplication appear coincident with the CMEE. The apparent coincidence of duplication and CMEE implies that the paralog should be considered a candidate primary gene, especially if criterion 1 also applies. Note that this could be considered a special case of criterion 2, and perhaps also of criterion 3.

6.4.3 Possible primary genes in flower evolution

Can candidate primary genes be identified for flower evolution, assuming that the Mostly Male theory is correct? The critical morphological–evolutionary event is clear: it is the appearance of ectopic ovules on the male structures of the gymnosperm ancestor. *FBP11* fulfills criterion 6 by creating ectopic ovules, though these ectopic ovules are not formed on the male structures of *Petunia* (Colombo *et al.*, 1995). Standard theories of petal origin suggest that petals were derived from stamens in many flowering plant lineages, which could make the ectopic development of these ovules more closely analogous to the CMEE. The analogous homeotic transformation has not been demonstrated in other plants.

Recent MADS-box gene trees show that *FBP11* is a member of the *AGAMOUS* subfamily of MADS genes, a subfamily that includes multiple paralogs in *Arabidopsis* and other plants. *FBP11* has a close paralog, *FBP7*, in *Petunia*. These two genes, with related genes in other angiosperms, attach to the tree *below* all other angiosperm genes in the *AGAMOUS* subfamily, and immediately above the gymnosperm AG-like genes (Theißen *et al.*, 2000, 2002). This suggests that the *FBP11* + *FBP7* subsubgroup of MADS genes is sufficiently ancient to satisfy criterion 2, above. This makes *FBP11* (or the gene ancestral to *FBP11* and *FBP7*) a candidate primary gene in the generation of ectopic ovules, according to the Mostly

Male theory (P. Engstrom, pers. comm.). This gene could also satisfy criterion 5, but so could any other gene that generates ectopic ovules, hence criterion 5 does not give further significant support to *FBP11* as a candidate primary gene.

Can candidate primary genes be identified for other major steps in the Mostly Male theory? Perhaps the second-most important event in the theory is loss of the pure female structure, after the ectopic ovules become efficient seed producers, and have the ability to generate some outcrossed seed. The Needle paralog of *FLO/LFY* satisfies criteria 1, 2 and most notably 7, as its loss might prevent development of the pure female structures; hence, it may be considered a candidate primary gene for this evolutionary event. However, the detailed scenario presented by Frohlich (2001) may argue against its primary status, as it may fail to fully satisfy criterion 5.

In the detailed scenario, the ectopic ovules become important seed producers long before any mechanism exists to avoid self-fertilization by pollen from the adjacent microsporophylls. I suggested (Frohlich, 2001) that the non-ectopic cupules may have produced a substantial fraction of outcrossed seed, hence they could have had an important function and been retained on the plant even if most seed was produced by the ectopic ovules. This presence of two types of seed-bearing structures would be analogous in function to the chasmogamous and cleistogamous flowers of some modern angiosperms, which produce seed that is, respectively, partly outcrossed versus all selfed. Many modern plants bear both types of flowers, showing that this is a stable evolutionary strategy. Hence, in the Mostly Male scenario the non-ectopic cupules may have been retained for a considerable period, until an outcrossing mechanism had evolved for the ectopic ovules in the protoflower. Long co-existence of both types of seed-bearing structures suggests that the plant would have perfected systems to control when, where and how many of each type of structure was borne. If the non-ectopic cupules gradually lost their value, then this sort of control mechanism could have gradually reduced their frequency close to zero. The likely existence of such regulatory systems provides a simpler method to eliminate the pure female structure; hence, these regulatory genes could well have been the primary genes for loss of the non-ectopic cupules, and loss of the Needle paralog would only achieve developmental homeostasis, preventing the pure female structure from ever forming.

6.5 Testing the Mostly Male theory

6.5.1 General considerations

The Mostly Male theory is subject to two general classes of tests. The discovery of fossil morphologies predicted by the theory, or of fossils contrary to expectations, could provide the strongest evidence to support or refute the theory. A detailed scenario that describes the full range of potential fossil intermediates consistent with the theory was presented by Frohlich (2001). While fossils may well be definitive, their discovery is unpredictable and often serendipitous.

Tests based on gene function and phylogeny do not depend on the vagaries of fossil discovery, so they provide the most immediately applicable tests. Tests based on gene evidence derive from the fundamental assertion of the Mostly Male theory regarding genes active in the flower: a higher proportion should be derived from genes formerly active in the male structure of the ancestral gymnosperm, than from genes active in the female structure. In contrast, genes specific to the ovule would not be expected to show this pattern.

The very long history of separately borne male and female structures in gymnosperms (Beck, 1988; Stewart and Rothwell, 1993; Taylor and Taylor, 1993), suggests that some genes likely became specialized to function only in the male or only in the female reproductive units *before* the divergences of extant seed plant lineages. Barring subsequent shifts of expression or function, such genes should show similar male versus female functional specializations in both the modern gymnosperms and in the gymnosperm ancestors of flowering plants, so predictions of the theory may be transferred to the genes of modern gymnosperms and modern flowering plants. Of the genes active in flowers, a higher proportion should be closely related to genes active in male gymnosperm structures than in female structures. If gene trees are sufficiently resolved to determine orthology, then more flowering plant genes should have their closest gymnosperm orthologs active in male reproductive units, not in female units.

Of course, this only applies in the absence of shifts in gene expression or function, which can rarely be judged *a priori* for any particular gene. Ergo, the most effective tests must be based on studies of many genes, so that confounding evidence from genes that changed expression or function can be recognized as anomalous.

6.5.2 Expressed sequence tag studies for testing the theory

The tools of modern genomics provide efficient and effective methods to clone large numbers of genes expressed in a particular organ. Messenger RNA obtained from the organ of interest is converted into cDNA and cloned to make a large library containing the expressed genes. Different expressed genes vary by several orders of magnitude in the relative abundance of their mRNAs, which is reflected in the relative abundances of their cDNA clones. Effective strategies exist to isolate both the abundant and very many of the low expressed genes from such libraries. These methods become practical due to recent great increases in the efficiency DNA sequencing. In the "cold plaque" method a moderate number (typically $c.1000$) of randomly selected cDNA clones is sequenced. The most highly expressed genes will be found during this first round of sequencing. These genes are pooled and used to generate a probe that can detect these same genes in other clones from the library. Further sequencing is done only on clones that do *not* bind probe from the first-round sequences, so one avoids re-sequencing the more abundant clones (Lanfranchi *et al.*, 1996; Nelson *et al.*, 1999). It is now practical to sequence as many as 10,000 cDNA clones that do not bind the abundant transcripts, which yields sequences from very many of the low-expressed genes. Typically, in such studies only one sequence is generated per clone, termed an "expressed sequence tag" (EST). The clones are archived, so cDNA clones of particular interest can be retrieved and fully sequenced.

To minimize effects of genes that have changed expression or function recently, one should study lineages of extant organisms that straddle the deepest nodes in the flowering plant tree and the deepest nodes in the gymnosperm tree. If lineages diverging from these deep nodes share the same expression patterns, then that pattern is more likely to have characterized plants at the basal nodes. Fortunately, a number of recent studies, especially those including multiple genes, converge on similar phylogenetic relationships near the base of the extant angiosperms, and may well represent the correct phylogeny. Relationships among the four extant gymnosperm groups are still in dispute, but Gnetales seem to be closer to conifers than to the other groups, and *Ginkgo* seems to be closer to cycads than to conifers or Gnetales.

The recently funded Floral Genome Project seeks to obtain the data that would most effectively test the Mostly Male theory, along with data pertinent to the subsequent evolution of angiosperm flowers (Soltis *et al.*, 2002). In particular, EST studies, as described above, will be performed (separately) on early developing male and female structures of both *Welwitschia mirabilis* in the Gnetales and on a species of *Zamia* in the Cycadales. ESTs will be obtained from early developing flowers of thirteen angiosperms, including members of the three apparently basal-most clades of angiosperms (*Amborella, Nuphar, Illicium*), as well as five representatives of other near-basal angiosperm clades (*Persea, Liriodendron, Acorus, Saruma, Papaver*). Expression studies of these ESTs will be carried out on a subset of these species, using microarray technology as well as other methods. Important genes that are not found among the ESTs, but that are expected to be present, will be sought using PCR and library probing. Several hundred cDNA clones per species will be completely sequenced. Gene phylogenies will be constructed for all the fully sequenced genes. Orthologous genes should show relationships that mirror the phylogenetic relationships of the organisms that contain them, allowing orthologs to be recognized, if the gene trees are sufficiently well resolved.

This ambitious study will provide exactly the data needed for a rigorous test of the Mostly Male theory. Further, the expression studies may allow reconstruction of some genetic control networks that specify organs and their cell types. It should then become possible to identify entire segments of the control network in flowers, which derive from systems active in male and/or female gymnosperm reproductive units, and to determine where the control of expression has shifted to meld the various elements together. This would allow a more profound understanding of the evolution of development, and could generate theories of flower evolutionary origins far more detailed and far more powerful than the Mostly Male theory.

ACKNOWLEDGMENTS

I thank Richard Bateman, Paul Kenrick and Christiane Anderson for comments on the manuscript, Y.-L. Qiu for calling to my attention the reports of ectopic *Ginkgo* ovules, Jer-Ming Hu for translating the Ma and Li (1991) paper, Florence S. Wagner for sharing specimens of *Ginkgo* leaves bearing ectopic ovules, Peter Engstrom for pointing out that *FBP11* appears sufficiently ancient to have possibly caused the ectopic ovule in the Mostly Male theory, Susan Reznicek for drawing Figure 6.5, and Paul Davies and David Lewis for hospitality at the NHM while studying the *Pteroma* fossils. This work is supported by National Science Foundation grant DEB-9974374.

REFERENCES

Akam, M. and Dawes, R. (1992) More than one way to slice an egg. *Current Biology*, 2, 395–398.

Axsmith, B. J., Taylor, E. L., Taylor, T. N. and Cuneo, N. R. (2000) New perspectives on the Mesozoic seed fern order Corystospermales based on attached organs from the Triassic of Antarctica. *American Journal of Botany*, 87, 757–768.

Bateman, R. M. and DiMichele, W. A. (2002) Generating and filtering major phenotypic novelties: neoGoldschmidtian saltation revisited, in *Developmental Genetics and Plant Evolution* (eds Q. C. B. Cronk, R. M. Bateman and J. A. Hawkins), Taylor & Francis, London, pp. 109–159.

Beck, C. B. (1988) *Origin and Evolution of Gymnosperms*. Columbia University Press, New York.

Blázquez, M., Soowal, L., Lee, I. and Weigel, D. (1997) *LEAFY* expression and flower initiation in *Arabidopsis*. *Development Supplement*, 124, 3835–3844.

Blázquez, M. A. and Weigel, D. (2000) Integration of floral inductive signals in *Arabidopsis*. *Nature*, 404, 889–892.

Bowe, L. M., Coat, G. and DePamphilis, C. W. (2000) Phylogeny of seed plants based on all three genomic compartments: extant gymnosperms are monophyletic and Gnetales' closest relatives are conifers. *Proceedings of the National Academy of Sciences USA*, 97, 4092–4097.

Busch, M. A., Bomblies, K. and Weigel, D. (1999) Activation of a floral homeotic gene in *Arabidopsis*. *Science*, 285, 585–587.

Chaw, S. M., Parkinson, C. L., Cheng, Y. C., Vincent, T. M. and Palmer, J. D. (2000) Seed plant phylogeny inferred from all three plant genomes: monophyly of extant gymnosperms and origin of Gnetales from conifers. *Proceedings of the National Academy of Sciences USA*, 97, 4086–4091.

Chen, X. M., Riechmann, J. L., Jia, D. X. and Meyerowitz, E. (2000) Minimal regions in the *Arabidopsis PISTILLATA* promoter responsive to the *APETALA3/PISTILLATA* feed back control do not contain a CArG box. *Sexual Plant Reproduction*, 13, 85–94.

Coen, E. S., Romero, J. M., Doyle, S., Elliott, R., Murphy, G. and Carpenter, R. (1990) Floricaula – a homeotic gene required for flower development in *Antirrhinum majus*. *Cell*, 63, 1311–1322.

Colombo, L., Franken, J., Vanwent, J., Angenent, H. J. M. and Vantunen, A. J. (1995) The *Petunia* MADS box gene *FBP11* determines ovule identity. *Plant Cell*, 7, 1859–1868.

Crane, P. R. (1985) Phylogenetic analysis of seed plants and the origin of angiosperms. *Annals of the Missouri Botanical Garden*, 72, 716–793.

Cronquist, A. (1981) *An Integrated System of Classification of Flowering Plants*. Columbia University Press, New York.

Darwin, F. and Seward, A. C. (eds) (1903) *More Letters of Charles Darwin: A Record of his Work in a Series of Hitherto Unpublished Letters*, vol. 2, p. 20. John Murray, London.

Donoghue, M. J. and Doyle, J. A. (2000) Seed plant phylogeny: demise of the anthophyte hypothesis? *Current Biology*, 10, R106–R109.

Doyle, J. A. (1978) Origin of angiosperms. *Annual Review of Ecology and Systematics*, 9, 365–392.

Doyle, J. A. (1994) Origin of the angiosperm flower: a phylogenetic perspective. *Plant Systematics and Evolution Supplement*, 8, 7–29.

Doyle, J. A. (1996) Seed plant phylogeny and the relationships of Gnetales. *International Journal of Plant Sciences*, 157, S3–S39.

Doyle, J. A. and Donoghue, M. J. (1986) Seed plant phylogeny and the origin of angiosperms: an experimental cladistic approach. *Botanical Review*, 52, 321–431.

Doyle, J. A. and Donoghue, M. J. (1987) The origin of angiosperms: a cladistic approach, in *The Origin of Angiosperms and their Biological Consequences* (eds E. M. Friis, W. G. Chaloner and P. R. Crane), Cambridge University Press, Cambridge, pp. 17–50.

Endress, P. K. (1996) Structure and function of female and bisexual organ complexes in Gnetales. *International Journal of Plant Science*, 157, S113–S125.

Fenster, C. B. and Galloway, L. F. (1997) Developmental homeostasis and floral form: evolutionary consequences and genetic basis. *International Journal of Plant Science*, 158, S121–S130.

Frohlich, M. W. (2001) A detailed scenario and possible tests of the Mostly Male theory of flower evolutionary origins, in *Beyond Heterochrony: the Evolution of Development* (ed. M. Zelditch), John Wiley, New York, pp. 59–104.

Frohlich, M. W. and Meyerowitz, E. M. (1997) The search for flower homeotic gene homologs in basal angiosperms and Gnetales: a potential new source of data on the evolutionary origin of flowers. *International Journal of Plant Sciences*, 158, S131–S142.

Frohlich, M. W. and Parker, D. S. (2000) The Mostly Male theory of flower evolutionary origins: from genes to fossils. *Systematic Botany*, 25, 155–170.

Fujii, K. (1896) On the different views hitherto proposed regarding the morphology of the flowers of *Ginkgo biloba* L. *Botanical Magazine (Tokyo)*, 10, 7–8, 13–15, 104–110.

Gifford, E. M. and Foster, A. S. (1989) *Morphology and Evolution of Vascular Plants* (3rd edn). W. H. Freeman, New York.

Gocal, G. F. W., King, R. W., Blundell, C. A., Schwartz, O. M., Andersen, C. H. and Weigel, D. (2001) Evolution of floral meristem identity genes. Analysis of *Lolium temulentum* genes related to *APETALA1* and *LEAFY* of *Arabidopsis*. *Plant Physiology*, 125, 1788–1801.

Harris, T. M. (1964) *The Yorkshire Jurassic Flora. II. Caytoniales, Cycadales and Pteridosperms*. British Museum (Natural History), London.

Hempel, F. D., Weigel, D., Mandel, M. A., Ditta, G., Zambryski, P. C., Feldman, L. J. and Yanofsky, M. (1997) Floral determination and expression of floral regulatory genes in *Arabidopsis*. *Development*, 124, 3845–3853.

Hofer, J., Turner, L., Hellens, R., Ambrose, M., Matthews, P., Michael, A. and Ellis, N. (1997) Unifoliata regulates leaf and flower morphogenesis in pea. *Current Biology*, 7, 581–587.

Hufford, L. (1996) The morphology and evolution of male reproductive structures of Gnetales. *International Journal of Plant Sciences*, 157, S113–S125.

Kyozuka, J., Konishi, S., Nemoto, K., Izawa, T. and Shimamoto, K. (1998) Down-regulation of RFL, the Flo/*LFY* homolog of rice, accompanied with panicle branch initiation. *Proceedings of the National Academy of Sciences USA*, 95, 1979–1982.

Lanfranchi, G., Muraro, T., Caldara, F., Pacchioni, B., Pallavicini, A., Pandolfo, D., Toppo, S., Trevisan, S., Scarso, S. and Valle, G. (1996) Identification of 4370 expressed sequence tags from a 3′-end-specific cDNA library of human skeletal muscle by DNA sequencing and filter hybridization. *Genome Research*, 6, 35–42.

Ma, F.-S. and Li, J.-X. (1991) *Ginkgo biloba*: the ovuliferous leaf and its phylogenetic implication. *Acta Phytotaxonomica Sinica*, 29, 187–189.

McLellan, T., Shephard, H. L. and Ainsworth, C. (2002) Identification of genes involved in evolutionary diversification of leaf morphology, in *Developmental Genetics and Plant Evolution* (eds Q. C. B. Cronk, R. M. Bateman and J. A. Hawkins), Taylor & Francis, London, pp. 315–329.

Mellerowicz, E. J., Horgan, K., Walden, A., Coker, A. and Walter, C. (1998) PRFLL – a *Pinus radiata* homologue of *FLORICAULA* and *LEAFY* is expressed in buds containing vegetative shoot and undifferentiated male cone primordia. *Planta*, 206, 619–629.

Moller, A. P. (1997) Developmental stability and fitness – a review. *American Naturalist*, 149, 916–932.

Mouradov, A., Glassick, T., Murphy, L., Fowler, B., Majla, S. and Teasdale, R. D. (1998) *NEEDLY*, a *Pinus radiata* ortholog of *FLORICAULA/LEAFY* genes, expressed in both reproductive and vegetative meristems. *Proceedings of the National Academy of Sciences USA*, 95, 6537–6542.

Nelson, P. S., Hawkins, V., Schummer, M., Bumgarner, R., Ng, W. L., Ideker, T., Ferguson, C. and Hood, L. (1999) Negative selection: a method for obtaining low-abundance cDNAs using high-density cDNA clone arrays. *Genetic Analysis–Biomolecular Engineering*, 15, 209–215.

Osborn, J. M. (1994) The morphology and ultrastructure of *Caytonanthus*. *Canadian Journal of Botany*, 72, 1519–1527.

Raff, R. (1996) *The Shape of Life: Genes, Development, and the Evolution of Animal Form*. University of Chicago Press, Chicago.

Retallack, G. J. and Dilcher, D. L. (1988) Reconstructions of selected seed ferns. *Annals of the Missouri Botanical Garden*, 75, 1010–1057.

Running, M. P. and Hake, S. (2001) The role of floral meristems in patterning. *Current Opinion in Plant Biology*, 4, 69–74.

Sakisaka, M. (1929) On the seed-bearing leaves of *Ginkgo*. *Japanese Journal of Botany*, 4, 219–236, plates XXIII–XXV.

Salazar-Ciudad, R. V., Solé, R. V. and Newman, S. A. (2001) Phenotypical and dynamical transitions in model genetic networks. II. Application to the evolution of segmentation mechanisms. *Evolution and Development*, 3, 95–103.

Samigullin, T. Kh., Martin, W. F., Troitsky, A. V. and Antonov, A. T. (1999) Molecular data from the chloroplast *rpoC1* gene suggest a deep and distinct dichotomy of contemporary spermatophytes into two monophyla: gymnosperms (including Gnetales) and angiosperms. *Journal of Molecular Evolution*, 49, 310–315.

Soltis, D. E., Soltis, P. S., Albert, V. A., Oppenheimer, D., dePamphilis, C. W., Ma, H., Frohlich, M. W. and Theißen, G. (2002) Missing links: the genetic architecture of the flower and floral diversification. *Trends in Plant Science*.

Stebbins, G. L. (1974) *Flowering Plants: Evolution Above the Species Level*. Harvard University Press, Cambridge, MA.

Stern, D. L. (1998) A role of Ultrabithorax in morphological differences between *Drosophila* species. *Nature*, 396, 463–466.

Stewart, W. N. and Rothwell, G. W. (1993) *Paleobotany and the Evolution of Plants* (2nd edn). Cambridge University Press, Cambridge.

Taylor, T. N. and Archangelsky, S. (1985) The Cretaceous pteridosperms *Ruflorinia* and *Ktalenia* and implications on cupule and carpel evolution. *American Journal of Botany*, 72, 1842–1853.

Taylor, T. N. and Taylor, E. L. 1993. *The Biology and Evolution of Fossil Plants*. Prentice Hall, Englewood Cliffs, NJ.

Theißen, G., Becker, A., Di Rosa, A., Kanno, A., Kim, J. T., Munster, T., Winter, K. U. and Saedler, H. (2000) A short history of MADS-box genes in plants. *Plant Molecular Biology*, 42, 115–149.

Theißen, G., Becker, A., Winter, K., Münster, T., Kirchner, C. and Saedler, H. (2002) How the land plants learned their floral ABCs: the role of MADS-box genes in the evolutionary origin of flowers, in *Developmental Genetics and Plant Evolution* (eds Q. C. B. Cronk, R. M. Bateman and J. A. Hawkins), Taylor & Francis, London, pp. 173–205.

Tomlinson, P. B., Braggins, J. E. and Rattenbury, J. A. (1997) Contrasted pollen capture mechanisms in Phyllocladaceae and certain Podocarpaceae (Coniferales). *American Journal of Botany*, 84, 214–223.

Wagner, D., Sablowski, R. W. M. and Meyerowitz, E. M. (1999) Transcriptional activation of *APETALA1* by *LEAFY*. *Science*, 285, 582–584.

Weigel, D., Alvarez, J., Smyth, D. R., Yanofsky, M. F. and Meyerowitz, E. M. (1992) *LEAFY* controls floral meristem identity in *Arabidopsis*. *Cell*, 69, 843–859.

Yang, Z. H. and Bielawski, J. P. (2000) Statistical methods for detecting molecular adaptation. *Trends in Ecology and Evolution*, 15, 496–503.

Yao, X., Taylor, T. N. and Taylor, E. L. (1995) The corystosperm pollen organ *Pteruchus* from the Triassic of Antarctica. *American Journal of Botany*, 82, 535–546.

Chapter 7

Generating and filtering major phenotypic novelties: neoGoldschmidtian saltation revisited

Richard M. Bateman and William A. DiMichele

ABSTRACT

Further developing a controversial neoGoldschmidtian paradigm that we first published in 1994, we here narrowly define saltational evolution as a genetic modification that is expressed as a profound phenotypic change across a single generation and results in a potentially independent evolutionary lineage termed a prospecies (the 'hopeful monster' of Richard Goldschmidt). Of several saltational and parasaltational mechanisms previously discussed by us, the most directly relevant to evolutionary–developmental genetics is dichotomous saltation, which is driven by mutation within a single ancestral lineage. It can result not only in instantaneous speciation but also in the simultaneous origin of a profound phenotypic novelty more likely to be treated as a new supraspecific taxon. Saltational events form unusually long branches on morphological phylogenies, which follow a punctuated equilibrium pattern, but at the time of origin they typically form zero length branches on the contrastingly gradualistic molecular phylogenies. Our chosen case-studies of heterotopy (including homeosis) and heterochrony in fossil seed-ferns and extant orchids indicate that vast numbers of prospecies are continuously generated by mutation of key developmental genes that control morphogenesis. First principles suggest that, although higher plant mutants are more likely to become established than higher animals, the fitness of even plant prospecies is in at least most cases too low to survive competition-mediated selection. The establishment of prospecies is most likely under temporary release from selection in environments of low biotic competition for resources, followed by honing to competitive fitness by gradual reintroduction to neoDarwinian selection. Unlike neoDarwinism alone, this two-phase evolutionary paradigm is consistent with the recent results of (a) whole-genome sequencing, which have revealed a surprisingly small total number of genes per model species, and (b) of Quantitative Trait Locus analyses, which indicate that major phenotypic features are determined by one or two homeotic genes of major phenotypic effect and only a handful of genes of lesser effect. Evolutionary–developmental genetics has already proved beyond reasonable doubt the credibility of the initial 'generative phase' of the neoGoldschmidtian hypothesis (though further investigation is needed of the effects of canalisation and epigenesis following both gain and loss of function). Unfortunately, far fewer data are currently available to test the subsequent 'establishment phase'; this deficiency places a premium on monitoring the ecological progress of many prospecies, in both artificial and natural

In *Developmental Genetics and Plant Evolution* (2002) (eds Q. C. B. Cronk, R. M. Bateman and J. A. Hawkins), Taylor & Francis, London, pp. 109–159.

habitats. Saltation is superior to classical neoDarwinian selection in removing the highly improbable requirement to drive alleles to fixation in large populations, explaining sympatric speciation, allowing lineages to cross otherwise lethal fitness valleys, giving key evolutionary roles to pre-adaptation and exaptation rather than adaptation, and providing a direct causal explanation of the qualitatively different levels of morphological divergence that underpin the Linnaean hierarchy.

7.1 Introduction

> On the theory of natural selection we can clearly understand the full meaning of that old canon in natural history, 'Natura non facit saltum.' This canon, if we look only to the present inhabitants of the world, is not strictly correct, but if we include all those of past times, it must by my theory be strictly true.
>
> (Jones, 1999: 189)

In the summer of 1993, we attempted to synthesise a new 'supraDarwinian' paradigm in the evolutionary biology of vascular plants for an edited volume entitled *Shape and Form in Plants and Fungi* (Bateman and DiMichele, 1994a). Our wide-ranging discussion primarily concerned the credibility of, and the evidence for, non-gradual evolution of plant form that did not rely on classic neoDarwinian mechanisms – in other words, on directional or disruptive selection acting via adaptively-driven fixation of mutant alleles and traced via Hardy–Weinberg equilibria within large panmictic populations (for clear expositions of these principles see Ridley, 1996; Patterson, 1999). We ultimately reached the following conclusions (p. 91):

> It has long been accepted that the fundamental unit of evolutionary change is the gene but that such changes are mediated via the phenotype of the host organism (the replicators and interactors respectively of Dawkins, 1982, 1986, 1989). Recent studies of [homeotic] gene expression in plants (e.g. Coen, 1991; Coen and Carpenter, 1992; ...) provide a vital causal link between genotype and phenotype – replicator and interactor – that allows reciprocal illumination between these two contrasting manifestations of the evolutionary process. [Attributes of homeotic] genes can be coded cladistically in order to make the crucial distinction between primitive and derived states and, as we have shown, the resulting cladograms can be used to test competing hypotheses of underlying evolutionary mechanisms. Despite recent advances (Chasan, 1993), biologists have been surprisingly slow to combine relevant concepts of gene expression, developmental control, phylogeny reconstruction, ecological filtering of phenotypes, and evolutionary theory into a truly integrated evolutionary synthesis. This problem has been exacerbated by over-enthusiastic generalisation from parochial studies of a few 'flagship' species to all-embracing evolutionary theories. Nonetheless, we are confident that future syntheses will confirm our opinion that plants have their own distinct approach to the evolution of shape and form.

These deliberately provocative statements, and our subsequent reformulations of various aspects of the paradigm (Bateman, 1994, 1996, 1999a, b; Bateman and

DiMichele, 1994b; DiMichele and Bateman, 1996; Bateman *et al.*, 1998) attracted disappointingly few commentaries (e.g. Rutishauser, 1995; Erwin, 2000; Tucker, 2000, 2001) prior to this volume (Baum and Donoghue, 2002; Cronk, 2002; Theissen *et al.*, 2002). Our 1994 synthesis was very much a product (or perhaps a victim?) of its time. Despite a few eloquent dissidents, often from the palaeobiological community (most notably Gould and Lewontin, 1979; cf. Morris, 2001), evolutionary theory was (and is) dominated by strict neoDarwinians whose perspectives differed only in fine detail: all relied on the mathematically modelled gradual spread through populations of alleles that engendered subtle modifications of phenotype, increasing in frequency through directional or disruptive selection mediated primarily by competition (for recent examples, the latter extensively quoted in this chapter, see Dawkins, 1986, 1989; Maynard Smith, 1989; Maynard Smith and Szathmáry, 1995; Ridley, 1996; Jones, 1999). Adaptation was used, often unthinkingly, as a null hypothesis for most specific evolutionary transitions (Bateman and DiMichele, 1994a) and for evolutionary radiations (Bateman, 1999a), however limited and ambiguous the evidence for levels of selection capable of inducing genuine speciation. And botanists had surprisingly little to say on such lofty matters, preferring to co-opt, with at best minor modifications, theories that were essentially zoocentric.

However, in the field of developmental genetics, genuine breakthroughs in understanding the control of metazoan body plans through the homeotic Hox genes (e.g. Slack *et al.*, 1993; Valentine *et al.*, 1999; Mindell and Meyer, 2001) had been accompanied by the elucidation of the classic ABC model of angiosperm flower development through studies of *Arabidopsis* and *Antirrhinum* (e.g. Coen and Meyerowitz, 1991; Theissen *et al.*, 1996, 2002; Cubas, 2002). Also, the increasingly perceived possibility of harnessing these emerging insights in developmental genetics for evolutionary studies prompted the benchmark plant evo-devo conference at Taos, New Mexico in the summer of 1993 (summarised by Chasan, 1993, as cited by us in the above quote).

Eight years on, it seems appropriate to assess how much (or indeed how little) progress has been made toward the crucial goal of formulating a holistic evolutionary paradigm that encompasses the origin, phylogenetic context and ecologically-mediated fates of mutations in key morphogenetic genes.

7.2 SupraDarwinian evolutionary mechanisms

7.2.1 NeoGoldschmidtian saltation

> Geneticists were once so impressed with mutation as to suggest that new forms of life arise not through the accumulation of small changes but in great leaps. Evolution was due to the instability of genes and genetics had, perhaps, destroyed Darwin's idea. It had not: mutation is the fuel rather than the engine of biological advance. The process involves mechanisms undreamed of in the science's first days.
>
> (Jones, 1999: 147)

In his notorious evolutionary tome *The Material Basis of Evolution*, the Berkeley-based developmental zoo-geneticist Richard Goldschmidt (1940) argued that 'systemic mutations' (large-scale chromosomal rearrangements) altered early developmental trajectories to generate, across a single generation, 'hopeful monsters' – teratological lineages of phenotypes radically different from their parents. By chance, some hopeful monsters possessed high levels of fitness that enabled their persistence as new lineages of great evolutionary significance. It was not difficult for aggressive advocates of the neoDarwinian *New Synthesis* to fatally undermine Goldschmidt's thesis by discrediting his concept of macromutations and mathematically 'proving' the improbability of hopeful monsters instantaneously acquiring competitively high fitness. Consequently, occasional attempts to resurrect various aspects of Goldschmidt's saltational paradigm (e.g. Waddington, 1957; Croizat, 1962; van Steenis, 1976; Gould, 1982; Arthur, 1984; Orr, 1991; Bateman and DiMichele, 1994a), or to re-assess the generally overlooked contributions of Goldschmidt's intellectual predecessors (e.g. see the discussion of the views of influential German plant morphologist W. Troll as portrayed by Kaplan, 2001b), attracted relatively little attention.

As few evolutionary biologists provided explicit definitions of saltational (or 'saltatory') evolution, the term has been used to encompass a wide range of often conflicting concepts. In our previous attempt to define (and thus recognise) saltational evolutionary events (Bateman and DiMichele, 1994a), we immediately excluded from consideration all non-genetic ecophenotypic and ontogenetic variation within species. We also emphasised the related difference between the concept of a *teratos* – an individual possessing a radically different morphology from its immediate ancestor(s) irrespective of the underlying causal mechanism – and Goldschmidt's concept of a '*hopeful monster*', where a genetic cause of the morphological discontinuity is assumed and non-heritable causes are specifically excluded (Table 7.1).

Saltation requires a substantial change in phenotype between ancestor and descendant. As developmental genetic studies have conclusively demonstrated that

Table 7.1 Formal definitions of key terms relating to exceptionally rapid speciation (modified after Bateman and DiMichele, 1994a: 66–67)

Saltation: a genetic modification that is expressed as a profound phenotypic change across a single generation and results in a potentially independent evolutionary lineage.

Parasaltation: a genetic modification that is expressed as a profound phenotypic change across two to several generations and results in a potentially independent evolutionary lineage.

Dichotomous saltation: saltation driven by mutation of at least one gene within a single ancestral lineage.

Reticulate saltation: saltation driven by allopolyploidy and thus blending the entire genomes of two ancestral lineages.

Teratos: an individual showing a profound phenotypic change from its parent(s) irrespective of whether the underlying cause is genetic or ecophenotypic (plural, *terata*).

Hopeful monster: an individual showing a profound phenotypic change from its parent(s) that demonstrably reflects a genetic modification.

Taxonomic species: one or more (typically many more) populations separated from all other comparable populations by phenotypic, and putative genotypic, discontinuities that are believed to reflect one or more isolating mechanisms operating over a considerable period of time.

Prospecies: a putatively recently evolved lineage possessing the essential intrinsic properties of a taxonomic species but yet to achieve levels of abundance and especially of longevity acceptable to practising taxonomists.

magnitude of phenotypic change engendered by a modification of a specific gene is decoupled from the magnitude of the change in the genome itself, quantification of the degree of phenotypic change is clearly the most appropriate criterion for saltation. Increased overall complexity is definitely not a requirement; indeed, current evidence suggests that saltational events which suppress developmental genes and consequently reduce morphological complexity are far more common than saltational events which increase overall complexity (cf. Kellogg, 2002).

In 1994 we abrogated our responsibility to define the profoundness of morphological transition necessary to pass the threshold of saltation, and we continue to do so here. The primary expectation is of a high degree of phenotypic divergence in a well-sampled phylogenetic (and perhaps phenetic: DiMichele *et al.*, 2001) study that allows reasonably accurate comparison of degrees of morphological divergence, which should be considerable, and of sequence divergence, which should be minimal in the immediate aftermath of the saltation event but, unlike morphology, changes progressively and gradually thereafter (Bateman, 1999a).

Rate of evolutionary change is a second key criterion for saltation. Some authors (e.g. Ayala and Valentine, 1979) chose to define saltation as a period when the temporal rate of evolution (change/time) is substantially greater than the long-term average within the lineage. However, this definition is more appropriately discussed in the context of evolutionary radiations (Bateman *et al.*, 1998; Bateman, 1999a), as it encompasses not only our saltation *s.s.* but also several contrasting mechanisms of rapid evolutionary change that we prefer to collectively term 'parasaltational' (see Section 7.2.2). We believe that saltation is better defined by generation time than absolute time, and hence argue that a saltational change must occur across a single generation. Although hopeful monsters will have a much greater likelihood of retaining their novel phenotype if they become reproductively isolated from the parental population, particularly if the isolation reflects intrinsic properties of the monsters rather than mere *ad hoc* spatial separation (i.e. sympatry rather than allopatry), new lineages can in theory become established even in the absence of reproductive isolation (Arthur, 1984). This criterion is therefore ancillary to, rather than inherent in, *saltation*, which can now be defined more precisely as a genetic modification that is expressed as a profound phenotypic change across a single generation and results in a potentially independent evolutionary lineage (Table 7.1).

A single hopeful monster has by definition a minimal geographical and ecological distribution; if also reproductively isolated, in principle it fulfils the criteria required for a biological species *sensu* Mayr (1963). However, in practice the hopeful monster is unlikely to be awarded specific rank by a taxonomist, who requires a *taxonomic species* to demonstrate a degree of historical tenacity by establishing a sizeable population that persists through many generations. This is a reasonable *modus operandi*, as most lineages resulting from saltation undoubtedly fail to survive beyond a single generation, and very few exceed ten or a hundred generations; these typically ephemeral entities are better described as *prospecies* (Table 7.1). Note that there is no intrinsic biological distinction between prospecies and taxonomic species; taxonomic species can only be distinguished retrospectively, on the basis of their far greater temporal continuity and spatial extent. Also, an ageist bias is evident in most taxonomic treatments. A species inevitably has an origin, an acme and ultimately an extinction that are defined both temporally and spatially (Levin, 2000, 2001).

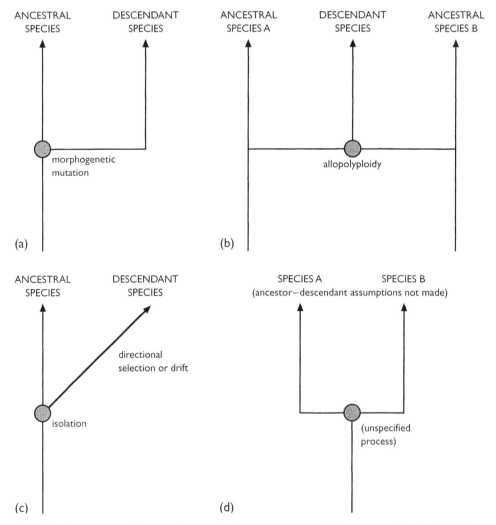

Figure 7.1 Comparison of contrasting evolutionary patterns. Dichotomous saltation (a) shows instantaneous divergence of descendant from ancestor via a mutation in a key gene controlling morphogenesis. Reticulate saltation (b) also occurs instantaneously via allopolyploidy but blends two parental genomes. Directional selection or drift (c) cause gradual divergence of the descendant following its isolation. Cladistic representations of speciation (d) assume no ancestor–descendant relationships and specify no underlying process.

However, in practice, taxonomists are more reluctant to recognise as a full species a rare but putatively recent, expanding population than a rare but putatively long-lived, senescent population that represents far greater historical continuity.

Thus far, we have considered only mutationally-driven saltation. Other modes of genotypic change rely on mixing pre-existing genes from individuals of two species (hybridisation) or on duplicating a complete set of pre-existing genes in a single individual (autopolyploidy), but neither phenomenon generates a new reproductively

Table 7.2 Examples of parasaltational evolutionary mechanisms, with relevant literary sources

(a) Reticulate processes that combine formerly disparate lineages

1. Hybridisation*	Stace, 1989, 1993; Goodnight, 1995, 1999; Rieseberg et al., 1995; Wendel et al., 1995; Rieseberg, 1997; Vogel et al., 1999; Rieseberg and Burke, 2001
2. Endosymbiosis	Margulis, 1993; Martin and Schnarrenburger, 1997

(b) Processes intrinsic to the behaviour of the genome

3. Homeosis*	Raff and Kauffman, 1983; Arthur, 1984, 2000; Kauffman, 1993; Wray, 1995; Albert et al., 1998
4. Neutral theory	Kimura, 1983, 1991
5. Nearly neutral theory	Ohta, 1992, 1995
6. Adaptive mutation	Foster and Cairns, 1992; Shapiro, 1997, 2002
7. Molecular drive	Dover, 1982, 2000
8. Meiotic drive	Pomiankowski and Hurst, 1993; McVean and Hurst, 1997
9. Multiple codes	Trifonov, 1997
10. Chromosomal rearrangements	Prescott and DuBois, 1996; Hoffman and Prescott, 1997
11. Transposon-induced deactivation	Wessler et al., 1995; Federoff, 2000; Walbot, 2002

(c) Processes emergent from genetic control of development

12. Epigenesis	Goodwin and Saunders, 1992; Jablonka et al., 1992; Goodwin, 1994; Jablonka, 1994; Jablonka and Lamb, 1995
13. Epistasis	Papers in Wolf et al., 2000

(d) Processes directly linked to the ecological spread of genetic novelties

14. Drift	Templeton, 1989; Gillespie, 1991; Barrett and Pannell, 1999; Patterson, 1999
15. Shifting balance	Wade, 1992; Goodnight, 1995; Whitlock, 1997; Wade and Goodnight, 1998; Mallet and Joron, 1999
16. Correlated selection	Price et al., 1993
17. Species selection	Gould, 1986; Eldredge, 1989
18. Clade selection	Williams, 1992

Note: * = events that are also potentially saltational.

isolated evolutionary lineage across a single generation. However, by first mixing genes of two lineages and then duplicating the entire heterogeneous genome, thus restoring fertility, allopolyploidy does immediately generate a novel lineage that is often also reproductively isolated (Stebbins, 1971; Stace, 1989, 1993; Thompson and Lumaret, 1992; Wendel *et al.*, 1995; Rieseberg, 1997; Vogel *et al.*, 1999; Rieseberg and Burke, 2001; Wolfe, 2001). Similarly, the endosymbiotic origins of mitochondria and plastids (Margulis, 1993; Martin and Schnarrenberg, 1997) can only realistically be viewed as unusually profound reticulate saltation events.

Thus, Bateman and DiMichele (1994a) concluded that two distinct modes of saltation exist (Table 7.1, Figure 7.1). In *dichotomous saltation* (Figure 7.1a), hopeful monsters originate by mutation; one new daughter lineage diverges instantaneously from the ancestral lineage, thereby forming a dichotomous pattern that can in theory be resolved cladistically (as can the gradual divergences implicit in directional selection or drift: Figure 7.1c). However, in *reticulate saltation* (Figure 7.1b), allopolyploidy combines elements from two ancestral lineages; the resulting reticulate pattern cannot be adequately accommodated in a dichotomous cladogram (note that the cladistic method is neutral regarding both ancestor–descendant and process-based interpretations: Figure 7.1d). The general absence of mutation in reticulate saltation restricts the potential range of immediate phenotypic innovation, so that speciation events are less likely to coincide with the origins of supraspecific taxa than is the case in dichotomous saltation (cf. van Steenis, 1976; Arthur, 1984; Stace, 1993; Wolfe, 2001). Thus, dichotomous saltation is of greater relevance to evolutionary–developmental genetics, though reticulate saltation also remains an important evolutionary process as each such event is on average more likely to generate viable phenotypes capable of establishing long-lived lineages.

7.2.2 The diversity of parasaltational mechanisms

> Quite how [mimetic] insects traversed the valley of death – in a sudden leap, with a single gene pushing them most of the way, or by small changes getting together by accident – is not clear.
>
> (Jones, 1999: 160)

The narrowness of our definition of saltation excludes several evolutionary mechanisms, most under-explored, which are capable of causing speciation events that are exceptionally rapid but not instantaneous. These are more appropriately described as *parasaltational* (Table 7.1).

First, the stringent requirement for both genotypic and phenotypic change across a single generation excludes from strict saltation most mutations of recessive alleles; here, the genotypic change can only be expressed in the F1 generation in rare cases where a recessive mutation in a germ cell precursor is followed by self-fertilisation involving two gametes, each of which carries the mutation (Arthur, 1984). Hybridisation *per se* is similarly excluded (cf. Abbott, 1992; Rieseberg *et al.*, 1995).

Specifying instantaneous speciation also rules out evolutionary scenarios that focus on populations of small effective sizes, typically due to reduction induced by various forms of environmental stress, by a marginalisation event leading to parapatry, or by a vicariance event leading to allopatry (Levinton, 1988). The neutral theory (Kimura, 1983, 1991) and subsequent nearly neutral theory (Ohta, 1992, 1995) predict that random sampling effects alone can lead to allele fixation or extinction in small populations, largely independent of selective advantage. Various reformulations of Wright's (1932, 1968) shifting balance theory (Lewis, 1962, 1966, 1969; Levin, 1970, 1993, 2000; Templeton, 1982, 1989; Carson, 1985; Lande, 1986; Wade, 1992; Whitlock *et al.*, 1995; Whitlock, 1997; Wade and Goodnight, 1998; Mallet and Joron, 1999; Wolf *et al.*, 2000; cf. Coyne *et al.*, 1997) predict that random genetic drift in small populations can temporarily override selective pressures on alleles, thereby allowing populations to cross non-lethal valleys on the adaptive landscape to the slopes of another peak, which is then climbed by classic neoDarwinian selection. Drift is in theory expressed most profoundly when it disrupts and destabilises developmental homeostasis (Levin, 1970; see also Patterson, 1999). Although most such populations fail, this process provides an occasional opportunity to substantially re-organise the developmental programming under conditions of low infraspecific competition and high physical stress ('catastrophic selection' *sensu* Lewis, 1962, 1966, 1969; Carson, 1985).

Shifting balance scenarios are consistent with evolutionary patterns that were termed 'punctuated equilibria' by Eldredge and Gould (1972) – long periods of stasis followed by brief periods of rapid phenotypic change. Vermeij (1987) extended the ecological component of these scenarios, arguing that periods of stasis reflect neoDarwinian processes and are punctuated by ecosystem disruptions that locally reduce selection pressure, species diversity and population sizes. Each such disruption allows escalation – a brief interval of intense competition to fill the vacated niches that increases the fitness of the competitors. Many of these observations apply equally well to populations that are very small, not because they have recently declined into an apocryphal 'bottleneck' but because they have just evolved by saltation – we will return to them later. Other explanations of punctuational patterns require differential survival of lineages, focusing on selection among species (Gould, 1986; Levinton, 1988; Gould and Eldredge, 1993) or even among clades (Williams, 1992).

7.3 Cladistic tests of evolutionary hypotheses

> Cladistics, a German invention, has strict rules and a complex vocabulary. It can, if not carefully used, give erratic results and is still filled with argument about just what should be plugged into its analyses. It has, nevertheless, transformed our view of the world.
>
> (Jones, 1999: 371)

In 1994, we felt obliged to outline the basic principles and methodology of cladistics before discussing its relevance to saltation theory. Presenting this background is no longer necessary, given the pre-eminence since achieved by cladistic techniques for reconstructing phylogenies. However, it is worthwhile restating our suggested use of cladograms for falsifying saltational hypotheses, and reviewing cladistic falsification of adaptation and exaptation in the light of increasing use of mapped quantitative variables for ecological interpretations (see review volumes by Harvey and Pagel, 1991; Harvey *et al.*, 1996; Silvertown *et al.*, 1997). Also, the rapid and profound switch during the last decade from cladistic analyses based on morphology to those based on increasingly profligate DNA sequence data requires further consideration (cf. Bateman, 1999a; Chase *et al.*, 2000). Phylogenies can in practice be reconstructed using sequence data alone, but evolution can be understood only by relating genotype explicitly to phenotype.

7.3.1 Falsification of adaptation and exaptation

The number of morphological cladistic analyses of plants leading to interpretations of underlying causal mechanisms remains surprisingly small; numbers began to rise during the 1990s, but then plateauxed as sequence matrices replaced morphology. Most such examples focused on adaptation (for early examples see Coddington, 1988; Donoghue, 1989; Maddison, 1990; for reviews see Harvey and Pagel, 1991; Harvey *et al.*, 1996; Silvertown *et al.*, 1997), and require three important codicils:

1   Many traits are likely to be *adaptive* (increase the perceived overall fitness of the organism) but far fewer are identifiable as *adaptations* that evolved via natural selection to fulfil a specific function. Despite many published criticisms, this key distinction is still often overlooked.
2   Morphological cladistic analyses by definition employ 'form' as characters, but rarely include explicit functions (cf. Lauder, 1990). Fortunately, this is not a serious handicap to interpretation, as particular functions can be plotted on a cladogram *a posteriori* (we discuss this procedure, strictly termed 'mapping', in more detail below).
3   When attempting to infer evolutionary process from cladistic pattern, it is only possible to state that a particular evolutionary process is *consistent with* a particular phylogenetic pattern. *Demonstrating* such a correlation requires additional biological data that are not appropriate for coding into the original cladistic matrix.

To be consistent with a hypothesis of adaptation, a particular form (represented as one or more character-state transitions on the morphological cladogram) and a particular function (mapped *a posteriori*) must evolve on the same branch of the cladogram. A form appearing below the postulated function on the cladogram is consistent with a hypothesis of exaptation; the form either evolved non-adaptively, or evolved adaptively but for a different function, only later acquiring its present function. A form appearing above the putative function on the cladogram refutes the hypothesis of positive correlation and thus of any causal relationship. Arrangements of form and function consistent with adaptation or exaptation are not positive proof of such hypotheses; rather, the value of the cladograms is negative, allowing falsification of postulated correlations.

7.3.2 Falsification of transference of function

This logic of transitional correlation also underpins the test of transference of function proposed by Baum and Donoghue (2002; also D. Baum, pers. comm. 2001), where the phylogenetic distribution of a particular function is mapped relative to two or more non-homologous structures known to fulfil the function across the clade under scrutiny. Polymorphism of function in the hypothetical ancestor, or the presence of species where neither structure fulfils the function in question, both refute the hypothesis of transfer of that function from one structure to the other during the evolution of the clade. As with tests of adaptation and exaptation, the *a priori* hypothesis is conclusively refuted by a negative correlation, but is not proven by a positive correlation.

7.3.3 Falsification of saltation

Bateman and DiMichele (1994a) adopted and amended the logic of phylogenetic falsification to develop a cladistic test of non-adaptive, saltational hypotheses. The emphasis switches from demonstrating the simultaneous origin of a character state and its presumed adaptive function to demonstrating the simultaneous origin of several developmentally correlated character states. This, in turn, focuses attention on long branches – those supported by several morphological character-state transitions – and requires a literal (and thus somewhat philosophically controversial) interpretation of the cladogram as an evolutionary history. In this scenario, potentially developmentally correlated characters changing simultaneously on the cladogram are assumed to have changed simultaneously during evolution, most probably as the direct or indirect consequence of a single mutation event. In other words, saltation is regarded as the null hypothesis to explain particular long branches in morphological cladograms.

The credibility of this test is heavily dependent on the density of species sampling. Ideally, all known species in the chosen clade, both extant and extinct, should be sampled and coded in order to maximise the probability of dissociating multiple character-state transitions on a single branch. A more pragmatic approach would be to analyse the readily obtained species initially and then add other relevant species as they become available, as a secondary test of the initial saltational hypothesis (Bateman and DiMichele, 1994a, Figure 3).

There are several potential difficulties with using long branches in morphological trees as circumstantial evidence of saltation events:

1   Some manuals of morphological cladistic analysis state that morphological characters which are potentially developmentally correlated fail the cladistic requirement for independence and hence should be reduced to a single character prior to tree-building. If effectively implemented, this procedure would eliminate any potential insights into developmental correlation. However, the complexities of developmental genetics mean that this is not a practical recommendation; many phenotypic characters are demonstrably under highly polygenic control, while single key genes frequently regulate many features of an organism (pleiotropy). The best approach to coding a species for morphological cladistic analysis is to describe as many features as possible that can be delimited and can reasonably be assumed to be under genetic control, leaving developmental correlation to be assessed *a posteriori*.

2   Long branches tend to be especially mutually attractive during tree-building, often yielding incorrect topologies (long-branch effects: e.g. Hendy and Penny,

1989); in such cases the long branch will no longer be present to be examined for evidence of saltation, having been substantially shortened in order to insert the long-branch taxon into an incorrect position on the cladogram.

3   Even if sampling of known species is maximised, long morphological branches could still reflect at least two non-saltational evolutionary scenarios. First, they could be due to the absence (most commonly by extinction) of several phylogenetically intermediate species. Second, it could conceivably reflect the acquisition of several phenotypic characters by a lineage in the absence of any divergent speciation (i.e. by anagenesis). Although there is limited direct evidence that anagenetic evolution takes place (but see Section 7.7.2), equally it is difficult to demonstrate that the observed character-state transitions accompanied a single speciation event.

7.4 Punctuated equilibria in morphology, plus phyletic gradualism in DNA sequences, equals evolution

That morphological and genetic changes are not always parallel seems entirely reasonable at first, but this statement must be tempered by the fact that establishing the extent of 'great morphological change' is highly subjective. Like 'key' characters, it is likely to be influenced by a single novelty that we view as significant because it epitomises a highly successful group of plants. The importance of such single novelties must be viewed against the backdrop of the majority of traits still shared by all taxa due to overall genetic similarity ... Statements like 'rapid radiations are from first principles better tracked by morphology than molecules' (Bateman, 1999: 432) are based on the assumption that specific novelties are adaptive and successful from the onset. Individual novelties only appear important because we know that particular groups are successful today. At the time when they first evolved, such novelties would not be highly significant because the overall genetic environment would still have been similar to that of their close relatives.

(Chase *et al.*, 2000: 166–167)

The 1990s were also characterised by ongoing debates regarding whether the punctuational pattern of evolution perceived and widely popularised by S. J. Gould, N. Eldredge and others (Eldredge and Gould, 1972; Gould and Lewontin, 1979; Eldredge, 1992; Gould and Eldredge, 1993) was real, rather than an artefact of the absence from the fossil and living records of a myriad of phylogenetically intermediate species ('ghost' species that would reflect non-preservation due to rapid extinction, persistent rarity and/or relatively poor preservation potential). Recent reviews of the completeness of the fossil record strongly suggest that the proportion of animal species that ever existed that are preserved in the fossil record is greater than all but the most optimistic palaeontologists had predicted (Donovan and Paul, 1998); it is less clear whether this statement applies equally well to plants.

The best-documented case studies of morphological evolution in the animal fossil record (e.g. Wray, 1995; Jackson and Cheetham, 1999) give great credibility to the reality of the rectangular, phenogram-like pattern inherent in punctuated equilibrium (Figure 7.2a). Furthermore, review of the recent literature suggests that the

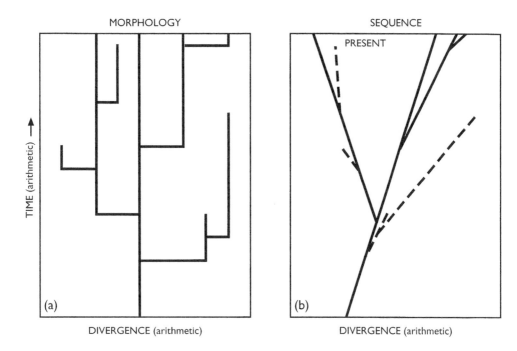

Figure 7.2 A simple hypothetical phylogeny of eight species, four extant and four extinct, all ultimately derived from a single ancestor. Evolutionary patterns are contrasted for morphological data (a), showing geologically instantaneous (punctuational) morphological
divergence, and sequences from non-morphogenetic regions of the genome (b), showing
constant, clock-like sequence divergence. In this example both speciation events and
extinction events are roughly evenly spaced and the magnitude of morphological divergence is random through time; features that maximise the likelihood of correctly reconstructing the phylogeny of the clade (Bateman, 1999a). Note that molecular data cannot
determine the relationships of the four extinct species, nor can they separate the most
recently divergent of the four extant species from its sister-species (in this example,
there has been insufficient time for the novel species to acquire molecular autapomorphies). Also, had the earlier divergences of the three molecularly distinguishable extant
lineages been more closely spaced in time (i.e. had they constituted a *bona fide* radiation),
we would have been far more likely to satisfactorily resolve their relationships through
morphological than through sequence data (modified after Bateman, 1999a, Figure 19.2a,
b; see also Bateman, 1996, 1999b).

reality of long periods of evolutionary stasis in specific lineages (i.e. 'equilibrium') is
increasingly accepted by both neoDarwinian and supraDarwinian researchers.
Bateman (1999a) argued that the most obvious explanation for stasis is that, in most
ecological circumstances, neoDarwinian natural selection enhances phenotypic
stability. In other words, the background mode of natural selection is stabilising
selection, which inhibits evolutionarily meaningful morphological change. The relatively rapid intervening periods of evolution and speciation (i.e. 'punctuation') may
occur under the relatively high directional or disruptive selection pressures that
underpin neoDarwinian microevolution, or alternatively they may occur even more
rapidly under the relatively low selection pressures that are more characteristic of

the various supraDarwinian macroevolutionary scenarios outlined above. In either case, the morphological divergence appears geologically instantaneous.

However, non-fossil phylogenetics was increasingly dominated through the 1990s by DNA sequence phylogenies for regions of the genome that are either non-coding or code for various biochemical pathways that are not morphologically expressed. There are reasons to assume that such regions of the genome have broadly clock-like properties, even if the clock is somewhat unreliable (e.g. Avise, 1994; Sanderson, 1997, 1998). It has become conventional wisdom that, at least in most circumstances, phylogenetically favoured regions of the genome such as plastid genes (Chase and Albert, 1998) and nuclear ribosomal genes (Hershkovitz *et al.*, 1999) accumulate non-lethal mutations with the semi-regularity of a Geiger counter (admittedly, evidence continually accrues of clear contraventions of this 'steady state'). Thus, first principles suggest that these sequence data change by something approaching phyletic gradualism (Figure 7.2b).

Hence, rather than being alternative patterns of evolutionary change that justifiably engender keenly-fought arguments among evolutionary theorists, evolution typically occurs via both punctuated equilibria (morphology) *and* phyletic gradualism (DNA sequences: Figure 7.2). As concluded by Bateman (1999a: 446),

> if morphological evolution follows a punctuational pattern (dictated by long periods of stabilising selection that are only occasionally broken by temporary release from selection and consequent speciation) and thus there is no morphological clock, but if in contrast genomic mutation is broadly clock-like, then in phylogenetic terms the vast majority of morphological character-state transitions occur *during* speciation events and the vast majority of molecular character-state transitions occur *between* them.

This important conclusion primarily contrasted morphological phylogenetic data with sequence data for the regions of the plant genome used routinely to infer plant phylogeny, which lack morphological expression. The key morphogenetic genes at the heart of this chapter are also likely to mutate in a broadly clock-like fashion (e.g. Möller and Cronk, 2001) but there the similarity ends. Contrary to the quote from Chase *et al.* (2000) that began this section, there is no requirement for instantaneous, well-honed adaptation, nor for the key gene to wait (at least, not for long) for its cohort of attendant genes to follow its evolutionary lead in order to create a novel 'overall genetic environment'. Genetic control of morphogenesis is not a democracy; rather, key characters reflect key transitions in key genes that are capable of autocratically altering the environment of gene expression.

Before exploring these concepts further, we will present two case studies to illustrate the morphological elements that underpin saltation theory.

7.5 Recognising and interpreting terata in extant and extinct species

> [E]veryone admits that there are at least individual differences in species under nature. But, besides such differences, all naturalists have admitted the existence of varieties, which they think sufficiently distinct to be worthy of record in

systematic works. No one can draw any clear distinction between individual differences and slight varieties; or between more plainly marked varieties and subspecies, and species.

(Jones, 1999: 447)

Hereafter we shall be compelled to acknowledge that the only distinction between species and well-marked varieties is that the latter are known, or believed, to be connected at the present day by intermediate gradations, whereas species were formerly thus connected ... It is quite possible that forms now generally acknowledged to be merely varieties may hereafter be thought worthy of specific names.

(Jones, 1999: 463)

Bateman and DiMichele (1994a) used two detailed examples to illustrate saltation. Reticulate saltation (little discussed in this chapter) was exemplified by the phylogenetic study of the asterid genus *Montanoa* (Funk and Brooks, 1990), which showed multiple origins of polyploids, each intimately related to similar transitions in ecological preferences (like many studies initially investigated by morphological cladistics alone, this interpretation required some revision following the acquisition of sequence data).

Dichotomous saltation (the focus of this chapter, as it is mutationally driven) was illustrated using frequent and profound architectural transitions within the largely fossil clade of rhizomorphic lycopsids (Bateman *et al.*, 1992; Bateman, 1994). Although a dominant element in Palaeozoic floras (Phillips and DiMichele, 1992), this clade has left only a single extant genus, *Isoetes* (including *Stylites*), which is highly morphologically reduced and ecologically specialised. This paucity of extant descendants renders the clade as a whole immune to both molecular phylogenetics and comparative evolutionary–developmental genetics. The arguments in favour of a strong evolutionary–developmental underpinning to the remarkable morphological diversification of the clade were therefore of necessity based on indirect, circumstantial evidence, which leaves the case unproven, however credible the underlying logic and biological inferences.

Here we will briefly review two other case studies that illustrate various aspects of dichotomous saltation: the first is based on a wholly extinct group of Palaeozoic gymnosperms, whereas the second concerns extant terrestrial orchids, particularly those of the subtribe Orchidinae.

7.5.1 *Palaeozoic lyginopterid pteridosperms: transient loss of gender segregation*

The earliest seed-bearing plants were the lyginopterid pteridosperms, which probably evolved from a single progymnospermous ancestor in the late Devonian (e.g. Rothwell and Scheckler, 1988). The pattern of morphological character acquisition within the group demonstrates progressive increase in reproductive sophistication (Retallack and Dilcher, 1988; Rothwell and Scheckler, 1988), but there is strong circumstantial evidence that even the earliest pteridosperms segregated male reproductive organs (clusters of increasingly synangial pollen-organs producing proximally germinating pre-pollen) from female reproductive organs (increasingly cupulate clusters of increasingly integumented ovules). The most common architectural model shows strong developmental similarity between male and female architectures; both were borne in dichotomously and three-dimensionally branching structures in the forks of complexly branched pinnate fronds. The most complete known specimen of such a plant, *Diplopteridium holdenii*, bears several fertile fronds, and each frond is uniformly female (Rowe, 1988). Thus, the available evidence strongly suggests that the plants were either dioecious or at best sequentially monoecious, presumably to ensure cross-pollination.

The only known exception to this rule is a single barely hermaphrodite specimen of the early Carboniferous lyginopterid ovulate cupule *Pullaritheca longii* (Long, 1977a, b; Rothwell and Wight, 1989); all other 38 *Pullaritheca* cupules recovered from two small adjacent outcrops in Southeast Scotland are uniformly ovulate (Bateman and Rothwell, 1990). Each cupule ('hemicupule' *sensu* Long, 1977a) is a pendulous bowl 7–10 mm long and 5–8 mm in diameter, consisting of six to twelve fleshy finger-like lobes surrounding a central discoid placenta (Figures 7.3a and 7.4a). The placenta bears twenty to thirty ovules that terminate in a characteristic chambered ('hydropteridalean') pollen-receiving apparatus, formed from the nucellus and well adapted to accommodate wind-borne pre-pollen grains retracted into the chamber via a pollen-drop mechanism (Figure 7.3b: Rothwell and Scheckler, 1988). The ovules expanded and matured into viable seeds only once they had been pollinated, eventually being shed from the cupule; consequently, most of the ovules remaining in the *Pullaritheca* cupules are unexpanded and abortive, presumably because they had not been pollinated (Figures 7.3a and 7.4a).

The exceptional hermaphrodite cupule was described in detail by Long (1977a), prompting him to indulge in a rare venture into biological speculation. Specifically, it allowed Long (1977b) to resurrect with greater conviction his earlier (Long, 1966) cupule–carpel theory, which suggested that angiosperms could have evolved directly from derived Mesozoic pteridosperms such as *Caytonia* (e.g. Harris, 1964; Retallack and Dilcher, 1988) by retention of the ovules within fully fused sterile cupules. This

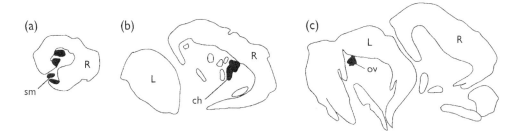

Figure 7.3 An example of a fossil putative teratos from the Lower Carboniferous of Oxroad Bay, East Lothian, Scotland. Longitudinal sections through the margin of an exceptional specimen of the early seed-fern cupule *Pullaritheca longii* (Lyginopteridaceae: Pteridospermales), showing a transition from (c) regulated expression of ovules to (a) atavistic expression of microsporangia via (b) non-functional structures of indeterminate gender. (a–c) modified from Long, 1977a, Figure 1c, e, i). Labels: L, left hemisphere; R, right hemisphere; ov, ovule; sm, microsporangium; ch, chimaeric structure possessing features of both ovule and microsporangium. (×4.8: after Bateman and DiMichele, 1994b, Figure 9.)

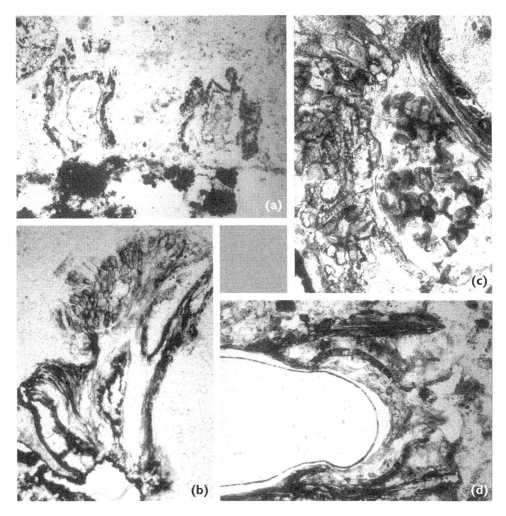

Figure 7.4 An example of a fossil putative teratos from the Lower Carboniferous of Oxroad Bay (see also Figure 7.3). (a) Portion of one of thirty-eight recorded dehisced specimens of the lobed placental cupule *Pullaritheca longii* that exclusively bear *Hydrasperma longii* ovules, showing retention of a few small abortive ovules attached to the vascularised placenta. (b–c) Two teratological structures first discovered by Long (1977a) at the margin of a single atypical cupule, suggesting ectopic expression of microsporangia; (b) largely resembles an abortive ovule but has undergone exceptional proliferation of the distal nucellar tissue that is normally adapted for capturing pre-pollen grains; (c) shows similar nucellar proliferation to (b), but contains many poorly-formed microspores rather than the expected single permanently retained megaspore. (d) The typical pollen-receiving chamber of *Hydrasperma* ovule, compressed by development of the ovum following successful fertilisation (see also Long, 1977a, b; Rothwell and Wight, 1987; Bateman and Rothwell, 1990; Bateman and DiMichele, 1994b). Magnifications: (a, b, d) x 63, (c) x 125. Photographs by RMB.

theory was dealt an apparently fatal blow by well-known morphological phylogenetic analyses of the 1980s and 1990s (beginning with Crane, 1985; Doyle and
Donoghue, 1986). These phylogenies consistently resurrected and promoted the
Anthophyte hypothesis, which interpolated between paraphyletic pteridosperms and
monophyletic angiosperms various putatively derived gymnosperms (conifers and
their relatives, pentoxylaleans, bennettites, and gnetaleans), rendering impossible the
direct transition from pteridosperm to angiosperm.

However, this conclusion has been seriously challenged by some arguments that
revised morphological homologies place the bennettites as sister to the cycads (W.
Crepet, pers. comm. 2001) and, more importantly, recent polygenic phylogenies that
placed the gnetophytes as a derived clade within a now monophyletic clade of extant
gymnosperms (cf. Mathews and Donoghue, 1999; Qiu *et al.*, 1999; Chaw *et al.*,
2000; Frohlich and Parker, 2000; Graham and Olmstead, 2000). This topology was
supported by evolutionary–developmental genetic evidence of synapomorphic loss of
the Needle copy of *lfy* from angiosperms (Frohlich, 2001, 2002), prompting the
development of the Mostly Male theory of angiosperm origin (Frohlich and Parker,
2000; Frohlich, 2001, 2002), which requires the ectopic expression of ovules on formerly male sporophylls. Does the hermaphrodite *Pullaritheca* cupule have any
bearing on the credibility of these important but highly speculative hypotheses?

1   The existence of thirty-eight wholly ovulate cupules demonstrates that the one
    recorded hermaphrodite cupule is an exceptional occurrence that has taken
    place within a routinely unisexual species (in other words, the hermaphrodite
    cupule is a developmental 'accident'), especially when viewed in the context of
    the fact that all other Palaeozoic pteridosperms have reliably unisexual repro-
    ductive clusters.
2   The near-radial symmetry of the cupule contrasts strongly with the localisation
    of the sporangia along a small, marginal portion of the placenta (Long, 1977a)
    (Figure 7.3). More symmetrical segregation of the genders, most probably with
    the male sporangia distributed along the entire periphery of the placenta, would
    be expected from a stable, genetically controlled hermaphrodite cupule.
3   The presence of two phenotypically intermediate structures at the junction of
    the ovulate and microsporangiate zones of the placenta also casts doubt on the
    stability of the hermaphrodite phenotype. One of these structures (Figure 7.4b)
    largely resembles the abortive *Hydrasperma* ovules shown in Figure 7.4a, retain-
    ing the tentacle-like distal lobes of the integument that give this ovule-genus its
    name. However, the nucellus-derived pollen chamber (Figure 7.4d) is absent,
    being replaced by an asymmetric, 'cancerous' outgrowth of the nucellus. This
    nucellar proliferation is also evident in the second structure (Figure 7.4c), which
    otherwise more closely resembles a microsporangium, even producing within the
    nucellar envelope a large number of incompletely formed pre-pollen grains.

Taken together, these three observations suggest to us that the developmental
anomaly is the product of a local physiological perturbation rather than being genetically determined. The two phenotypically 'hybrid' structures imply that there was a
breakdown in gender control across the placenta and that the mechanism of gender
control was expressed clinally. This accords with gender control of the unisexual

cones of extant conifers, where nutritional clines separate pollen-bearing cones from the better resourced ovulate cones. In the case of the hermaphrodite *Pullaritheca* cupule, one possible explanation lies in the apparently relatively poor vasculature supplying the microsporangium-bearing margin of the cupule (R. Bateman, unpubl. obs.).

Thus, the cupule is undoubtedly a teratos, reflecting an instantaneous phenotypic shift within an individual, but if the above interpretation is correct, it is not a *bona fide* hopeful monster (Table 7.1), as the phenotypic shift is not heritable. Nonetheless, the aberrant cupule provides a useful insight into the control of gender expression in pteridosperms, which in turn suggests that a genuinely genetically controlled mutation could indeed allow ectopic expression of male structures on a fundamentally female structure to produce a flower precursor. It also suggests the feasibility of the converse phenomenon, namely ectopic expression of female structures on a fundamentally male structure, that is required for the Mostly Male theory (Frohlich, 2002). And lastly, it indicates the potential contribution to plant evolution of heterotopy.

Strictly, *heterotopy* is a spatial shift in a developmental programme and its resulting phenotypic structure across the bauplan of an organism (cf. Bateman, 1994; Crane and Kenrick, 1997; Frohlich and Parker, 2000; Baum and Donoghue, 2002; Cronk, 2002; Kellogg, 2002; Rudall and Buzgo, 2002). In the utilitarian terminology of Baum and Donoghue (2002), a shift of a structure to a new location is termed a 'neoheterotopy', whereas the at least partial replacement of a pre-existing structure is a 'homeoheterotopy' (complete replacement of the structure constitutes homeosis *s.s.*: see also the following section).

Given their pivotal position in the phylogeny of seed-plants, it is especially unfortunate that there are no extant pteridosperms available to be subjected to the full panoply of evolutionary–developmental genetic techniques. In order to learn more of the evolutionary power of heterotopy, we will now consider a strongly contrasting clade that has an exceptionally poor fossil record (e.g. Mehl, 1986) but is remarkably species-rich in the extant flora.

7.5.2 Floral symmetry and speciation in orchids: many attempts yield few successes

7.5.2.1 Background and terminology

The literal blossoming of plant evolutionary–developmental genetics was spearheaded by studies of heterotopy *sensu lato* in relatively derived eudicot angiosperms (notably the model genera *Antirrhinum* and *Arabidopsis*), allowing elucidation of the basic ABC model of organ identity (e.g. Coen and Meyerowitz, 1991) that is now becoming far more complex as new data constantly demand revisions and amendments (e.g. Cubas, 2002; Theissen *et al.*, 2002). It was inevitable that this would soon lead to reconsideration of an old evolutionary chestnut, floral symmetry (e.g. Coen, 1999; Endress, 1999, 2001; Cubas *et al.*, 2001; Cubas, 2002; Knapp, 2002; Rudall and Bateman, 2002).

To classical morphologists the traditional primary distinction in floral symmetry separates 'regular' radial symmetry (actinomorphy) from 'irregular' bilateral symmetry (zygomorphy), though this fundamental dichotomy requires amendment to take into account of irregular flowers that lack any recognisable symmetry (asymmetric *sensu* Endress, 2001; anartiomorphic *sensu* R. Bateman and L. McCook, unpubl. obs.). Actinomorphy and zygomorphy can be defined largely on recognisable numbers of mirror planes (planes that divide the flower into two mirror images when viewed perpendicular to the pedicel); there are at least two and usually more in actinomorphy, and typically one in zygomorphy (a few zygomorphic flowers possessing four mutually perpendicular sepals and/or petals arranged in opposing pairs of unequal size have two unequal mirror planes: Rudall and Bateman, 2002).

Sadly, this well-established botanical terminology has been mutated by developmental geneticists to create a tension that is clearly evident in this volume; actinomorphic flowers have been termed 'symmetrical' and zygomorphic flowers 'asymmetrical', prompting Endress (1999, 2001) to redescribe actinomorphic flowers as polysymmetric and zygomorphic flowers as monosymmetric. Here, we retain the standard botanical terminology, noting also that (a) with regard to the vertical axis in zygomorphic flowers the botanical preference for the term 'dorsiventral' is more linguistically correct than the prevalent developmental genetic preference for the term 'dorsoventral' (cf. Brown, 1956), and (b) the term dorsoventral has a contrasting meaning in the developmental anatomy of animals (Kaplan, 2001a).

A third complication is that these descriptions of symmetry tend to be used to characterise the whole flower, when the symmetries of each of the four fundamental whorls can in theory be different (often, perianth symmetry is prioritised and stamen and carpel symmetry are essentially ignored: Bateman, 1985). For example, the flowers of *Antirrhinum* are zygomorphic in all four whorls, whereas those of *Arabidopsis* appear actinomorphic until one observes the characteristic but subtle bilateralism evident in stamen insertion and early sepal development. It is therefore preferable to consider the symmetry of each of the four whorls separately, rather than attempt to summarise the symmetry of an entire flower in a single aggregate term.

7.5.2.2 A survey of floral terata in terrestrial orchids

Bateman (1985) reviewed occurrences of floral terata among British orchids, focusing on examples of peloria: any transition from zygomorphy to actinomorphy or vice versa (see also Leavitt, 1909; Theißen, 2000). Noting that the gynostemium (a fusion of the male and female whorls) always retained a degree of bilateralism, he focused on transitions in floral symmetry of the three sepals plus the three petals, taking into account the usual strong morphological differentiation of the median petal. This petal, termed the labellum, acts as a landing stage for pollinating insects in most species; hence, although it is developmentally uppermost, in most cases it is spatially lowermost; in erect inflorescences this is due to a 180° torsion of the ovary that is termed 'resupination' (Ernst and Arditti, 1994; Rudall and Bateman, 2002).

Three categories of morphological transition were evident: type A peloria, when the two lateral petals are replaced by two additional labella (cf. Figure 7.5a, b), the less common type B peloria, when the labellum is replaced by an additional lateral petal, and a third category of more heterogeneous morphological transitions that were collectively termed pseudopeloria by Bateman (1985). Here, the modified

(a)

(b)

(c)

(d)

Figure 7.5 Two examples of extant teratos and putative hopeful monsters from the orchid subtribe Orchidinae. (a) is a typical flower of the insect-pollinated orchid *Ophrys insectifera* from Hampshire, England, that shows complex adaptations for transfer of pollinia via cephalic pseudocopulation by male solitary wasps. (b) is a flower of an adjacent individual showing type A perianthic peloria *sensu* Bateman (1985); the two mimicked 'antennae' of (a) have been homeotically replaced by additional 'bodies'. (c) is a typical flower of *Platanthera chlorantha* from Perthshire, Scotland, that shows a long, nectariferous spur, fragrance and white coloration adapted to attract night-flying moths as pollinators. (d) is the flower of an adjacent individual showing pseudopeloria *sensu* Bateman (1985); the petals of the normal flower, including the elaborate labellum and spur, have been replaced by three additional sepal-like structures (see also Figure 10 of Rudall and Bateman, 2002). Magnifications: (a) x 2.5, (b) x 1.9, (c) x 2.9, (d) x 2.6. (a–c) by RMB, (d) courtesy of R. Bush.

labellum becomes less distinctive than the typical labellum but can still be differentiated from the lateral petals and so confers a degree of zygomorphy to the petal whorl (cf. Figure 7.5c, d; Plate 1a, b). Often, the modified labellum more closely resembles the sepals than either the lateral petals or the normal labellum, and hence it is frequently termed 'sepaloid' in the literature. Similar floral transitions occur in other plant families but are rarely described, perhaps because they are less immediately obvious than genuinely peloric morphs. Adding a fourth (and rarer) phenomenon to the classification of Bateman, Horsman (1990) recognised type C peloria, wherein all three petals are apparently replaced by sepals, generating two near-identical whorls of three perianth segments (see also Lang, 2001).

Given the above definitions, one might interpret both type A and type B peloria in orchids as lateral homeoheterotopy (i.e. a change of organ identity confined to a single whorl, namely the three petals) and type C peloria and pseudopeloria as vertical homeoheterotopy (a change of organ identity between two whorls; in this case, apparent expression of three sepals and one sepal respectively in the relatively acropetal petal whorl). However, in most cases of pseudopeloria it is also possible to view the 'sepaloid' labellum as reflecting one or more changes in the timing of development (*heterochrony*); specifically, its simplicity relative to the typical labellum could in theory reflect a labellum whose development had either slowed (neotenic) or ceased abnormally early (progenetic) to yield a mature structure that resembles the juvenile stage of the homologous structure in a notional ancestral organism (paedomorphic: all terminology pertaining to heterochrony follows Alberch *et al.*, 1979; see also McKinney and McNamara, 1991; Zelditch and Fink, 1996).

Systematic and causal interpretation of such terata depends strongly on the level of comparison of the 'homeotic' (*s.l.*) morph with the normal 'wild type' morph. In the most obvious case, bimodal variation in floral symmetry within a single individual (e.g. Darwin, 1859; analogous to the hermaphroditic *Pullaritheca* cupule described above) is likely to reflect a non-genetic cause or a somatic mutation, and thus be of no evolutionary potential. Nonetheless, interpretation is simplified by the fact that the conspecificity of both morphs is assured, and the identity of the abnormal morph can therefore be determined by comparison with other conspecific individuals bearing only one floral morph. Examples occur widely within the orchid family; the most common manifestation is the duplication of the labellum to produce a perianth of seven segments, a phenomenon that typically affects only the lowermost or the uppermost flowers of a (usually unbranched) orchid inflorescence.

Determining the conspecificity of the homeotic and normal floral morphs becomes more challenging when they are borne on separate individuals. In most cases, both morphs co-exist in a single local population, and resemble each other in all other phenotypic characters. However, distinguishing the normal morph from the homeotic morph (i.e. determining the polarity of the morphological transition) can be challenging, usually relying on the homeotic morph being appreciably less frequent than the normal morph in the populations within which it occurs or, even less convincingly, in its absence from other closely related species. This was the case with the type A peloric individual of *Ophrys insectifera* shown in Figure 7.5b, which was the only such morph observed in an estimated population of 6,000 flowering plants. However, this assumption of relative rarity is not always justified, especially in populations of orchid species that are autogamous; for example, *Epipactis phyllanthes* var.

phyllanthes (Young, 1952; Bateman, 1985) often forms uniformly pseudopeloric populations, presumably because their self-pollination leads to very low genetic diversity (Hollingsworth *et al.*, subm.). Under such circumstances, arguments for transitional polarity generally rely on the existence of many populations that wholly lack the putatively homeotic morph (as is the case for *E. phyllanthes s.l.*: Young, 1952). In such cases, the homeotic morph is likely to reflect a mutation (or combination of mutations) that is by definition polymorphic within the populations under scrutiny.

More usefully, previous unsubstantiated assertions of conspecificity or lack of conspecificity between the two contrasting floral morphs can now be readily tested using molecular markers. For example, a wide range of markers failed to detect any differences between normal and pseudopeloric *E. phyllanthes* (P. Hollingsworth, R. Bateman *et al.*, unpubl. obs.). Molecular data are especially valuable for refuting previous accusations of hybrid origins for homeotic (especially pseudopeloric) morphs. McKean (1982) interpreted the orchid shown in Figure 7.5d as a bigeneric hybrid between *Platanthera chlorantha* and *Pseudorchis albida*, whereas Bateman (1985) argued that it was probably a pseudopeloric morph of *P. chlorantha* (Figure 7.5c). This species is frequent at the locality yielding the supposed hybrid, where the vegetation includes several teratologies that are thought to reflect the high heavy metal content of the underlying spoil heaps. Nuclear rDNA (ITS) sequences revealed a large phylogenetic disparity between *Platanthera* and *Pseudorchis* (Bateman *et al.*, 1997; Pridgeon *et al.*, 1997; Bateman, 1999a, 2001) and further studies using molecular markers have conclusively demonstrated that the contrasting plants shown in Figure 7.5c and d are, in fact, both assignable to *P. chlorantha*.

Homeosis becomes of greatest interest to systematists when the putative homeotic morph regularly forms fairly uniform populations and hence becomes recognised as a distinct taxonomic species (cf. Bateman and DiMichele, 1994a; Rudall and Bateman, 2002). In other words, the teratology in question definitely reflects mutation, and that mutation has become fixed in the populations in question. Under these circumstances, we can usefully apply the full panoply of morphological and molecular phylogenetic techniques to the problem.

In some such cases the mutant orchid species is assigned to a pre-existing genus, a good example being the type B peloric cypripedioid *Phragmipedium lindenii* (e.g. Pridgeon *et al.*, 1999). In other cases, particularly in the Orient, the florally simplified mutant has been controversially recognised not only as a distinct species but also as a monotypic genus that is assumed to be phylogenetically primitive. Of these supposed genera (most of which are lower epidendroids: for reviews see Chen, 1982; Rudall and Bateman, 2002), *Tangtsinia* closely resembles *Cephalanthera*, *Sinorchis* appears attributable to either *Cephalanthera* or, more likely, *Aphyllorchis*, and *Diplandrorchis* and the polyspecific *Archineottia* resemble *Neottia*. Within the orchidoids, *Aceratorchis* is a more widely distributed *bona fide* species that resembles the co-occurring genus *Galearis*. Rudall and Bateman (2002) suggested that, far from being primitive, DNA sequence data will demonstrate that these 'genera' are nested within other more species-rich genera that possess much more strongly differentiated labella. Available data are insufficient to determine whether the morphs show types B or C peloria or pseudopeloria.

A similar example is evident among South African orchidoids of the *Satyrium* group, where two 'subactinomorphic' (pseudopeloric) species possessing reduced labella and elongate gynostemia have been assigned to the genus *Pachites*. It is not clear whether these two species are sisters, nor whether they are the sister group, and potentially ancestral to, *Satyrium s.s.* or nested within *Satyrium* as secondarily reduced morphs (Linder and Kurzweil, 1999). Although molecular phylogenies have begun to be constructed for South African orchids (Douzery *et al.*, 1999; Bateman *et al.*, subm.), this putative genus has not yet been sequenced.

7.5.2.3 Possible post-saltational radiations

In yet other cases, the prospecies not only successfully establishes a new taxonomic species and genus but also subsequently radiates to generate several descendant species, often via polyploidy and/or a transition from allogamy to autogamy. The type B or C peloric genus from Australasia, *Thelymitra*, has been demonstrated using sequence data to be monophyletic (Kores *et al.*, 2001) and contains *c.* fifty species, many of which are autogamous (Bower, 2001; Bateman and Rudall, 2002).

Here, however, we have chosen to focus on the pseudopeloric European genus *Nigritella*, as we have access to both molecular and morphological phylogenetic data and so can conduct a cross-matrix comparison (Figure 7.6: cf. Bateman, 1999a). *Nigritella* contains about fifteen species, most reflecting relatively recent polyploidy events and/or transitions from allogamy to autogamy (Hedrén *et al.*, 2000). ITS sequence data (Figure 7.6b) imply that *Nigritella* is nested phylogenetically within a paraphyletic *Gymnadenia s.s.* (consequently, *Nigritella* was sunk into *Gymnadenia* by Bateman *et al.*, 1997), and that both putative genera are appreciably divergent from their sister genus, *Dactylorhiza*. The morphological cladistic analysis (Figure 7.6a) similarly suggests that *Nigritella* is nested within *Gymnadenia*, but identifies a contrasting sister-species. More interestingly, it also reveals that the equal longest morphological branch in the tree separates *Gymnadenia s.s.* from *Nigritella*. Although the two putative genera are very similar vegetatively, *Gymnadenia s.s.* has far more strongly zygomorphic and fully resupinate flowers that possess long nectar-iferous spurs and are pollinated by lepidoptera (Figure 7.7a–c). In contrast, *Nigritella* lacks resupination and has only a simplified labellum and vestigial spur (Figure 7.7d–f); these are classic characteristics of pseudopeloria (Bateman *et al.*, 1997, subm.; Pridgeon *et al.*, 1997; Bateman, 1999a, 2001; Rudall and Bateman, 2002).

Interestingly, there is no measurable sequence divergence or qualitative morpho-logical divergence among the several (arguably over-split) species of *Nigritella*, prompting Bateman (1999a) to suggest that the radiation of this group, which is confined to alpine and boreal habitats in Europe, may reflect a species-level radia-tion that occurred since the Pleistocene ice retreated at the close of the Loch Lomond Stadial (Younger Dryas), about 10,000 years ago. However, as noted by M. Frohlich (pers. comm. 2001), the present distributions could be the relics of a genus that was more widespread during the Pleistocene glacials, when terrains suit-able for the plants would have been more extensive. In either case, much of the driving force for that radiation appears to have been polyploidy and at least one transition to autogamy, suggesting that the clade is 'attempting' to maximise its fitness within the 'undesirable' constraint of a floral morphology that can no longer support the ancestral lepidopteran pollination syndrome. It is regrettable that greater progress has not been made in the evolutionary–developmental genetics of

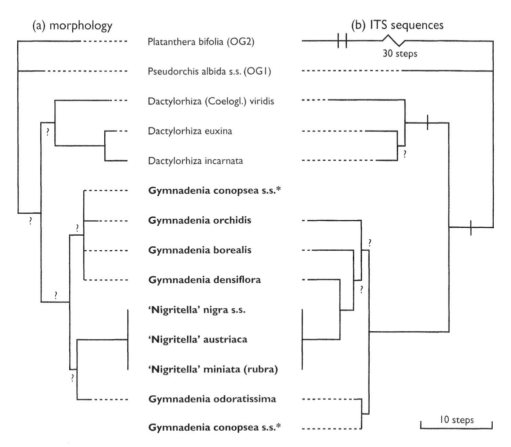

(a) morphology   (b) ITS sequences

Platanthera bifolia (OG2)

30 steps

Pseudorchis albida s.s. (OGI)

Dactylorhiza (Coelogl.) viridis

Dactylorhiza euxina

Dactylorhiza incarnata

Gymnadenia conopsea s.s.*

Gymnadenia orchidis

Gymnadenia borealis

Gymnadenia densiflora

'Nigritella' nigra s.s.

'Nigritella' austriaca

'Nigritella' miniata (rubra)

Gymnadenia odoratissima

Gymnadenia conopsea s.s.*

10 steps

Figure 7.6 Comparison of one of two most-parsimonious cladograms for morphology (a) and one of several most-parsimonious cladograms for ITS sequences (b) for the *Gymnadenia* alliance (boldface), its sister-group the *Dactylorhiza* alliance, and *Pseudorchis* plus *Platanthera* as a paraphyletic outgroup (all Orchidinae: Orchidaceae). Branch lengths are proportional to the number of steps under Acctran optimisation. Cross-bars indicate unambiguous indels; nodes bearing question marks have less than 80 per cent bootstrap support. Note the incongruent positions between the two trees of *Gymnadenia conopsea s.s.* (asterisked).(a) from R. M. Bateman, I. Denholm and P. M. Hollingsworth (unpubl. obs.), (b) from R. M. Bateman, P. M. Hollingsworth and J. Preston (unpubl. obs.).

orchids (for summary see Johansen and Frederiksen, 2002), as we suspect that a single mutation in a key developmental gene triggered the origin of *Nigritella* from within a species of *Gymnadenia*, stochastically forcing subsequent evolution in the lineage along a very different path from the plant–pollinator co-evolution that had previously driven evolution in the lineage.

7.5.2.4 Radiation of the orchid family

Reviewing the origin and initial radiation of the orchid family, Rudall and Bateman (2002) noted the importance of heterochrony, heterotopy, organ suppression and

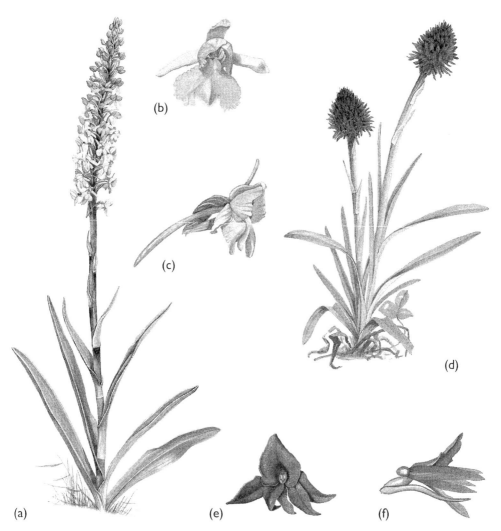

Figure 7.7 A putative example of a homeotic transformation leading to the immediate origin of a new orchid genus, based on the phylogenies depicted in Figure 7.6 which suggest that *Gymnadenia densiflora* (a–c) is the most likely ancestor of the *Nigritella* alliance, here represented by *Nigritella austriaca* (d–f). The two putative genera are very similar vegetatively but differ strongly in floral morphology: compared with *Gymnadenia s.s.*, *Nigritella* has a more compact pyramidal inflorescence and non-resupinate flowers bearing a poorly differentiated labellum and a vestigial spur that lacks nectar (a–f from Landwehr, 1977). Magnifications: (a) × 0.30, (b–c) × 1.5, (d) × 0.59, (e–f) × 2.1.

organ fusion in underpinning gross morphological divergence; processes that typically leave the two distal floral whorls of the orchid flower fused and at least the three distal floral whorls bilaterally symmetrical. The origin of the family was probably facilitated by epigyny and the mycorrhizal association that encouraged the maturation of many small seeds per ovary, but strictly is delimited by bilaterally symmetrical suppression of at least half of the six ancestral stamens, with subsequent multiple reductions to two functional stamens and a single reduction to one stamen (each group possessing at least some species with one or more relictual staminodes, and occasional atavistic reversals to greater numbers of stamens and/or staminodes). Style–stamen fusion formed the gynostemium, and petal–stamen (or possibly petal–staminode) fusion may have formed the labellum. Differentiation of the labellum conferred bilateral symmetry on the whorl of three petals, which is inserted immediately distal to the more equidimensional whorl of three sepals. Finer-scale phylogenetic divergences in floral form tend to involve modifications of the size, shape and colour of the labellum and associated spur (where present).

Within the context of this evolutionary history, Rudall and Bateman described both type A and type B peloria as examples of lateral homeoheterotopy, with median petals (i.e. additional labella) replacing lateral petals and lateral petals replacing median petals, respectively. Both are also genuine examples of homeosis *sensu* Baum and Donoghue (2002), as they show positional translocation during evolution of an entire pre-existing structure. However, pseudopeloria was attributed not to vertical (in this case acropetal) homeoheterotopy (translocation of a sepal to the location normally occupied by the median petal) but to heterochrony; specifically, the median petal remains recognisable (albeit less morphologically distinct). The considerable variation in morphology evident among pseudopeloric individuals indicates that several genes (probably expressed downstream of the better-known homeobox genes) may be involved in such transitions. Pseudopeloria also appears more likely to disrupt the structure of the gynostemium than is peloria, tentatively indicating that separate sets of developmental genes control perianthic zygomorphy and gynostemial fusion. These morphologically-based inferences are now amenable to testing using evo-devo techniques.

7.6 Plant versus animal evolution

> Plants work to rules rather different from those of animals. They are more ready to accept a foreign mate, and, quite often, the offspring of such a liaison find themselves with combinations of genes that fit together so well that the new mixtures flourish. Hybridisation becomes a fast track to a new existence, rather than (as in animals) a crack through which DNA leaks to dilute the prospects of a hopeful species ... Plant species can, it seems, originate at some speed, with no need for the long probation in new forms of animal life. Their genes are more ready to cooperate than are those of animals.
>
> (Jones, 1999: 242)

Our 1994 paper included a forthright critique of zoocentrism. The above quote from Jones (1999) is a welcome indication that plants are at last entering into the thinking of some of the most effective popular exponents of evolutionary theory: influential scientists who are inevitably drawn from the zoological or palaeozoological communities. The list of major differences in both the generation and establishment phases of saltation presented by Bateman and DiMichele (1994a; see also Darley, 1990) can now be considerably expanded.

Perhaps the most important factor is the sedentary, autotrophic lifestyle of almost all higher plants and their consequent reliance on a root–shoot dichotomy and on the multiple localised meristems that allow open, additive growth and differentiation, and offer the potential for clonal vegetative reproduction. The majority of the component organs are, of necessity, ephemeral serial homologues (rootlets, leaves, leaf-derived floral organs of the four angiosperm floral whorls, pollen, ovules), and most of these terminate ontogenetic cascades, minimising the ontogenetic constraints inherent in the concept of 'burden' (Riedl, 1979) and maximising the potential for evolutionary mosaicism (Stebbins, 1983).

The history of green plants since their single putatively Silurian invasion of the land (Kenrick and Crane, 1997; Bateman *et al.*, 1998) owes much to diversification in the type, number, spatial arrangement and branching patterns of meristems, which along with dormancy substitute for the motility of animals (Darley, 1990). Moreover, the largely 'on–off' nature of meristematic control, and the inability of plant cells to translocate within the plant body once the cellulose cell wall has formed, reduce the potential for gradual transitions in form, either within specific organs or between adjacent organs.

Also, there is no juvenile sequestering of gametes in plants; any meristem can in theory undergo mutations that are subsequently passed on to meiotically-derived sets of 'sibling' gametes generated during the next reproductive phase of the plant. The regularly repeated phases of reproduction that characterise most plant life cycles also provide far greater opportunities for polyploidy, arguably enhanced by self-generated chemical spindle inhibitors and externally imposed physical shocks (Cronk, 2001); moreover, there is increasingly strong evidence that mutant embryos are far less frequently aborted in plants than animals.

7.7 Establishing hopeful monsters

[E]ach new species is formed by having had some advantage in the struggle for life over other and preceding forms. From the extraordinary manner in which European productions have recently spread over New Zealand we may believe that in the course of time a multitude of British forms would exterminate many of the natives. Under this point of view, the productions of Great Britain may be said to be higher than those of New Zealand. Yet the most skilful naturalist, from an examination of the species of the two countries, could not have foreseen this result.

(Darwin, 1859; rewritten by Jones, 1999: 303)

Evolution's progress (if such it is) is not uninterrupted. Catastrophe on a scale unknown to history has played a part. Whether it had a constructive, rather than a merely lethal, effect is another issue. Some claim that mass destruction led to biological explosions as the survivors evolved to fill the gaps. If they are right, cataclysms drive change as much as does slow modification.

(Jones, 1999: 292)

We argue in the following section that the field observations of many generations of natural historians (such as those discussed for the Orchidaceae in Section 7.5.2), now supported by detailed studies of model plants (e.g. Meyerowitz and Somerville, 1994) and causal explanations recently provided by various subdisciplines of molecular biology, constitute an irrefutable case for the widespread occurrence of macromutations. However, these are not Goldschmidt's concept of genetically extensive mutations, but rather are typically very small genomic changes of large phenotypic effect. Thus, we believe that future conceptual battles over the credibility of saltation theory are far more likely to be fought over the viability of the prospecies in natural habitats (Bateman and DiMichele, 1994a, b; DiMichele and Bateman, 1996; Bateman, 1999a) than over previous erroneous scepticism regarding whether macromutation occurs. Sadly, the profound advances in developmental genetic knowledge achieved since 1994 that have helped to conclusively demonstrate the ubiquity of macromutation (see below) have not been matched by corresponding new insights into the all-important ecological filtration of novel phenotypes. Nonetheless, we will now review (and, where appropriate, amend) our previous scenario.

7.7.1 The model: ecological filtration in primarily and secondarily vacant niches

Once it has originated as a single mutant or small colony of sibling mutants, the prospecies must be sufficiently functional to establish itself within its local environment and subsequently to reproduce (though note that there are several alternatives to sexual reproduction available to plants; initial sterility is not necessarily a bar to successful establishment). Considerable expansion in the size and range of the population is then required to protect the now fully-fledged taxonomic species from the vicissitudes of the environment and from competition (albeit indirect, environmentally-mediated competition for resources) with other plant species, not least the parental species (Bateman and DiMichele, 1994a, b; DiMichele and Bateman, 1996).

In our previous models we assumed lower fitness of the hopeful monster than its parents and hence emphasised (possibly over-emphasised) the benefit of the prospecies originating serendipitously in a habitat of low competition that allows at least temporary respite from selection pressures. In this context, there exists a strong contrast between examples of relatively low infra- and interspecific competition associated with (a) the initial invasion of a previously unoccupied habitat type by plants following their invasion of the land (Bateman et al., 1998) and (b) establishment in habitats that have become temporarily vacant over a specific geographical area due to one or more major environmental perturbations.

The greening of the terrestrial realm during the Palaeozoic is essentially a story of diversification into increasingly stressful (especially increasingly xeric) environments (Bateman et al., 1998; Stein, 1998; Kenrick, 2002; Schneider et al., 2002). Each successive phase involved radiations of one or more major taxa possessing a recognisable range of morphologies well adapted to the range of environments being colonised. Major morphological innovations tended to occur at the cutting edge of this wave of colonisation, with consolidation and specialisation characterising previously colonised habitat types (e.g. of lycopsids and sphenopsids in Devono-Carboniferous wetlands). Reverse colonisations, such as the post-Carboniferous invasion of previously marginal, xeromorphic conifers into the (by then relatively desiccated) adjacent wetlands, were possible only in the wake of severe environmental perturbations such as the equatorial warming and drying of the Early Permian (DiMichele and Aronson, 1992; Foote, 1996; DiMichele et al., 2001; DiMichele, subm.).

Cases involving the occupation of 'secondary' vacancies inevitably dominate the post-Palaeozoic history of the terrestrial flora (e.g. Wing and Boucher, 1998). Here, the identity of the re-occupying species is strongly influenced by three factors. If by good fortune the occupier grew adjacent to, but was unaffected by, the environmental perturbation, its expansion is caused purely by happenstance. If the occupier experienced the perturbation but survived there is ample scope for an adaptive explanation (e.g. fire-induced seed germination in species regularly affected by wildfires). However, where the environmental catastrophe is insufficiently frequent to be able to influence microevolution, the persistence of the occupier and its subsequent expansion in numbers can only reflect pre-existing properties of the organism; in other words, its success can be attributed to pre-adaptation.

More controversially, Bateman and DiMichele (1994a) also argued that in all these cases the successful invader then benefits from incumbent advantage, a theory that essentially states that mere physical occupancy of a particular site confers an advantage beyond that attributable to raw estimates of intrinsic fitness (e.g. Rosenzweig and McCord, 1991; DiMichele and Bateman, 1996). We further hypothesised that, because of the lack of direct Malthusian competition among individual plants, a much wider range of features of a typical higher plant are selectively neutral at any particular moment in time than are those of a typical higher animal, giving plants considerably more freedom for non-selective experimentation in form (see also van Steenis, 1976).

7.7.2 Relevance of phylogenetic patterns

Thus far, our 'armchair' test of this scenario has been weak and indirect. Specifically, we predicted that the overall pattern of plant morphological diversification through time would appear fractal, with the greatest phenotypic divergences between ancestral and descendant species characterising the earliest lineages in a clade and the least divergences characterising the most recent (DiMichele and Bateman, 1996; Bateman, 1999a, b; DiMichele et al., 2001; see also Arthur, 1984, 2000). In the few clades where strong data sets for both sequences and morphology are available, this does indeed appear to be the case (Bateman, 1999a). There are at least three possible explanations for such a pattern: increased developmental canalisation, increased niche saturation (both resulting from speciation) and preferential extinction of early-formed lineages (the antithesis of speciation).

Increased developmental integration and canalisation through time appear likely from first principles. It is becoming evident that most genes of profound phenotypic effect in higher plants have putative homologues in prokaryotes and early eukaryotes. The subsequent evolutionary history of plant form owes much to gene duplication events (including those engendered by polyploidy) followed by divergence of function of the paralogues, the increase in overall numbers of functional genes presumably being balanced to some degree by 'extinction' of specific genes deactivated through gene silencing or pseudogene formation via events such as transposon insertion (e.g. Wessler et al., 1995; Federoff, 2000; Walbot, 2002). The resulting enhanced complexity of the developmental control mechanisms is likely to increase the interdependency among specific genes. However, there are two opposing interpretations of the consequences of such interdependency; either the ever-more complex system of sequentially acting genes will become more prone to lethal collapse when faced with a strongly expressed mutation (thereby inducing progressive canalisation), or the probable proliferation of alternative systems, resulting in enhanced developmental homeostasis that can buffer the organism *against* lethal collapse (in which case canalisation is likely to prove to be a largely mythical phenomenon). This issue deserves more attention than it currently receives.

Transferring our emphasis from the generation phase to the establishment phase of saltation, an increased probability of niche saturation inevitably accompanied the well-documented progressive increase in species-level plant diversity through time (e.g. Valentine, 1980; Behrensmeyer *et al.*, 1992). As niche saturation is approached, competition for resources increases and the chances of a novel prospecies finding a low-competition niche in which to establish itself decrease correspondingly. In other words, the rate of generation of prospecies may remain high, but the more critical rate of successful establishment of those prospecies will on average decrease through time. To some extent, this effect may be mitigated by finer niche partitioning of specific habitats; Tertiary angiosperms in particular appear to have benefited from unusually strong co-evolutionary drivers for speciation. However, this exceptional species richness reflects typically trivial morphological differences among species, and is not mirrored in greater diversification of form at higher taxonomic levels (DiMichele *et al.*, 2001).

This observation does raise the thorny issue of whether selective extinction among major clades strongly affects perceptions of patterns of morphological divergence in land-plant phylogenies. It is abundantly clear that the survival of major class- and ordinal-level taxa (both monophyletic and paraphyletic) of pre-angiosperm radiations has been highly selective, with the most profound and troublesome gaps in the extant flora reflecting Siluro-Devonian pteridophytes (e.g. Rhyniopsida, Zosterophyllopsida, Trimerophytopsida, Progymnospermopsida) and Late Palaeozoic–Early Mesozoic seed-plants (e.g. Cordaitales, Pteridospermales, Bennettitales, most Ginkgoales). These gaps can, to some extent, be plugged by fossil taxa in morphological phylogenies, but they seriously undermine attempts to generate molecular phylogenies of land-plants suitable for testing saltation hypotheses. Exceptional sampling of many genes cannot compensate for poor sampling of taxa, and the current trend to eschew morphological phylogenies in favour of sequence data is becoming an increasingly severe handicap to making genuine progress in the understanding of major morphological transitions (cf. Bateman, 1999a; Hawkins, 2002).

Thus, in practice, the best opportunities for empirically testing saltation theory lie among extant angiosperms, tackling sets of closely related species that show recent origins, as inferred from short terminal branches on sequence trees, but considerable morphological divergence, as inferred from long terminal branches on morphological trees (Bateman, 1999a). Good 'mainland' examples occur in Orchidaceae (Bateman, 1999a), Zingiberaceae (Harris *et al.*, 2000) and Fabaceae (Richardson *et al.*, 2001a). Putative evolutionary radiations on geographic and physiographic islands have in recent years provided many such case studies of massive morphological divergence, imposing legendary status on groups such as the Hawaiian silver-swords (Baldwin *et al.*, 1998), Macaronesian Poteriaeae (Bateman, 1999a; Helfgott *et al.*, 2000) and South Atlantic Rhamnaceae (Richardson *et al.*, 2001b), but what has been lacking is the null hypothesis of equally careful studies of groups that have *failed* to radiate.

For example, the twelve orchid species currently recognised in Macaronesia represent eleven genera, and each species has a close analogue in the adjacent regions of the Mediterranean, apparently reflecting anagenetic speciation (R. Bateman, unpubl. obs.). Indeed, the majority of immigrants to oceanic islands over volcanic hot-spots have failed to dichotomously speciate, let alone radiate, suggesting that fresh terrain and low competition is a prerequisite for, rather than a driver of, radical innovation in form. In the case of the Macaronesian orchids, the potential for radiation is constrained not only by the need for fidelity from associated pollinating insects but also apparently from at least two cohorts of mycorrhizal fungi; the first is essential to initiate germination of the minute embryos in the air-borne seeds and the second is needed to supply nutrition to the mature tubers or rhizomes.

7.7.3 Urgent need for field data

The main problem with assessing the establishment phase of saltation (and indeed with demonstrating any evolutionary hypothesis among higher plants) is demonstrating the potential for speciation by making direct field observations. This challenge has been to a large degree abrogated by neoDarwinian theorists on the grounds that the period of time required for speciation via microevolution is too long for observation to be feasible. Also, most supposed cases of microevolution (including the much-vaunted textbook example of the melanic form of the moth *Biston betularia*) do not ultimately lead to speciation, but rather diminish once the directional selection pressure proves transient and subsides or are introgressed back into the fold by the 'parental' populations. By contrast, in the case of saltation, the observer can easily demonstrate the origin (often multiple origins) of a particular morph, as in the orchid floral examples discussed in Section 7.5.2, and by deploying modern evo-devo techniques can (albeit with much effort) identify the causal mutation. Rather, what is lacking is a concerted effort to monitor such lineages in the field in the hope of identifying one of the rare examples of successful establishment.

A few empirical projects have been pursued in experimental plots, most notably the intriguing effects of epidermal cell mutation on pollinator choice in snapdragon flowers discussed by Glover and Martin (2002). However, in these cases the mutation under scrutiny is usually too phenotypically subtle and hence questionably constitutes a *bona fide* macromutation (indeed, most fail to attract the attention of evo-devo researchers at all). Moreover, the massed individuals carefully planted in serried ranks do not experience the environmentally-induced traumas or the challenging cut-and-thrust of interspecific competition that occur in natural ecosystems.

One possible source of stronger field data comes from the sudden intense interest among environmentalists in determining the behaviour of transgenic crop plants in natural and semi-natural ecosystems, which has emerged in the wake of the recent GM controversies (cf. Ellstrand *et al.*, 1999). Detailed monitoring of the possible spread of transgenes from the intended 'host' to other individuals, of both that and other species growing in the habitat under scrutiny, should give greater insights into the integration and dissemination of specific genetic novelties.

7.8 Generating hopeful monsters

> However Darwinians may protest (and they do), millions of generations of inertia scarcely fit his image of life as poised for an instant response to any challenge. The argument between supporters of evolution as unhurried Victorian progress and those who hold the modern view of history as boredom mitigated by panic is unresolved.
>
> (Jones, 1999: 294)

7.8.1 A rapidly expanding body of data guided by phylogenies

The examples of teratology discussed in Section 7.5 emphasise the value of the distinction between the general concept of a sudden morphological shift between parent and offspring (of whatever cause) inherent in terata and the narrower concept of a heritable shift reflecting a mutation or epimutation that defines a prospecies ('hopeful monster'). Our 1994 discussion of the generation of hopeful monsters centred on homeotic (notably Hox and MADS-box) genes and their relatives, and a survey of the current volume will demonstrate that this remains the focus of evolutionary–developmental genetics. Nonetheless, the range of genes receiving detailed attention is progressively broadening.

At that time the empirical evidence for the genetic underpinnings of such transitions lay in organisms that were undesirably morphologically simple. Studies of higher plants primarily concerned *Arabidopsis* and *Antirrhinum*, with floral morphogenesis perhaps better understood than vegetative morphogenesis (cf. Poethig, 1990; Coen and Meyerowitz, 1991). Although thale-cress and snapdragon flowers remain at the forefront of plant developmental genetics, much progress has been made on all fronts, and the range of model organisms is now rapidly diversifying. In addition, the crucial choice of such organisms is being carefully guided by increasingly robust phylogenies (e.g. Angiosperm Phylogeny Group, 1998; Pryer *et al.*, 2000). Perhaps the greatest remaining gap in sampling of extant green plants lies in the vast panoply of cryptogamic plants. Studies in progress of model moss *Physcomitrella* and model fern *Ceratopteris* will no doubt yield many new insights (Cronk, 2000), though a study of a long-extinct Devonian rhyniophyte such as *Aglaophyton*, with its near-isomorphic sporophyte and gametophyte generations, would have provided a fascinating exploration of the importance of buffering of gene expression in diploid relative to haploid (Kenrick, 1994; Bateman, 1996).

We do not deny that the view, long-held by virtually all evolutionary biologists, that almost all hopeful monsters are in practice hopeless is correct; such mutations are either physiologically lethal or structurally highly dysfunctional. However, it is also clear from case-studies such as the orchid terata described above (Rudall and Bateman, 2002) that the production rate of profound morphological mutations capable of at least transient existence in natural habitats is extremely high. Rare successful establishment of such mutants is arguably sufficient to explain much of evolution, just as rare fixation of mutant alleles in neoDarwinian populations arguably suffices. Moreover, it is also clear that indistinguishable morphological variants can be generated frequently and repeatedly within a single 'parental' species. Many different mechanisms of mutation and epimutation are now recognised. Also, the decreasing number of key developmental genes now thought to exist in plant genomes as single copies offers more opportunities for divergence of function (e.g. Cronk, 2002). Thus, there are no guarantees that these multiple origins of similar morphs are genuinely homologous at the molecular level; indeed, it seems likely that there are multiple routes to engendering most profound morphological transitions (cf. Hawkins, 2002). Such causal diversity can only enhance the probability of successful establishment of the prospecies.

Nonetheless, constraints on potential saltational transitions are also evident. Some of the most interesting potential saltations that occur frequently in the land-plant phylogeny involve radical changes in life history or ecological preference, such as terrestrial to aquatic existence (e.g. Bateman, 1996), homospory to heterospory (Bateman and DiMichele, 1994b), autotrophy to mycoheterotrophy (e.g. Nickrent *et al.*, 1998) and allogamy to autogamy (e.g. Hollingsworth *et al.*, subm.). We have not encountered in the phylogenetic literature any unequivocal examples of successful reversals of these transitions. It seems likely that, after a very brief period of time, the genetic framework necessary for producing terrestrial adaptations, or hermaphroditic gametophytes, or photosynthesis, or effective incompatibility mechanisms, or reliable relationships with specific pollinators, cannot be resurrected. Indeed, in the case of mycoheterotrophs there is circumstantial evidence of increased mutation rates in physiologically expressed genes *prior* to the transition to mycoheterotrophy, suggesting that the transition may constitute the only remaining evolutionary option for the survival of the affected lineage rather than reflecting the finessed craftsmanship inherent in increased fitness through directional or disruptive selection.

This conclusion offers an insight into the concept of contingency promoted by Gould (1989), who argued that replaying a specific evolutionary radiation, even under similar conditions, would yield dissimilar products. Here, we have demonstrated the iterative origination of very similar floral morphs within single orchid species, and of multiple lineages of mycoheterotrophic orchids. Thus, the real contingency may in fact lie not in the origin of the novel morphology but in the environment in which it originates; it is the post-origination fate of each mutant that is decidedly unpredictable.

7.8.2 Loss of features

One can view the history of plant evolution as one of accumulating an optimal repertoire of phenotypic features underpinned by genetic structures sufficiently rigid to preserve advantageous elements of the phenotype but sufficiently flexible to allow change. The diversity of features and underlying genomes strongly influence both flexibility and complexity.

We have already noted the distinction between the ecological processes that operated during the initial period of land-plant colonisation, when the focus lay in previously unoccupied habitats, and subsequent periods of expansion, when the primarily challenge was (and is) locating habitats temporarily vacated by their incumbents. Interestingly, a similar distinction can be drawn between the gradual accumulation of increasingly complex and sophisticated morphological features during the Palaeozoic, and the subsequent challenges of altering form in the face of increased canalisation within lineages and decreased probability of ecological establishment of the products of macromutation. In such circumstances, re-organising and in many cases eliminating pre-existing phenotypic elements is more likely to lead to radical new body plans (cf. Bateman, 1996; Teotónio and Rose, 2001; Kellogg, 2002). As we noted previously (Bateman and DiMichele, 1994a: 83),

> Saltation breaks canalization, toppling the hopeful monster from the adaptive optimum of its parents but also freeing the potential lineage for radical reorganisation of form ... Superficially, such character losses appear improbable agents for innovative evolutionary change. However, by breaking canalisation and simplifying development, they clear the evolutionary palette for future adaptive innovation. The 'developmental ratchet' (Vermeij, 1987; Levinton, 1988) is reset at a lower level, leaving a combination of adaptation and contingency to define a new evolutionary trajectory for the lineage should it survive the establishment bottleneck.

This statement was followed by several empirical examples of simplification of form revealed by phylogenetic analyses, to which could be added the above discussions of simplification associated with mycoheterotrophy. Some of these examples included discussion of vestigial structures, such as secondary thickening in the aquatic pteridophyte *Isoetes*, that presumably represent unbreakable canalisation. However, as noted by Cronk (2001), the *absence* of a vestigial structure in a putative descendant relative to an ancestor is even more indicative of non-adaptive evolution, since it is more convincingly explained by a sudden catastrophic loss, forcing transfer of function (if any), rather than by gradual diminution in size through micromutation following loss of function. Many examples of changes in number, position, appearance and degrees of fusion among floral organs in angiosperms, such as those outlined for orchids by Rudall and Bateman (2002) and for legumes by Tucker (2000, 2001), are far more parsimoniously explained by outright suppression or duplication of structures through specific mutation or epimutation events small in genetic scale but large in phenotypic effect. This inevitably leads to reconsideration of the potential significance of epigenesis, epistasis, pleiotropy, heterotopy and heterochrony relative to the underlying macromutation(s) – subjects that merit greater discussion than is feasible in this chapter.

7.8.3 Recent advances in core evo-devo data

Which genes are crucial to saltation, how are they modified and how are they expressed? In 1994, no single plant chromosome had been wholly sequenced. At the time of writing (summer 2001), the *Arabidopsis* genome has been sequenced in its entirety (Arabidopsis Genome Initiative, 2000; Walbot, 2000) and several grass genera will soon follow (Adam, 2000). How will these advances aid evolutionary understanding? (cf. Singh and Krimbas, 2000).

Arabidopsis, chosen for its small, compact overall nuclear genome and small number of chromosomes ($n = 5$), has proved to harbour about 120 m bases that include 26,000 genes of, at most, 15,000 different kinds, thus resembling in number *Drosophila* and *Caenorhabditis* (Walbot, 2000) and approaching that of *Homo*. The remarkable differences must therefore lie in the precise nature of those genes (though it is clear that the origins of many of the more important genes preceded the divergence of plants, animals and fungi: Meyerowitz and Somerville, 1994) and their mode of expression. Not surprisingly, the sedentary lifestyle of plants is reflected in their genetic complement; a much higher proportion of plant genes influence cell-wall formation, water and hormonal transport in plants, whereas animals specialise in signal transduction (Walbot, 2000). Plants contain an apparent superfluity of some transcription factors; for example, *Arabidopsis* maintains an order of magnitude more of the still enigmatic MYB transcription factors than does *Homo* (Cronk, 2001). On average, five times as many plant nuclear genes communicate with organellar genomes in plants than in animals or fungi (perhaps not surprising, given that plants uniquely possess plastids), and the much wider range of secondary compounds suggests much more complex biosynthetic pathways (Walbot, 2000) that may be more readily switched among closely related species (Bateman, 1999a).

In terms of pinpointing the genes that underpin evolution, a fascinating cautionary tale was summarised by zoologists Marshall *et al.* (1999), who noted that the high-level Hox gene *Ubx* is able to regulate the fundamental distinction between two-winged and four-winged insects. However, comparison of a two-winged fly with a four-winged butterfly pointed the finger at one or more genes located downstream from the apparently highly conserved *Ubx*. They further noted the demonstration by Averof and Patel (1997) that striking differences in body plan among the major groups of crustaceans primarily reflect the precise location of the initial embryonic expression boundary of *Ubx* and *abd-A*. Also, some intermediate morphologies show reduced levels or mosaic patterns of Hox gene expression, suggesting that morphogenesis may be gradational and reflect gradual accumulation of mutations that modify the degree and location of Hox gene expression. Moreover, during insect evolution, first *Ubx* and then *abd-A* were co-opted to repress limb formation (Palopoli and Patel, 1998).

Many of the contributions to this volume discuss interactions among low copy-number genes of large phenotypic effect. Originally termed 'single-copy' genes, this categorisation has been revised to 'low copy number' genes on the basis of more recent research. Gene duplication, both *en masse* through polyploidy (Walbot, 2000; Wolfe, 2001) and individually, may immediately provide a larger amount of transcript or be followed by a 'gain-of function' mutation. The gain-of-function mutation must occur sufficiently rapidly to prevent loss-of-function mutations from generating a non-expressed pseudogene. The result can be different forms of the same functional protein with subtly different properties, increasing the flexibility of response of the 'host' organism to the environment and potentially resulting in fixed heterozygosity (Cronk, 2001). One great potential benefit of the genomics revolution in general, and comparative whole-genome sequencing in particular, is stronger tests of orthology versus paralogy for these powerful classes of gene.

The most popular examples of interactions among such genes are the interaction of the MADS-IKC genes that lead to the ABC model of floral organogenesis. Comparison of early (e.g. Coen and Meyerowitz, 1991) and more recent (e.g. Kramer and Irish, 1999, 2000; Cubas, 2002; Frohlich, 2002; Gillies *et al.*, 2002; Glover and Martin, 2002; Theißen *et al.*, 2002) accounts of this model show elaborations of the original 'three primary factors generate four whorls' model but no major refutations of the cornerstones of the original hypothesis. The *CYC* and *DICH* genes that confer zygomorphy on many angiosperm flowers (see discussion of Orchidaceae in Section 7.5.2) apparently represent a relatively recent gene duplication event among TCP/R transcription factors (Cubas *et al.*, 1999; Luo *et al.*, 1999; Cronk, 2001, 2002; Cubas, 2002; Gillies *et al.*, 2002), with *DICH* having a similar function to, but accentuating, *CYC*; moreover, an additional duplication of this gene family has been reported in Gesneriaceae (Citerne *et al.*, 2000). In the vegetative realm, the key meristem-control *KNOTTED*-like genes are divisible into two distinct classes in gene trees (Bharathan *et al.*, 1999), reflecting a pre-bryophyte duplication and suggesting a range of possible interactions between duplicates. Meristematic control by *KNOX*-like genes is down-regulated by *PHAN*-like genes, which are in turn negatively regulated by the *STM*-like genes that ensure that *PHAN* expression is restricted to developing leaves (cf. Harrison *et al.*, 2002; Langdale *et al.*, 2002; McLellan *et al.*, 2002; Tsiantis *et al.*, 2002). And in the antithesis of gene duplication and mutual interaction, the apparent loss of function of one of the two copies of *LFY* during the transition from gymnosperms to angiosperms may have resulted from the transfer of the 'female' function to the still-functional 'male' copy of *LFY*, prompting the origin of the hermaphroditic organisation that characterises most angiosperm flowers (Frohlich and Parker, 2000; Frohlich, 2001, 2002).

Much interesting debate currently surrounds the most likely properties of the genes involved in those morphogenetic cascades that are most pivotal to evolution. Current evidence, though sparse, indicates that genes of a relatively narrow range of expression are less likely to be strongly conserved (and thus evolutionarily recalcitrant) than are those with strongly pleiotropic expression across the bauplan (e.g. Doebley *et al.*, 1997; Doebley and Lukens, 1998) or those that provide the focal point of several biosynthetic pathways (Cronk, 2001). The more focused effects of regulatory genes in general, and transcription factors in particular, make them more probable drivers of evolution, with the emphasis currently on mutations in *cis*-regulatory regions (Marshall *et al.*, 1999; Wang *et al.*, 1999; Cronk, 2001). This view partly reflects the observation that the origin of teosinte from maize involved a selective sweep (rapid incorporation of favourable mutations allowing simultaneous fixation of linked genes irrespective of their contribution to fitness) of the flanking regions of the maize allele of the key developmental gene *TB1*, demonstrating intense selection of the *cis*-regulatory regions but no corresponding selection of the coding region of the gene (Wang *et al.*, 1999). Mutations in control regions that subtly influence spatio-temporal expression patterns are strongly implicated in gain-of-function mutations. One potentially important distinction surprisingly rarely discussed is whether the optimal conditions for success for a typical 'gain-of-function' mutation are substantially different from those needed for a successful 'loss-of-function' mutation.

The relatively few well-founded studies of developmental gene interactions can only have revealed the tip of the iceberg; the gene expression and regulation scenarios surrounding specific morphogenetic transitions will undoubtedly appear more complex with increased knowledge. These are likely to reinforce the distinction made by McLellan *et al.* (2002) between primary genes, whose mutation generates novel morphs, and secondary genes, whose expression patterns are modified as a result of mutations in upstream primary genes. In this context, none of the above discussion covers the genes of modest phenotypic effect that constitute the bulk of the genes in any particular QTL survey, yet in these genes resides the potential for subtle speciation *without* the simultaneous origination of higher taxa (e.g. the modes of speciation that, for example, allow co-evolutionary switching of partners among closely related orchids). We suspect that these phenotypic transitions – essentially quantitative rather than qualitative in nature – will prove more recalcitrant to causal interpretation than the larger-scale morphological transitions that are the raw materials of saltational evolution. Here, epistatic interactions among genes, often generating multiple component 'supergenes', may prove to have considerable explanatory power (papers in Wolf *et al.*, 2000), especially if considered in the context of the epigenetic environment of expression (Jablonka *et al.*, 1992; Jablonka, 1994; Jablonka and Lamb, 1995).

7.9 Integrating the developmental genetics of plants into mainstream evolutionary syntheses

> I have now recapitulated the chief facts and considerations which have thoroughly convinced me that species have been modified, during a long course of descent, by the preservation or the natural selection of successive slight favourable variation ... Why, it may be asked, have all the most eminent living naturalists and geologists rejected this view of the mutability of species? ...
>
> A few naturalists, endowed with much flexibility of mind, and who have already begun to doubt on the immutability of species, may be influenced by this volume; but I look with confidence to the future, to young and rising naturalists, who will be able to view both sides of the question with impartiality. Whoever is led to believe that species are mutable will do good service by conscientiously expressing his conviction; for only thus can the load of prejudice by which this subject is overwhelmed be removed.
>
> (Darwin, 1859: 480–482; rewritten by Jones, 1999: 459–460)

The above observations lead us to the nub of the (arguably artificial) divide between macromutation and micromutation, namely whether evolutionary transitions reflect modification of many genes of small effect, consistent with neoDarwinism, or modification of very few genes of profound phenotypic effect, consistent with saltation? After reviewing current zoological data, Shubin and Marshall (2000: 324) concluded that 'it appears that relatively few genetic changes may be responsible for most of the phenotypic differences among species'; thus, initial results appear favourable to saltation theory, but may require reinterpretation.

For example, Marshall *et al.* (1999) reported the work of Doebley and Wang (1997) which demonstrated that the now well-known transition from teosinte to maize originally pursued by Iltis (1983) involved a very small number of genes and was dominated by a single homeotic gene, *TB1* (cf. Kellogg, 2002). They further noted that recent quantitative trait locus (QTL) studies (e.g. Orr, 1998; Goodnight, 2000) among closely related species suggest that major morphological features tend to be controlled by a modest but nonetheless significant number of genes (e.g. the eighteen to seventy-four suggested for leaf shape control, as reviewed by McLellan *et al.*, 2002), results that are superficially intermediate between the predictions of neo-Darwinian (many more) and saltation theory (even fewer). However, this raw figure disguises an exponential distribution of numbers of genes versus degree of pheno-typic effect, with very few genes of large phenotypic effect (indeed, they are of far larger effect than was predicted as viable by Fisher, 1930) and much greater numbers of genes of lesser phenotypic effect (Orr, 1998; McLellan et al., 2002). Marshall et al. validly used these observations to criticise the Fisherian concept of evolution being confined to panmictic populations carrying many mutations of small phenotypic effect (Fisher, 1930), and reiterated arguments advanced by Kimura (1983, 1991) that each of the many small mutations is extremely vulnerable to random loss from the population while still very rare (this is, for example, evident from frequent loss of highly adaptive self-incompatibility alleles from angiosperm lineages).

However, Marshall et al. then chose to set their insightful observations in the context of neoDarwinian adaptation, in turn criticising Kimura's model on the grounds that it assumed a single step to a new adaptive optimum following a pre-sumed environmental change, rather than a more complex 'adaptive walk' consis-tent with neoDarwinian models.

There are models that are neither strictly neoDarwinian nor strictly saltational that are consistent with the above observations, falling within our broad category of parasaltation (see Section 7.2.2). The most relevant relate to unusually rapid spread of novel alleles through populations that are small, morphologically and genetically atypical of the 'parental' species, and often allopatric or parapatric: these theories include drift (Barrett and Pannell, 1999), shifting balance (Wright, 1932, 1968; Waddington, 1957; Whitlock et al., 1995; Mallet and Joron, 1999), and crash–flush–founder (Carson and Templeton, 1984; Carson, 1985; Slatkin, 1996), each capable of reforming adaptive gene complexes.

However, many of the data gradually being revealed by both QTL and candidate gene approaches to evo-devo (e.g. Baum, 2002) are most parsimoniously explained by the saltation model of major evolutionary transitions in phenotype. The initial qualitative transition is prompted by mutations that modify or suppress the expression of one or at most very few genes of profound phenotypic effect, conferring instant iso-lation from its parents; thus, saltation is essentially the sympatric equivalent of the aforementioned parasaltational mechanisms. Then, the resulting radically altered phe-notype is gradually honed by natural selection to a phenotype that is subtly better adapted for its environment, resulting in modest changes of expression in a much greater (though still modest) number of genes of far less profound, quantitative influ-ence on phenotype (note that their subtle effects are likely to escape the eyes of both field biologists and experimental geneticists scrutinising mutant screens).

We therefore conclude that saltation theory offers the best explanation for several key issues that have long perplexed evolutionary biologists. Most notably, saltation:

1   removes the troublesome mathematical challenge inherent in natural selection of driving an initially very rare mutant allele to fixation in a large panmictic popu-lation, by causing instant isolation from the 'parental' population (cf. Bateman, 1999a; Marshall et al., 1999; Shubin and Marshall, 2000);

2   thereby provides a simple and credible explanation for the equally mathematically improbable phenomenon of true sympatric speciation without introgression;

3   permits a lineage to cross valleys in the fitness landscape that would be lethal to a lineage attempting to make the evolutionary journey gradually, thus immediately reaching at least the higher slopes of another peak and giving natural selection a subordinate role of subsequently assisting that novel phenotype towards the crest of its new-found fitness peak (Bateman, 1999a; Cronk, 2001); in short, saltation is a pre-requisite for radical re-organisation of form;

4   accommodates a genuine qualitative difference in degrees of morphological divergence among phylogenetically delimited taxa of contrasting taxonomic ranks (cf. Kellogg, 2002); some newly originated species are trivially morphologically distinct and hence are readily assigned to pre-existing genera, whereas other species originate with sufficiently profound morphological transitions to simultaneously represent previously unknown higher taxa.

The overall contrast between neoDarwinism and supraDarwinian saltation theory (together with the more heterogeneous panoply of parasaltational mechanisms only trivially explored in this chapter but of considerable explanatory power: Table 7.2) is to place greater evolutionary (and systematic) emphasis on the origination of a genetically determined phenotypic novelty and less on its subsequent dissemination through populations; rather, the genetic–phenotypic novelty is seen as constituting an equally novel population. The fate of this prospecies is determined at least in part by the likelihood that the phenotypic novelty will offer one or more serendipitous features that function as pre-adaptations or exaptations (or allow transfer of an essential function) in the environment initially occupied by the novel organism(s).

Whether this changed emphasis constitutes a genuine paradigm shift (Kuhn, 1962), as we predicted in 1994, remains to be seen. For the present, we will simply echo Darwin by appealing for greater 'flexibility of mind' in the light of these fascinating recent developments, not least among the more evangelistic of Darwin's modern disciples.

ACKNOWLEDGEMENTS

We thank the late Derek Turner Ettlinger for drawing our attention to the peloric *Ophrys insectifera*, Gar Rothwell and the late Albert Long for helpful discussions on *Pullaritheca* cupules, and the Hancock Museum, Newcastle for long-term loan of related specimens. Peter Hollingsworth generously granted permission to use unpublished data on the molecular phylogenetics of the *Gymnadenia* alliance. Michael Frohlich, Julie Hawkins and Paula Rudall kindly provided useful comments on the manuscript.

REFERENCES

Abbott, R. J. (1992) Plant invasions, interspecific hybridisation, and the evolution of new plant taxa. *Trends in Ecology and Evolution*, 7, 401–405.

Adam, D. (2000) Now for the hard ones. *Nature*, 408, 792–793.

Alberch, P., Gould, S. J., Oster, G. F. and Wake, D. B. (1979) Size and shape in ontogeny and phylogeny. *Paleobiology*, 5, 296–317.

Albert, V. A., Gustafsson, M. H. G. and Di Laurenzio, L. (1998) Ontogenetic systematics, molecular developmental genetics, and the angiosperm petal, in *Molecular Systematics of Plants 2* (eds D. E. Soltis, P. S. Soltis and J. J. Doyle), Chapman & Hall, London, pp. 349–374.

Angiosperm Phylogeny Group (1998) An ordinal classification for the families of flowering plants. *Annals of the Missouri Botanical Garden*, 85, 531–553.

Arabidopsis Genome Initiative (2000) Analysis of the genome sequence of the flowering plant *Arabidopsis thaliana*. *Nature*, 408, 796–815.

Arthur, W. (1984) *Mechanisms of Morphological Evolution*. Wiley, New York.

Arthur, W. (2000) The concept of developmental reprogramming and the quest for an inclusive theory of evolutionary mechanisms. *Evolution and Development*, 2, 49–57.

Averof, M. and Patel, N. H. (1997) Crustacean appendage evolution associated with changes in Hox gene expression. *Nature*, 388, 682–686.

Avise, J. C. (1994) *Molecular Markers, Natural History, and Evolution*. Chapman and Hall, London.

Ayala, F. J. and Valentine, J. W. (1979) *Evolving: The Theory and Processes of Organic Evolution*. Benjamin–Cummings, Menlo Park, CA.

Baldwin, B. G., Crawford, D. J., Francisco-Ortega, J., Kim, S.-C., Sang, T. and Steussy, T. F. (1998) Molecular phylogenetic insights on the origin and evolution of oceanic island plants, in *Molecular Systematics of Plants 2* (eds D. E. Soltis, P. S. Soltis and J. J. Doyle), Chapman & Hall, London, pp. 410–441.

Barrett, S. C. H. and Pannell, J. R. (1999) Metapopulation dynamics and mating-system evolution in plants, in *Molecular Systematics and Plant Evolution* (eds P. M. Hollingsworth, R. M. Bateman and R. J. Gornall), Taylor & Francis, London, pp. 74–100.

Bateman, R. M. (1985) Peloria and pseudopeloria in British orchids. *Watsonia*, 15, 357–359.

Bateman, R. M. (1994) Evolutionary–developmental change in the growth architecture of fossil rhizomorphic lycopsids: scenarios constructed on cladistic foundations. *Biological Reviews*, 69, 527–597.

Bateman, R. M. (1996) Non-floral homoplasy and evolutionary scenarios in living and fossil land-plants, in *Homoplasy and the Evolutionary Process* (eds M. J. Sanderson and L. Hufford), Academic Press, London, pp. 91–130.

Bateman, R. M. (1999a) Integrating molecular and morphological evidence for evolutionary radiations, in *Molecular Systematics and Plant Evolution* (eds P. M. Hollingsworth, R. M. Bateman and R. J. Gornall), Taylor & Francis, London, pp. 422–471.

Bateman, R. M. (1999b) Architectural radiations cannot be optimally interpreted without morphological and molecular phylogenies, in *The Evolution of Plant Architecture* (eds M. H. Kurmann and A. R. Hemsley), Royal Botanic Gardens Kew, pp. 221–250.

Bateman, R. M. (2001) Evolution and classification of European orchids: insights from molecular and morphological characters. *Journal Europäischer Orchideen* 33: 33–119.

Bateman, R. M., Crane, P. R., DiMichele, W. A., Kenrick, P., Rowe, N. P., Speck, T. and Stein, W. E. (1998) Early evolution of land plants: phylogeny, physiology, and ecology of the primary terrestrial radiation. *Annual Review of Ecology and Systematics*, 29, 263–292.

Bateman, R. M. and DiMichele, W. A. (1994a) Saltational evolution of form in vascular plants: a neoGoldschmidtian synthesis, in *Shape and Form in Plants and Fungi* (eds D. S. Ingram and A. Hudson), Academic Press, London, pp. 63–102.

Bateman, R. M. and DiMichele, W. A. (1994b) Heterospory: the most iterative key innova-
    tion in the evolutionary history of the plant kingdom. *Biological Reviews*, 69, 345–417.
Bateman, R. M., DiMichele, W. A. and Willard, D. A. (1992) Experimental cladistic analyses of
    anatomically preserved arborescent lycopsids from the Carboniferous of Euramerica: an essay
    in paleobotanical phylogenetics. *Annals of the Missouri Botanical Garden*, 79, 500–559.
Bateman, R. M., Hollingsworth, P. M., Preston, J., Luo, Y.-B., Pridgeon, A. M. and Chase,
    M. W. (subm.) Molecular phylogenetics of the Orchidinae and selected 'Habenariinae'
    (Orchidaceae). *Botanical Journal of the Linnean Society*.
Bateman, R. M., Pridgeon, A. M. and Chase, M. W. (1997) Phylogenetics of subtribe Orchidinae
    (Orchidoideae, Orchidaceae) based on nuclear ITS sequences. 2. Infrageneric relationships and
    taxonomic revision to achieve monophyly of *Orchis sensu stricto*. *Lindleyana*, 12, 113–141.
Bateman, R. M. and Rothwell, G. W. (1990) A reappraisal of the Dinantian floras at Oxroad
    Bay, East Lothian, Scotland. 1. Floristics and the development of whole-plant concepts.
    Transactions of the Royal Society of Edinburgh B, 81, 127–159.
Baum, D. A. (2002) Identifying the genetic causes of phenotypic evolution: a review of experi-
    mental strategies, in *Developmental Genetics and Plant Evolution* (eds Q. C. B. Cronk, R.
    M. Bateman and J. A. Hawkins), Taylor & Francis, London, pp. 493–507.
Baum, D. A. and Donoghue, M. J. (2002) Transference of function, heterotopy and the evolu-
    tion of plant development, in *Developmental Genetics and Plant Evolution* (eds Q. C. B.
    Cronk, R. M. Bateman and J. A. Hawkins), Taylor & Francis, London, pp. 52–69.
Behrensmeyer, A. K., Damuth, J. D., DiMichele, W. A., Potts, R., Sues, H.-D. and Wing, S. L.
    (1992) *Terrestrial Ecosystems Through Time*. Chicago University Press.
Bharathan, G., Janssen, B. J., Kellogg, E. A. and Sinha, H. (1999) Phylogenetic relationships
    and evolution of the *KNOTTED* class of plant homeodomain proteins. *Molecular Biology
    and Evolution*, 16, 553–563.
Bower, C. C. (2001) *Thelymitra*: pollination, in *Genera Orchidacearum 2: Orchidioideae Part
    1* (eds A. M. Pridgeon, P. J. Cribb, M. C. Chase and F. N. Rasmussen), Oxford University
    Press, Oxford, pp. 208–213.
Brown, R. W. (1956) *Composition of Scientific Words*. Smithsonian Institution Press, Wash-
    ington, DC.
Carson, H. L. (1985) Unification of speciation theory in plants and animals. *Systematic
    Botany*, 10, 380–390.
Carson, H. L. and Templeton, A. R. (1984) Genetic revolutions in relation to speciation phenom-
    ena: the founding of new populations. *Annual Review of Ecology and Systematics*, 15, 97–131.
Chasan, R. (1993) Meeting report: evolving developments [Taos]. *Plant Cell*, 5, 363–369.
Chase, M. W. and Albert, V. A. (1998) A perspective on the contribution of plastid *rbcL*
    sequences to angiosperm phylogenetics, in *Molecular Systematics of Plants 2* (eds D. E.
    Soltis, P. S. Soltis and J. J. Doyle), Chapman & Hall, London, pp. 488–507.
Chase, M. W., Fay, M. F. and Savolainen, V. (2000) Higher-level classification in the
    angiosperms: new insights from the perspective of DNA sequence data. *Taxon*, 49, 685–704.
Chaw, S.-M., Parkinson, C. L., Cheng, Y., Vincent, T. M. and Palmer, J. D. (2000) Seed-plant
    phylogeny inferred from all three plant genomes: monophyly of extant gymnosperms and
    origins of Gnetales from conifers. *Proceedings of the National Academy of Sciences of the
    USA*, 97, 4086–4091.
Chen, S.-C. (1982) The origin and early differentiation of the Orchidaceae. *Acta Phytotaxo-
    nomica Sinica*, 20, 1–20.
Citerne, H. L., Möller, M. and Cronk, Q. C. B. (2000) Diversity of *cycloidea*-like genes in
    Gesneriaceae in relation to floral symmetry. *Annals of Botany*, 86, 167–176.
Clements, M. A. (2001) Diurideae: phylogeny, in *Genera Orchidacearum 2. Orchidoideae,
    Part 1* (eds A. M. Pridgeon, P. L. Cribb, M. W. Chase and F. N. Rasmussen), Oxford
    University Press, pp. 61–63.

Coddington, J. A. (1988) Cladistic tests of adaptational hypotheses. *Cladistics*, 4, 3–22.

Coen, E. S. (1991) The role of homeotic genes in flower development and evolution. *Annual Review of Plant Physiology and Plant Molecular Biology*, 42, 241–279.

Coen, E. S. (1999) *The Art of the Genes: How Organisms Make Themselves*. Oxford University Press, Oxford.

Coen, E. and Carpenter, R. (1992) The power behind the flower. *New Scientist*, 134, 24–27.

Coen, E. and Meyerowitz, E. M. (1991) The war of the whorls: genetic interactions controlling flower development. *Nature*, 353, 31–37.

Coyne, J. A., Barton, N. H. and Turelli, M. (1997) A critique of Sewall Wright's Shifting Balance theory of evolution. *Evolution*, 51, 643–671.

Crane, P. R. (1985) Phylogenetic analysis of the seed plants and the origin of angiosperms. *Annals of the Missouri Botanical Garden*, 72, 716–793.

Crane, P. R. and Kenrick, P. (1997) Diverted development of reproductive organs: a source of morphological innovation in land plants. *Plant Systematics and Evolution*, 206, 161–174.

Croizat, L. (1962) *Space, Time, Form: The Biological Synthesis*. Author's publication, Caracas.

Cronk, Q. C. B. (2001) Plant evolution and development in a post-genomic context. *Nature Reviews, Genetics*, 2, 607–619.

Cronk, Q. C. B. (2002) Perspectives and paradigms in plant evo-devo, in *Developmental Genetics and Plant Evolution* (eds Q. C. B. Cronk, R. M. Bateman and J. A. Hawkins), Taylor & Francis, London, pp. 1–14.

Cubas, P. (2002) Role of TCP genes in the evolution of morphological characters in angiosperms, in *Developmental Genetics and Plant Evolution* (eds Q. C. B. Cronk, R. M. Bateman and J. A. Hawkins), Taylor & Francis, London, pp. 247–266.

Cubas, P., Coen, E. and Martinez-Zapater, J. M. (2001) Ancient asymmetries in the evolution of flowers. *Current Biology*, 11, 1050–1052.

Cubas, P., Vincent, C. and Coen, E. (1999) An epigenetic mutation responsible for natural variation in floral symmetry. *Nature*, 401, 157–161.

Darley, W. M. (1990) The essence of 'plantness'. *American Biology Teacher*, 52, 354–357.

Darwin, C. (1859) *The Origin of Species by Means of Natural Selection*. Murray, London.

Dawkins, R. (1982) *The Extended Phenotype: The Gene as a Unit of Selection*. Freeman, San Francisco.

Dawkins, R. (1986) *The Blind Watchmaker*. Longman, Harlow.

Dawkins, R. (1989) *The Selfish Gene* (2nd edn). Oxford University Press, Oxford.

Dietrich, M. R. (1992) Macromutation, in *Keywords in Evolutionary Biology* (eds E. F. Keller and E. A. Lloyd), Belknap Press, Harvard, pp. 194–201.

DiMichele, W. A. (subm.) Landscape position and the evolution of innovation in vascular plants. *Paleobiology*.

DiMichele, W. A. and Aronson, R. B. (1992) The Pennsylvanian–Permian vegetational transition: a terrestrial analogue to the onshore–offshore hypothesis. *Evolution*, 46, 807–824.

DiMichele, W. A. and Bateman, R. M. (1996) Plant paleoecology and evolutionary inference: two examples from the Paleozoic. *Review of Palaeobotany and Palynology*, 90, 223–247.

DiMichele, W. A., Stein, W. E. and Bateman, R. M. (2001) Ecological sorting of vascular plant classes during the Paleozoic evolutionary radiations, in *Evolutionary Paleoecology: The Ecological Context of Macroevolutionary Change* (eds W. D. Allmon and D. J. Bottjer), Columbia University Press, NY, pp. 285–335.

Doebley, J. and Lukens, L. (1998) Transcriptional regulators and the evolution of plant form. *Plant Cell*, 10, 1075–1082.

Doebley, J., Stec, A. and Hubbard, L. (1997) The evolution of apical dominance in maize. *Nature*, 386, 485–488.

Doebley, J. and Wang, R. L. (1997) Genetics and the evolution of plant form: an example from maize. *Cold Spring Harbor Symposia on Quantitative Biology*, 62, 361–367.

Donoghue, M. J. (1989) Phylogenies and the analysis of evolutionary sequences, with examples from seed plants. *Evolution*, 43, 1137–1156.

Donovan, S. K. and Paul, C. R. C. (eds) (1998) *The Adequacy of the Fossil Record*. Wiley, New York.

Douzery, E. J. P., Pridgeon, A. M., Kores, P., Linder, H. P., Kurzweil, H. and Chase, M. W. (1999) Molecular phylogenetics of disease (Orchidaceae): a contribution from nuclear ribosomal ITS sequences. *American Journal of Botany*, 86, 887–899.

Dover, G. A. (1982) Molecular drive: a cohesive mode of species evolution. *Nature*, 299, 111–117.

Dover, G. A. (2000) How genomic and developmental dynamics affect evolutionary processes. *BioEssays*, 22, 1153–1159.

Doyle, J. A. and Donoghue, M. J. (1986) Seed plant phylogeny and the origin of angiosperms: an experimental cladistic approach. *Botanical Review*, 52, 321–431.

Eldredge, N. (1989) *Macroevolutionary Dynamics: Species, Niches, and Adaptive Peaks*. MacGraw-Hill, New York.

Eldredge, N. (ed.) (1992) *Systematics, Ecology, and the Biodiversity Crisis*. Columbia University Press, New York.

Eldredge, N. and Gould, S. J. (1972) Punctuated equilibria: an alternative to phyletic gradualism, in *Models in Paleobiology* (ed. T. J. M. Schopf), Freeman, San Francisco, pp. 82–115.

Ellstrand, N. C., Prentice, H. C. and Hancock, J. F. (1999) Gene flow and introgression from domesticated plants into their wild relatives. *Annual Review of Ecology and Systematics*, 30, 539–563.

Endress, P. K. (1999) Symmetry in flowers: diversity and evolution. *International Journal of Plant Science*, 160 (6 suppl.), S3–S23.

Endress, P. K. (2001) Evolution of floral symmetry. *Current Opinion in Plant Biology*, 4, 86–91.

Ernst, R. and Arditti, J. (1994) Resupination, in *Orchid Biology: Reviews and Perspectives, VII* (ed. J. Arditti), Wiley, New York, pp. 135–188.

Erwin, D. H. (2000) Macroevolution is more than repeated rounds of microevolution. *Evolution and Development*, 2, 78–84.

Federoff, N. V. (2000) Transposons and genome evolution in plants. *Proceedings of the National Academy of Sciences of the USA*, 97, 7002–7007.

Fisher, R. A. (1930) *The Genetical Theory of Natural Selection*. Oxford University Press, Oxford.

Foote, M. (1996) Models of morphological diversification, in *Evolutionary Paleobiology* (eds D. Jablonski, D. H. Erwin and J. H. Lipps), Chicago University Press, Chicago, pp. 62–86.

Foster, P. L. and Cairns, J. (1992) Mechanisms of directed mutation. *Genetics*, 131, 783–789.

Frohlich, M. W. (2001) A detailed scenario and possible tests of the Mostly Male theory of flower origins, in *Beyond Heterochrony: The Evolution of Development* (ed. M. Zelditch), Wiley, New York, pp. 59–104.

Frohlich, M. W. (2002) The Mostly Male theory of flower origins: summary and update regarding the Jurassic pteridosperm *Pteroma*, in *Developmental Genetics and Plant Evolution* (eds Q. C. B. Cronk, R. M. Bateman and J. A. Hawkins), Taylor & Francis, London, pp. 85–108.

Frohlich, M. W. and Parker, D. S. (2000) The mostly male theory of flower evolutionary origins. *Systematic Botany*, 25, 155–170.

Funk, V. A. and Brooks, D. R. (1990) *Phylogenetic Systematics as the Basis of Comparative Biology*. Smithsonian Institution Contributions to Botany 73, Washington, DC.

Gillespie, J. H. (1991) *The Causes of Molecular Evolution*. Oxford University Press, Oxford.

Gillies, A. C. M., Cubas, P., Coen, E. S. and Abbott, R. J. (2002) Making rays in the Asteraceae: genetics and evolution of radiate versus discoid flower heads, in *Developmental Genetics and Plant Evolution* (eds Q. C. B. Cronk, R. M. Bateman and J. A. Hawkins), Taylor & Francis, London, pp. 237–246.

Glover, B. J. and Martin, C. (2002) Evolution of adaptive petal cell morphology, in *Develop-

mental Genetics and Plant Evolution (eds Q. C. B. Cronk, R. M. Bateman and J. A. Hawkins), Taylor & Francis, London, pp. 160–172.

Goldschmidt, R. (1940) *The Material Basis of Evolution.* Yale University Press, New Haven.

Goodnight, C. J. (1995) Epistasis and the increase in additive genetic variance: implications for phase 1 of Wright's shifting balance process. *Evolution*, 49, 502–511.

Goodnight, C. J. (2000) Quantitative trait loci and gene interaction: the quantitative genetics of metapopulations. *Heredity*, 84, 587–598.

Goodwin, B. C. (1994) Homology, development and hierarchies, in *Homology: The Hierarchical Basis of Comparative Biology* (ed. B. K. Hall), Academic Press, London, pp. 229–247.

Goodwin, B. C. and Saunders, P. (eds) (1992) *Theoretical Biology: Epigenetic and Evolutionary Order From Complex Systems.* Johns Hopkins Press, Baltimore.

Gould, S. J. (1982) The uses of heresy: introduction to reprint of R. Goldschmidt, 1940, *The Material Basis of Evolution.* Yale University Press, New Haven, xiii–xlii.

Gould, S. J. (1986) Punctuated equilibrium at the third stage. *Systematic Zoology*, 35, 143–148.

Gould, S. J. (1989) *Wonderful Life: The Burgess Shale and the Nature of History.* Norton, New York.

Gould, S. J. and Eldredge, N. (1993) Punctuated equilibrium comes of age. *Nature*, 366, 223–227.

Gould, S. J. and Lewontin, R. C. (1979) The spandrels of San Marco and the Panglossian paradigm: a critique of the adaptationist program. *Proceedings of the Royal Society of London, B*, 205, 581–598.

Graham, S. W. and Olmstead, R. G. (2000) Utility of 17 chloroplast genes for inferring the phylogeny of the basal angiosperms. *American Journal of Botany*, 87, 1712–1730.

Harris, D. J., Poulsen, A. D., Frimodt-Moller, C., Preston, J. and Cronk, Q. C. B. (2000) Rapid radiation in *Aframomum* (Zingiberaceae): evidence from nuclear ribosomal DNA internal transcribed spacer (ITS) sequences. *Edinburgh Journal of Botany*, 57, 377–395.

Harris, T. M. (1964) *The Yorkshire Jurassic Flora. II. Caytoniales, Cycadales and Pteridosperms.* British Museum (Natural History), London.

Harrison, C. J., Cronk, Q. C. B. and Hudson, A. (2002) An overview of seed plant leaf evolution, in *Developmental Genetics and Plant Evolution* (eds Q. C. B. Cronk, R. M. Bateman and J. A. Hawkins), Taylor & Francis, London, pp. 395–403.

Harvey, P. H., Leigh Brown, A. J., Maynard Smith, J. and Nee, S. (eds) (1996) *New Uses for New Phylogenies.* Oxford University Press, Oxford.

Harvey, P. H. and Pagel, M. D. (1991) *The Comparative Method in Evolutionary Biology.* Oxford University Press, Oxford.

Hawkins, J. A. (2002) Evolutionary developmental biology: impact on systematic theory and practice, and the contribution of systematics, in *Developmental Genetics and Plant Evolution* (eds Q. C. B. Cronk, R. M. Bateman and J. A. Hawkins), Taylor & Francis, London, pp. 32–51.

Hedrén, M., Klein, E. and Teppner, H. (2000) Evolution of polyploids in the European orchid genus *Nigritella*: evidence from allozyme data. *Phyton*, 40, 239–275.

Helfgott, D. M., Francisco Ortega, J., Santos-Guerra, A., Jansen, R. K. and Simpson, B. B. (2000) Biogeography and breeding system evolution of the woody *Bencomia* alliance (Rosaceae) in Macaronesia based on ITS sequence data. *Systematic Botany*, 25, 82–97.

Hendy, M. D. and Penny, D. (1989) A framework for the quantitative study of evolutionary trees. *Systematic Zoology*, 38, 297–309.

Hershkovitz, M. A., Zimmer, E. A. and Hahn, W. J. (1999) Ribosomal DNA sequences and angiosperm systematics, in *Molecular Systematics and Plant Evolution* (eds P. M. Hollingsworth, R. M. Bateman and R. J. Gornall), Taylor & Francis, London, pp. 268–326.

Hoffman, D. C. and Prescott, D. M. (1997) Evolution of internal eliminated segments and scrambling in the micronuclear gene encoding DNA polymerase A in two *Oxytricha* species. *Nucleic Acids Research*, 25, 1883–1889.

Hollingsworth, P. M., Squirrell, J. and Bateman, R. M. (subm.) Taxonomic complexity and breeding system transitions: conservation genetics of the *Epipactis leptochila* complex (Orchidaceae: Neottieae). *Molecular Ecology*.

Horsman, F. (1990) Peloria in *Dactylorhiza*. *BSBI News*, 55, 16–18.

Iltis, H. H. (1983) From teosinte to maize: the catastrophic sexual transmutation. *Science*, 222, 886–894.

Jablonka, E. (1994) Inheritance systems and the evolution of new levels of individuality. *Journal of Theoretical Biology*, 170, 301–309.

Jablonka, E., Lachmann, M. and Lamb, M. J. (1992) Evidence, mechanisms and models for the inheritance of acquired characters. *Journal of Theoretical Biology*, 158, 245–268.

Jablonka, E. and Lamb, M. J. (1995) *Epigenetic Inheritance and Evolution*. Oxford University Press, Oxford.

Jackson, J. B. C. and Cheetham, A. H. (1999) Tempo and mode of speciation in the sea. *Trends in Ecology and Evolution*, 14, 72–77.

Johansen, B. and Frederiksen, S. (2002) Orchid flowers: evolution and molecular development, in *Developmental Genetics and Plant Evolution* (eds Q. C. B. Cronk, R. M. Bateman and J. A. Hawkins), Taylor & Francis, London, pp. 206–219.

Jones, J. S. (1999). *Almost Like a Whale: the* Origin of Species *Updated*. Doubleday, London.

Kaplan, D. R. (2001a) Fundamental concepts of leaf morphology and morphogenesis: a contribution to the interpretation of molecular genetic mutants. *International Journal of Plant Sciences*, 162, 465–474.

Kaplan, D. R. (2001b) The science of plant morphology: definition, history, and role in modern biology. *American Journal of Botany*, 88, 1711–1741.

Kauffman, S. A. (1993) *The Origins of Order: Self-Organization and Selection in Evolution*. Oxford University Press, Oxford.

Kellogg, E. A. (2002) Are macroevolution and microevolution qualitatively different? Evidence from Poaceae and other families, in *Developmental Genetics and Plant Evolution* (eds Q. C. B. Cronk, R. M. Bateman and J. A. Hawkins), Taylor & Francis, London, pp. 70–84.

Kenrick, P. (1994) Alternation of generations in land plants: new phylogenetic and palaeobotanical evidence. *Biological Reviews*, 69, 293–330.

Kenrick, P. (2002) The Telome Theory, in *Developmental Genetics and Plant Evolution* (eds Q. C. B. Cronk, R. M. Bateman and J. A. Hawkins), Taylor & Francis, London, pp. 365–387.

Kenrick, P. and Crane, P. R. (1997) *The Origin and Early Diversification of Plants on Land: A Cladistic Study*. Smithsonian Institution Press, Washington, DC.

Kimura, M. (1983) *The Neutral Theory of Molecular Evolution*. Cambridge University Press.

Kimura, M. (1991) Recent development of the neutral theory viewed from the Wrightian tradition of theoretical population genetics. *Proceedings of the National Academy of Sciences of the USA*, 88, 5969–5973.

Knapp, S. (2002) Floral diversity and evolution in the Solanaceae, in *Developmental Genetics and Plant Evolution* (eds Q. C. B. Cronk, R. M. Bateman and J. A. Hawkins), Taylor & Francis, London, pp. 267–297.

Kores, P. J., Molvray, M., Weston, P. H., Hopper, S. D., Brown, A. P., Cameron, K. C. and Chase, M. C. (2001) A phylogenetic analysis of Duirideae (Orchidaceae) based on plastid DNA sequence data. *American Journal of Botany*, 88, 1903–1914.

Kramer, E. M. and Irish, V. F. (1999) Evolution of genetic mechanisms controlling petal development. *Nature*, 399, 144–148.

Kramer, E. M. and Irish, V. F. (2000) Evolution of the petal and stamen developmental programs: evidence from comparative studies of the lower eudocts and basal angiosperms. *International Journal of Plant Sciences*, 161, S29–S40.

Kuhn, T. S. (1962) *The Structure of Scientific Revolutions*. Chicago University Press, Chicago.

Lande, R. (1986) The dynamics of peak shifts and the pattern of morphological evolution. *Paleobiology*, 12, 343–354.

Landwehr, J. (1977) *Wilde Orchideeën Van Europa*. Amsterdam.

Lang, D. C. (2001) A new variant of *Ophrys apifera* in Britain. *BSBI News*, 88, 40–46.

Langdale, J. A., Scotland, R. W. and Corley, S. B. (2002) A developmental perspective on the evolution of leaves, in *Developmental Genetics and Plant Evolution* (eds Q. C. B. Cronk, R. M. Bateman and J. A. Hawkins), Taylor & Francis, London, pp. 388–394.

Lauder, G. V. (1990) Functional morphology and systematics: studying functional patterns in a historical context. *Annual Review of Ecology and Systematics*, 21, 317–340.

Leavitt, R. G. (1909) A vegetative mutant, and the principle of homeosis in plants. *Botanical Gazette*, 47, 30–68.

Levin, D. A. (1970) Developmental instability and evolution in peripheral isolates. *Evolution*, 104, 343–353.

Levin, D. A. (1993) Local speciation in plants: the rule not the exception. *Systematic Botany*, 18, 197–208.

Levin, D. A. (2000) *The Origin, Expansion, and Demise of Plant Species*. Oxford University Press, Oxford.

Levin, D. A. (2001) Fifty years of plant speciation. *Taxon*, 50, 69–91.

Levinton, J. (1988) *Genetics, Paleontology, and Macroevolution*. Cambridge University Press, Cambridge.

Lewis, H. (1962) Catastrophic selection as a factor in speciation. *Evolution*, 16, 257–271.

Lewis, H. (1966) Speciation in flowering plants. *Science*, 152, 167–172.

Lewis, H. (1969) Speciation. *Taxon*, 18, 21–25.

Linder, H. P. and Kurzweil, H. (1999) *Orchids of Southern Africa*. Balkema, Rotterdam.

Long, A. G. (1966) Some Lower Carboniferous fructifications from Berwickshire, together with a theoretical account of the evolution of ovules, cupules and carpels. *Transactions of the Royal Society of Edinburgh B*, 66, 345–375.

Long, A. G. (1977a) Some Lower Carboniferous pteridosperm cupules bearing ovules and microsporangia. *Transactions of the Royal Society of Edinburgh B*, 70, 1–11.

Long, A. G. (1977b) Lower Carboniferous pteridosperm cupules and the origin of the angiosperms. *Transactions of the Royal Society of Edinburgh B*, 70, 13–35.

Luo, D., Carpenter, R., Copsey, L., Vincent, C., Clark, J. and Coen, E. (1999) Control of organ asymmetry in flowers of *Antirrhinum*. *Cell*, 99, 367–376.

McKean, D. R. (1982) × *Pseudanthera breadalbanensis* McKean: a new integeneric hybrid from Scotland. *Watsonia*, 14, 129–131.

McKinney, M. L. and McNamara, K. J. (1991) *Heterochrony: The Evolution of Ontogeny*. Plenum, New York.

McLellan, T., Shephard, H. L. and Ainsworth, C. (2002) Identification of genes involved in evolutionary diversification of leaf morphology, in *Developmental Genetics and Plant Evolution* (eds Q. C. B. Cronk, R. M. Bateman and J. A. Hawkins), Taylor & Francis, London, pp. 315–329.

McVean, G. T. and Hurst, L. T. (1997) Evidence for a selectively favourable reduction in the mutation rate of the X chromosome. *Nature*, 386, 388–392.

Maddison, W. P. (1990) A method for testing the correlated evolution of two binary characters: are gains or losses concentrated on certain branches of a phylogenetic tree? *Evolution*, 44, 539–557.

Mallet, J. and Joron, M. (1999) Evolution of diversity and warning colour and mimicry: polymorphisms, shifting balance, and speciation. *Annual Review of Ecology and Systematics*, 30, 201–233.

Margulis, L. (1993). *Symbiosis in Cell Evolution: Microbial Communities in the Archean and Proterozoic Eons* (2nd edn). Freeman, New York.

Marshall, C. R., Orr, H. A. and Patel, N. H. (1999) Morphology innovation and developmental genetics. *Proceedings of the National Academy of Sciences of the U.S.A.*, 96, 9995–9996.

Martin, W. and Schnarrenberger, C. (1997) The evolution of the Calvin cycle from prokaryotic to eukaryotic chromosomes: a case study of functional redundancy in ancient pathways through endosymbiosis. *Current Genetics*, 32, 1–18.

Mathews, S. and Donoghue, M. J. (1999) The root of angiosperm phylogeny inferred from duplicate phytochrome genes. *Science*, 282, 947–950.

Maynard Smith, J. (1989) *Evolutionary Genetics.* Oxford University Press, Oxford.

Maynard Smith, J. and Szathmáry, E. (1995) *The Major Transitions of Evolution.* Freeman, Oxford.

Mayr, E. (1963) *Animal Species and Evolution.* Belknap Press, Harvard.

Mehl, J. (1986) Die fossile Dokumentation der Orchideen. *Journal Berlin Naturwissenschaften Verein Wuppertal*, 39, 121–133.

Meyerowitz, E. M. and Somerville, C. R. (1994) *Arabidopsis.* Cold Spring Harbor Laboratory Press, New York.

Mindell, D. P. and Meyer, A. (2001) Homology evolving. *Trends in Ecology and Evolution*, 16, 434–440.

Möller, M. and Cronk, Q. C. B. (2001) Evolution of morphological novelty: a phylogenetic analysis of growth patterns in *Streptocarpus* (Gesneriaceae). *Evolution*, 55, 918–929.

Morris, R. (2001) *The Evolutionists: The Struggle for Darwin's Soul.* Freeman, San Francisco.

Nickrent, D. L., Duff, R. J., Colwell, A. E., Wolfe, A. D., Young, N. D., Steiner, K. E. and dePamphilis, C. W. (1998) Molecular phylogenetic and evolutionary studies of parasitic plants, in *Molecular Systematics of Plants 2* (eds D. E. Soltis, P. S. Soltis and J. J. Doyle), Chapman & Hall, London, pp. 211–241.

Ohta, T. (1992) The nearly neutral theory of molecular evolution. *Annual Review of Ecology and Systematics*, 23, 263–286.

Ohta, T. (1995) Synonymous and nonsynonymous substitutions in mammalian genes and the nearly neutral theory. *Journal of Molecular Evolution*, 40, 56–63.

Orr, H. A. (1991) Is single gene speciation possible? *Evolution*, 45, 764–769.

Orr, H. A. (1998) The population genetics of adaptation: the distribution of factors fixed during adaptive evolution. *Evolution*, 52, 935–949.

Palopoli, M. F. and Patel, N. H. (1998) Evolution of the interaction between *Hox* genes and a downstream target. *Current Biology*, 8, 587–590.

Patterson, C. (1999) *Evolution* (2nd edn). Natural History Museum, London.

Phillips, T. L. and DiMichele, W. A. (1992) Comparative ecology and life-history biology of arborescent lycopsids in Late Carboniferous swamps of Euramerica. *Annals of the Missouri Botanical Garden*, 79, 560–588.

Poethig, R. S. (1990) Phase change and the regulation of shoot morphologies in plants. *Science*, 250, 923–930.

Pomiankowski, A. and Hurst, L. D. (1993) Siberian mice upset Mendel. *Nature*, 363, 396–397.

Prescott, D. M. and DuBois, M. L. (1996) The mercurial germ-line genome of hypotrichous ciliates, in *Genomes of Plants and Animals* (eds J. P. Gustafson and R. B. Flavell), Plenum, New York, pp. 271–279.

Price, T., Turelli, M. and Slatkin, M. (1993) Peak shifts produced by correlated response to selection. *Evolution*, 47, 280–290.

Pridgeon, A. M., Bateman, R. M., Cox, A. V., Hapeman, J. R. and Chase, M. W. (1997) Phylogenetics of subtribe Orchidinae (Orchidoideae, Orchidaceae) based on nuclear ITS sequences. 1. Intergeneric relationships and polyphyly of *Orchis sensu lato*. *Lindleyana*, 12, 89–109.

Pridgeon, A. M., Cribb, P. J., Chase, M. W. and Rasmussen, F. N. (eds) (1999) *Genera Orchidacearum. 1. General introduction, Apostasioideae, Cypripedioideae*. Oxford University Press, Oxford.

Pryer, K. M., Schneider, H., Smith, A. R., Cranfill, R., Wolf, P. G., Hunt, J. S. and Sipes, S. D. (2000) Horsetails and ferns are a monophyletic group and the closest living relatives of seed plants. *Nature*, 409, 618–622.

Qiu, Y.-L., Lee, J., Bernasconi-Quadroni, F., Soltis, D. E., Soltis, P. S., Zanis, M., Zimmer, E. A., Chen, Z.-D., Savolainen, V. and Chase, M. W. (1999) The earliest angiosperms: evidence from mitochondrial, plastid and nuclear genomes. *Nature*, 402, 404–407.

Raff, R. A. and Kauffman, T. C. (1983) *Embryos, Genes, and Evolution*. Macmillan, New York.

Retallack, G. J. and Dilcher, D. L. (1988) Reconstructions of selected seed ferns. *Annals of the Missouri Botanical Garden*, 75, 1010–1057.

Richardson, J. E., Pennington, R. T., Pennington, T. D. and Hollingsworth, P. M. (2001a) Rapid diversification of a species-rich genus of Neotropical rain forest trees. *Science*, 293, 2242–2245.

Richardson, J. E., Weitz, F. M., Fay, M. F., Cronk, Q. C. B., Linder, H. P., Reeves, G. and Chase, M. W. (2001b) Rapid and recent origin of species richness in the Cape flora of South Africa. *Nature*, 412, 181–183.

Ridley, M. (1996) *Evolution* (2nd edn). Blackwell, Oxford.

Riedl, R. (1979) *Order in Living Organisms*. Wiley, New York.

Rieseberg, L. H. (1997) Hybrid origins of plant species. *Annual Review of Ecology and Systematics*, 28, 359–389.

Rieseberg, L. H. and Burke, J. M. (2001) The biological reality of species: gene flow, selection, and collective evolution. *Taxon*, 50, 47–67.

Rieseberg, L. H., Van Fossen, C. and Desrochers, A. M. (1995) Hybrid speciation accompanied by genomic reorganisation in wild sunflowers. *Nature*, 375, 313–316.

Rosenzweig, M. L. and McCord, R. D. (1991) Incumbent replacement: evidence for long-term evolutionary progress. *Paleobiology*, 17, 202–213.

Rothwell, G. W. and Scheckler, S. E. (1988) Biology of ancestral gymnosperms, in *Origin and Evolution of Gymnosperms* (ed. C. B. Beck), Columbia University Press, New York, pp. 85–134.

Rothwell, G. W. and Wight, D. C. (1989) *Pullaritheca longii* gen. nov. and *Kerryia mattenii* gen. et spec. nov., Lower Carboniferous cupules with ovules of the *Hydrasperma tenuis* type. *Review of Palaeobotany and Palynology*, 60, 295–309.

Rowe, N. P. (1988) New observations on the Lower Carboniferous pteridosperm *Diplopteridium* Walton and an associated synangiate organ. *Botanical Journal of the Linnean Society*, 97, 125–158.

Rudall, P. J. and Bateman, R. M. (2002) Roles of synorganisation, zygomorphy and heterotopy in floral evolution: the gynostemium and labellum of orchids and other lilioid monocots. *Biological Reviews*.

Rudall, P. J. and Buzgo, M. (2002) Evolutionary history of the monocot leaf, in *Developmental Genetics and Plant Evolution* (eds Q. C. B. Cronk, R. M. Bateman and J. A. Hawkins), Taylor & Francis, London, pp. 431–458.

Rutishauser, R. (1995) Developmental patterns of leaves in Podostemaceae compared with more typical flowering plants: saltational evolution and fuzzy morphology. *Canadian Journal of Botany*, 73, 1305–1317.

Sanderson, M. J. (1997) A nonparametric approach to estimating divergence times in the absence of rate constancy. *Molecular Biology and Evolution*, 14, 1218–1231.

Sanderson, M. J. (1998) Reappraising adaptive radiation. *American Journal of Botany*, 85, 1650–1655.

Schneider, H., Pryer, K. M., Cranfill, R., Smith A. R. and Wolf, P. G. (2002) Evolution of vascular plant body plans: a phylogenetic perspective, in *Developmental Genetics and Plant Evolution* (eds Q. C. B. Cronk, R. M. Bateman and J. A. Hawkins), Taylor & Francis, London, pp. 330–364.

Shapiro, J. A. (1997) Genome organization, natural genetic engineering, and adaptive mutation. *Trends in Genetics*, 13, 98–104.

Shapiro, J. A. (2002) A 21st Century view of evolution. *Journal of Biological Physics* (in press).

Shubin, N. H. and Marshall, C. R. (2000) Fossils, genes, and the origins of novelty. *Paleobiology*, 26, 324–340.

Silvertown, J., Franco, M. and Harper, J. L. (eds) (1997) *Plant Life Histories: Ecology, Phylogeny and Evolution*. Cambridge University Press, Cambridge.

Singh, R. S. and Krimbas, C. B. (eds) (2000) *Evolutionary Genetics: From Molecules to Morphology*. Cambridge University Press, Cambridge.

Slack, J. M. W., Holland, P. W. H. and Graham, C. F. (1993) The zootype and the phylotypic stage. *Nature*, 361, 490–492.

Slatkin, M. (1996) In defense of founder–flush theories of speciation. *American Naturalist*, 147, 493–505.

Stace, C. A. (1989) *Plant Taxonomy and Biosystematics* (2nd edn). Arnold, London.

Stace, C. A. (1993) The importance of rare events in polyploid evolution, in *Evolutionary Patterns and Processes* (eds D. R. Lees and D. Edwards), Academic Press, New York, pp. 159–169.

Stebbins, G. L. (1971) *Chromosomal Evolution in Higher Plants*. Arnold, London.

Stebbins, G. L. (1983) Mosaic evolution: an integrating principle for the modern synthesis. *Experientia*, 39, 823–834.

Stein, W. E. (1998) Developmental logic: establishing a relationship between developmental process and phylogenetic pattern in primitive vascular plants. *Review of Palaeobotany and Palynology*, 102, 15–42.

Templeton, A. R. (1982). Genetic architecture of speciation, in *Mechanisms of Speciation* (ed. C. Barigozzi), Liss, New York, pp. 105–121.

Templeton, A. R. (1989) The meaning of species and speciation: a genetic perspective, in *Speciation and its Consequences* (eds D. Otte and J. A. Endler), Sinauer, Sunderland, MA, pp. 3–27.

Teotónio, H. and Rose, M. R. (2001) Reverse evolution (Perspective). *Evolution*, 55, 653–660.

Theißen, G. (2000) Evolutionary developmental genetics of floral symmetry: the revealing power of Linnaeus' monstrous flower. *BioEssays*, 22, 209–213.

Theißen, G., Becker, A., Winter, K.-U., Münster, T., Kirchner, C. and Saedler, H. (2002) How the land plants learned their floral ABCs: the role of MADS-box genes in the evolutionary origin of flowers, in *Developmental Genetics and Plant Evolution* (eds Q. C. B. Cronk, R. M. Bateman and J. A. Hawkins), Taylor & Francis, London, pp. 173–205.

Theißen, G., Kim, J. T. and Saedler, H. (1996) Classification and phylogeny of the MADS-box gene subfamilies in the morphological evolution of eukaryotes. *Journal of Molecular Evolution*, 10, 484–516.

Thompson, J. D. and Lumaret, R. (1992) The evolutionary dynamics of polyploid plants: origins, establishment and persistence. *Trends in Ecology and Evolution*, 7, 302–307.

Trifonov, E. N. (1987) Genetic sequences as products of compression by inclusive superposition of many codes. *Molecular Biology*, 31, 759–767.

Tsiantis, M., Hay, A., Ori, N., Kaur, H., Holtan, H., McCormick, S. and Hake, S. (2002) Developmental signals regulating leaf form, in *Developmental Genetics and Plant Evolution* (eds Q. C. B. Cronk, R. M. Bateman and J. A. Hawkins), Taylor & Francis, London, pp. 418–430.

Tucker, S. C. (2000) Organ loss and its possible relationship to saltatory evolution, in *Advances in Legume Systematics 9* (eds P. Herendeen and A. Bruneau), Royal Botanic Gardens Kew, pp. 107–120.

Tucker, S. C. (2001) The ontogenetic basis for missing petals in *Crudia* (Leguminosae: Cae-salpinioideae: Detarieae). *International Journal of Plant Science*, 162, 83–89.

Valentine, J. W. (1980) Determinants of diversity in higher taxonomic categories. *Paleobiology*, 6, 444–450.

Valentine, J. W., Jablonski, D. and Erwin, D. H. (1999) Fossils, molecules, and embryos: new perspectives on the Cambrian explosion. *Development*, 126, 851–859.

Van Steenis, C. G. G. J. (1976) Autonomous evolution in plants: differences in plant and animal evolution. *Gardens' Bulletin, Singapore*, 29, 103–126.

Vermeij, G. J. (1987) *Evolution and Escalation.* Princeton University Press, NJ.

Vogel, J. C., Barrett, J. A., Rumsey, F. J. and Gibby, M. (1999) Identifying multiple origins in polyploid homosporous pteridophytes, in *Molecular Systematics and Plant Evolution* (eds P. M. Hollingsworth, R. M. Bateman and R. J. Gornall), Taylor & Francis, London.

Waddington, C. H. (1957) *The Strategy of the Genes.* Allen & Unwin, London.

Wade, M. J. (1992) Sewall Wright: gene interaction and the shifting balance theory, in *Oxford Surveys in Evolutionary Biology 8* (eds D. Futuyma and J. Antonovics), Oxford University Press, Oxford, pp. 35–62.

Wade, M. J. and Goodnight, C. J. (1998) The theories of Fisher and Wright in the context of metapopulations: when nature does many small experiments (Perspective). *Evolution*, 52, 1537–1553.

Walbot, V. (2000) A green chapter in the book of life. *Nature*, 408, 794–795.

Walbot, V. (2002) Impact of transposons on plant genomes, in *Developmental Genetics and Plant Evolution* (eds Q. C. B. Cronk, R. M. Bateman and J. A. Hawkins), Taylor & Francis, London, pp. 15–31.

Wang, R.-L., Stec, A., Hey, A., Lukens, L. and Doebley, J. (1999) The limits of selection during maize domestication. *Nature*, 398, 236–239.

Wendel, J. F., Schnabel, A. and Seelanan, T. (1995) Bidirectional interlocus concerted evolution following allopolyploid speciation in cotton (*Gossypium*). *Proceedings of the National Academy of Sciences of the USA*, 92, 280–284.

Wessler, S. R., Bureau, T. E. and White, S. E. (1995) LTR-retrotransposons and MITEs: important players in the evolution of plant genomes. *Current Opinion on Genetics and Development*, 5, 814–821.

Whitlock, M. C. (1997) Founder effects and peak shifts without genetic drift: adaptive peak shifts occur easily when environments fluctuate slightly. *Evolution*, 51, 1044–1048.

Whitlock, M. C., Phillips, P. C., Moore, F. B.-G. and Tonsor, S. J. (1995) Multiple fitness peaks and epistasis. *Annual Review of Ecology and Systematics*, 26, 601–629.

Williams, G. C. (1992) *Natural Selection: Domains, Levels and Challenges.* Oxford University Press, Oxford.

Wing, S. L. and Boucher, L. (1998) Ecological aspects of Cretaceous plant radiation. *Annual Review of Earth and Planetary Sciences*, 26, 379–421.

Wolf, J. B., Brodie, E. D. III and Wade, M. J. (eds) (2000) *Epistasis and the Evolutionary Process.* Oxford University Press, Oxford.

Wolfe, K. H. (2001) Yesterday's polyploids and the mystery of diploidization. *Nature Reviews, Genetics*, 2, 333–341.

Wray, G. A. (1995) Punctuated evolution of embryos. *Science*, 267, 1115–1116.

Wright, S. (1932) The roles of mutation, inbreeding, crossbreeding and selection in evolution. *Proceedings of the Sixth International Congress of Genetics*, 1, 356–366.

Wright, S. (1968) *Evolution and the Genetics of Populations.* Chicago University Press, Chicago.

Young, D. P. (1952) Studies in the British *Epipactis*, III, IV. *Watsonia*, 2, 253–276.

Zelditch, M. L. and Fink, W. L. (1996) Heterochrony and heterotopy: stability and innovation in the evolution of form. *Paleobiology*, 22, 247–250.

Chapter 8

Evolution of adaptive petal cell morphology

Beverley J. Glover and Cathie Martin

ABSTRACT

The speed and extent of angiosperm radiations owe much to the evolution of animal-attracting floral parts, particularly petals. Understanding the ways in which attractive features of petals evolved is thus an essential component of understanding angiosperm diversity. One extremely common adaptation which increases petal attractiveness is specialised epidermal cell shape. Conical-papillate epidermal cells may attract pollinators by enhancing colour, sparkle, scent, intrafloral temperature or by providing tactile cues. This specialised cell form has been studied in three eudicot species. It appears to be produced by a common developmental programme in *Antirrhinum* and *Petunia*, but to have evolved independently in *Arabidopsis*. Further work on a wider range of angiosperm species is necessary for a fuller understanding of this adaptive cell morphology.

8.1 Introduction

Darwin famously described the speed and extent of the radiation of the angiosperms as 'an abominable mystery'. The sheer diversity of morphology, biochemistry and life history within the angiosperms is astonishing, and they represent approximately one-sixth of the estimated species diversity of all life on Earth. However, it was the speed of angiosperm radiation, above all, that perturbed Darwin. The angiosperms first appeared in the fossil record in the Early Cretaceous, about 130 million years ago, and then radiated very quickly over the next 40 million years to give most of the huge diversity that we see today (Kenrick, 1999). The explanation for this phenomenal radiation almost certainly lies in the evolution of the angiosperm flower. The development of animal-attracting flowers allowed diversification to fill the niches provided by the variety of potentially pollinating animals, which themselves underwent co-evolution along with the flowers. Since it is the petals which function as the attractant in most flowers, advertising the presence of a floral reward, the radiation of the angiosperms can thus be attributed almost entirely to the evolution of adaptive features of petals. This chapter will examine the evolution of one particular character which enhances the attractiveness of a petal to a pollinating animal – the presence of conical-papillate epidermal cells. This specialised cell morphology may act to increase the pollination success of a plant in a number of ways, which are described below. The molecular genetic control of the cellular differenti-

In *Developmental Genetics and Plant Evolution* (2002) (eds Q. C. B. Cronk, R. M. Bateman and J. A. Hawkins), Taylor & Francis, London, pp. 160–172.

ation programme leading to conical-papillate petal epidermal cells is analysed and the conservation of this control mechanism across the angiosperms is discussed.

8.1.1 Petal evolution

The first petals to evolve are thought to have served a protective function. In primitive plants they appear to have developed from protective leaves or bracts surrounding the sporangia (Weberling, 1992). However, petals have developed in flowering plants usually from the stamens rather than from the bracts. Stamen-derived petals are found in the eudicots and monocots, while earlier angiosperm clades have petals derived variously from stamens and bracts (Endress, 1993). Recent evidence for the derivation of petals from stamens has been provided by the molecular genetic analysis of flower development. Both petal and stamen development are partially specified by common genes, the 'B function' genes, which encode MADS-box transcription factors regulating the expression of organ specific traits (Theißen et al., 2002). Expression of these B function genes in stamens is conserved throughout all angiosperms tested, but their expression in petals is variable, both spatially and during development. This has led to the suggestion that the conserved pattern in stamens may reflect their single origin, whereas the diverse pattern in petals may be due to their multiple evolutionary origins (Kramer and Irish, 1999). The evolution of petals into attractive signalling devices is thought to have occurred in conjunction with the evolution of insects as pollinators, and there is certainly a temporal correlation between the radiation of the angiosperms and the even more dramatic radiation of the insects. There is an almost infinite variety of ways in which petals can attract pollinators, including use of their size, form, smell, colour and pattern. In addition, petals may also form landing platforms and traps for pollinating animals, and in some flowers they serve as sites for nectar synthesis and scent production.

8.1.2 Enhancing petal attractiveness

The most commonly adapted features which enhance the attractiveness of the corolla and thus increase pollinator visitation are shape and colour. Corolla shape can be modified by alterations of petal size or shape and by alterations of corolla structure and symmetry. Since a petal develops from a primordium of dividing cells, its size can be increased by starting development earlier so that the cells undergo more divisions, or by prolonging cell division in the developing organ, or by enhancing the speed of cell division in the developing organ. The shape of the petal can be altered by manipulation of the rate of cell division in different parts of the developing structure and by regulation of the distribution of regions of maximum cell division during development.

Irregular flowers, in which radial symmetry is lost to bilateral symmetry, provide an adaptation to insect pollination (and bird pollination) that appears to have evolved independently on numerous occasions (Coen et al., 1995). The model Arabidopsis flower, like many common flowers, is radially symmetrical. The Antirrhinum majus flower, in common with those of many orchids, legumes and other families, has bilateral symmetry. This gives it clear dorsiventral asymmetry, or dorsiventrality. The advantage of dorsiventrality is that it confers position on each petal. When each developing petal has a specific position then it is possible to modify their development independently so that a corolla with a more specialised structure is produced. Several genes associated with the establishment of bilateral symmetry have been identified in *Antirrhinum* (*CYCLOIDEA, DICHOTOMA, RADIALIS*) (Luo *et al.*, 1995; Cubas, 2002). The best characterised of these mutants is *cycloidea*, in which lateral petals are converted into ventral petals, and dorsal petals are morphologically somewhere between dorsal and lateral petals. The *dichotoma* mutant produces flowers with a similar phenotype. The double mutant of *cycloidea* and *dichotoma* has radially symmetrical flowers, with all the petals fully ventralised. The genes are expressed only in the dorsal region of the floral meristem and confer a dorsal identity on the organ primordia developing in that region. *Antirrhinum* belongs to a large monophyletic group, the Lamiales, which all have irregular flowers, suggesting that a common mechanism involving related genes may underlie bilateral symmetry in all member species. Other genetic mechanisms may dictate bilateral symmetry in unrelated species. In some regular flowers bilateral symmetry may have been lost, which would mean that irregularity evolved on fewer independent occasions.

Flower colour and pattern also show remarkable diversity within petals, although they usually result in a contrasting appearance with associated vegetative tissues. As Darwin put it:

> Flowers rank amongst the most beautiful productions of nature; but they have been rendered conspicuous on contrast with the green leaves, and in consequence at the same time beautiful, so that they may be easily observed by insects. I have come to this conclusion from finding it an invariable rule that when a flower is fertilised by the wind it never has a gaily coloured corolla
>
> (Darwin, 1859)

The understanding of colour formation is greatest for the anthocyanins, where colour is dependent not only on the pigments themselves but also on the presence of related compounds (co-pigments) which may modify their colour in solution and on the pH and metal ion content of the cell (Martin and Gerats, 1993). Other associated compounds, the flavones and flavonols, absorb strongly in the ultraviolet (UV) range, and can be seen as colours by some pollinators, such as bees. The presence of other pigments, such as aurones and carotenoids, can also affect the final flower colour (Martin and Gerats, 1993). Carotenoids are widely dispersed pigments responsible for the colour of most yellow and orange flowers. They accumulate in plastids and are lipid-soluble. Combinations of these different pigments result in different coloured corollas, as mixing two water-soluble pigments together gives a different final colour than a water-soluble pigment against a background of a lipid-soluble pigment. Interactions between anthocyanins and metal ions can also alter the final colour observed and enhance the attractiveness of the corolla (Martin and Gerats, 1993). For example, the bright blue of cornflowers stems from an interaction between the anthocyanin delphinidin and the metal iron. A change in vacuolar pH is another way that petal colour can be adapted. For example, the light blue petals of *Ipomoea tricolor* (morning glory) owe their colour to the effect of a high vacuolar pH on anthocyanins with a specialised structure (Yoshida *et al.*, 1995).

8.2 Role of conical-papillate epidermal cells

A very subtle way in which the attractiveness of a corolla can be enhanced is by distribution of specialised cell types in the epidermis. In one study that examined the surface structures of 201 species of flower from sixty families (Kay *et al.*, 1981), 79 per cent were found to exhibit some form of conical or papillate cells on the adaxial epidermis (the epidermis oriented towards potential pollinators; Figure 8.1a). The frequency of this specialised cell morphology within the angiosperms, and its almost universal restriction to the petal, argue for an adaptive explanation involving the function of the petal in pollinator attraction. The leaves of some shade-adapted plants also develop conical-papillate cells but it is not known whether these have the same or a similar developmental origin.

Antirrhinum majus is one of the majority of species which produce conical-papillate petal epidermal cells (Figure 8.1b). However, the *mixta* mutant of *Antirrhinum* fails to develop this specialised cell form and instead has flat petal cells (Figure 8.1c). This mutant has been used to test the hypothesis that conical-papillate petal cells are

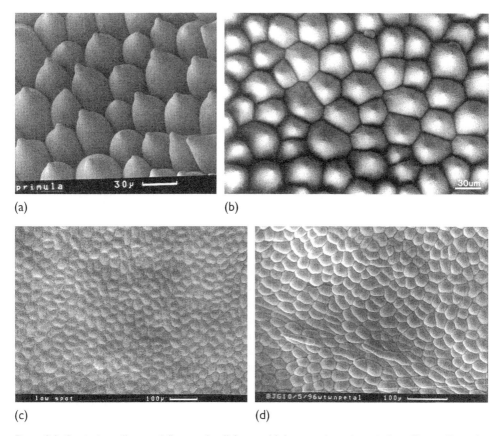

(a)

(b)

(c)

(d)

Figure 8.1 Conical-papillate and flat petal cell forms: (a) Lemon-shaped conical-papillate cells in the petal epidermis of *Primula*. (b) Conical-papillate cells in a wild-type *Antirrhinum* petal epidermis. (c) Flat cells in a *mixta* mutant *Antirrhinum* petal epidermis. (d) Flat cells in the petal epidermis of *Solanum dulcamara*.

adaptive and enhance the pollination success of the plant, and also to investigate how they might do so. Isogenic plants of the wild type and *mixta* mutant genotypes were planted out in a field plot and flowers were emasculated before dehiscence to prevent self-pollination. Fruit set was then scored one month later and the presence of a fruit used as an indicator of a pollinator visit. The shape of the petal cells had a highly significant effect on the likelihood of fruit set (and thus a pollinator visit) in all replicate plots. Flat-celled flowers set significantly fewer fruits than conical-papillate-celled flowers (Glover and Martin, 1998). Interestingly, a similar result was observed when conical-papillate-celled acyanic *Antirrhinum* flowers were compared with double mutant flat-celled/acyanic flowers. These lines were produced by crossing the isogenic lines described above to the *nivea* mutant of *Antirrhinum* which contains a deletion of the gene encoding chalcone synthase, the first committed step in anthocyaninin production. Although conical-papillate and flat-celled surfaces were identical to the human eye in the absence of pigmentation, the flat-celled flowers experienced reduced fruit set (Glover and Martin, 1998) and a reduction in pollinator visits (Comba *et al.*, 2000). These data indicate that enhanced colour is unlikely to be the main explanation for the evolutionary success of conical-papillate cells.

Four explanations for the function of conical-papillate cells in enhancing the attractiveness of the corolla have been proposed: depth of colour and sparkle, scent, temperature, and tactility.

8.2.1 Colour and sparkle

It was first proposed that conical-papillate cells increased the amount of light absorbed by the floral pigments, enhancing the perceived colour of the petal (Kay *et al.*, 1981). This would result in a more brightly coloured flower which would contrast strongly with vegetative tissue. Conical-papillate cells would also scatter light reflected back from the mesophyll more evenly than would flat or lenticular (rounded) cells, resulting in a sparkling effect, or velvety texture to the petal (Kay *et al.*, 1981; Kay, 1988).

Initially, the only evidence to support these hypotheses was the observation that petals of species with conical-papillate cells are usually brighter and more deeply coloured than the petals of species with flat cells (compare flat-celled *Solanum dulcamara* (Figure 8.1d) with conical-papillate-celled *Antirrhinum*). However, the isolation of the *mixta* mutant of *Antirrhinum* allowed the influence of petal cell shape to be assessed in isogenic material with no other variables between the petals. In fact, the *mixta* mutant was originally identified in a screen of mutagenised plants because it was paler in colour and believed to be a putative pigmentation mutant (Noda *et al.*, 1994). The mutant petal also has a matt texture, unlike the velvety sparkle of the wild-type petal.

By comparing the ability of epidermal cells to focus light in the wild-type and *mixta* mutant lines, Gorton and Vogelmann (1996) demonstrated how conical-papillate cells act to enhance visible pigmentation. They found that conical-papillate cells focused light approximately twice as well as the mutant flat cells, and that this focusing was at a depth of 45 to 52 μm, well within the pigment-containing vacuoles of the epidermal cells (which are around 60 μm tall). However, the flat cells of the *mixta* mutant line produced maximal focal intensification at a depth of 60 to 75 μm, below the 25 μm-thick flat epidermis, in the unpigmented mesophyll tissue. The conical cells of the wild-type genotype therefore appear deeper in colour as a result of the focusing of the captured light into the pigmented epidermal cells, rather than below them. Gorton and Vogelmann (1996) also showed that the wild-type *Antirrhinum* petals reflected significantly less light away from the flower than did *mixta* mutant petals, and absorbed significantly more light. These differences can be attributed to the focusing of the light onto the light-absorbing anthocyanin pigments in the vacuoles of the epidermal cells, and to the reduction in reflection of light at low angles of incidence, resulting in the greater depth of colour of wild-type conical-celled flowers. Thus, a combination of biophysical analysis and molecular genetics has confirmed the model of Kay *et al.* (1981), and demonstrated that conical-papillate cells do enhance visible pigmentation by directing light.

8.2.2 Tactile cues

An alternative, although not mutually exclusive, explanation for the specialised shape of petal epidermal cells is that the texture of a flower may provide tactile cues to pollinator position on the flower, sensed through the sensilla trichodea on the tips of the antennae of insects (Kevan and Lane, 1985). Moreover, the shape of petal epidermal cells may be used as a tactile cue by insects in their discrimination between particular flowers. In support of this hypothesis, Kevan and Lane (1985) trained bees to recognise the petals of different species by the shape of the cells. Bees were provided with a food reward when they touched epidermal layers composed of certain cell types, but no reward when they touched other shaped cells. They learned very quickly to associate the reward with the texture of the petal and would search for food only when presented with the epidermal tissue which usually accompanied the reward. To investigate the influence of the texture of epidermal tissues on pollinator behaviour, the response of bumblebees to the wild-type and *mixta* mutant lines was analysed in a field experiment (Comba *et al.*, 2000). Analysis of bee behaviour indicated that bees distinguished between the two genotypes both before and after landing. It is therefore unlikely that epidermal cell shape is used here solely as a tactile cue (as was suggested by Kevan and Lane, 1985), since cell shape evidently provides information that is available before landing. However, this conclusion does not exclude the possibility that cell shape is also used by pollinators as a tactile cue post-landing.

8.2.3 Temperature

A third way in which conical-papillate cells might be perceived as adaptations to enhance the attractiveness of the corolla is through a role in regulating intrafloral microclimate. The temperature within a flower has been shown to influence nectar secretion rate, nectar evaporation rate, nectar standing crop and nectar concentration (e.g. Corbet, 1990). There is also some evidence to suggest that a high intrafloral temperature is itself attractive to insects, particularly in cooler climates and at dawn (e.g. Corbet *et al.*, 1993). Analysis of intrafloral temperature using fine thermocouples demonstrated a striking difference between the wild-type and *mixta* mutant lines of *Antirrhinum*. The intrafloral temperature of wild-type, conical-celled flowers increased directly with increase in the warming effect of direct solar radiation. This was measured by subtracting ambient temperature from the temperature of a ball of black 'Blu-tak' placed in direct sunlight (globe temperature excess). Although intrafloral temperature of conical-papillate celled flowers increased directly with globe temperature excess, the intrafloral temperature of mutant flat-celled flowers remained constant, however strong the warming effect of solar radiation (Comba *et al.*, 2000).

8.2.4 Scent

A further possible advantage to a plant in producing conical-papillate cells has recently been demonstrated by Kolosova *et al.* (2001). The emission of scent from a petal requires the expression of genes encoding enzymes involved in volatile production. Benzoate methyl transferase, necessary for scent production in *Antirrhinum*, is expressed specifically in conical-papillate cells and the heads of trichomes in the corolla tube. Conical-papillate cells may influence the directionality of volatile emission, either through the sloping sides or, in combination with their influence on microclimate, through microgradients of temperature. It is therefore possible that conical-papillate petal epidermal cells are of adaptive significance in a particular angiosperm species through any one or a combination of effects on colour, texture, temperature or scent.

8.3 Molecular genetics of conical-papillate cells in *Antirrhinum*

The identification of the *mixta* mutant of *Antirrhinum* allowed the isolation of the first gene having a role in the development of conical-papillate petal cells. This mutant phenotype is the result of insertion of a transposon, *Tam4*, into the *MIXTA* gene. The gene was cloned using the transposon as a tag and shown to encode an MYB-related transcription factor (Noda *et al.*, 1994). The MYBs are a large family of transcription factors in plants which are believed to regulate expression of a wide variety of target genes (Martin and Paz Ares, 1997).

The *MIXTA* gene has a very limited expression pattern, consistent with a gene whose only role is in the differentiation of a specialised cell type found in only one tissue. Northern analysis demonstrated that the gene is expressed only in petal tissue and in no other organ of the plant (Noda *et al.*, 1994). *In situ* hybridisation showed that this petal-specific expression is limited to the adaxial epidermis of the corolla lobe, and occurs in no other part of the petal. Furthermore, *MIXTA* expression is limited temporally to relatively mature petals, and is found only after mitotic cell divisions in the epidermis have ceased and predictive cell cycle marker genes, such as *CYCLIN D3b*, are no longer expressed (Glover *et al.*, 1998).

This precise expression pattern indicates that the *MIXTA* gene is itself regulated by a number of other genes. To date, the only one of these known is the B function gene, *DEFICIENS*, which is required for petal and stamen development in *Antirrhinum*. *MIXTA* expression is absent from weak *deficiens* alleles (Perez-Rodriguez and Martin, unpubl. obs.).

The MYB transcription factor encoded by *MIXTA* presumably activates a set of target genes involved in the differentiation of the conical-papillate cell. These cells appear to develop through the localised production of specialised cell wall material. It is therefore likely that putative target genes will include those involved in the synthesis of cell wall components, enzymes required for cell wall loosening, and components of the cytoskeleton which must necessarily be reorganised to produce the specialised cell structure. The identification of these target genes will be very important in understanding how conical-papillate cells develop and ultimately how that development evolved.

Within *Antirrhinum* two genes with significant sequence similarity to *MIXTA* have been isolated from a cDNA library made from *mixta* mutant petal tissue. These genes, *AmMYBML1* (*ML1*) and *AmMYBML2* (*ML2*), may also play roles in the specification of adaptive cell forms (Glover *et al.*, 1998; C. Martin, unpubl. obs.). Both are expressed in petal tissue, *ML1* early in petal development and *ML2* later in petal development. They may thus play roles in reinforcing the function of *MIXTA* or in the production of other specialised cell types such as the hairs of the throat of the corolla tube or the proliferating cells which generate the lip of the *Antirrhinum* flower. The identification of plants with mutations in these genes will be essential in understanding their functions.

8.4 Comparative development of conical-papillate cells and trichomes

To investigate whether the *MIXTA* gene is sufficient by itself to activate conical-papillate cell differentiation, it was ectopically expressed in tobacco. The development of conical-papillate cells on organs such as leaves, stem and sepals demonstrated that the gene is sufficient to regulate this developmental pathway. However, the ectopic development of excess long-stalked multicellular trichomes on all organs, including normally hair-free organs, indicated that conical cell formation and multicellular trichome formation share common components (Figure 8.2). The formation of either conical-papillate cells or trichomes was strongly correlated with the timing of expression of *MIXTA* in relation to the developmental stage of the tissue, particularly its competence for further cell divisions. Leaf tissue expressing *MIXTA* relatively late (maximum expression when the cells had finished dividing) differentiated cones, whereas epidermal leaf tissue that had high *MIXTA* expression while the cells were still mitotically active, differentiated predominantly long-stalked trichomes (Glover *et al.*, 1998).

Therefore, trichome and conical-papillate cell development in *Antirrhinum* and tobacco appear to share parts of a common developmental programme. Separation of the two cellular differentiation programmes may have arisen by gene duplication. Differentiation of expression of two genes with a common function, particularly in relation to the progress of cell division, may thus result in different forms of cellular morphogenesis. A likely candidate for a gene regulating trichome development in flowers (i.e. the related cellular differentiation pathway) is the *ML1* gene, which is expressed relatively early in petals at the same time as genes predictive of cell division such as *CYCLIN D3b* (Glover *et al.*, 1998).

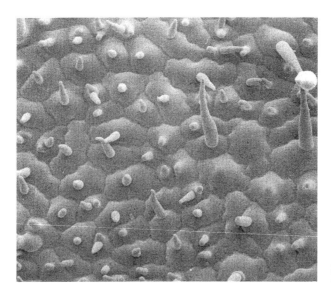

30 μm

Figure 8.2 Conical-papillate cells and excess trichomes in the leaf epidermis of a tobacco plant ectopically expressing *MIXTA*. A complete range of cell structures from rounded cell through to multicellular trichome can be seen, underlining the extent to which these two cell forms are produced by similar developmental programmes.

8.5 Conical-papillate cells in other species

Conical-papillate petal epidermal cells are found throughout the angiosperms (Kay *et al.*, 1981). Almost all families contain some species with conical-papillate cells, and this observation is true not only of the eudicots but also of other clades. The monocots *Freesia* and *Iris*, for example, both produce conical-papillate cells (Kay *et al.*, 1981). The prevalence of the conical-papillate cell form may not be surprising in light of the variety of ways in which these cells can function as an adaptation to enhance pollination success. It is clear that the acquisition of the developmental programme leading to this type of cellular differentiation may present a plant with a strong selective advantage in a number of different ways. A selective advantage which acts to increase pollinator visits and thus seed set is likely to spread very rapidly throughout a population. This observation alone makes it unlikely that those angiosperms which do not produce conical-papillate petal cells have lost such a selectively favourable developmental programme, and indicates that a likelier explanation is that this cell form has evolved multiple times. To investigate this hypothesis it is necessary to analyse the mechanism of conical-papillate cell development in a range of species. However, it should be borne in mind that demonstrating that similar genes act to control this process in different species does not necessarily point to a single evolutionary origin, as the same genes and developmental programmes may be recruited more than once.

The conical-papillate cells of *Petunia hybrida* strongly resemble those of *Antirrhinum*. A mutant with flattened petal epidermal cells has been identified and the mutated gene, *PhMYBPH1*, has strong sequence similarity to *MIXTA* (van Houwelingen *et al.*, 1998). However, if both *Petunia* and *Antirrhinum* contain multicellular trichome developmental programmes regulated by orthologous master genes likely to be similar in sequence to the *MIXTA* MYB genes, then it is also plausible that duplication of this gene occurred separately in the two lineages. The result of this multiple duplication would be different but similar genes in both species involved in the differentiation of conical-papillate cells. In either case it is likely that the downstream target genes activated are similar, as the processes of cellular morphogenesis are clearly conserved. It is worth noting that not all members of the Scrophulariaceae or the Solanaceae produce petals with conical-papillate cells. Unless adaptive explanations for the loss of this cell form in some species can be found, it may be simpler to postulate that the same pathways have evolved multiple times. One such adaptive explanation might revolve around the functions of the petals. *Solanum dulcamara* has lenticular (rounded) petal epidermal cells, but does not use its petals to advertise a reward of nectar. Instead it is pollinated by pollen-gathering bees which release the pollen from the poricidal anthers by grasping them and vibrating ('buzz pollination'). It is therefore possible that conical-papillate cells are of no tactile or temperature-regulating use to this plant, and their role as a visual attractant may also be insignificant in this particular case.

All members of the Brassicaceae so far studied produce conical-papillate petal epidermal cells. However, investigations with *Arabidopsis* have suggested that the developmental programme leading to this specialised cell form is not the same as that operating in *Antirrhinum* and *Petunia*. Ectopic expression of the *MIXTA* gene of *Antirrhinum* in *Arabidopsis* did not result in the formation of conical-papillate cells on any organ (Glover *et al.*, 1998; Payne *et al.*, 1998). Although it is possible that this result simply represents divergence in protein sequence between quite distantly related species of eudicots, analysis of the MYB genes of *Arabidopsis* has reinforced the idea that the developmental programmes are different. Functional analysis of those MYB genes with greatest similarity to the *MIXTA* subgroup has indicated that, although they may play roles in cellular morphogenesis programmes, it does not appear that there is an orthologue of the *MIXTA* gene (Jin and Martin, unpubl. obs.). This suggests that, although the developmental pathway which produces conical-papillate cells may have evolved once in the Brassicaceae, this may be a separate evolutionary origin from that in *Antirrhinum* (and possibly *Petunia*). Interestingly, trichome formation in *Arabidopsis* requires an MYB gene, *GL1*. *Arabidopsis* also expresses a very similar gene to *GL1*, *AtMYB23*, which also promotes epidermal cell differentiation. It is possible that this gene might serve a role in conical cell formation in *Arabidopsis*. No orthologues of *GL1* or *AtMYB23* have been found in species outside the Brassicaceae.

There has been little analysis of conical-papillate petal cell development in any other taxonomic group. However, it is clear that petal development varies between different clades of angiosperms. Although B function genes may be involved in petal development in all species, their expression patterns vary. Kramer and Irish (1999) showed that certain members of the Ranunculaceae, a very primitive eudicot family, have B function genes which are only expressed in the early stages of petal primordium development. In *Antirrhinum*, expression of the B function genes *DEFICIENS* and *GLOBOSA* is necessary throughout petal development or the tissue reverts and develops sepaloid features (Coen and Carpenter, 1992). Expression of these B function genes is also necessary for the expression of *MIXTA* late in petal development and thus for the differentiation of conical-papillate cells. However, if phylogenetically distant species do not have this continuous B function expression then they must either develop conical-papillate cells through a different route or, if they do use a *MIXTA* type developmental programme, control the activation of that programme through different regulatory factors. It would be of great interest to investigate the molecular genetic control of conical-papillate petal cell development in species with limited B function expression in monocots and in more basal angiosperms.

8.6 Conclusions

Investigation of the evolution of adaptive petal cell morphology is at a very early stage. Only one species has been analysed in detail, and two species have been studied to a lesser degree. All three species are eudicots. However, it is already abundantly clear that the selective pressures favouring the evolution of the conical-papillate petal cell form are very varied. These cells influence fruit set and thus fecundity through roles in visual attraction of pollinators, regulation of intrafloral temperature and thus reward availability, and possibly also through tactile cues. The strength of this selective pressure makes the fixing of a gene or developmental programme specifying such a cell form in a population likely to occur extremely rapidly.

It is probable that specialised petal cell morphologies have evolved more than once. Petals themselves are believed to have evolved more than once, and there are a number of variations on the theme of conical-papillate cells within the angiosperms. It is certainly easier to postulate multiple evolutionary origins of a developmental programme which can be acquired by modification of an existing programme (the multicellular trichome pathway) than it is to explain the serial loss of such an advantageous feature in many different angiosperm families. Although there may be adaptive explanations for the loss of conical-papillate cells in certain species, in the majority of cases such a loss would be counter-intuitive. Further evidence for the idea that conical-papillate petal cells may have evolved multiple times comes from the molecular genetic evidence which suggests that multiple mechanisms, sharing some common features, may exist in different species.

If radiations of the angiosperms can be explained in large part by the evolution of animal-attracting flowers then understanding the evolution of the adaptive cell forms on the petals of those flowers is critical. Comparative molecular and developmental analysis of the petal cells of different species should shed further light on this question during the next few years.

ACKNOWLEDGEMENTS

We thank Maria Perez Rodriguez, Hailing Jin and Kiran Bhatt for unpublished data and John Parker for helpful discussion. Work in our laboratories on this subject is funded by the BBSRC, the Royal Society, the Nuffield Foundation and the Gatsby Charitable Trust.

REFERENCES

Coen, E. S. and Carpenter, R. (1992) The power behind the flower. *New Scientist*, 1818, 24–27.

Coen, E. S., Nugent, J. M., Luo, D., Bradley, D., Cubas, P., Chadwick, M., Copsey, L. and Carpenter, R. (1995) Evolution of floral symmetry. *Philosophical Transactions of the Royal Society, Series B*, 350, 35–38.

Comba, L., Corbet, S., Hunt, H., Outram, S., Parker, J. S. and Glover, B. J. (2000) The role of genes influencing the corolla in pollination of *Antirrhinum majus*. *Plant, Cell and Environment*, 23, 639–647.

Corbet, S. A. (1990) Pollination and the weather. *Israel Journal of Botany*, 39, 13–30.

Corbet, S. A., Fussell, M., Ake, R., Fraser, A., Gunson, C., Savage, A. and Smith, K. (1993) Temperature and the pollinating activity of social bees. *Ecological Entomology*, 18, 17–30.

Cubas, P. (2002) The role of TCP genes in the evolution of key morphological characters in Angiosperms, in *Developmental Genetics and Plant Evolution* (eds Q. C. B. Cronk, R. M. Bateman and J. A. Hawkins), Taylor & Francis, London, pp. 247–266.

Darwin, C. (1859) *The Origin of Species by Means of Natural Selection* (1st edn). Murray, London.

Endress, P. K. (1993) Floral structure and the evolution of primitive angiosperms: recent advances. *Plant Systematics and Evolution*, 192, 79–97.

Glover, B. J. and Martin, C. (1998) The role of petal cell shape and pigmentation in pollination success in *Antirrhinum majus*. *Heredity*, 80, 778–784.

Glover, B. J., Perez-Rodriguez, M. and Martin, C. (1998) Development of several epidermal cell types can be specified by the same MYB-related plant transcription factor. *Development*, 125, 3497–3508.

Gorton, H. L. and Vogelmann, T. C. (1996) Effects of epidermal cell shape and pigmentation on optical properties of *Antirrhinum* petals at visible and ultraviolet wavelengths. *Plant Physiology*, 112, 879–888.

Kay, Q. O. N. (1988) More than the eye can see: the unexpected complexity of petal structure. *Plants Today* (July–August), 109–114.

Kay, Q. O. N., Daoud, H. S. and Stirton, C. H. (1981) Pigment distribution, light reflection and cell structure in petals. *Botanical Journal of the Linnean Society*, 83, 57–84.

Kenrick, P. (1999) The family tree flowers. *Nature*, 402, 358–359.

Kevan, P. G. and Lane, M. A. (1985) Flower petal microtexture is a tactile cue for bees. *Proceedings of the National Academy of Sciences USA*, 82, 4750–4752.

Kolosova, N., Shermon, D., Karlson, D. and Dudareva, N. (2001) Cellular and subcellular localisation of S-adenosyl-L-methionine: benzoic acid carboxyl methyltransferase, the enzyme responsible for biosynthesis of the volatile ester methylbenzoate in snapdragon flowers. *Plant Physiology*, 126, 956–964.

Kramer, E. M. and Irish, V. F. (1999) Evolution of genetic mechanisms controlling petal development. *Nature*, 399, 144–148.

Luo, D., Carpenter, R., Vincent, C., Copsey, L. and Coen, E. (1995) Origin of floral asymmetry in *Antirrhinum. Nature*, 383, 794–799.

Martin, C. and Gerats, T. (1993) The control of flower colouration, in *The Molecular Biology of Flowering* (ed. B. R. Jordan), CAB International, Wallingford, pp. 219–255.

Martin, C. and Paz Ares, J. (1997) MYB transcription factors in plants. *Trends in Genetics*, 13, 67–73.

Noda, K. I., Glover, B. J., Linstead, P. and Martin, C. (1994) Flower colour intensity depends on specialized cell shape controlled by a Myb-related transcription factor. *Nature*, 369, 661–664.

Payne, T., Clement, J., Arnold, D. and Lloyd, A. (1998) Heterologous myb genes distinct from GL1 enhance trichome production when overexpressed in *Nicotiana tabacum*. *Development*, 126, 671–682.

Theißen, G., Becker, A., Kirchner, C., Munster, T., Saedler, H. and Winter, K.-U. (2002) How the plants on land learned the floral ABCs: the role of MADS-box genes in the evolutionary origin of flowers, in *Developmental Genetics and Plant Evolution* (eds Q. C. B. Cronk, R. M. Bateman and J. A. Hawkins), Taylor & Francis, London, pp. 173–205.

van Houwelingen, A., Souer, E., Spelt, K., Kloos, D., Mol, J. and Koes, R. (1998) Analysis of flower pigmentation mutants generated by random transposon mutagenesis in *Petunia hybrida*. *Plant Journal*, 13, 39–50.

Weberling, F. (1992) *Morphology of Flowers and Inflorescences*. Cambridge University Press, Cambridge.

Yoshida, K., Kondo, T., Okazaki, Y. and Katou, K. (1995) Cause of blue petal colour. *Nature*, 373, 291.

Chapter 9

How the land plants learned their floral ABCs: the role of MADS-box genes in the evolutionary origin of flowers

Günter Theißen, Annette Becker, Kai-Uwe Winter, Thomas Münster, Charlotte Kirchner and Heinz Saedler

ABSTRACT

Despite dramatic progress in the reconstruction of seed plant phylogeny, understanding the evolutionary origin of flowers has remained one of the most intractable problems of botany and evolutionary biology. Novel attempts to solve the "abominable mystery" of flower origin are currently being made by evolutionary–developmental genetics. Since the structure of the flower is sculpted by floral organ identity genes encoding MADS-domain transcription factors, clarifying the evolution of these genes should in theory provide us with valuable clues concerning flower origin. Current evidence suggests that non-seed land plants such as ferns and mosses lack orthologs of floral organ identity genes. In contrast, extant gymnosperms, being the sister group of the angiosperms, have orthologs of genes that specify (1) petal and stamen or (2) stamen, carpel and ovule identity in flowering plants (class B and class C and D floral organ identity genes, respectively). In addition, the class B genes have a recently identified sister group in both gymnosperms and angiosperms; these B_{sister} genes are involved in the development of ovules and carpels. The presence of orthologs of B, B_{sister}, and C and D genes in diverse gymnosperms and angiosperms suggests that the genetic system for the specification of reproductive organ identity in the angiosperm flower was recruited during evolution from an ancestral and functionally related system, rather than being established *de novo*. This system was already present in the last common ancestor of all extant seed plants about 300 million years ago and may have been involved in sex determination and the specification of reproductive organs. At the molecular genetic level, therefore, flower origin seems far less "abominable" than morphology and paleontology had suggested for more than a century.

A general function of the expression of class B genes (and possibly B_{sister} genes) in seed plants may be to distinguish between male reproductive organs, such as microsporophylls and stamens (where B gene expression is on, and B_{sister} gene expression is off), and female reproductive organs, such as ovules and carpels (where B gene expression is off, but B_{sister} gene expression is on). Differential expression of class B genes (and possibly B_{sister} genes) may thus represent the ancestral sex determination mechanism of all seed plants. Changes in class B gene expression could therefore have been crucial for the establishment of reproductive structures bearing male and female organs on the same growth axis, a key event in flower origin.

A first simple, perianth-less, hermaphroditic flower may have originated from

In *Developmental Genetics and Plant Evolution* (2002) (eds Q. C. B. Cronk, R. M. Bateman and J. A. Hawkins), Taylor & Francis, London, pp. 173–205.

either a male or a female cone of a gymnosperm ancestor, by reduction of class B gene expression in the apical region of the cone, by ectopic expression of class B genes in the basal region of a female cone, or by antagonistic changes in B_{sister} genes. The hermaphroditic arrangement of reproductive organs, together with carpel closure (another key structural innovation of the flowering plants), may have facilitated outcrossing by animal pollinators. Improved outcrossing may have been the initial reason why hermaphroditic "hopeful monster" flowers were positively selected. Co-evolution with their pollinators may have catalyzed the first wave of radiation of the angiosperms, which eventually allowed them to dominate the vast majority of terrestrial ecosystems. The perianth may have originated later than hermaphroditism. Derivation of class A genes (specifying sepals and petals) from floral meristem identity genes, and the recruitment of these and of class B genes to superimpose sepal, petal or tepal identity on a leaf developmental program, may have been crucial processes during perianth origin.

Molecular events that may have led to the critical changes in gene expression and the resulting morphological shifts are discussed. Genes that control class B gene (and B_{sister} gene) expression, and the cis-regulatory elements of the B genes (and B_{sister} genes), are identified as putative key elements of flower evolution.

9.1 Introduction

9.1.1 Enigmas of macroevolution: the Cambrian explosion and an abominable mystery

While we are approaching a fairly good understanding of microevolution (Singh and Krimbas, 2000), the mechanisms of macroevolution – here meaning all evolution above species level – have remained elusive. The origin of complex evolutionary novelties, such as new body plans, has been especially difficult to explain. For example, the fossil record suggests that the body plans of almost all animals, both extant and extinct, were generated within a few million years in the early Cambrian, about 540 million years ago (MYA). This phenomenon has become widely known as the "Cambrian explosion" (Valentine et al., 1999; Conway Morris, 2000). Since animal body plans are largely genetically determined one wonders how the genetic information for an enormous diversity of animals was established in a remarkably short period of time.

Another striking example of the sudden origin of new and diverse forms is the appearance of the flowering plants (angiosperms) in the fossil record of the early Cretaceous, about 100 MYA. This "Cretaceous explosion" has become better known as the "abominable mystery" since Charles Darwin used that often-repeated expression in a letter to his close friend, the botanist Joseph Hooker (Crepet, 1998, 2000; Frohlich, 1999, 2002; Frohlich and Parker, 2000; Ma and dePamphilis, 2000). Darwin was worried about the fact that the angiosperms appear in the fossil record suddenly and in considerable diversity, but without obvious antecedents. This was difficult to reconcile with his idea that evolution should proceed in a countless number of very small steps, a concept since termed "gradualism." More than a century later, the "abominable mystery" remains unsolved (Crepet, 1998, 2000; Frohlich, 1999; Frohlich and Parker, 2000; Ma and dePamphilis, 2000). Consequently, the evolutionary origin of flowers, the prominent and well-known reproductive structures of the angiosperms, has also remained enigmatic (for reviews see Crane *et al.*, 1995; Crepet, 1998, 2000; Frohlich, 1999).

About 300,000 flowering plant species exist on earth today, from which we obtain (directly or indirectly) almost all of our food and a myriad of other benefits, from medicine to cut flowers and lumber. It seems strange that we have nothing more than very vague hypotheses as to how a structure originated that is so familiar and important to us such as the flower. What makes flowering plant origin such a great mystery? For a long time, it seemed that problems understanding the origin of flowering plants were due mainly to an incomplete fossil record and taxonomic uncertainties. The closest relatives of the angiosperms are the gymnosperms, which can be subdivided into four extant groups: conifers, gnetophytes, cycads and *Ginkgo* (with conifers possibly not being monophyletic: Bowe *et al.*, 2000; Chaw *et al.*, 2000). Angiosperms and gymnosperms together constitute a clade termed seed plants (spermatophytes: organisms that reproduce by seeds). However, although angiosperms and gymnosperms are clearly relatively closely related, they are quite different in many respects (Frohlich and Parker, 2000). Among the most important reasons that have been suggested why the origin of flowering plants has remained mysterious are:

a   the large morphological gap between angiosperms and both the living and fossil gymnosperms, which makes homology assessments extremely difficult; and

b   uncertainties in relationships among the major groups of seed plants, and among major groups within the angiosperms, which impede the use of phylogenetic reconstruction to suggest the series of early morphological steps that led to flowers.

The most diagnostic synapomorphies of angiosperms are male reproductive organs (stamens) with two pairs of pollen sacs, and a carpel enclosing one or more ovules (Crane *et al.*, 1995). In contrast, male and female reproductive organs united in a hermaphroditic structure, and the presence of an elaborate perianth surrounding the reproductive organs, are characteristic but not unique features of flowering plants. Gymnosperms typically have unisexual reproductive axes, with male and female cones borne either on the same (conifers) or on different individual plants, and lack a perianth surrounding their sporophylls. An exception are Gnetales, which was one of the reasons for the popular hypothesis that gnetophytes are more closely related to the angiosperms than to the other groups of gymnosperms (Doyle and Donoghue, 1986; Doyle, 1994a, 1996, 1998; Donoghue and Doyle, 2000).

The difficult question of flower origin may usefully be divided:

a   How did carpel, stamen and perianth organs (sepal, petal or tepal) originate?

b   How did male and female organs come together in one intimate structure?

This second question is the long-standing one as to whether flowers were derived from a simple branch with attached sporophylls, or from multiple branches ("euanthial" versus "pseudanthial" scenario: Doyle, 1994a; Crane *et al.*, 1995; Frohlich and Parker, 2000). Modern versions of these scenarios are represented by the "Anthophyte" and the "Neo-Pseudanthial" hypotheses (*sensu* Frohlich and Parker, 2000). Both hypotheses assume a close relationship between gnetophytes and angiosperms. In the Neo-Pseudanthial hypothesis an entire gnetalean reproductive structure, comprising many fertile shoots, was reduced to form a flower (Doyle, 1994a). The Anthophyte hypothesis suggests that the flower-like organization of reproductive structures – sterile bracts or sepals surround male reproductive units, which surround female reproductive organs positioned in the centre – evolved very early, and that this arrangement is homologous (though sometimes reduced) in the members of the Anthophytes (angiosperms, gnetophytes and the wholly fossil gymnosperms Bennettitales). Later, the wholly fossil group Caytoniales, although known not to have male and female structures grouped together, was added to the Anthophytes (Doyle, 1994a; Frohlich and Parker, 2000). Nevertheless, the Anthophyte hypothesis maintains a euanthial scenario of flower origin.

9.1.2 Recent progress in reconstructing angiosperm phylogeny

Our understanding of seed plant phylogeny has dramatically improved due to recent phylogeny reconstructions employing molecular markers from all three plant genomes. A series of studies identified *Amborella trichopoda* as sister to all other angiosperms, and Nymphaeales (water lilies), followed by a clade uniting Illiciaceae, Schisandraceae, Trimeniaceae and Austrobaileyaceae, as the next two successive branches (Mathews and Donoghue, 1999; Parkinson *et al.*, 1999; Qiu *et al.*, 1999; Soltis *et al.*, 1999). The generally basal nature of these "ANITA" clades has already found wide acceptance (for review see Kuzoff and Gasser, 2000); nevertheless, the basalmost position of *Amborella* may be uncertain because of a number of systematic errors during phylogeny reconstructions (Barkman *et al.*, 2000; Graham and Olmstead, 2000). In any case, the new data can be used to reconstruct the flowers of the last common ancestor of extant angiosperms: they probably had an undifferentiated perianth with organs arranged in more than two cycles or a spiral. The carpels in these flowers were probably urn-shaped (ascidiate), were not attached to one another (apocarpous), and possessed margins that did not fuse completely but were closed at maturity by secretion. Differentiated sepals and petals and complete carpel closure probably evolved later in angiosperm evolution (Kuzoff and Gasser, 2000).

Another remarkable recent result concerns relationships among the diverse seed plant groups. The considerable morphological difference between gnetophytes, conifers, cycads and *Ginkgo* has long been taken as evidence for the gymnosperms being paraphyletic. More and more molecular data, however, strongly suggest that extant gymnosperms are in fact monophyletic (Hasebe *et al.*, 1992; Goremykin *et al.*, 1996; Chaw *et al.*, 1997, 2000; Samigullin *et al.*, 1999; Bowe *et al.*, 2000; Frohlich and Parker, 2000) and separated from the lineage that led to angiosperms as early as 300 MYA (Wolfe *et al.*, 1989; Savard *et al.*, 1994; Goremykin *et al.*, 1997). This means that the clade of all extant gymnosperms has to be considered as the sister group of the angiosperms.

9.1.3 The mystery of flower origin: more abominable than ever

During recent years, morphological studies had consistently identified gnetophytes as the extant sister group of the angiosperms, a presumed relationship that gave rise to the Anthophyte hypothesis and was used as a major basis for understanding character evolution leading to flowering plants (for reviews see Doyle, 1994a, 1996,

1998; Donoghue and Doyle, 2000). The now strongly supported hypotheses that gnetophytes are more closely related to conifers than to angiosperms (Hansen *et al.*, 1999; Winter *et al.*, 1999), and that extant gymnosperms are a monophyletic group, suggest that many morphological similarities between angiosperms and gnetophytes (e.g. double fertilization, flower-like reproductive structures) arose independently. This implies that, essentially, all well-established theories of flower origin should be abandoned, because they explicitly assume a close relationship between gnetophytes and angiosperms (Doyle, 1994a; Donoghue and Doyle, 2000). Obviously, new models of flower origin need to be developed.

This leaves us in a paradox: despite the fact that uncertainties in relationships among the different groups of seed plants and flowering plants have often been considered as a major obstacle to understanding angiosperm and flower origin, recent clarifications of seed and flowering plant phylogeny have been of little help in solving the mystery of flower origin. On the contrary, with the structurally diverse gymnosperms as a clade on the one hand, and a relatively elaborate flower at the base of extant angiosperms on the other hand, the apparent morphological gap between gymnosperms and angiosperms has substantially increased, and the mystery of angiosperm origin seems more abominable than ever.

How do we escape from this frustrating conundrum?

9.2 Flower origin studies go evo-devo

One can always hope, of course, that fossils will be found that close the morphological gap between angiosperms and gymnosperms. But there is no guarantee that this will ever happen. The recently discovered *Archaefructus liaoningensis*, claimed to represent the "oldest fossil angiosperm," has such an unusual combination of characters that its phylogenetic position, and thus its relevance for understanding flower origin, is highly ambiguous (Crepet, 1998; Sun *et al.*, 1998). One problem is certainly an incomplete fossil record, but the situation could be even worse: the apparently rapid origin of the angiosperms may reflect a real phenomenon, not just an artefact of a patchy fossil record. Although usually sharply criticized by microevolutionists and population geneticists, ideas about saltational, non-gradualistic evolution at the morphological level are again slowly gaining ground (Bateman and DiMichele, 1994a, 2002; Theißen *et al.*, 2000). But if angiosperms really originated and diversified very rapidly, intermediates between angiosperms and non-angiosperms may be difficult to find.

It seems justified, therefore, to try some alternative approaches to gain a better understanding of flower origin. One of the most promising concepts in that respect is evolutionary developmental genetics ("evo-devo" or "evodevotics"). The rationale of an application of evo-devo to plant evolution, and especially to the problem of flower origin, has already been outlined elsewhere (Theißen and Saedler, 1995; Theißen *et al.*, 2000). In brief, evo-devo assumes that there is a close interrelationship between developmental and evolutionary processes (Gould, 1992; Gilbert *et al.*, 1996). This is because even the most complex organisms are renewed in each generation by developmental processes that generally start with a single cell, the fertilized egg-cell (zygote). In the case of multicellular organisms, evolution of form is thus always the evolution of developmental processes. Since development is largely under genetic control, changes in developmental control genes may be a major starting point for evolutionary changes in morphology (Theißen and Saedler, 1995; Theißen *et al.*, 2000). Understanding the phylogeny of developmental control genes should thus contribute significantly to understanding the evolution of plant and animal form.

Recent research shows that the key developmental control genes are often members of a very limited number of multigene families which encode transcription factors. The paradigm for such a gene family are the homeobox genes (Gehring, 1992), which play key roles in the specification of the animal body plan in both development and evolution (Slack *et al.*, 1993). A subfraction of the homeobox genes, the *HOX* genes, are crucial for differentiating different body regions from each other along the anterior–posterior axis of the developing animal. *HOX* genes are characteristically arranged in genomic clusters. *HOX* genes encode transcription factors which bind to regulatory DNA sequences of their "target genes," and either activate or repress these genes as appropriate for the development of the respective segment of the animal. These "target genes," or genes further downstream, function as "realizator genes," meaning that they encode the enzymes and structural proteins needed for the respective body segment of the animal to develop its characteristic identity.

Thus, in order to apply the rationale of evo-devo to the problem of flower origin we first have to identify those genes that play a role in flower development similar to the *HOX* genes in establishing the animal body plan. To achieve that goal we obviously need some understanding of flower developmental genetics.

9.3 Genetics of flower development

Current knowledge of the genetic basis of flower development largely reflects research on two model plants: thale cress (*Arabidopsis thaliana*) (Figure 9.1) and snapdragon (*Antirrhinum majus*) (for reviews see Coen and Meyerowitz, 1991; Meyerowitz, 1994; Weigel and Meyerowitz, 1994; Theißen and Saedler, 1999; Theißen *et al.*, 2000; Theißen, 2001a, b; Cubas, 2002; Glover and Martin, 2002). Flower development in these species can be subdivided into several major steps, such as floral induction, floral meristem formation and floral organ development. Genetic control of the different steps of flower development is achieved by a hierarchy of interacting regulatory genes, most of which encode transcription factors (Figure 9.1).

Close to the top of the gene hierarchy that controls flower development are so-called "flowering time" genes, which are triggered by developmental cues such as plant age, and by environmental factors such as day length and temperature. Upon floral induction, flowering time genes mediate the switch from vegetative to reproductive development by activating certain classes of "meristem identity genes." These control the transition from vegetative to inflorescence and floral meristems and work as upstream regulators of "floral organ identity genes." In *Arabidopsis* and *Antirrhinum*, floral meristems arise at the flanks of the inflorescence meristems at the shoot apices. Two key floral meristem identity genes are responsible for the transition from inflorescence to floral meristems and the specification of floral meristem identity in *Antirrhinum*, *FLORICAULA* (*FLO*) and *SQUAMOSA* (*SQUA*). The putative orthologs and functional equivalents from *Arabidopsis* are *LEAFY* (*LFY*) and *APETALA1* (*AP1*), respectively. The function of these genes is indicated by the phenotype of loss-of-function mutants. In these mutants, floral meristems often fail to form; secondary inflorescences form instead, indicating that the transition from inflorescence to floral meristems does not occur. Three other floral meristem identity genes support the specification of floral meristem identity in *Arabidopsis*: *APETALA2 (AP2)*, *FRUITFULL (FUL)* and *CAULIFLOWER (CAL)*.

When the transition from inflorescence to floral meristems has occurred, floral organs arise at defined positions from within these meristems under the control of different types of genes. In *Arabidopsis*, "floral meristem size" genes such as *CLAVATA1 (CLV1)*, *CLV2*, *CLV3* and *WIGGUM (WIG = ERA1)* regulate the size of the floral meristem and also influence floral organ number. "Cadastral" genes like *LEUNIG (LUG)*, *AP2* and *AGAMOUS (AG)* are involved in establishing the boundaries of floral organ identity gene functions. "Floral organ pattern" genes such as *PERIANTHIA (PAN)* act to establish floral organ primordia in specific numbers and positions. These primordia develop into the different types of floral organs under the control of specific homeotic selector genes, termed "floral organ identity" genes. Combinatorial interactions of these genes specify the identity of the different floral organs by activating organ-specific "realizator" genes. How organ identity is realized is not exactly clear, because the downstream targets of the "floral homeotic" genes are largely unknown.

The function of floral organ identity genes was recognized during the study of homeotic mutants in which the identity of floral organs is changed. In *Arabidopsis* and *Antirrhinum* such mutants can be categorized in three classes: A, B and C. Ideal Class A mutants have carpels in the first whorl instead of sepals, and stamens in the second whorl instead of petals. Class B mutants have sepals rather than petals in the second whorl, and carpels rather than stamens in the third whorl. Class C mutants have petals instead of stamens in the third whorl, and replacement of the carpels by sepals in the fourth whorl. In addition, these mutants are indeterminate (i.e. there is continued production of mutant floral organs inside the fourth whorl).

Based on these classes of mutants and all combinations of double and triple mutants, the "ABC" model proposes three classes of combinatorially acting floral organ identity genes, called A, B and C, with A specifying sepals in the first floral whorl, A + B specifying petals in the second whorl, B + C specifying stamens in the third whorl, and C specifying carpels in the fourth whorl (Figure 9.1) (Coen and Meyerowitz, 1991; Weigel and Meyerowitz, 1994). The model also maintains that the Class A and Class C genes negatively regulate each other. Based on studies in petunia (*Petunia hybrida*), the ABC model was later extended by Class D genes, specifying ovules (Angenent and Colombo, 1996). Meanwhile, it has been demonstrated by a reverse genetic approach that yet another class of floral organ identity genes, termed Class E genes (Theißen, 2001a), is involved in specifying petals, stamens and carpels (Pelaz *et al.*, 2000).

The floral organ identity genes can be interpreted as acting as major developmental switches that activate the entire genetic program for a particular organ. Combinatorial interactions between the floral organ identity genes are used for the manifestation of positional information ("molecular address labels") during subsequent organogenesis. As outlined in more detail elsewhere (Theißen and Saedler, 1995; Meyerowitz, 1997), the function of the floral organ identity genes is thus quite similar to the function of the *HOX* genes during animal development, although both sets of genes have an independent evolutionary origin (i.e. are not homologous). The floral organ identity genes are thus key players in the drama of

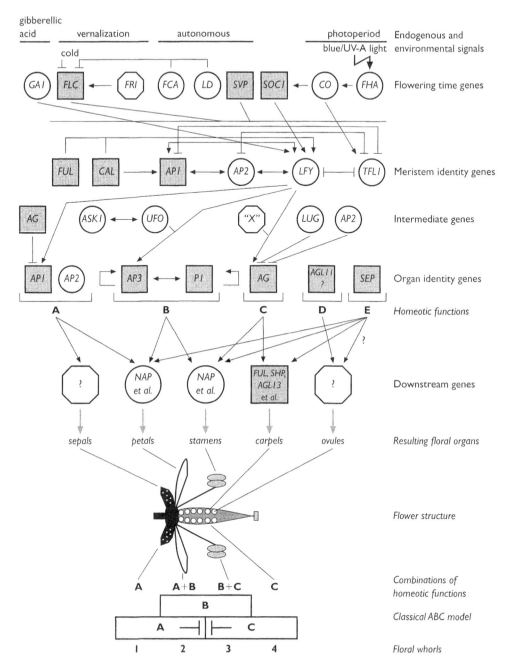

gibberellic acid | vernalization | autonomous | photoperiod | Endogenous and environmental signals
blue/UV-A light

cold

GA1 | FLC ← FRI | FCA | LD | SVP | SOC1 ← CO ← FHA | Flowering time genes

FUL | CAL → AP1 ← AP2 ← → LFY ⊢─┤ TFL1 | Meristem identity genes

AG | ASK1 ← → UFO | "X" | LUG | AP2 | Intermediate genes

AP1 | AP2 | AP3 ← → PI | AG | AGL11 ? | SEP | Organ identity genes

A | B | C | D | E | Homeotic functions

? | NAP et al. | NAP et al. | FUL, SHP, AGL13 et al. | ? | Downstream genes

sepals | petals | stamens | carpels | ovules | Resulting floral organs

Flower structure

A | A+B | B+C | C | Combinations of homeotic functions

B

A ┤├ C | Classical ABC model

1 | 2 | 3 | 4 | Floral whorls

Figure 9.1 A simplified and preliminary depiction of the genetic hierarchy that controls flower development in *Arabidopsis thaliana*. Examples for the different types of genes within each hierarchy level are shown. "Gibberellic acid," "vernalization," "autonomous" and "photoperiod" refer to the different promotion pathways of floral induction, which respond to developmental cues or environmental factors. "Intermediate genes" summarizes a functionally diverse class of genes including "cadastral genes." MADS-box genes are

continued

flower evolution and, therefore, the valuable tools we need in order to better understand the "abominable mystery."

In *Arabidopsis*, Class A genes comprise *APETALA1* (*AP1*) and *APETALA2* (*AP2*). The Class B genes are represented by *APETALA3* (*AP3*) and *PISTILLATA* (*PI*), and the Class C gene is *AGAMOUS* (*AG*). In *Antirrhinum*, the Class B genes comprise *DEFICIENS* (*DEF*) and *GLOBOSA* (*GLO*), and the Class C gene is *PLENA* (*PLE*). Thus far, Class D genes have been recognized only in *Petunia*, where they have been termed *FLORAL BINDING PROTEIN7* (*FBP7*) and *FBP11*. The Class E genes in *Arabidopsis* comprise *SEPALLATA1* (*SEP1*), *SEP2* and *SEP3*, which have highly redundant functions (i.e. any *SEP* gene can functionally replace the two other *SEP* genes).

Recently, the molecular mode of interaction of floral organ identity genes was tentatively identified (Honma and Goto, 2001). There is evidence that these genes encode proteins that assemble to four different ternary or quaternary complexes, one for each type of floral organ (Figure 9.2). These protein complexes, assumed to represent transcription factors, may exert their function by binding to the promoters of target genes, which they either activate or repress as appropriate for the development of the identities of the different floral organs. According to this "quartet model" (Figure 9.2: Theißen, 2001a; Theißen and Saedler, 2001), two protein dimers of each tetramer recognize two different DNA sites (termed CArG-boxes: consensus sequence 5'-CC(A/T)$_6$GG-3') which are brought into close vicinity by DNA bending.

Among the regulators of the floral organ identity genes in *Arabidopsis* is the transcription factor LFY (Parcy *et al.*, 1998). LFY alone can induce expression of the Class A gene *AP1*; other, flower- or region-specific co-regulators are not needed. In contrast, the Class B gene *AP3* and the Class C gene *AG* are activated by LFY in region-specific patterns within flowers, depending on other factors such as the F-box gene *UNUSUAL FLORAL ORGANS* (*UFO*) in case of *AP3* and an unknown

shown as squares and are highlighted by shading, non-MADS-box genes as circles, and genes whose sequence has not previously been reported as octagons. Some regulatory interactions between the genes are symbolized by arrows (activation), double arrows (synergistic interaction) or barred lines (inhibition, antagonistic interaction). Only a few selected examples of the known genes and interactions involved in flower development are shown. In the case of downstream genes, just one symbol is shown for every type of floral organ, though whole cascades of many direct target genes and further downstream genes are probably activated in each organ of the flower. A generic flower structure is shown in the lower region of the figure, and at the bottom of the figure, the classical "ABC model" of flower organ identity is depicted. According to this model, flower organ identity is specified by three classes of floral organ identity genes that provide homeotic functions A, B and C, which are each active in two adjacent whorls. Function A alone specifies sepals in whorl 1; the combined activities of A + B specify petals in whorl 2; B + C specify stamens in whorl 3; and C alone specifies carpels in whorl 4. The activities A and C are mutually antagonistic, as indicated by barred lines: A prevents the activity of C in whorls 1 and 2, and C prevents the activity of A in whorls 3 and 4. Abbreviations of gene names used: *AG, AGAMOUS; AGL, AGAMOUS-LIKE GENE; AP, APETALA; ASK1, ARABIDOPSIS SKP1-LIKE1; CAL, CAULIFLOWER; CO, CONSTANS; FLC, FLOWERING LOCUS C; FRI, FRIGIDA; FUL, FRUITFULL; LD, LUMINIDEPENDENS; LFY, LEAFY; LUG, LEUNIG; NAP, NAC-LIKE, ACTIVATED BY AP3/PI; PI, PISTILLATA; SEP, SEPALLATA; SHP, SHATTERPROOF; SOC1, SUPPRESSOR OF OVEREXPRESSION OF CO1; SVP, SHORT VEGETATIVE PHASE; UFO, UNUSUAL FLORAL ORGANS; TFL1, TERMINAL FLOWER1.*

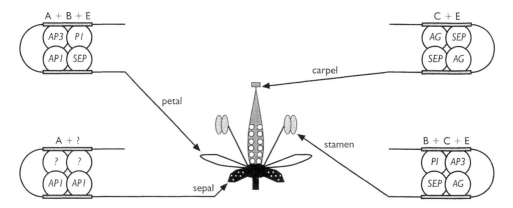

Figure 9.2 The "quartet model" of floral organ identity in *Arabidopsis*. In the centre is shown a generic flower structure with four types of organs: sepals, petals, stamens and carpels. The identity of these organs is, according to the "quartet" model (Theißen, 2001a), determined by four unique combinations of floral homeotic proteins, depicted around the flower. The experimental basis for at least parts of the model has recently been published (Egea-Cortines *et al.*, 1999; Pelaz *et al.*, 2000; Honma and Goto, 2001). The protein quartets, assumed to represent transcription factors, may exert their function by binding to the promoters of target genes, which they either activate or repress as appropriate for the development of the identities of the different floral organs. Above the quartets, the combinations of classes of floral organ identity proteins represented by these complexes is shown, with AP1 being the Class A protein, AP3 and PI being the Class B proteins, AG being the Class C protein, and SEP symbolizing the Class E proteins. According to this model two protein dimers of each tetramer recognize two different DNA sites (termed CArG-boxes; consensus sequence 5'-CC(A/T)$_6$GG-3'), shown here as gray boxes, which are brought into close vicinity by DNA bending. Note that the exact structures of the multimeric complexes of MADS-domain proteins controlling the identity of flower organs are still hypothetical. Question marks denote components whose identity is especially uncertain. For a more detailed description of the model see Theißen (2001a). Abbreviations for protein names used: AG, AGAMOUS; AP1, APETALA1; AP3, APETALA3; PI, PISTILLATA; SEP, SEPALLATA.

factor "X" in case of *AG* (Figure 9.1). Recently, it was shown that *AP1* and *AG* are direct downstream targets of LFY (Busch *et al.*, 1999; Wagner *et al.*, 1999). *LFY* and *FLO*, the *LFY* ortholog from *Antirrhinum*, are members of a family termed *FLO*-like genes (Theißen, 2000a).

Many of the genes encoding transcription factors within the gene hierarchy contributing to flower development are MADS-box genes (Figure 9.1). This gene type is well represented at the levels of flowering time genes, meristem identity genes, intermediate genes and downstream genes. MADS-box genes are even dominant among the floral organ identity genes; cloning of all the floral organ identity genes mentioned above revealed that all but *AP2* belong to the MADS-box gene family. Therefore, the origin of floral organ identity genes can only be understood in the context of MADS-box gene phylogeny. And since floral organ identity strictly depends on the activity of the floral organ identity genes, an understanding of flower origin also critically depends on insights into MADS-box gene phylogeny. So what are MADS-box genes, and how did they evolve?

9.4 MADS-box genes: a primer

The defining characteristic of all MADS-box genes is the presence of a highly conserved DNA sequence, approximately 180 nucleotides long, that is termed the MADS-box. It encodes the DNA-binding domain of the respective MADS-domain transcription factors (for recent reviews of MADS-box genes in plants, see Theißen *et al.*, 1996, 2000; Riechmann and Meyerowitz, 1997). "MADS" is an acronym for the four founder proteins <u>M</u>CM1 (from brewer's yeast, *Saccharomyces cerevisiae*), <u>A</u>GAMOUS (from *Arabidopsis*), <u>D</u>EFICIENS (from *Antirrhinum*), and <u>S</u>RF (from humans), on which the definition of this gene family is based (Schwarz-Sommer *et al.*, 1990). MADS-box genes control many developmental processes in flowering plants during both vegetative and reproductive growth.

MADS-domain proteins, like many other eukaryotic transcription factors, have a modular structural organization. MADS-box genes in total are structurally and functionally very diverse (Theißen *et al.*, 2000), but the vast majority of plant MADS-box genes known so far belong to a single clade of genes with a conserved structural organization, the so-called MIKC-type domain structure (reviewed by Becker *et al.*, 2000; Theißen *et al.*, 2000), including a MADS (M-), intervening (I-), keratin-like (K-) and C-terminal (C-) domain.

The MADS-domain is by far the most highly conserved region of the proteins (Purugganan *et al.*, 1995). In most cases, it is found at the N-terminus of the putative proteins, although some plant proteins contain additional residues N-terminal to the MADS-domain (termed NMIKC-type proteins). The MADS-domain is the major determinant of DNA-binding, but it also performs dimerization and accessory factor binding functions. Part of it folds into an unprecedented structural motif for DNA interaction, an antiparallel coiled coil of α-helices that lies flat on the DNA minor groove (Pellegrini *et al.*, 1995). In line with the conserved nature of their DNA-binding domain, MADS-domain proteins bind to similar DNA sites based on the consensus sequence $CC(A/T)_6GG$, which is called a "CArG-box" (for "CC-A rich-GG").

The I-domain, directly downstream of the MADS-domain, typically comprises approximately thirty amino acids, but is quite variable in length. It is only relatively weakly conserved among plant MADS-domain proteins (Purugganan *et al.*, 1995; Becker *et al.*, 2000). In some *Arabidopsis* MADS-domain proteins, the I-domain constitutes a key molecular determinant for the selective formation of DNA-binding dimers (Riechmann *et al.*, 1996; Riechmann and Meyerowitz, 1997). The K-domain is characterized by a conserved, regular spacing of hydrophobic residues, which is hypothesized to allow for the formation of an amphipathic helix. It is assumed that such an amphipathic helix interacts with that of another K-domain containing protein to promote dimerization (Riechmann and Meyerowitz, 1997). The K-domain is absent from any of the animal and fungal MADS-domain proteins known so far, but also from many plant proteins (Fischer *et al.*, 1995; Alvarez-Buylla *et al.*, 2000b; Riechmann *et al.*, 2000). The most variable region is the C-domain, located at the C-terminal end of the MADS-domain proteins. In some MADS-domain proteins it is involved in transcriptional activation (Cho *et al.*, 1999), or in the formation of ternary or quaternary complexes (Egea-Cortines *et al.*, 1999).

In contrast to the *HOX* genes of animals, which are organized in genomic clusters, the MADS-box genes of plants are scattered throughout the plant genomes (Fischer *et al.*, 1995; Alvarez-Buylla *et al.*, 2000a; Riechmann *et al.*, 2000).

9.5 Specification of floral organ identity is a MADS-box gene (sub)family business

According to the reasoning of evo-devo, understanding the origin and evolution of flower development depends on an understanding of the origin and evolution of the gene network governing the ontogeny of the flower. Changes in gene number, expression and interaction could all have contributed to the evolution of flowers. Since MADS-box genes play such an important role in the network of flower development (Figure 9.1), understanding the phylogeny of MADS-box genes might strongly improve our understanding of flower evolution. Since the identity of the floral organs is wholly dependent on the activity of the floral organ identity genes, the origin and evolution of these genes is of special interest.

Phylogeny reconstructions demonstrated that the MADS-box gene family is composed of several well-defined gene clades (Doyle, 1994b; Purugganan *et al.*, 1995; Theißen and Saedler, 1995; Theißen *et al.*, 1996, 2000; Hasebe and Banks, 1997; Münster *et al.*, 1997). Most clade members share highly related functions and similar expression patterns (Figure 9.3). For example, the Class A, B and C floral homeotic genes each fall into separate clades, namely *SQUAMOSA*- (Class A), *DEFICIENS*- or *GLOBOSA*- (Class B), and *AGAMOUS*-like genes (Class C) (Doyle, 1994b; Purugganan *et al.*, 1995; Theißen *et al.*, 1996) (for rules to allow consistent naming of MADS-box gene clades, see Theißen *et al.*, 1996). The Class D genes determining ovule identity (Angenent and Colombo, 1996) also belong to the

---

Figure 9.3 Recruitment of MADS-box genes during evolution of the flower. The phylogeny of selected plant MADS-box genes (a) focuses on floral genes from *Arabidopsis* and some other genes for which the function is known. More detailed trees have been published elsewhere (Hasebe, 1999; Becker *et al.*, 2000; Theißen *et al.*, 2000). At the terminal branches gene names, genus names of species of origin, and the function of the respective genes (if known) are indicated. "A," "B," "C," "D" and "E" symbolize the different classes of floral organ identity genes involved in specifying sepal and petal (A), petal and stamen (B), stamen and carpel (C), ovule (D), or sepal, stamen and carpel (E) identity, respectively (for reviews see Meyerowitz, 1994; Weigel and Meyerowitz, 1994; Angenent and Colombo, 1996; Theißen *et al.*, 2000; Theißen, 2001a, b). The gymnosperms *Picea*, *Pinus* and *Gnetum* are highlighted by grey boxes; all other taxa are flowering plants. Only a few of the many known members of each clade are shown (Theißen *et al.*, 2000). The separation of gymnosperm and angiosperm clade members about 300 MYA is indicated by black dots (topology of the *GGM13*-like genes may be incorrect; compare with Figure 9.5). The functions of the *Arabidopsis* genes involved in flower development are mapped on a scheme of a generic flower (b). Abbreviations for genes used: ABS, *ARABIDOPSIS BSISTER GENE*; AeAP3-2; *ASARUM EUROPAEUM AP3-LIKE GENE2* (a misnomer); AG, *AGAMOUS*; AGL, *AGAMOUS-LIKE* (sensu Ma et al., 1991); AP, *APETALA*; CAL, *CAULIFLOWER*; DAL, *DEFICIENS/AGAMOUS-LIKE*; DAPI, *DELPHINIUM AJACIS PI GENE*; DEFH, *DEFICIENS-HOMOLOG*; FBP, *FLORAL BINDING PROTEIN*; FUL, *FRUITFULL*; GGM, *GNETUM GNEMON MADS*; GLO, *GLOBOSA*; OSMADS2, *ORYZA SATIVA MADS2*; PI, *PISTILLATA*; PLE, *PLENA*; PRDGL, *PINUS RADIATA DEFICIENS GLOBOSA-LIKE*; SHP, *SHATTERPROOF*; SQUA, *SQUAMOSA*; TM6, *TOMATO GENE6*; ZMM17, *ZEA MAYS MADS17*.

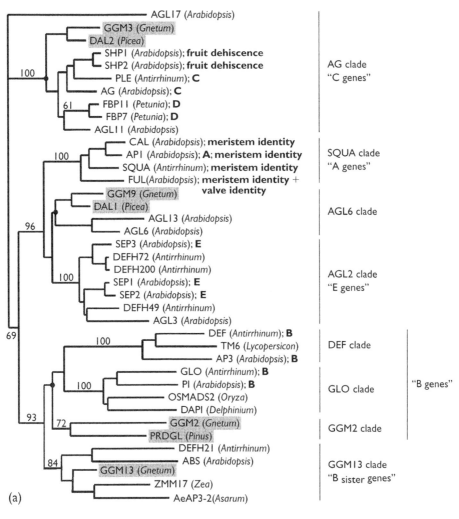

(a)

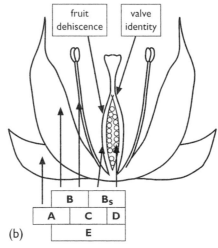

(b)

clade of *AGAMOUS*-like genes (Theißen *et al.*, 1996). Class E genes belong to a clade termed *AGL2*-like genes (Theißen, 2001a).

These findings strongly suggest that the establishment of the aforementioned gene clades was an important pre-requisite for the establishment of the floral homeotic genes and their functions (Theißen *et al.*, 1996). To answer the key question as to when these gene clades arose during evolution and how some of their members were transformed into floral homeotic genes, the phylogeny of MADS-box genes had to be reconstructed and superimposed on credible phylogenies of land plants (e.g. Figure 9.4). Toward that goal, MADS-box genes had to be studied in phylogenetically informative taxa, including the putative sister group of the angiosperms, the gymnosperms, but also in non-seed plants.

9.6 Origin of MADS-box gene subfamilies in the evolution of land plants

Since all floral organ identity genes that belong to the MADS-box gene family have a MIKC-type domain structure the following discussion focuses on this clade of genes.

The isolation of several MADS-box genes from the model moss *Physcomitrella patens* (Krogan and Ashton, 2000; Henschel, Münster and Theißen, unpubl. obs.) strongly suggests that the most recent common ancestor of mosses and vascular plants, which existed about 425 MYA, already possessed at least one MADS-box gene with a MIKC-type domain structure (Figure 9.4).

A detailed characterization of the MADS-box gene family in the pteridophytes *Ceratopteris richardii* and *C. pteroides* (leptosporangiate ferns) and *Ophioglossum pedunculosum* (eusporangiate fern) (Kofuji and Yamaguchi, 1997; Münster *et al.*, 1997; Hasebe *et al.*, 1998; Theißen *et al.*, 2000; Münster, Faigl and Theißen, unpubl. obs.) suggests that, by about 400 MYA, the most recent common ancestor of ferns and seed plants already contained at least two different MIKC-type MADS-box genes (Münster *et al.*, 1997; Theißen *et al.*, 2000). These genes probably had expression patterns and functions that were more ubiqitous than those of the highly specialized floral organ identity genes from extant flowering plants, as suggested by the expression patterns of genes from extant ferns. However, no evidence has yet been obtained that orthologs of floral homeotic genes occur in ferns. Thus, the last common ancestors of extant ferns and seed plants probably lacked orthologs of floral homeotic genes (Theißen *et al.*, 2000). The floral homeotic gene lineages must therefore have been established in the lineage that led to the flowering plants after its separation from the fern lineage, but most probably before the radiation of the flowering plants. The only extant taxa whose lineage branched off from the lineage that led to flowering plants during that critical time interval are the gymnosperms (Figure 9.4).

Extensive characterizations of the MADS-box gene families in the gnetophytes (*Gnetum* species), conifers (*Picea* and *Pinus* species) and *Ginkgo* (Tandre *et al.*, 1995, 1998; Mouradov *et al.*, 1998b, 1999; Rutledge *et al.*, 1998; Hasebe, 1999; Shindo *et al.*, 1999; Sundström *et al.*, 1999; Winter *et al.*, 1999; Becker *et al.*, 2000) revealed that the MADS-box gene families of extant gymnosperms and angiosperms are of similar complexity (Becker *et al.*, 2000). Phylogeny reconstructions allowed the identification of gene clades containing putative orthologs from both angiosperm and gymnosperm lineages. Thus, the minimum number of MADS-box genes already

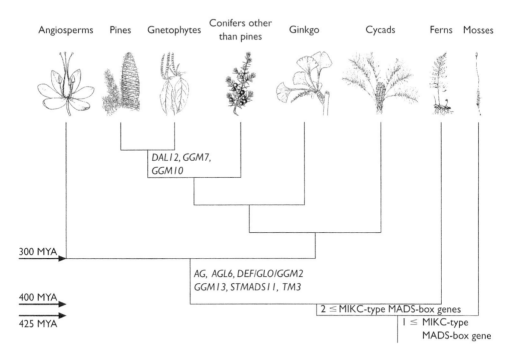

Figure 9.4 The ancestry of some clades of MIKC-type MADS-box genes in the evolution of land plants. A phylogenetic tree of some major taxa of land plants is shown. "Ferns" is used in a broad sense, including also horsetails and whisk ferns (but not clubmosses, which are not shown here). The topology of the tree, which is still controversial, has been compiled from some recent publications (Bowe *et al.*, 2000; Chaw *et al.*, 2000; Pryer *et al.*, 2001). The ages (in MYA) given at three nodes of the tree are approximations. The gene names beside internal branches denote gene clades, not single genes. These clades have been established during the time interval represented by the respective branches of the phylogenetic tree, at the latest. This could be concluded from the distribution of respective clade member genes among extant taxa. For example, *AG-, AGL6-, DEF/GLO/GGM2-, GGM13-, STMADS11-* and *TM3*-like genes have already been isolated from angiosperms and gymnosperms, but not from ferns (Becker *et al.*, 2000). "*DEF/GLO/GGM2*" denotes the B gene superclade. Gene clades that have been found so far only in one of the taxonomic groups represented are not shown here. "2 ≤ MIKC-type genes" denotes that the last common ancestor of ferns and seed plants already possessed at least two MIKC-type MADS-box genes (Münster *et al.*, 1997). "1 ≤ MIKC-type gene" indicates that the last common ancestor of mosses and vascular plants already possessed at least one MIKC-type MADS-box gene (Krogan and Ashton, 2000).

present in the last common ancestor of extant gymnosperms and angiosperms can be estimated. Taken together, the available data suggest that there were already at least six different MADS-box genes present at the base of extant seed plants about 300 MYA (Figure 9.4: Becker *et al.*, 2000). Despite several serious attempts, obvious orthologs from ferns or more basal plants have not yet been isolated. It therefore seems that the respective gene clades were established 300–400 MYA (Becker *et al.*, 2000), although molecular clock estimates indicate a considerable greater age for at least some of the gene clades (Purugganan *et al.*, 1995; Purugganan, 1997).

Comparative expression studies involving pairs of putatively orthologous genes revealed a diversity of patterns in both vegetative and reproductive organs that obviously has been largely conserved since the time when the angiosperm and gymnosperm lineages diverged (Becker *et al.*, 2000). Ancestral clade members are likely to have already acquired diverse functions.

9.7 On the origin and ancestral function of floral organ identity genes

Among the gene clades that were found in both gymnosperms and angiosperms are the *AG*-like genes (*sensu* Theißen *et al.*, 1996): floral homeotic Class C/Class D genes or their orthologs (Tandre *et al.*, 1995; Rutledge *et al.*, 1998; Hasebe, 1999; Winter *et al.*, 1999), and the *DEF/GLO/GGM2*-like genes, being Class B genes or orthologs thereof (Mouradov *et al.*, 1999; Sundström *et al.*, 1999; Winter *et al.*, 1999) (Figure 9.4). For simplicity, members of the *AG* clade will henceforth be termed "C genes," and *DEF-*, *GLO-* and *GGM2*-like genes will be called "B genes." This notation refers to the genealogy of these genes, and not necessarily to their function; it contrasts with the term "Class C gene," which denotes a specific type of "floral organ identity gene."

Arabidopsis plants ectopically expressing the C genes from conifers or *Gnetum* resemble plants ectopically expressing *AG* itself, suggesting that the conifer genes can substitute at least some functional aspects of the Class C floral homeotic genes in a flowering plant background (Rutledge *et al.*, 1998; Tandre *et al.*, 1998; Kirchner and Theißen, unpubl. obs.). Similarly, the "B gene" *GGM2* from *Gnetum* can substitute at least parts of the function of the endogenous B genes from *Arabidopsis* (Winter, 2000; Winter and Theißen, unpubl. obs.). The fact that a gene is able to (partially) substitute the function of another gene in another species does not always identify the function of that gene in its species of origin. Class C genes of flowering plants specify stamens and carpels, and Class B genes specify stamens and petals (Figures 9.1–9.3), but what could be the function of orthologs of these floral organ identity genes in taxa that do not form flowers, and thus do not form petals, stamens and carpels?

Based on expression studies, some educated guesses can be made. Since the Class C and D genes of angiosperms specify the identity of reproductive organs (stamens and carpels, or ovules, respectively), and since their putative orthologs from gymnosperms are also expressed in both male and female reproductive units but not in vegetative organs (Tandre *et al.*, 1995, 1998; Rutledge *et al.*, 1998; Winter *et al.*, 1999), the function of the expression of Class C and D gene orthologs ("C genes") in gymnosperms might be to distinguish between reproductive organs (where expression is on) and non-reproductive organs (where expression is off), and to specify ovules. Due to the fact that the Class B genes of angiosperms specify stamens (male organs), but not carpels (female organs), and because the orthologs from *Gnetum* and conifers are also exclusively expressed in male reproductive units (Mouradov *et al.*, 1999; Sundström *et al.*, 1999; Winter *et al.*, 1999), it might be the function of the expression of Class B gene orthologs in gymnosperms to distinguish between male reproductive organs (where expression is on) and female reproductive organs (where expression is off) (Theißen *et al.*, 2000). Differential expression of B genes may thus represent the ancestral sex-determination mechanism of all seed plants (Winter *et al.*, 1999). Genes that control B gene expression and the cis-regulatory elements of the B genes may, therefore, be key elements of flower evolution.

Homologs of the *LFY* gene, the best characterized upstream regulator of the floral homeotic genes, have already been isolated from conifers (Mellerowicz *et al.*, 1998; Mouradov *et al.*, 1998a). But when and how did the floral homeotic Class A genes come into play?

The fact that the Class A gene function is involved in specifying organ identity within the organs of the floral perianth and that gymnosperms are unlikely to have homologous organs suggests a relatively recent origin of the Class A gene function at the base of the flowering plants. In line with this conclusion, putative orthologs of MADS-type Class A genes (*SQUA*-like genes) have not yet been isolated from extant gymnosperms and thus may be absent. Class A function may have been established independently several times during angiosperm evolution (Theißen *et al.*, 2000). In the case of *Arabidopsis*, the two Class A genes (*AP1* and *AP2*) also function as meristem identity genes, and we suggest that, at least in some cases, the Class A gene function could be derived from the function that determines meristem identity (Theißen *et al.*, 2000). In line with this interpretation, Class A genes or their orthologs (such as *SQUAMOSA* from *Antirrhinum*) are often expressed in several whorls of the flower and in non-floral organs, not just in sepals and petals (Huijser *et al.*, 1992; Jofuku *et al.*, 1994). Expression studies of the non-MADS Class A gene *AP2* suggest an even more ancestral function in ovule or seed development (Jofuku *et al.*, 1994). Thus, orthologs of *AP2* may have evolved step by step from genes involved in seed formation, via genes that are also involved in specifying inflorescence and floral meristem identity to genes that also play a role in specifying organ identity in the perianth of the flower (Theißen *et al.*, 2000).

The presence of orthologs of Class B and Class C genes and their upstream regulators in diverse gymnosperms suggests that the system for the specification of reproductive organ identity in angiosperms, rather than being established from scratch, was recruited from a similar system (involved in sex determination and the specification of reproductive organs) that was already present in the last common ancestor of all extant seed plants, about 300 MYA (Theißen *et al.*, 2000). At the molecular genetic level, therefore, flower origin seems far less abominable than morphological studies had suggested for more than a century.

This evolutionary scenario can be compared with the establishment of some (at least economically) successful computer programs that have been derived from more ancient versions rather than being developed from first principles. While this way of establishing novelties has some well-known technical disadvantages, evolutionary novelties (unlike human inventions) cannot start from scratch, because the thread of life (the successive chain of life histories) must not be broken at any time if the lineage is to survive.

Interestingly, during flower origin and diversification, the above-mentioned clades from the prolific MADS-box gene family did not only provide the floral organ identity genes, but also genes with other functions during flower development, such as genes specifying substructures of floral organs, or those controlling seed dehiscence (Figure 9.3: Liljegren *et al.*, 2000; Theißen, 2000b).

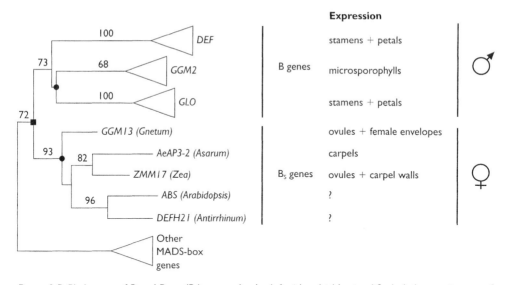

Figure 9.5 Phylogeny of B and B$_{sister}$ (B$_s$) genes. At the left side a highly simplified phylogenetic tree of the *DEF-*, *GLO-*, *GGM2*-(B) and *GGM13*-like (B$_s$) genes is shown, with "*DEF*," "*GLO*" and "*GGM2*" symbolizing gene clades, not individual genes. The complete tree involving a diverse set of plant MADS-box genes known is available via the world wide web (http://www.mpiz-koeln.mpg.de/mads/). Numbers adjacent to some nodes are bootstrap percentages. Genus names of species from which the B$_s$ genes were isolated are given in parentheses after the gene names. The separation of B and B$_s$ genes into orthologous gene lineages in gymnosperms and angiosperms, caused by the split of gymnosperm and angiosperm lineages about 300 MYA, is symbolized by black dots. The split of an ancestral gene lineage into B and B$_s$ genes, assumed to have occurred 300–400 MYA, is indicated by a black square. Predominant sites of expression of the different types of genes are summarized at the terminal branches of the tree. The GenBank/EMBL/DDBJ accession number of the unpublished *Arabidopsis* gene *ABS* (*Arabidopsis* B$_{sister}$) is AB007648.

9.8 The B$_{sister}$ hypothesis

The tale about the floral homeotic genes in seed plants recently acquired yet another new twist. It started with the characterization of MADS-box genes from the gnetophyte gymnosperm *Gnetum gnemon* (Winter *et al.*, 1999; Becker *et al.*, 2000). In phylogeny reconstructions one of these genes, termed *GGM13*, appeared to be quite closely related to the B genes (Figure 9.5: Winter *et al.*, 1999; Becker *et al.*, 2000; Theißen *et al.*, 2000). In addition, two other features strongly supported a close relationship between *GGM13* and the B genes. Sequence alignments with other MADS-domain proteins revealed that the conceptual GGM13 protein, like all B proteins, has an I-domain that is characteristically shorter than that of other MADS-domain proteins (Becker *et al.*, 2002). Moreover, the GGM13 protein contains a subterminal "PI Motif" sequence and a terminal "PaleoAP3 Motif" at its C-terminal end (Figure 9.6). PI motifs and derived sequences have so far been found only in B proteins, and PaleoAP3 motifs only in a subset of B proteins from both gymnosperms and angiosperms (Figure 9.6: Kramer *et al.*, 1998; Mouradov *et al.*, 1999; Sundström *et al.*, 1999; Winter *et al.*, 1999).

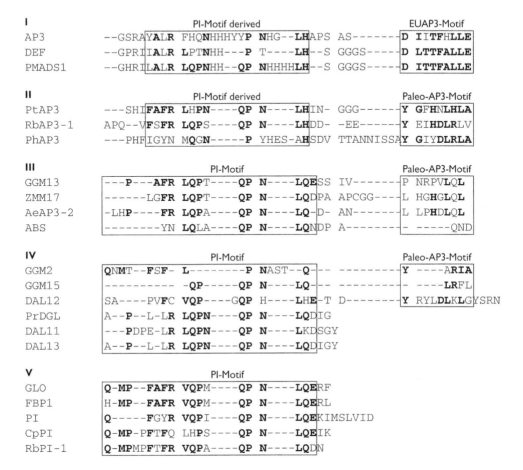

Figure 9.6 B$_{sister}$ proteins share conserved sequence motifs in their C-terminal regions with B proteins. An alignment of C-terminal regions of some B and B$_{sister}$ proteins from angiosperms and gymnosperms is shown. (I) DEF-like proteins from higher eudicotyledonous flowering plants (from top to bottom: AP3, *Arabidopsis thaliana*, DEF, *Antirrhinum majus*, PMADS1, *Petunia hybrida*). (II) DEF-like proteins from lower eudicots (PtAP3, *Pachysandra terminalis*; RbAP3-1, *Ranunculus bulbosus*) or a basal angiosperm/magnoliid dicot (PhAP3, *Peperomia hirta*). (III) B$_{sister}$ proteins, from a gymnosperm (GGM13, *Gnetum gnemon*), a monocot (ZMM17, *Zea mays*), a basal angiosperm (AeAP3-2, *Asarum europaeum*), and a higher eudicot (ABS, *Arabidopsis thaliana*). (IV) B proteins from gymnosperms (GGMn, *Gnetum gnemon*; DALn, *Picea abies*; PrDGL, *Pinus radiata*). (V) GLO-like proteins from higher eudicots (GLOBOSA, *Antirrhinum majus*; FBP1, *Petunia hybrida*; PI, *Arabidopsis thaliana*) or lower eudicots, respectively (CpPI, *Caltha palustris*; RbPI-1, *Ranunculus bulbosus*). Gene names follow Kramer and Irish (2000). Regions defining motifs identified by Kramer *et al.* (1998) have been boxed. Within boxes of the alignment, the most frequent amino acids at any position are highlighted in bold.

The fact that the B genes were obviously important for the evolution of male reproductive organs, that is microsporophylls and stamens (Kramer et al., 1998; Mouradov et al., 1999; Sundström et al., 1999; Winter et al., 1999; Theißen et al., 2000), increased our interest in the conservation and evolutionary importance of *GGM13* and its putative close relatives. Isolation of *ZMM17* by screening of a maize (*Zea mays* subsp. *mays*) cDNA library representing mRNAs from immature ears soon revealed that *GGM13*-like genes are also present in angiosperms (Figure 9.5), and suggested that the most recent common ancestor of angiosperms and gymnosperms already possessed such a gene 300 MYA (Becker et al., 2002). *ZMM17* is a single-copy gene on the small arm of maize chromosome 5, in a region where no developmental mutation has been localized thus far (Werth and Theißen, unpubl. obs.).

Sequence comparisons and phylogeny reconstructions revealed that *GGM13* and *ZMM17* constitute a well-supported gene clade (Figure 9.5). This clade of *GGM13*-like genes represents the sister clade (i.e. the closest known relatives) of the B genes and was therefore termed B_{sister} (B_s) genes (Figure 9.5). In line with this, the ZMM17 protein shows also the gap in the I-domain (Becker et al., 2002) and the PI Motif-Derived and Paleo-AP3 Motif sequences (Figure 9.6) that characterize B proteins and GGM13. We thus interpret a relatively short I-domain and the presence of PI Motif-Derived and PaleoAP3 Motif sequences (Figure 9.6) as synapomorphies of members of the B plus B_s genes "superclade." Although the function of the sequence motifs in the C-domain remains unknown, there is evidence that the I-domain is a key molecular determinant for the selective formation of DNA-binding dimers (Riechmann et al., 1996). It seems possible, therefore, that both B and B_s proteins form dimers only with members of the B plus B_s genes superclade, but not with other MADS-domain proteins. Additional dimerization specificities evolved within the B proteins, because DEF- and GLO-like proteins from eudicots form heterodimers but do not homodimerize (Davies et al., 1996), unlike the B proteins from gymnosperms and monocot GLO-like proteins (Winter et al., 2002). As most of the other MADS-domain proteins that have been tested can homodimerize, the capacity to homodimerize probably represents the ancestral state.

A specific B_s gene, termed *AeAP3-2*, was also reported from the relatively primitive extant angiosperm *Asarum europaeum*, but was misclassified as a *DEF*-like gene (*AP3* is the *DEF* ortholog from *Arabidopsis*), implying that *AeAP3-2* was considered a B rather than a B_s gene (Kramer and Irish, 2000). Ironically, B_s genes from the eudicotyledonous model plant species *Antirrhinum* (*DEFH21*) and *Arabidopsis* (*ABS*) (Becker et al., 2002) (Figure 9.5) were identified only after genes from a gymnosperm, a monocot and a basal angiosperm had already been characterized.

Surprising (but probably quite revealing) aspects of the B_s genes are their expression patterns. Whereas their closest relatives, the B genes, are predominantly expressed in male reproductive organs (and angiosperm petals), B_s genes are predominantly expressed in female reproductive organs, specifically ovules and organs surrounding the ovules (Figure 9.5: Kramer and Irish, 2000; Becker et al., 2002).

Our phylogeny reconstructions (Figure 9.5) suggest that the first B and B_s genes were generated by a gene duplication event somewhere in the lineage that led to extant seed plants, a considerable period of time before the existence of the last common ancestor of extant seed plants at about 300 MYA (because gymnosperms as well as angiosperms have both B and B_s genes), but after the split of the seed plant lineage from the fern lineage at about 400 MYA (because there is no evidence for the presence of either B or B_s genes in ferns). Their conservation for at least 300 million years suggests an important function for B_s genes. Most well-characterized MADS-box genes from plants are faithfully expressed where they exert their function. For example, floral organ identity genes are generally transcribed in the organs whose identity they specify (Weigel and Meyerowitz, 1994; Theißen and Saedler, 1995; Angenent and Colombo, 1996; Theißen et al., 1996, 2000; Riechmann and Meyerowitz, 1997). It seems likely, therefore, that B_s genes influence the development of female reproductive organs; they may be involved in the specification of ovules or carpels, or substructures thereof. In the classic ABC model of floral organ specification (Coen and Meyerowitz, 1991; Weigel and Meyerowitz, 1994), Class C genes are involved in specifying both male and female reproductive organs (stamens and carpels), and the Class B genes distinguish between male reproductive organs (where B gene expression is on) and female reproductive organs (where B gene expression is off). The data presented here make it conceivable, however, that female organs are not simply defined by "C on, B off," but rather by "C on, B_s on" (in contrast to the male condition "C on, B on"). This B_{sister} hypothesis prompts us to suggest an extended ABC model, which also incorporates the Class D and Class E genes that have been defined since the ABC model was first developed (Figure 9.7).

However, it remains possible that B_s genes, rather than being involved in specifying complete female organs, could have more restricted functions, such as determining aspects of ovule formation. We eagerly await the outcome of reverse genetic approaches currently underway to determine the function of B_s genes in genetic model plants.

Our present knowledge is compatible with the view that the precursor of B and B_s genes had a general function in reproductive organ development of the most recent fern- or progymnosperm-like ancestor of seed plants, which still had sporophylls that were not yet differentiated into micro- and megasporophylls; this plant may have existed in the Devonian or Carboniferous between 400 and 300 MYA (Taylor and Taylor, 1993; Bateman and DiMichele, 1994b). We do not yet know whether the gene duplication that gave rise to distinct B and B_s gene lineages occurred before or after the origin of micro- and megasporophylls in the lineage that led to extant seed plants. Our current preferred hypothesis is that the dichotomy into a male-specific B and a female-specific B_s gene lineage was a pre-requisite for the evolutionary establishment of distinct male microsporophylls and female megasporophylls within the lineage that led to extant seed plants. During or after the origin of the flower between 300 and 200 MYA, expression of B_s genes expanded into one of the key structures of angiosperms, the carpel, while B gene expression and function expanded into the petal, another evolutionary novelty of the flowering plants. Thus, the establishment of B and B_s genes via gene duplication, and changes in the expression and function of these genes, may all have contributed to the origin and evolution of seed plant reproductive structures.

However, it is possible that the gene duplication which gave rise to distinct B and B_s lineages occurred after micro- and megasporophylls had already been established in the seed plant lineage; if so, the last common ancestor of these genes could have been expressed in male, or female, or both types of floral organs. In the latter case,

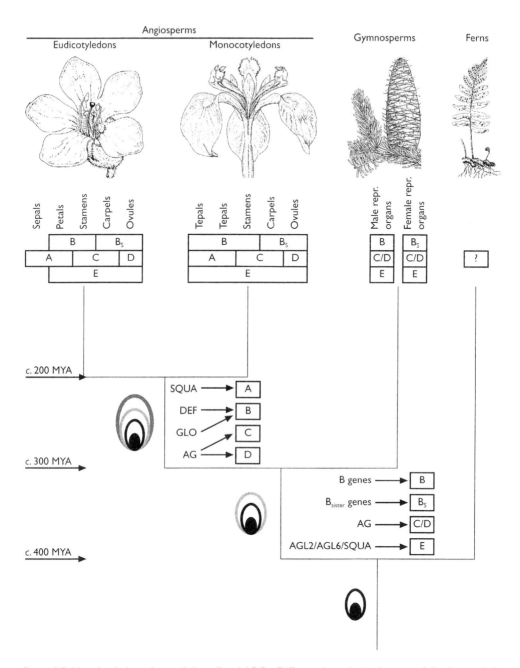

Figure 9.7 How land plants learned their floral ABCs. Different hypothetical states of the "extended ABC model" ("ABB$_s$CDE model") of floral organ specification are plotted onto a phylogenetic tree of major taxa of vascular plants. The organs specified by the different classes of homeotic genes are indicated above the models. The ages (in MYA) given at nodes of the tree are rough estimates. At the left side of the root and some internal branches of the tree three important stages in the evolution of the megasporangium are schematically depicted. From bottom to top: a sporangium that is not covered by an integument, a

restriction to either male or female organs would have occurred after the split into B and B$_s$ gene lineages. Thus, the early female expression observed for some Class B genes in developing angiosperm flowers (reviewed by Münster *et al.*, 2001) might either be a molecular rudiment or an atavism, depending on the evolving expression patterns of the ancestors of extant B genes.

9.9 Models of flower origin: past, present and future

9.9.1 Dead as a dodo? Previous models of flower origin

As indicated above, studies on orthologs of MADS-type floral organ identity genes from gymnosperms, and many other molecular phylogenetic studies, have seriously challenged a close relationship between gnetophytes and angiosperms, and have thus undermined well-established models of flower origin such as the euanthial Antho-phyte Hypothesis and the Neo-Pseudanthial Theory (Doyle and Donoghue, 1986; Doyle, 1994a; Donoghue and Doyle, 2000; Frohlich and Parker, 2000).

Expression patterns of the orthologs of floral organ identity genes in the gymnosperm *Gnetum gnemon* are also incompatible with organ homologies implied by some of the contemporary models of flower origin. One version of the euanthial model of flower origin suggested by Doyle (1994a), for example, assumes that the organs of the outer envelope of *Gnetum* reproductive units are homologous with angiosperm petals. Since petals express (and are specified by) Class B floral organ identity genes, but do not express Class C genes, one would also expect to find similar expression patterns for the *Gnetum* envelope organs and floral organ identity gene orthologs. In reality, however, the opposite pattern was observed: the outer envelope organs of *Gnetum* reproductive units express *GGM3*, the putative ortholog of Class C and Class D floral organ identity genes, but not the putative Class B gene ortholog *GGM2* (Winter *et al.*, 1999). Indeed, the organs of the outer envelope of *Gnetum* show expression patterns that are more similar to those of carpels, or the integuments of angiosperm ovules (perhaps indicating respective homologies) and, therefore, do not appear to be homologous with angiosperm petals, an observation that is incompatible with the euanthial model *sensu* Doyle (1994a).

---

condition still found in extant ferns; a sporangium that is covered by an integument (ovule); and a sporangium that, in addition, is surrounded by a carpel. At the right side of some internal branches of the tree, gene clades rather than individual genes are indicated (e.g. "*SQUA*" means the clade of *SQUA*-like genes; "*AGL2/AGL6/SQUA*" means a super-clade of *AGL2*-, *AGL6*- and *SQUA*-like genes; "B genes" means a superclade of *DEF*-, *GLO*- and *GGM2*-like genes). The relationships between representatives of these gene clades and organ identity functions (boxed) are symbolized by arrows. For example, a *SQUA*-like gene (i.e. *AP1*) provides the Class A organ identity function (specifying sepals and petals) in *Arabidopsis*. (Note that *SQUA* itself is not a Class A gene!) The different relationships have been established during the time interval represented by the respective branches of the phylogenetic tree, at the latest. Additional abbreviations used: A, B, B$_s$, C, D, E, functions provided by floral organ identity genes; *AG*, *AG*-like genes; C/D, a precursor of Class C and D floral organ identity genes; *DEF*, *DEF*-like genes; *GLO*, *GLO*-like genes; *SQUA*, *SQUA*-like genes.

9.9.2 Characterizing the first flower

With the previous models of flower origin refuted, we are obliged to develop new ones. Towards that goal, speculations concerning the morphology of the first flower may be helpful, because that at least partly defines the key morphological transitions. The conclusion based on recent results in phylogeny reconstruction, that the last common ancestor of extant angiosperms had hermaphroditic flowers with an elaborate perianth (see above), should not be confused with statements about the first floral structure that ever existed. For example, the last common ancestor of extant angiosperms may well have possessed antecedents bearing structurally strongly diverging flowers. Except the angiosperms, however, all its descendants may have died out, so that we have no living evidence for these diverse structures.

Some models assume that the most ancestral flower already possessed a perianth. One hypothesis suggests that the ancestral condition was a single, petaloid whorl expressing Class A and Class B floral organ identity genes (Albert et al., 1998). According to this hypothesis, the calyx whorl, expressing only Class A genes, was later added externally to protect flower buds from predation. Other ancestral ABC models assume that the basal flower had one or more sepaloid perianth whorls specified by Class A genes. Petals, and thus the distinction between corolla and calyx, could have evolved later by the outward extension of Class B gene expression into the inner of two perianth whorls (Baum, 1998).

In contrast, since the identity of the organs of perianth-less flowers could be completely specified by organ identity genes that may have been present already in the last common ancestor of angiosperms and gymnosperms (Figure 9.7), we have argued that such simple flowers should be seriously considered as plausible models for the archetypic first flower (Theißen et al., 2000). Flowers with a perianth may well have evolved later, prompted by the establishment of Class A genes (which could be derivatives of floral meristem identity genes; see above). In fact, perianth-less flowers may help to bridge the enormous morphological gap between gymnosperm and angiosperm reproductive structures. It also seems that the fossil evidence, in contrast to traditional views, increasingly supports simple, perianth-less flowers as being ancestral (Crepet, 1998; Sun et al., 1998).

These inferences do not, however, mean that the key features of the flower are not functionally correlated and originated at very different time points during evolution. On the contrary, it seems likely that both the association of female and male organs on the same axis, and the development of nectaries and of attractive, colored organs (such as petals), increased the effectiveness of animal pollination and thus of outcrossing (Dilcher, 2000). The closed carpel, the most characteristic key feature of the angiosperms, and biochemical incompatibility, may have evolved in parallel to serve as a plant's control mechanism to guarantee outcrossing (Long, 1977; Dilcher, 2000). Thus, the evolution of the closed carpel and of the hermaphroditic flower may have occurred at about the same time. It may have been through the success of co-evolution between animal pollinators and angiosperms that the latter became dominant in the vegetation of most ecosystems on land (Behrensmeyer et al., 1992). However, early hermaphroditic flowers may have attracted pollinators by attractive features of the reproductive organs itself; if so, they could well have been perianth-less.

How can the origin of ancestral, perianth-less flowers be explained in molecular terms? We believe that, with respect to the characteristic structural features of the flower, MADS-box genes are the most informative genes so far identified for explaining the hermaphroditic condition, therefore we focus on that topic below.

9.9.3 Boys meet girls: molecular hypotheses of flower origin

We have argued above that differential expression of class B floral organ identity genes (perhaps in antagonistic interaction with B_s genes) may represent the primary sex-determination mechanism of all seed plants (Winter *et al.*, 1999). If this assumption is correct, switching from male to female organ identity depends on changes in the activity of just a few genes (or even only one). This allows euanthial scenarios explaining the origin of floral hermaphroditism that begins with truly unisexual axes, as observed in extant and fossil gymnosperms. Such scenarios may have seemed unlikely to classical morphologists, because without knowledge about homeotic organ identity genes the genetic and molecular requirements for such a process may have been over-estimated. Anyway, the assumption that unisexuality rather than hermaphroditism represents the ancestral state is justified by fossil evidence (for review see Taylor and Taylor, 1993).

Obviously, we now have two contrasting options: we can start with either a male or a female cone (Figure 9.8). The "out-of-male" hypothesis assumes that hermaphroditic flowers with their basal (outer) whorl(s) of male organs and their upper (inner) whorl of female organs originated from a male gymnosperm cone. The "out-of-female" hypothesis assumes that flowers originated from a female cone. In the out-of-male scenario, reduction of B gene expression (or ectopic expression of B_s genes) in the upper region of the cone led to the development of female rather than male reproductive units. In the out-of-female scenario, ectopic expression of B genes

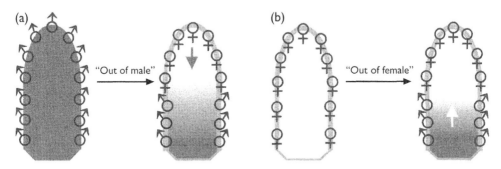

Figure 9.8 Transitions from unisexual gymnosperm cones (a) to hermaphroditic flowers (b). The "out-of-male" model assumes that hermaphroditic flowers with their basal arrangements of male and their upper arrangements of female organs originated from a male gymnosperm cone. The "out-of-female" scenario assumes that flowers originated from a female cone. In the out-of-male scenario, reduction of B gene expression (or ectopic expression of B_s genes) in the upper region of the cone led to the development of female rather than male reproductive units. In the out-of-female scenario, ectopic expression of B genes (or reduction of B_s gene expression) in the basal region of the cone led to the development of male rather than female reproductive units. B gene expression is shown in dark grey.

(or reduction of B$_s$ gene expression) in the basal region of the cone led to the development of male rather than female reproductive units. In both cases, a perianth-less flower-like structure with male reproductive units in the basal region and female reproductive units in the apical region would have been established (Figure 9.8). This arrangement of organs may have facilitated outcrossing by animal pollinators, which may have become attracted by pollination droplets that were secreted by the ovules (as in gnetophytes: Endress, 1996; Hufford, 1996), by nectaries, scent or the colour of the pollen.

This raises the question as to which molecular changes may have caused the required modifications in B or B$_s$ gene expression. For simplicity, we focus the following discussion on B genes, although analogous arguments could also be made for their sister (B$_s$) genes.

In general, changes in genes that control B gene expression (encoding so-called "trans-acting factors"), or changes in the cis-regulatory elements of the B genes themselves may have caused changes in the expression patterns of B genes and thus may be key elements in flower evolution. It has been argued that changes in cis-regulatory elements of genes encoding transcription factors are of special importance for the evolution of plant morphology (Doebley and Lukens, 1998), but far too few case studies have yet been published to test that hypothesis adequately.

Changes in cis-regulatory elements, located in the promoter region of a respective gene or in introns, may have rendered B gene expression more susceptible to an apical–basal gradient already present within the cone. Such a gradient could be based on a low molecular weight compound such as a phytohormone or on a protein (presumably another transcription factor). The underlying molecular changes within the gene could range from a simple single nucleotide change to major rearrangements at the B gene locus. Changes caused by the activity of transposable elements are especially intriguing possibilities (see reviews by Lönnig and Saedler, 1997; Walbot, 2002).

Among the regulators of the Class B genes in *Arabidopsis* and *Antirrhinum* are the orthologous genes *UNUSUAL FLORAL ORGANS* (*UFO*) and *FIMBRIATA* (*FIM*), which are involved in activating the B genes and establishing the boundaries of B gene expression (Ingram *et al.*, 1997) and as such are prime candidates for involvement in changes in B gene expression domains. Complexes between FIM or UFO and other proteins may act by promoting the degradation of transcriptional inhibitors of B genes (Ingram *et al.*, 1997).

Another important key regulator in *Arabidopsis* of the floral homeotic genes, including the B genes, is the floral meristem identity protein LFY (Parcy *et al.*, 1998), which directly binds to the Class A and C genes (Busch *et al.*, 1999; Wagner *et al.*, 1999), and possibly also to the Class B genes. Remarkably, extant gymnosperms have two paralogous *LFY*-like genes, which seem to specialize in either male or female cone development, respectively (Mellerowicz *et al.*, 1998; Mouradov *et al.*, 1998a). Even more remarkable is the fact that angiosperms have only one *LFY*-like gene, which seems to be the ortholog of the gymnosperm gene influencing male cone development (Frohlich and Parker, 2000). This has been taken as evidence that, at the genetic level, angiosperm flowers have more in common with male than with female gymnosperm cones (Frohlich and Parker, 2000). This "mostly male" theory thus resembles our "out-of-male" hypothesis. However, there are differences with respect to the molecular mechanism assumed to underlie the morphological changes. The "mostly male" theory maintains that female organs (ovules) are ectopic in male cones, caused for example by the ectopic expression of ovule identity genes (such as Class D genes). In our "out-of-male" hypothesis, spatial changes in B (or B_s) gene expression caused the initial morphological change. Under this hypothesis, changes in the spatiotemporal pattern of the expression of *LFY*-like genes, or the interaction between LFY-like proteins and B (or B_s) genes, may have contributed to flower evolution.

The classical scenarios of flower origin, such as the "Anthophyte" and "Neo-Pseudanthial" hypotheses, were based on morphological considerations, but obviously could not make predictions that could be rigorously tested by morphological data alone. The attractive feature of the more recent hypotheses is that they can be tested directly by the methods of molecular biology. For example, according to the novel models, the gene networks that control flower development should be more similar to either those controlling male cone development or to those governing female cone development, depending on which of these hypotheses is correct. Comparative studies comprising all groups of gymnosperms and diverse, phylogenetically informative angiosperms (magnoliid angiosperms as well as monocots and eudicots) may be needed to obtain a reliable answer. Valid tests may require gene cloning, determination of gene expression patterns, and establishment of gene interactions, all at large scale, and thus would represent a major research effort. However, comparative analyses of the gene networks that control the development of angiosperm flowers and of male and female cones of gymnosperms may be feasible now by employing the efficient technologies of genomics and proteomics, such as DNA microarrays for expression studies and the yeast two-hybrid system for investigations of protein–protein interactions at large scale. In our view, such monumental efforts would be justifiable in order to free biology from one of its most "abominable" paradoxes: that the origin of something so common as flowers is so difficult to explain.

ACKNOWLEDGEMENTS

We thank Kerstin Kaufmann for help with the B_{sister} project, Andreas Freialdenhoven for bringing some relevant unpublished data to our attention, and Britta Grosardt and Susanne Werth for technical assistance. We are indebted to Richard Bateman for his rigorous proof-reading. Financial support from the DFG to G.T. (grant Th 417/3-2) and to A.B. (Graduiertenkolleg "Molekulare Analyse von Entwicklungsprozessen bei Pflanzen") is gratefully acknowledged.

REFERENCES

Albert, V. A., Gustafsson, M. H. G. and Di Laurenzio, L. (1998) Ontogenetic systematics, molecular developmental genetics, and the angiosperm petal, in *Molecular Systematics of Plants II* (eds D. E. Soltis, P. S. Soltis and J. J. Doyle), Kluwer Academic Publishers, Boston, USA, pp. 349–374.

Alvarez-Buylla, E. R., Liljegren, S. J., Pelaz, S., Gold, S. E., Burgeff, C., Ditta, G. S., Vergara-Silva, F. and Yanofsky, M. F. (2000a) MADS-box gene evolution beyond flowers: expression in pollen, endosperm, guard cells, roots and trichomes. *Plant Journal*, 24, 457–466.

Alvarez-Buylla, E. R., Pelaz, S., Liljegren, S. J., Gold, S. E., Burgeff, C., Ditta, G. S., de Pouplana, L. R., Martinez-Castilla, L. and Yanofsky, M. F. (2000b) An ancestral MADS-box gene duplication occurred before the divergence of plants and animals. *Proceedings of the National Academy of Sciences of the United States of America*, 97, 5328–5333.

Angenent, G. C. and Colombo, L. (1996) Molecular control of ovule development. *Trends in Plant Science*, 1, 228–232.

Barkman, T. J., Chenery, G., McNeal, J. R., Lyons-Weiler, J., Ellisens, W. J., Moore, G., Wolfe, A. D. and dePamphilis, C. W. (2000) Independent and combined analyses of sequences from all three genomic compartments converge on the root of flowering plant phylogeny. *Proceedings of the National Academy of Sciences of the United States of America*, 97, 13166–13171.

Bateman, R. M. and DiMichele, W. A. (1994a) Saltational evolution of form in vascular plants: a neoGoldschmidtian synthesis, in *Shape and Form in Plants and Fungi* (eds D. S. Ingram and A. Hudson), Academic Press, London, pp. 61–100.

Bateman, R. M. and DiMichele, W. A. (1994b) Heterospory: the most iterative key innovation in the evolutionary history of the plant kingdom. *Biological Reviews*, 69, 345–417.

Bateman, R. M. and DiMichele, W. A. (2002) Generating and filtering major phenotypic novelties: neoGoldschmidtian saltation revisited, in *Developmental Genetics and Plant Evolution* (eds Q. C. B. Cronk, R. M. Bateman and J. A. Hawkins), Taylor & Francis, London, pp. 109–159.

Baum, D. A. (1998) The evolution of plant development. *Current Opinion in Plant Biology*, 1, 79–86.

Becker, A., Kaufmann, K., Freialdenhoven, A., Vincent, C., Li, M. A., Saedler, H. and Theißen, G. (2002) A novel MADS-box gene subfamily with a sister-group relationship to class B homeotic genes. *Molecular Genetics and Genomics* (in press).

Becker, A., Winter, K.-U., Meyer, B., Saedler, H. and Theißen, G. (2000) MADS-box gene diversity in seed plants 300 million years ago. *Molecular Biology and Evolution*, 17, 1425–1434.

Behrensmeyer, A. K., Damuth, J. D., DiMichele, W. A., Potts, R., Sues, H.-D. and Wing, S. L. (1992) *Terrestrial Ecosystems Through Time*. Chicago University Press, Chicago.

Bowe, L. M., Coat, G. and dePamphilis, C. W. (2000) Phylogeny of seed plants based on all three genomic compartments: extant gymnosperms are monophyletic and Gnetales' closest relatives are conifers. *Proceedings of the National Academy of Sciences of the United States of America*, 97, 4092–4097.

Busch, M. A., Bomblies, K. and Weigel, D. (1999) Activation of a floral homeotic gene in *Arabidopsis. Science*, 285, 585–587.

Chaw, S.-M., Parkinson, C. L., Cheng, Y., Vincent, T. M. and Palmer, J. D. (2000) Seed plant phylogeny inferred from all three plant genomes: monophyly of extant gymnosperms and origin of Gnetales from conifers. *Proceedings of the National Academy of Sciences of the United States of America*, 97, 4086–4091.

Chaw, S.-M., Zharkikh, A., Sung, H.-M., Lau, T.-C. and Li., W.-H. (1997) Molecular phylogeny of extant gymnosperms and seed plant evolution: analysis of nuclear 18S rRNA sequences. *Molecular Biology and Evolution*, 14, 56–68.

Cho, S., Jang, S., Chae, S., Chung, K. M., Moon, Y.-H., An, G. and Jang, S. K. (1999) Analysis of the C-terminal region of *Arabidopsis thaliana APETALA1* as a transcription activation domain. *Plant Molecular Biology*, 40, 419–429.

Coen, E. S. and Meyerowitz, E. M. (1991) The war of the whorls: genetic interactions controlling flower development. *Nature*, 353, 31–37.

Conway Morris, S. (2000) The Cambrian "explosion": slow-fuse or megatonnage? *Proceedings of the National Academy of Sciences of the United States of America*, 97, 4426–4429.

Crane, P. R., Friis, E. M. and Pedersen, K. R. (1995) The origin and early diversification of angiosperms. *Nature*, 374, 27–33.

Crepet, W. L. (1998) The abominable mystery. *Science*, 282, 1653–1654.

Crepet, W. L. (2000) Progress in understanding angiosperm history, success, and relationships: Darwin's abominable "perplexing phenomenon." *Proceedings of the National Academy of Sciences of the United States of America*, 97, 12939–12941.

Cubas, P. (2002) Role of TCP genes in the evolution of morphological characters in angiosperms, in *Developmental Genetics and Plant Evolution* (eds Q. C. B. Cronk, R. M. Bateman and J. A. Hawkins), Taylor & Francis, London, pp. 247–266.

Davies, B., Egea-Cortines, M., de Andrade Silva, E., Saedler, H. and Sommer, H. (1996) Multiple interactions amongst floral homeotic MADS box proteins. *EMBO Journal*, 15, 4330–4343.

Dilcher, D. (2000) Toward a new synthesis: major evolutionary trends in the angiosperm fossil record. *Proceedings of the National Academy of Sciences of the United States of America*, 97, 7030–7036.

Doebley, J. and Lukens, L. (1998) Transcriptional regulators and the evolution of plant form. *The Plant Cell*, 10, 1075–1082.

Donoghue, M. J. and Doyle, J. A. (2000) Seed plant phylogeny: demise of the anthophyte hypothesis? *Current Biology*, 10, R106–R109.

Doyle, J. A. (1994a) Origin of the angiosperm flower: a phylogenetic perspective. *Plant Systematics and Evolution* (Suppl.), 8, 7–29.

Doyle, J. A. (1996) Seed plant phylogeny and the relationships of Gnetales. *International Journal of Plant Science*, 157 (6 Suppl.), S3–S39.

Doyle, J. A. (1998) Molecules, morphology, fossils, and the relationship of angiosperms and Gnetales. *Molecular Phylogenetics and Evolution*, 9, 448–462.

Doyle, J. A. and Donoghue, M. J. (1986) Seed plant phylogeny and the origin of angiosperms – an experimental cladistic approach. *Botanical Reviews*, 52, 321–431.

Doyle, J. J. (1994b) Evolution of a plant homeotic multigene family: towards connecting molecular systematics and molecular developmental genetics. *Systematic Biology*, 43, 307–328.

Egea-Cortines, M., Saedler, H. and Sommer, H. (1999) Ternary complex formation between the MADS-box proteins SQUAMOSA, DEFICIENS and GLOBOSA is involved in the control of floral architecture in *Antirrhinum majus*. *EMBO Journal*, 18, 5370–5379.

Endress, P. K. (1996) Structure and function of female and bisexual organ complexes in Gnetales. *International Journal of Plant Science*, 157 (6 Suppl.), S113–S125.

Fischer, A., Baum, N., Saedler, H. and Theißen, G. (1995) Chromosomal mapping of the MADS-box multigene family in *Zea mays* reveals dispersed distribution of allelic genes as well as transposed copies. *Nucleic Acids Research*, 23, 1901–1911.

Frohlich, M. W. (1999) MADS about Gnetales. *Proceedings of the National Academy of Sciences of the United States of America*, 96, 8811–8813.

Frohlich, M. W. (2002) The Mostly Male theory of flower origins: summary and update regarding the Jurassic pteridosperm *Pteroma*, in *Developmental Genetics and Plant Evolution* (eds Q. C. B. Cronk, R. M. Bateman and J. A. Hawkins), Taylor & Francis, London, pp. 85–108.

Frohlich, M. W. and Parker, D. S. (2000) The mostly male theory of flower evolutionary origins: from genes to fossils. *Systematic Botany*, 25, 155–170.

Gehring, W. J. (1992) The homeobox in perspective. *Trends in Biochemical Science*, 17, 277–280.

Gilbert, S. F., Opitz, J. M. and Raff, R. A. (1996) Resynthesizing evolutionary and developmental biology. *Developmental Biology*, 173, 357–372.

Glover, B. J. and Martin, C. (2002) Evolution of adaptive petal cell morphology in angiosperms, in *Developmental Genetics and Plant Evolution* (eds Q. C. B. Cronk, R. M. Bateman and J. A. Hawkins), Taylor & Francis, London, pp. 160–172.

Goremykin, V., Bobrova, V., Pahnke, J., Troitsky, A., Antonov, A. and Martin, W. (1996) Noncoding sequences from the slowly evolving chloroplast inverted repeat in addition to *rbcL* data do not support gnetalean affinities of angiosperms. *Molecular Biology and Evolution*, 13, 383–396.

Goremykin, V., Hansmann, S. and Martin, W. F. (1997) Evolutionary analysis of 58 proteins encoded in six completely sequenced chloroplast genomes: revised molecular estimates of two seed plant divergence times. *Plant Systematics and Evolution*, 206, 337–351.

Gould, S. J. (1992) Ontogeny and phylogeny – revisited and reunited. *BioEssays*, 14, 275–279.

Graham, S. W. and Olmstead, R. G. (2000) Utility of 17 chloroplast genes for inferring the phylogeny of the basal angiosperms. *American Journal of Botany*, 87, 1712–1730.

Hansen, A., Hansmann, S., Samigullin, T., Antonov, A. and Martin, W. (1999) *Gnetum* and the angiosperms: molecular evidence that their shared morphological characters are convergent, rather than homologous. *Molecular Biology and Evolution*, 16, 1006–1009.

Hasebe, M. (1999) Evolution of reproductive organs in land plants. *Journal of Plant Research*, 112, 463–474.

Hasebe, M. and Banks, J. A. (1997) Evolution of MADS gene family in plants, in *Evolution and Diversification of Land Plants* (eds K. Iwatsuki and P. H. Raven), Springer-Verlag, Tokyo, Japan, pp. 179–197.

Hasebe, M., Kofuji, R., Ito, M., Kato, M., Iwatsuki, K. and Ueda, K. (1992) Phylogeny of gymnosperms inferred from *rbcL* gene sequences. *Botanical Magazine of Tokyo*, 105, 673–679.

Hasebe, M., Wen, C.-K., Kato, M. and Banks, J. A. (1998) Characterization of MADS homeotic genes in the fern *Ceratopteris richardii*. *Proceedings of the National Academy of Sciences of the United States of America*, 95, 6222–6227.

Honma, T. and Goto, K. (2001) Complexes of MADS-box proteins are sufficient to convert leaves into floral organs. *Nature*, 409, 525–529.

Hufford, L. (1996) The morphology and evolution of male reproductive structures of Gnetales. *International Journal of Plant Science*, 157 (6 Suppl.), S95–S112.

Huijser, P., Klein, J., Lönnig, W.-E., Meijer, H., Saedler, H. and Sommer, H. (1992) Bracteomania, an inflorescence anomaly, is caused by the loss of function of the MADS-box gene *squamosa* in *Antirrhinum majus*. *EMBO Journal*, 11, 1239–1249.

Ingram, G. C., Doyle, S., Carpenter, R., Schultz, E. A., Simon, R. and Coen, E. S. (1997) Dual role for *fimbriata* in regulating floral homeotic genes and cell division in *Antirrhinum*. *EMBO Journal*, 16, 6521–6534.

Jofuku, K. D., den Boer, B. G. W., Van Montagu, M. and Okamuro, J. K. (1994) Control of *Arabidopsis* flower and seed development by the homeotic gene *APETALA2*. *Plant Cell*, 6, 1211–1225.

Kofuji, R. and Yamaguchi, K. (1997) Isolation and phylogenetic analysis of MADS genes from the fern *Ceratopteris richardii*. *Journal of Phytogeography and Taxonomy*, 45, 83–91.

Kramer, E. M., Dorit, R. L. and Irish, V. F. (1998) Molecular evolution of genes controlling petal and stamen development: duplication and divergence within the *APETALA3* and *PISTILLATA* MADS-box gene lineages. *Genetics*, 149, 765–783.

Kramer, E. M. and Irish, V. F. (2000) Evolution of the petal and stamen developmental programs: evidence from comparative studies of the lower eudicots and basal angiosperms. *International Journal of Plant Science*, 161 (6 Suppl.), S29–S40.

Krogan, N. T. and Ashton, N. W. (2000) Ancestry of plant MADS-box genes revealed by bryophyte (*Physcomitrella patens*) homologues. *New Phytologist*, 147, 505–517.

Kuzoff, R. K. and Gasser, C. S. (2000) Recent progress in reconstructing angiosperm phylogeny. *Trends in Plant Science*, 5, 330–336.

Liljegren, S. J., Ditta, G. S., Eshed Y., Savidge, B., Bowman, J. L. and Yanofsky, M. F. (2000) *SHATTERPROOF* MADS-box genes control seed dispersal in *Arabidopsis*. *Nature*, 404, 766–770.

Long, A. G. (1977) Some Lower Carboniferous pteridosperm cupules bearing ovules and microsporangia. *Transactions of the Royal Society of Edinburgh*, B, 70, 1–11.

Lönnig, W.-E. and Saedler, H. (1997) Plant transposons: contributors to evolution? *Gene*, 205, 245–253.

Ma, H. and dePamphilis, C. (2000) The ABCs of floral evolution. *Cell*, 101, 5–8.

Ma, H., Yanofsky, M. F. and Meyerowitz, E. M. (1991) *AGL1-AGL6*, an *Arabidopsis* gene family with similarity to floral homeotic and transcription factor genes. *Genes and Development*, 5, 484–495.

Mathews, S. and Donoghue, M. J. (1999) The root of angiosperm phylogeny inferred from duplicate phytochrome genes. *Science*, 286, 947–950.

Mellerowicz, E. J., Horgan, K., Walden, A., Coker, A. and Walter, C. (1998) *PRFLL* – a *Pinus radiata* homologue of *FLORICAULA* and *LEAFY* is expressed in buds containing vegetative shoot and undifferentiated male cone primordia. *Planta*, 206, 619–629.

Meyerowitz, E. M. (1994) The genetics of flower development. *Scientific American*, 271, 40–47.

Meyerowitz, E. M. (1997) Plants and the logic of development. *Genetics*, 145, 5–9.

Mouradov, A., Glassick, T., Hamdorf, B., Murphy, L., Fowler, B., Marla, S. and Teasdale, R. D. (1998a) *NEEDLY*, a *Pinus radiata* ortholog of *FLORICAULA/LEAFY* genes, expressed in both reproductive and vegetative meristems. *Proceedings of the National Academy of Sciences of the United States of America*, 95, 6537–6542.

Mouradov, A., Glassick, T. V., Hamdorf, B. A., Murphy, L. C., Marla, S. S., Yang, Y. and Teasdale, R. (1998b) Family of MADS-box genes expressed early in male and female reproductive structures of Monterey pine. *Plant Physiology*, 117, 55–61.

Mouradov, A., Hamdorf, B., Teasdale, R. D., Kim, J. T., Winter, K.-U. and Theißen, G. (1999) A *DEF/GLO*-like MADS-box gene from a gymnosperm: *Pinus radiata* contains an ortholog of angiosperm B class floral homeotic genes. *Developmental Genetics*, 25, 245–252.

Münster, T., Pahnke, J., Di Rosa, A., Kim, J. T., Martin, W., Saedler, H. and Theissen, G. (1997) Floral homeotic genes were recruited from homologous MADS-box genes preexisting in the common ancestor of ferns and seed plants. *Proceedings of the National Academy of Sciences of the United States of America*, 94, 2415–2420.

Münster, T., Wingen, L. U., Faigl, W., Werth, S., Saedler, H. and Theißen, G. (2001) Characterization of three *GLOBOSA*-like MADS-box genes from maize: evidence for ancient paralogy in one class of floral homeotic B-function genes of grasses. *Gene*, 262, 1–13.

Parcy, F., Nilsson, O., Busch, M. A., Lee, I. and Weigel, D. (1998) A genetic framework for floral patterning. *Nature*, 395, 561–566.

Parkinson, C. L., Adams, K. L. and Palmer, J. D. (1999) Multigene analyses identify the three earliest lineages of extant flowering plants. *Current Biology*, 9, 1485–1488.

Pelaz, S., Ditta, G. S., Baumann, E., Wisman, E. and Yanofsky, M. F. (2000) B and C floral organ identity functions require SEPALLATA MADS-box genes. *Nature*, 405, 200–203.

Pellegrini, L., Tan, S. and Richmond, T. J. (1995) Structure of serum response factor core bound to DNA. *Nature*, 376, 490–498.

Pryer, K. M., Schneider, H., Smith, A. R., Cranfill, R., Wolf, P. G., Hunt, J. S. and Sipes, S. D. (2001) Horsetails and ferns are a monophyletic group and the closest living relatives to seed plants. *Nature*, 409, 618–622.

Purugganan, M. D. (1997) The MADS-box floral homeotic gene lineages predate the origin of seed plants: phylogenetic and molecular clock estimates. *Journal of Molecular Evolution*, 45, 392–396.

Purugganan, M. D., Rounsley, S. D., Schmidt, R. J. and Yanofsky, M. (1995) Molecular evolution of flower development: diversification of the plant MADS-box regulatory gene family. *Genetics*, 140, 345–356.

Qiu, Y.-L., Lee, J., Bernasconi-Quadroni, F., Soltis, D. E., Soltis, P. S., Zanis, M., Zimmer, E. A., Chen, Z., Savolainen, V. and Chase, M. W. (1999) The earliest angiosperms: evidence from mitochondrial, plastid and nuclear genomes. *Nature*, 402, 404–407.

Riechmann, J. L., Heard, J., Martin, G., Reuber, L., Jiang, C.-Z., Keddie, J., Adam, L., Pineda, O., Ratcliffe, O. J., Samaha, R. R., Creelman, R., Pilgrim, M., Broun, P., Zhang, J. Z., Ghandehari, D., Sherman, B. K. and Yu, G.-L. (2000) *Arabidopsis* transcription factors: genome-wide comparative analysis among eukaryotes. *Science*, 290, 2105–2110.

Riechmann, J. L., Krizek, B. A. and Meyerowitz, E. M. (1996) Dimerization specificity of *Arabidopsis* MADS domain homeotic proteins APETALA1, APETALA3, PISTILLATA, and AGAMOUS. *Proceedings of the National Academy of Sciences of the United States of America*, 93, 4793–4798.

Riechmann, J. L. and Meyerowitz, E. M. (1997) MADS domain proteins in plant development. *Biological Chemistry*, 378, 1079–1101.

Rutledge, R., Regan, S., Nicolas, O., Fobert, P., Coté, C., Bosnich, W., Kauffeldt, C., Sunohara, G., Séguin, A. and Stewart, D. (1998) Characterization of an *AGAMOUS* homologue from the conifer black spruce (*Picea mariana*) that produces floral homeotic conversions when expressed in *Arabidopsis*. *Plant Journal*, 15, 625–634.

Samigullin, T. K., Martin, W. F., Troitsky, A. V. and Antonov, A. S. (1999) Molecular data from the chloroplast *rpoC1* gene suggest a deep and distinct dichotomy of contemporary spermatophytes into two monophyla: gymnosperms (including Gnetales) and angiosperms. *Journal of Molecular Evolution*, 49, 310–315.

Savard, L., Li, P., Strauss, S. H., Chase, M. W., Michaud, M. and Bousquet, J. (1994) Chloroplast and nuclear gene sequences indicate Late Pennsylvanian time for the last common ancestor of extant seed plants. *Proceedings of the National Academy of Sciences of the United States of America*, 91, 5163–5167.

Schwarz-Sommer, Z., Huijser, P., Nacken, W., Saedler, H. and Sommer, H. (1990) Genetic control of flower development by homeotic genes in *Antirrhinum majus*. *Science*, 250, 931–936.

Shindo, S., Ito, M., Ueda, K., Kato, M. and Hasebe, M. (1999) Characterization of MADS genes in the gymnosperm *Gnetum parvifolium* and its implication on the evolution of reproductive organs in seed plants. *Evolution and Development*, 1, 180–190.

Singh, R. S. and Krimbas, C. B. (eds) (2000) *Evolutionary Genetics: From Molecules to Morphology*. Cambridge University Press, Cambridge.

Slack, J. M. W., Holland, P. W. H. and Graham, C. F. (1993) The zootype and the phylotypic stage. *Nature*, 361, 490–492.

Soltis, P. S., Soltis, D. E. and Chase, M. W. (1999) Angiosperm phylogeny inferred from multiple genes as a tool for comparative biology. *Nature*, 402, 402–404.

Sun, G., Dilcher, D. L., Zheng, S. and Zhou, Z. (1998) In search of the first flower: a Jurassic angiosperm, *Archaefructus*, from northeast China. *Science*, 282, 1692–1695.

Sundström, J., Carlsbecker, A., Svensson, M. E., Svenson, M., Johanson, U., Theißen, G. and Engström, P. (1999) MADS-box genes active in developing pollen cones of Norway spruce (*Picea abies*) are homologous to the B-class floral homeotic genes in angiosperms. *Developmental Genetics*, 25, 253–266.

Tandre, K., Albert, V. A., Sundas, A. and Engström, P. (1995) Conifer homologues to genes that control floral development in angiosperms. *Plant Molecular Biology*, 27, 69–78.

Tandre, K., Svenson, M., Svensson, M. E. and Engström, P. (1998) Conservation of gene structure and activity in the regulation of reproductive organ development of conifers and angiosperms. *Plant Journal*, 15, 615–623.

Taylor, T. N. and Taylor, E. L. (1993) *The Biology and Evolution of Fossil Plants*. Prentice Hall, Englewood Cliffs, New Jersey.

Theißen, G. (2000a). *FLO*-like meristem identity genes: from basic science to crop plant design. *Progress in Botany*, 61, Springer-Verlag, Berlin Heidelberg, pp. 167–183.

Theißen, G. (2000b) Plant biology: shattering developments. *Nature*, 404, 711–713.

Theißen, G. (2001a) Development of floral organ identity: stories from the MADS house. *Current Opinion in Plant Biology*, 4, 75–85.

Theißen, G. (2001b) Flower development, Genetics of, in *Encyclopedia of Genetics* (eds S. Brenner and J. H. Miller), Academic Press, London (in press).

Theißen, G., Becker, A., Di Rosa, A., Kanno, A., Kim, J. T., Münster, T., Winter, K.-U. and Saedler, H. (2000) A short history of MADS-box genes in plants. *Plant Molecular Biology*, 42, 115–149.

Theißen, G., Kim, J. T. and Saedler, H. (1996) Classification and phylogeny of the MADS-box multigene family suggest defined roles of MADS-box gene subfamilies in the morphological evolution of eukaryotes. *Journal of Molecular Evolution*, 43, 484–516.

Theißen, G. and Saedler, H. (1995) MADS-box genes in plant ontogeny and phylogeny: Haeckel's "biogenetic law" revisited. *Current Opinion in Genetics and Development*, 5, 628–639.

Theißen, G. and Saedler, H. (1999) The Golden Decade of molecular floral development (1990–1999): a cheerful obituary. *Developmental Genetics*, 25, 181–193.

Theißen, G. and Saedler, H. (2001) Floral quartets. *Nature*, 409, 469–471.

Valentine, J. W., Jablonski, D. and Erwin, D. H. (1999) Fossils, molecules and embryos: new perspectives on the Cambrian explosion. *Development*, 126, 851–859.

Wagner, D., Sablowski, R. W. M and Meyerowitz, E. M. (1999) Transcriptional activation of APETALA1 by LEAFY. *Science*, 285, 582–584.

Walbot, V. (2002) Impact of transposons on plant genomes, in *Developmental Genetics and Plant Evolution* (eds Q. C. B. Cronk, R. M. Bateman and J. A. Hawkins), Taylor & Francis, London, pp. 15–31.

Weigel, D. and Meyerowitz, E. M. (1994) The ABCs of floral homeotic genes. *Cell*, 78, 203–209.

Winter, K.-U. (2000) *Charakterisierung von Orthologen floraler homöotischer B Funktionsgene der Gymnosperme Gnetum gnemon* L. PhD thesis, University of Cologne, Germany.

Winter, K.-U., Becker, A., Münster, T., Kim, J. T., Saedler, H. and Theissen, G. (1999) MADS-box genes reveal that gnetophytes are more closely related to conifers than to flowering plants. *Proceedings of the National Academy of Sciences of the United States of America*, 96, 7342–7347.

Winter, K.-U., Weiser, C., Kaufmann, K., Bohne, A., Kirchner, C., Kanno, A., Saedler, H. and Theißen, G. (2002) Evolution of class B floral homeotic proteins: obligate heterodimerization originated from homodimerization. *Molecular Biology and Evolution*, 19 (in press).

Wolfe, K. H., Gouy, M., Yang, Y.-W., Sharp, P. M. and Li, W.-H. (1989) Date of the monocot–dicot divergence estimated from chloroplast DNA sequence data. *Proceedings of the National Academy of Sciences of the United States of America*, 86, 6201–6205.

Chapter 10

Orchid flowers: evolution and molecular development

Bo Johansen and Signe Frederiksen

ABSTRACT

Studies of MADS-box genes have expanded greatly during the last ten years. This increase in knowledge will help to place the study of flower development in an evolutionary context, but requires a change in focus from the few well-known model plants to include species representing the much greater floral variation observed among angiosperms. The pre-requisites for such evolutionary studies are phylogenetic hypotheses of all known MADS-box genes and of the chosen group of plants. Studies, predominantly of higher eudicots, have led to the ABC model of flower development. Within the monocots only well-known crop species such as *Oryza* and *Zea* have been thoroughly studied, but their highly reduced flowers make them unsuitable for more general flower developmental studies. Among the petaloid monocots Orchidaceae seem to form a more suitable study group. All expected whorls in the flowers are present, and evolution of the many special structures of the orchid flower is supported by several more or less congruent phylogenetic hypotheses. Thus, the Orchidaceae can be used to test the validity of the ABC model in the monocots and to study how MADS-box genes are involved in defining the different highly specialised structures in these flowers. Our studies on the expression pattern of an *AdOM1* orthologue in the epidendroid orchid *Cleisostoma racemiferum* indicate a function in connection with cell-wall modification in several different tissues. The *AdOM1* orthologue probably controls this function by activating other, presently unknown, MADS-box genes.

10.1 Introduction

During the last decade, the discovery of MADS-box genes and the study of model plants such as *Arabidopsis* and *Antirrhinum* have greatly improved our knowledge of flower development, but as studies are mainly restricted to these model plants it is impossible to place the newly gained knowledge in a rigorous evolutionary context. To understand how flowers evolved and develop, we should now focus on the enormous variation observed among flowers today. Some molecular biologists claim that the *in situ* approach is not really necessary as we can just pull the genes of interest out of any plant, put them into *Arabidopsis,* and observe the result. Although studies of transgenic *Arabidopsis* may seem to be a reasonable way to study slow-growing plants like many palms, it is not a viable alternative to studying the effect of

In *Developmental Genetics and Plant Evolution* (2002) (eds Q. C. B. Cronk, R. M. Bateman and J. A. Hawkins), Taylor & Francis, London, pp. 206–219.

gene expression in the plant itself. This is especially true for MADS-box proteins, as they all regulate transcription of other genes. It is therefore unlikely that a distant relative of *Arabidopsis* (for example, a palm) possesses exactly the same genetic environment that the chosen MADS-box gene can act upon. Thus, we cannot expect a gene governing primary thickening in palms to produce plants with this feature when transferred to an *Arabidopsis* genome. Moreover, MADS-box proteins often function as heterodimers, which means that more than one MADS-box gene has to be transferred to the host genome to observe their true function. As evolutionary biologists are interested in evolution of the structures that the genes govern, studies on transgenic *Arabidopsis* alone are not sufficient; rather we need to study expression in the plant itself.

The recent complete sequencing of the *Arabidopsis* genome shows that at least forty paralogous MADS-box genes reside in its genome (Alvarez-Buylla *et al.*, 2000), but so far only some of these genes have been studied in detail. Consequently we now know a great deal about A-, B- and C-class MADS-box genes but little about the rest (e.g. Ma *et al.*, 1991; Pnueli *et al.*, 1991; Davies *et al.*, 1996; Mandel and Yanofsky, 1998; Yu and Goh, 2000).

In order to understand how flowers evolved, it is necessary to elucidate from which MADS-box genes the ABC genes evolved, and the function of these genes in the plant. This requires phylogenetic study of as many MADS-box genes as possible, combined with mapping of transcription patterns across the phylogeny. Most published phylogenetic studies have been based on very restricted numbers of genes (e.g. Southerton *et al.*, 1998; Yu *et al.*, 1999; Yu and Goh, 2000). However, omitting large numbers of relevant genes may lead to an incorrect phylogenetic hypothesis. Moreover, many of these hypotheses are based on neighbour joining analysis rather than parsimony (e.g. Rounsley *et al.*, 1995; Mouradov *et al.*, 1999). Neighbour joining is, however, basically a cluster analysis based on similarity, and consequently the ability to distinguish between homology and homoplasy does not exist (Siebert, 1992). Although evolution is not necessarily parsimonious, a phylogenetic analysis should be based on the parsimony criterion, as the basis of all science is always to seek the simplest explanation that can account for all pertinent observations. As a robust phylogenetic hypothesis should be based on as many characters as possible, phylogenies should utilise nucleotide sequences rather than amino acid sequences. Furthermore, as most amino acids are coded for by more than one codon, amino acid sequences are subject to convergence that will affect the resulting trees, irrespective of whether they are produced by neighbour joining, maximum likelihood or maximum parsimony (Simmons, 2000).

The ABC model of flower development has been well established within the higher eudicots (e.g. Bowman *et al.*, 1991; Coen and Meyerowitz, 1991; Weigel and Meyerowitz, 1994), but apart from the important crop grasses *Zea* and *Oryza* (Chung *et al.*, 1995; Kang *et al.*, 1995; Mena *et al.*, 1995; Theißen *et al.*, 1995; Greco *et al.*, 1997; Lopez-Dee *et al.*, 1999; Moon *et al.*, 1999; Heuer *et al.*, 2000; Kyozuka *et al.*, 2000), little is known about the molecular development of monocot flowers. As the grass flower is morphologically very reduced and the homology of some of the elements of the flower is still problematic, petaloid monocots appear more suitable for studies of the MADS-box gene expression. They possess all the whorls expected in a monocot flower, the homology of the different elements is unquestionable, and both simple and more complex flowers are known. Within the petaloid monocots, the specialised flowers of the orchids present an excellent opportunity to both evaluate the universality of the ABC model and explore whether MADS-box genes are involved in development of the more derived structures in the orchid flower. Gene expression studies give unambiguous information on evolution when combined with a thorough understanding of the phylogeny of the group examined. A phylogeny of the Orchidaceae based on *rbcL* data was recently published by Cameron *et al.* (1999), and an almost congruent phylogeny based on morphological characters was published by Freudenstein and Rasmussen (1999). These phylogenies support earlier studies on the evolution of some of the special morphological structures found in orchid flowers, further suggesting that Orchidaceae are suitable for observing specialised functions of MADS-box genes. Few MADS-box genes are known from the orchids, and only one has been reported to be an A-class gene, the rest being genes of unknown function belonging to the *AGL2* clade (Yu and Goh, 2000).

This chapter gives a brief overview of orchid flower morphology and, based on published phylogenetic hypotheses, the evolution of some of the specialised structures is also studied. Developmental patterns of these structures are in accordance with the relatively few observations of MADS-box gene expression in orchid flowers.

10.2 Evolutionary trends in the orchid flower

Containing more than 20,000 species, the Orchidaceae is definitely the largest family within the monocots and one of the largest families of seed plants. Although variable with respect to vegetative architecture, the structure of the flower always discloses the family. Like other petaloid monocots, all orchids have six coloured tepals arranged in two whorls. In most orchid species the median tepal in the inner whorl, the labellum, generally differs from the rest of the tepals; thus, the perianth is zygomorphic. Furthermore, the style and stamens are always united to form the highly modified gynostemium or column. Typical petaloid monocots bear six stamens in two whorls, whereas in orchids the number of fertile stamens is reduced. Never are more than three found: the median stamen from the outer whorl and the two lateral stamens from the inner whorl (Figure 10.1a–d).

Orchidaceae may be divided into five subfamilies characterised in part by specialisations in the gynostemium. In Apostasioideae, two or three fertile stamens are formed (Figure 10.1a–b), and contrary to all other orchids only the basal part of each filament is united with the style (Figure 10.1e). In Cypripediodeae (slipper orchids), the median stamen of the inner whorl forms a large, plate-like staminodium, whereas the lateral stamens of the outer whorl are fertile (Figures 10.1c, 10.2a). These two subfamilies contain only a modest number of species; the great majority of orchids are monandrous, wherein only the median stamen of the inner whorl is fertile and the laterals are reduced (Figures 10.1d, 10.2c). The three monandrous subfamilies (Vanilloideae, Orchidoideae, Epidendroideae) are all monophyletic, and the Vandeae are a monophyletic clade within the Epidendroideae (Cameron *et al.*, 1999; Freudenstein and Rasmussen, 1999) (Figure 10.3). However, some phylogenetic analyses indicate that Vanilloideae is sister-group to Cypripedioideae, Orchidoideae and Epidendroideae (e.g. Cameron and Chase, 2000; Pridgeon *et al.*, 2001). Thus, the monandrous orchids may have evolved twice.

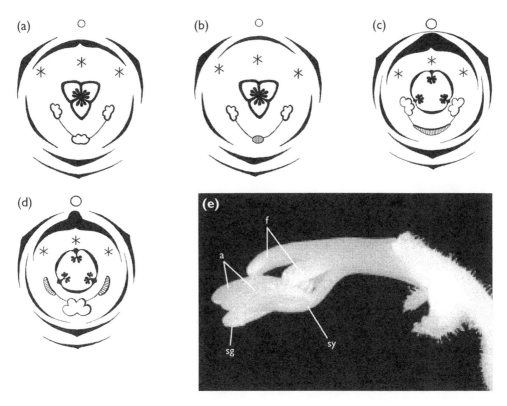

Figure 10.1 Flower diagrams (a–d) and details of gynostemium (e) from different subfamilies of Orchidaceae. (a) *Neuwiedia* (Apostasioideae). (b) *Apostasia* (Apostasioideae). (c) *Cypripedium* (Cypripedioideae). (d) Monandrous orchid. Asterisks: absent stamens; hatched area: staminodium. (e) Gynostemium of *Neuwiedia*, tepals removed. a: anther; f: filament; sg: stigma; sy: style. (a–d: after Eichler, 1875; E: photo: F. N. Rasmussen).

In Apostasioideae and Cypripedioideae, pollen is shed as single, sticky grains (Figure 10.2b), whereas in monandrous orchids pollen are fused together into pollinia (Figure 10.2c–d). Generally, one pollinium occurs in each theca. The pollinia in Vanilloideae and Orchidoideae are soft, whereas the pollinia in Epidendroideae (especially in the more derived Vandeae) are hard and difficult to squash. The pollinia are frequently attached to a pollinium stalk. In some species, especially of Orchidoideae, the stalk is composed of aborted pollen termed a caudicle (Dressler, 1981, 1993). In other species the pollinium stalk is formed from a strip of the median stigma lobe, a tegula (Rasmussen, 1982). The tegula is multilayered in many Epidendroideae but in all Vandeae it is unilayered (Freudenstein and Rasmussen, 1999) (Figure 10.2c–d). In a few genera of Epidendroideae, the entire apical part of the median stigma lobe breaks off and functions as a pollinium stalk called a hamulus (Rasmussen, 1986). When the pollinium stalk is formed as unilayered tegula, the epidermis of the abaxial surface of the stigma lobe is clearly modified; the anticlinal walls dissolve, and the tegula consists of the outer periclinal walls and cuticle.

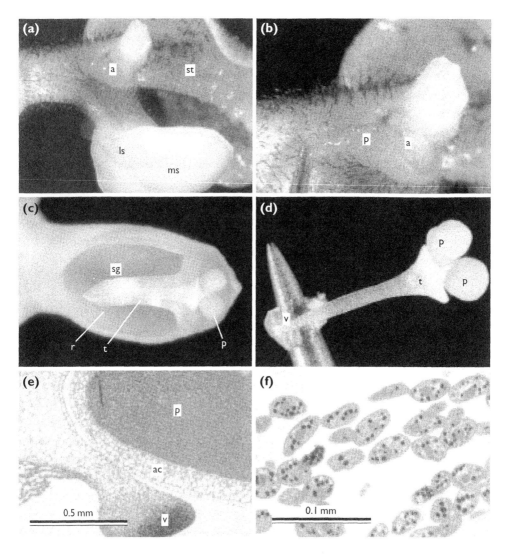

Figure 10.2 Details of gynostemium from Cypripedioideae (a–b) and Epidendroideae (c–f), (c–d) from Vandeae. (a) Gynostemium of *Paphiopedilum*, lateral view. (b) Anther of *Paphiopedilum* with a fluid mass of pollen. (c) Gynostemium of *Doritis*, front view, anther cap removed. Cavity represents the median stigma lobe. (d) Pollinarium of *Doritis*. (e) Longitudinal section through the apical part of the gynostemium of *Dendrobium* (note the lack of a pollinium stalk). (f) Magnification of detached stigma cells from *Dendrobium* stained in PAS-ABB. Black dots are starch grains. a: anther; ac: anther cap; p: pollen/pollinium; r: rostellum; sg: stigma; st: staminodium; t: tegula; v: viscidium.

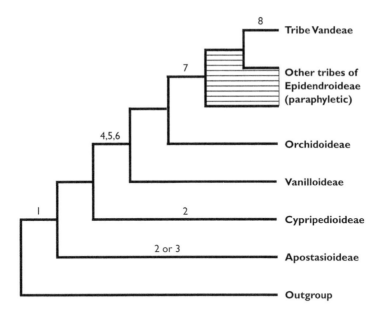

Figure 10.3 Phylogenetic tree of Orchidaceae (summarised from *rbcL* study of Cameron *et al.*, 1999). Mapping of 'advanced' characters on the tree. 1 = column, 2 = two stamens, 3 = three stamens, 4 = one stamen, 5 = pollinia, 6 = separated stigma cells, 7 = hard pollinia, 8 = unilayered tegula.

The phylogenetic hypothesis used here indicates evolutionary transitions from three to two or one fertile stamen (Figure 10.3). Reduction from three to two anthers has probably evolved twice: within the Apostasioideae and in the ancestor of the Cypripedioideae. Reduction from three to one stamen and the evolution of pollinia from solitary grains occurred in the ancestors of the monandrous orchids (Figure 10.3). The phylogenetic hypothesis also indicates that hard pollinia evolved from soft ones.

In the petaloid monocots the stigma consists of three nearly equal, solid stigma lobes. Three solid stigma lobes also characterise Apostasioideae and Cypripedioideae, but in the latter subfamily the median lobe is larger than the lateral lobes (Figure 10.2a).

In the monandrous orchids the cells lining the stylar canal, as well as the stigma cells, are highly specialised (Figure 10.3); in the mature flower they become more or less detached from each other and float in a mucilaginous matrix (Figure 10.2e–f). Furthermore, the apex of the median stigma lobe in the monandrous orchids is modified to form the viscidium, a structure that secretes an adhesive substance that glues the pollinia or the pollinium stalk to the pollinator (Figure 10.2d–e). The pollen dispersal unit including the viscidium, the pollinium and (if present) the pollinium stalk is termed a pollinarium (Figure 10.2d).

If the alternative hypothesis, that the monadrous orchids have evolved twice, reflects the true evolution of the orchids, then also pollinia, viscidium and the specialised stigma must have evolved twice.

10.3 Phylogenetic analysis of MADS-box genes

Based on a phylogenetic analysis, MADS-box genes were recently divided into two groups: Type I (including the vertebrate serum response factor, SRF) and Type II (including the monocyte enhancer factor, *MEF*, and most plant MADS-box genes) (Alvarez-Buylla *et al.*, 2000). The major difference between the plant MADS-box genes of interest here and the *MEF*-like genes is the special MIK structure, which means that in addition to the MADS-box *per se*, the genes include two other domains: a rather variable I-domain and a more conserved K-domain (Purugganan *et al.*, 1995; Lawton-Rauh *et al.*, 1999; Alvarez-Buylla *et al.*, 2000). A phylogenetic analysis with the SRF-like genes as outgroups, and a representative number of different plant MADS-box genes and *MEF*-like genes, showed that the *MEF*-like genes form a sister group to the plant genes with MIK-structure. Consequently, we have used the *MEF*-like genes as outgroup in the present phylogenetic analyses.

Genes with MIK-structure have very similar exon lengths (Goto and Meyerowitz, 1994; Barrier *et al.*, 1999; Krogan and Ashton, 2000), providing a tool for assessing positional homology during alignment. However, positional homology becomes uncertain in any gene that does not possess the fixed exon structure. Several such plant MADS-box genes, which have formerly been included in analyses without further discussion (e.g. Hasebe *et al.*, 1998; Becker *et al.*, 2000; Svensson *et al.*, 2000), have for that reason been excluded here.

A phylogenetic analysis based on the parsimony criterion and nucleotide sequences of 198 almost full-length MADS-box genes with fixed exon structure resulted in fifty-three equally parsimonious trees (Johansen *et al.*, in prep.). The structure of the lower branches is summarised on Figure 10.4a. Nothing is known about the expression pattern of many of the included genes but both of the successive sister clades to the rest, *JOINTLESS* and *PPM*, include genes with expression in vegetative organs. *JOINTLESS* is expressed in development of the abscission zone of the tomato flower (Mao *et al.*, 2000), whereas *PPM* was extracted from vegetative (gametophytic) tissue of a bryophyte (Krogan and Ashton, 2000). The *AGAMOUS* clade includes the C-class genes, the *PISTILLATA* and *APETALA3* clades include the B-class genes, and the *SQUAMOSA* clade includes the A-class genes. Several genes within these clades behave in accordance with the ABC model in flowers but are also expressed in other parts of the plant (e.g. Ma *et al.*, 1991; Huijser *et al.*, 1992; Yu *et al.*, 1999; Yu and Goh, 2000). The remaining clades, including the *AGL2* clade, contain genes with more diverse functions.

10.4 *In situ* PCR: a tool for studying molecular development

The principle in reversed transcribed *in situ* PCR (RT-ISPCR) studies is to make a reverse transcription followed by a PCR reaction directly in the tissue on the slide. Although the technique has mainly been used on animal material (for an overview see Gu, 1995), we now routinely perform RT-ISPCR on various types of plant material (Johansen, 1997; Petersen *et al.*, 2000; Mølhøj *et al.*, 2001). During thermal cycling a digoxigenin-(DIG-) labelled nucleotide is incorporated into the PCR product. The DIG-labelled PCR product can then be detected by immunological methods.

RT-ISPCR has several advantages when compared with ordinary *in situ* hybridisation. RT-ISPCR is much more sensitive (Teo and Shauniak, 1995; Nuovo, 1998); in theory, a single copy of mRNA can be detected. Furthermore, cross-reaction with highly similar mRNAs, which is a major problem during *in situ* hybridisation (or northern blot), is overcome as primers can be designed in such a way that they only amplify one mRNA, and as a control the integrity of the PCR product can be proven by sequencing. Cross-reaction is especially evident when using *in situ* hybridisation for studying MADS-box gene expression, as similar paralogues often exist (Hardenack *et al.*, 1994). We recently found five different paralogues of B-class genes in the orchid *Cleisostoma racemiferum* (Lindl.) Garay, and doubt whether a set of probes can be designed specific for all five of these genes.

As a positive control, ISPCR on *rbcL* is convenient, as most plant cells possess plastids and the signal is easily detected as dots in the cytoplasm (Johansen, 1997). As a negative control, the reverse transcription step can be omitted and no signal should be visible.

Further informations about RT-ISPCR and a viable protocol can be found on the Web at www.bot.ku.dk/staff/boj.htm.

10.5 MADS-box gene expression in flower buds of *Cleisostoma racemiferum*

MADS-box genes have so far been published from only two orchids. *AdOM1* was isolated from the supposed bigeneric hybrid *Aranda* Deborah (Lu *et al.*, 1993) and *DgMADS1*, *DgMADS2*, *DgMADS3* and *DgOTG7* from a *Dendrobium* hybrid (Yu and Goh, 2000). Northern blot analysis showed that *AdOM1* was expressed only in the flowers (Lu *et al.*, 1993), whereas the *Dendrobium* genes studied by *in situ* hybridisation show a more complex expression pattern (Yu and Goh, 2000). *DgMADS2* was found to be an A-class gene, *DgOTG7* is expressed during the transition from vegetative to inflorescence meristem, and *DgMADS1* and *DgMADS3* are expressed in the flower meristem and in different parts of the developing flower (Yu and Goh, 2000). *AdOM1* and *DgMADS1* were accepted as orthologues by Yu and Goh (2000), and in our phylogeny they are included in the *AGL2* clade with *DgMADS3* and *DgOTG7* (Figure 10.4b). Although the function of the genes in the *AGL2* clade are poorly known, they appear to be expressed in the flower, and many of them seem to play a role in flower initiation and/or during the seed development (Ma *et al.*, 1991; Pnueli *et al.*, 1994; Bonhomme *et al.*, 1997; Greco *et al.*, 1997; Mandel and Yanofsky, 1998; Southerton *et al.*, 1998; Sung *et al.*, 2000). Recently it was shown that several genes belonging to the *AGL2* clade possess both a glutamine-rich and a proline-rich transcription activation site (Cho *et al.*, 1999), and that B-class genes remain inactive without transcription of these *AGL2*-like genes (Honma and Goto, 2001).

Expression studies of MADS-box genes in *Cleisostoma racemiferum* (Epidendroideae: Vandeae) were started to find out if any of the known MADS-box genes are involved in development of the special structures of the orchid flower. *Cleisostoma* belongs to the Vandeae and thus possesses all the advanced characters mentioned earlier, including a tegula. Furthermore, *C. racemiferum* produce several branched inflorescences with hundreds of flowers of small size, making it ideal for studying molecular development.

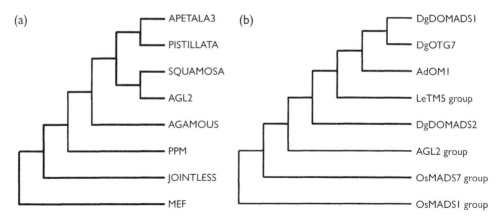

Figure 10.4 Strict consensus tree based on 198 almost full-length MADS-box genes with MIK structure. (a) Lower branches. (b) Details of the *AGL2* clade.

Expression of the orthologue of *AdOM1* was studied in buds of *Cleisostoma racemiferum* inflorescence apices, in an ontogenetic series from small to large buds at a stage just prior to anthesis. In the smallest buds, transcription was found throughout the bud; more intense staining occurred in the developing column than in the tepals (Figure 10.5a–b). Intensely stained regions were also found in developing strands of conducting tissue (procambrial strands), but no transcription was observed in the apical meristem of the inflorescence. Thus, the gene is not involved in maintaining meristem identity.

When the bud develops further expression becomes more specific, showing intense staining in the developing pollen and stigma and stylar canal (Figure 10.5c). Later, intense expression is visible in the tip of the rostellum where the viscidium will subsequently develop, and in cells on the upper (abaxial) side of the rostellum where the tegula will develop (Figure 10.5d). All expression ceases in buds approaching anthesis.

As the expression of the *AdOM1* orthologue decreases early in the developing perianth, *AdOM1* seems to have no role in determining tepal identity. The gene is expressed in the young column but expression diminishes throughout development, except for the areas already mentioned. In the stigma and stylar canal, strong expression is found in the cells that are later detached, but expression decreases before the cells actually separate. In the rostellum strong expression is found at later stages in the part that subsequently differentiates to the viscidium and in the cells later forming the tegula. These observations are in accordance with observations in the *AdOM1* orthologue from *Dendrobium*, but in *Dendrobium* no expression is found in the tegula area of the rostellum (Yu and Goh, 2000) because the genus lacks a pollinium stalk (Figure 10.2e).

AdOM1 and its orthologues clearly have functions in connection with the development of the stigma, pollinium, viscidium and tegula in the orchids examined. The cells of these different tissues all seem to have highly modified cell walls at maturity, and it is thus possible that *AdOM1* is involved in regulation of cell wall modifications. This corresponds with the fact that *AdOM1* is also transcribed in the procam-

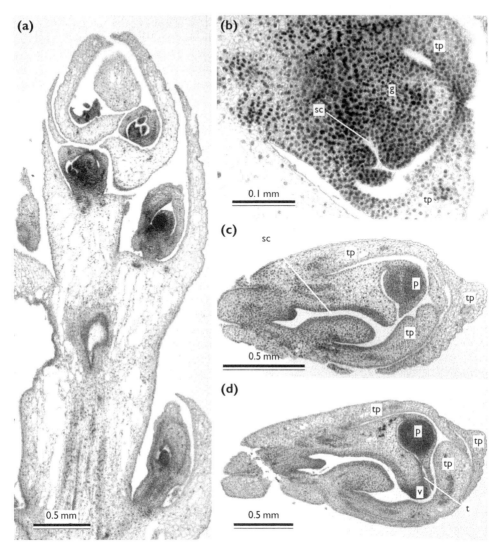

Figure 10.5 In situ PCR of AdOM1 in *Cleisostoma racemiferum* (Epidendroideae: Vandeae). (a) Longitudinal section of young inflorescence. (b) Detail of (a). (c) Bud less than 2 mm long. (d) Bud about 2 mm long. g: gynostemium; p: pollinium; sc: stylar canal; t: tegula; tp: tepal; v: viscidium.

bial strands, as at least the developing vessel members will have modified cell walls at maturity. As it is unlikely that a single MADS-box gene can regulate the development of all the above-mentioned tissues, we assume that other as yet unspecified MADS-box genes must be transcribed simultaneously with *AdOM1* in these tissues. Heterodimerisation may then be responsible for the different fates of the various cells, and *AdOM1* may simply function as an activator of other MADS-box genes, like other *AGL2*-like genes.

By studying MADS-box gene expression during flower development in a derived orchid, we have shown that MADS-box genes belonging to the *AGL2* clade are involved in defining the identity of specialised floral structures such as the viscidium, the tegula and the detached stigma cells. As these structures are only found in orchids, MADS-box genes are not only involved in defining the general floral architecture (the ABC model) but may also be responsible for the huge variation observed among orchid flowers.

ACKNOWLEDGEMENTS

Louise B. Pedersen and Martin Skipper are thanked for inspiring discussions, and Charlotte Hansen and Kate Jensen are acknowledged for their skilful technical assistance.

REFERENCES

Alvarez-Buylla, E., Pelaz, S., Liljegren, S. J., Gold, S. E., Burgeff, C. *et al.* (2000) An ancestral MADS-box gene duplication occurred before the divergence of plants and animals. *Proceedings of the National Academy of Sciences USA*, 97, 5328–5333.

Barrier, M., Baldwin, B. G., Robichaux, R. H. and Purugganan, M. D. (1999) Interspecific hybrid ancestry of a plant adaptive radiation: allopolyploidy of the Hawaiian silversword alliance (Asteraceae) inferred from floral homeotic gene duplications. *Molecular Biology and Evolution*, 16, 1105–1113.

Becker, A., Winther, K.-U., Meyer, B., Saedler, H. and Theißen, G. (2000) MADS-box gene diversity in seed plants 300 million years ago. *Molecular Biology and Evolution*, 17, 1425–1434.

Bonhomme, F., Sommer, H., Bernier, G. and Jacqmard, A. (1997) Characterization of SaMADS D from *Sinapis alba* suggests a dual function of the gene: in inflorescence development and floral organogenesis. *Plant Molecular Biology*, 34, 573–582.

Bowman, J. L., Smyth, D. R. and Meyerowitz, E. M. (1991) Genetic interactions among floral homeotic genes of *Arabidopsis*. *Development*, 112, 1–20.

Cameron, K. M. and Chase, M. W. (2000) Nuclear 18S rDNA sequences of Orchidaceae confirm the subfamilial status and circumscription of Vanilloideae, in *Monocots: Systematics and Evolution* (eds. K. L. Wilson and D. A. Morrison), CSIRO, Collingwood, Australia, pp. 457–464.

Cameron, K. M., Chase, M. W., Whitten, W. M., Kores, P. J., Jarrell, D. C. *et al.* (1999) A phylogenetic analysis of the Orchidaceae: evidence from *rbcL* nucleotide sequences. *American Journal of Botany*, 86, 208–224.

Cho, S., Jang, S., Chae, S., Chung, K. M., Moon, Y. H., An, G. and Jang, S. K. (1999) Analysis of the C-terminal region of *Arabidopsis thaliana* APETALA1 as a transcription activation domain. *Plant Molecular Biology*, 40, 419–429.

Chung, Y.-Y., Kim, S.-R., Kang, H.-G., Noh, Y.-S., Park, M. C. and An, G. (1995) Characterization of two rice MADS box genes homologous to GLOBOSA. *Plant Science*, 109, 45–56.

Coen, E. S. and Meyerowitz, E. M. (1991) The war of the whorls: genetic interactions controlling flower development. *Nature*, 353, 31–37.

Davies, B., Egea-Cortines, M., Andrade Silva, E. de, Saedler, H. and Sommer, H. (1996) Multiple interactions amongst floral homeotic MADS box proteins. *EMBO Journal*, 15, 4330–4343.

Dressler, R. L. (1981) *The Orchids: Natural History and Classification*. Harvard University Press, Cambridge, MA.

Dressler, R. L. (1993) *Phylogeny and Classification of the Orchid Family*. Cambridge University Press, Cambridge.

Eichler, A. W. (1875) *Blüthendiagramme. Part 1*. Wilhelm Engelmann, Leipzig.

Freudenstein, J. V. and Rasmussen, F. N. (1999) What does morphology tell us about orchid relationships? A cladistic analysis. *American Journal of Botany*, 86, 225–248.

Goto, K. and Meyerowitz, E. M. (1994) Function and regulation of the *Arabidopsis* floral homeotic gene *Pistillata*. *Genes and Development*, 8, 1548–1560.

Greco, R., Stagi, L., Colombo, L., Angenent, G. C., Sari-Gorla, M. and Pe, M. E. (1997) MADS box genes expressed in developing inflorescences of rice and sorghum. *Molecular and General Genetics*, 253, 615–623.

Gu, J. (1995) *In situ* PCR – an overview, in *In Situ Polymerase Chain Reaction and Related Technology* (ed. J. Gu), Birkhäuser, Boston, pp. 1–21.

Hardenack, S., Ye, D., Saedler, H. and Grant, S. (1994) Comparison of MADS box gene expression in developing male and female flowers of the dioecious plant white campion. *Plant Cell*, 6, 1775–1787.

Hasebe, M., Wen, C.-K., Kato, M. and Banks, J. A. (1998) Characterisation of MADS homeotic genes in the fern *Ceratopteris richardii*. *Proceedings of the National Academy of Sciences USA*, 95, 6222–6227.

Heuer, S., Loerz, H. and Dresselhaus, T. (2000) The MADS box gene ZmMADS2 is specifically expressed in maize pollen and during maize pollen tube growth. *Sexual Plant Reproduction*, 13, 21–27.

Honma, T. and Goto, K. (2001) Complexes of MADS-box proteins are sufficient to convert leaves into floral organs. *Nature*, 409, 525–529.

Huijser, P., Klein, J., Lonnig, W. E., Meijer, H., Saedler, H. and Sommer, H. (1992) Bracteomania, an inflorescence anomaly, is caused by the loss of function of the MADS-box gene *squamosa* in *Antirrhinum majus*. *EMBO Journal*, 11, 1239–1249.

Johansen, B. (1997) *In situ* PCR on plant material with sub-cellular resolution. *Annals of Botany*, 80, 697–700.

Kang, H. G., Noh, Y. S., Chung, Y. Y., Costa, M. A., An, K. and An, G. (1995) Phenotypic alterations of petal and sepal by ectopic expression of a rice MADS box gene in tobacco. *Plant Molecular Biology*, 29, 1–10.

Krogan, N. T. and Ashton, N. W. (2000) Ancestry of plant MADS-box genes revealed by bryophyte (*Physcomitrella patens*) homologues. *New Phytologist*, 147, 505–517.

Kyozuka, J., Kobayashi, T., Morita, M. and Shimamoto, K. (2000) Spatially and temporally regulated expression of rice MADS genes with similarity to *Arabidopsis* class A, B, C genes. *Plant Cell Physiology*, 41, 710–718.

Lawton-Rauh, A. L., Buckler, E. S. I. and Purugganan, M. D. (1999) Patterns of molecular evolution among paralogous floral homeotic genes. *Molecular Biology and Evolution*, 16, 1037–1045.

Lopez-Dee, Z. P., Wittich, P., Enrico Pe, M., Rigola, D., Del Buono, I. *et al.* (1999) OsMADS13, a novel rice MADS-box gene expressed during ovule development. *Developmental Genetics*, 25, 237–244.

Lu, Z.-X., Wu, M., Loh, C.-S., Yeong, C.-Y. and Goh, C.-J. (1993) Nucleotide sequence of a flower-specific MADS box cDNA clone from orchid. *Plant Molecular Biology*, 23, 901–904.

Ma, H., Yanofsky, M. F. and Meyerowitz, E. M. (1991) AGL1–AGL6, an *Arabidopsis* gene family with similarity to floral homeotic and transcription factor genes. *Genes and Development*, 5, 484–495.

Mandel, M. A. and Yanofsky, M. F. (1998) The *Arabidopsis* AGL9 MADS-box gene is expressed in young flower primordia. *Sexual Plant Reproduction*, 11, 22–28.

Mao, L., Begum, D., Chuang, H. W., Budiman, M. A., Szymkowiak, E. J. *et al.* (2000) JOINTLESS is a MADS-box gene controlling tomato flower abscission zone development. *Nature*, 406, 910–913.

Mena, M., Mandel, M. A., Lerner, D. R., Yanofsky, M. F. and Schmidt, R. J. (1995) A characterization of the MADS-box gene family in maize. *Plant Journal*, 8, 845–854.

Mølhøj, M., Johansen, B., Ulvskov, P. and Borkhardt, B. (2001) Expression of a membrane-anchored endo-1,4-β-glucanase from *Brassica napus*, orthologues to *KOR* from *Arabidopsis thaliana*, is inversely correlated to elongation in light-grown plants. *Plant Molecular Biology*, 45, 93–105.

Moon, Y. H., Jung, J. Y., Kang, H. G. and An, G. (1999) Identification of a rice APETALA3 homologue by yeast two-hybrid screening. *Plant Molecular Biology*, 40, 167–177.

Mouradov, A., Hamdorf, B., Teasdale, R. D., Kim, J. T., Winter, K. U. and Theißen, G. (1999) A DEF/GLO-like MADS-box gene from a gymnosperm: *Pinus radiata* contains an ortholog of angiosperm B class floral homeotic genes. *Developmental Genetics*, 25, 245–252.

Nuovo, G. J. (1998) *In situ* localization of PCR-amplified DNA and cDNA. *Molecular Biotechnology*, 10, 49–62.

Petersen, M., Brodersen, P., Naested, H., Andreasson, E., Lindhart, U., Johansen, B. *et al.* (2000) *Arabidopsis* MAP kinase 4 negatively regulates systemic acquired resistance. *Cell*, 103, 1111–1120.

Pnueli, L., Abu-Abeid, M., Zamir, D., Nacken, W., Schwarz-Sommer, Z. and Lifschitz, E. (1991) The MADS box gene family in tomato: temporal expression during floral development, conserved secondary structures and homology with homeotic genes from *Antirrhinum* and *Arabidopsis*. *Plant Journal*, 1, 255–266.

Pnueli, L., Hareven, D., Broday, L. and Hurwitz, C. (1994) The TM5 MADS box gene mediates organ differentiation in the three inner whorls of tomato flowers. *Plant Cell*, 6, 175–186.

Pridgeon, A. M., Cribb, P. J., Chase, M. W. and Rasmussen, F. N. (eds) (2001) *Genera Orchidacearum, vol. 2*. Oxford University Press, Oxford.

Puruggannan, M. D., Rounsley, S. D., Schmidt, R. J. and Yanofsky, M. F. (1995) Molecular evolution of flower development: diversification of the plant MADS-box regulatory gene family. *Genetics*, 140, 345–356.

Rasmussen, F. N. (1982) The gynostemium of the neottioid orchids. *Opera Botanica*, 65, 1–96.

Rasmussen, F. N. (1986) On the various contrivances by which pollinia are attached to viscidia. *Lindleyana*, 1, 21–32.

Rounsley, S. D., Ditta, G. S. and Yanofsky, M. F. (1995) Diverse roles for MADS box genes in *Arabidopsis* development. *Plant Cell*, 7, 1259–1269.

Siebert, D. J. (1992) Tree statistics; trees and 'confidence'; consensus trees; alternatives to parsimony; character weighting; character conflict and its resolution, in *Cladistics: a Practical Course in Systematics* (eds P. L. Forey, C. J. Humphries, I. L. Kitching, R. W. Scotland, D. J. Siebert and D. M. Williams), Oxford University Press, Oxford, pp. 72–88.

Simmons, M. P. (2000) A fundamental problem with amino-acid sequence characters for phylogenetic analysis. *Cladistics*, 16, 274–282.

Southerton, S. G., Marshall, H., Mouradov, A. and Teasdale, R. D. (1998) Eucalypt MADS-box genes expressed in developing flowers. *Plant Physiology*, 118, 365–372.

Sung, S. K., Yu, G. H., Nam, J., Jeong, D. H. and An, G. (2000) Developmentally regulated expression of two MADS-box genes, MdMADS3 and MdMADS4, in the morphogenesis of flower buds and fruits in apple. *Planta*, 210, 519–528.

Svensson, M. E., Johannesson, H. and Engström, P. (2000) The LAMB1 gene from the club-moss, *Lycopodium annotinum*, is a divergent MADS-box gene, expressed specifically in sporogenic structures. *Gene*, 253, 31–43.

Teo, I. A. and Shauniak, S. (1995) Polymerase chain reaction *in situ*: an appraisal of an emerging technique. *Histochemical Journal*, 27, 647–659.

Theißen, G., Strater, T., Fischer, A. and Saedler, H. (1995) Structural characterization, chromosomal localization and phylogenetic evaluation of two pairs of AGAMOUS-like MADS-box genes from maize. *Gene*, 156, 155–166.

Weigel, D. and Meyerowitz, E. M. (1994) The ABCs of floral homeotic genes. *Cell*, 78, 203–209.

Yu, D., Kotilainen, M., Pollanen, E., Mehto, M., Elomaa, P. *et al.* (1999) Organ identity genes and modified patterns of flower development in *Gerbera hybrida* (Asteraceae). *Plant Journal*, 17, 51–62.

Yu, H. and Goh, C. J. (2000) Identification and characterization of three orchid MADS-box genes of the AP1/AGL9 subfamily during floral transition. *Plant Physiology*, 123, 1325–1336.

Chapter 11

Involvement of non-ABC MADS-box genes in determining stamen and carpel identity in *Gerbera hybrida* (Asteraceae)

Teemu H. Teeri, Victor A. Albert, Paula Elomaa, Jaana Hämäläinen, Mika Kotilainen, Eija Pöllänen and Anne Uimari

ABSTRACT

The prevailing molecular viewpoint on flower organ determination is largely the result of resolving relationships between homeotic mutations and their corresponding genes in only a few model plant species. The outcome of these studies has been a simplified 'ABC' model, in which three homeotic functions, each with overlapping expression in two adjacent whorls of floral organs, determine the identity of organs in these whorls. According to this model, stamens and carpels develop when the C function is expressed. In mutants that are impaired in the C function, petals and sepals develop instead of stamens and carpels. According to the *Arabidopsis* ABC model, the C function is encoded by a single regulatory gene of the MADS-box family. The gene product is hypothesised to homodimerise before binding DNA. Our approach has been to study flower development outside of the common model species by using *Gerbera hybrida* (Asteraceae) as our experimental organism. Being transformable with foreign genes, *Gerbera* is a powerful system for functional gene analysis. Among the MADS-box regulatory genes, we have isolated the *Gerbera* orthologues that encode both the B and C functions. Through phylogenetic analysis, we have found that several other *Gerbera* MADS-box genes fall outside the A, B and C function clades. The functions of *Gerbera* genes that group with the *Arabidopsis AGL2* (or *SEPALLATA*)-like MADS-box genes are discussed here. These genes separately affect stamen and carpel development, and we propose that they are needed to fulfil the *Gerbera* C function because, when transgenically down-regulated, they phenocopy down-regulation of the *Gerbera AGAMOUS* orthologues. Involvement in the C function may take place through heterodimer formation with the classical C function MADS-box protein.

11.1 Introduction

Our current understanding of flower development relies heavily on analysis of developmental mutants in the model species of plant molecular genetics, *Arabidopsis thaliana*. The combination of genetic accessibility, compact genome size, diverse methods of genetic transformation, and the possibility to screen exhaustive pools of induced mutations rapidly, has led to an incomparable wealth of data from *Arabidopsis* concerning the genes involved in flower development, starting from flower induction to determination of organ identity within each flower (Yanofsky, 1995).

In *Developmental Genetics and Plant Evolution* (2002) (eds Q. C. B. Cronk, R. M. Bateman and J. A. Hawkins), Taylor & Francis, London, pp. 220–232.

For the latter, an elegantly simple combinatorial model has unfolded, in which three genetic 'functions' determine which of the four types of floral organs (sepals, petals, stamens and carpels) will develop in each whorl of the flower (Coen and Meyerowitz, 1991). Floral homeotic mutations, in which floral organ identity is changed, nearly always affect two adjacent whorls. Based on this observation, it was proposed that three genetic functions, labelled A, B and C, each cover two adjacent whorls: A is expressed in whorls one and two, B in whorls two and three, and C in whorls three and four. Thus, sepals develop where A acts alone, petals where A and B are present together, stamens where B is combined with C, and finally carpels where C only is expressed. With the assumption that A and C are negative regulators of each other (Bowman *et al.*, 1991), not only could all three types of homeotic mutations be explained, but also the phenotypes of double and triple mutants became logical. In the triple mutant, all whorls are occupied by leaf-like organs; however, the basic concentrically-whorled structure of the flower remains unchanged, and must therefore be determined by genes other than those that code for the A, B and C functions (Meyerowitz *et al.*, 1991).

Although the central elements of the *Arabidopsis* ABC model for determination of floral organ identity have been shown to apply widely (see also Theißen *et al.*, 2002), variants do exist. The most thorough and, in fact, temporally parallel studies on the genetic basis of flower development were accomplished in *Antirrhinum majus*, the snapdragon, but the spectrum of homeotic floral mutations in *Antirrhinum* conspicuously lacks a true A type mutation in which sepals and petals are converted to carpels and stamens, respectively. In *Antirrhinum*, such a phenotype does occur, but only with the dominant *ovulata* mutation, which leads to ectopic expression of the C function (Bradley *et al.*, 1993). Analysis of genes responsible for the A, B and C functions in *Petunia*, maize and *Gerbera* (e.g. Kater *et al.*, 1998; Yu *et al.*, 1999; Ambrose *et al.*, 2000) further demonstrate that, although the ABC model might be generally applicable to all flowering plants, each species has its own variations. For example, clear A-type genetic factors seem to be specific for *Arabidopsis*, whereas they become entangled with genetic functions that determine flower meristem identity in other plants (and, indeed, to a certain extent in *Arabidopsis* as well). Furthermore, it is becoming clear that the spectrum of floral mutants that can be obtained in *Arabidopsis* points out only a subset of the genes involved in determination of organ identity in the flower. Many players never show up due to genetic redundancy; they have no apparent phenotype unless several genes are simultaneously affected by mutation (Pelaz *et al.*, 2000).

The molecular cloning of the genes responsible for the A, B and C functions began with *Antirrhinum*, when the *DEFICIENS* gene (B function) was isolated and characterised (Sommer *et al.*, 1990). *DEFICIENS* encodes a protein that has high sequence similarity to transcription factors isolated from yeast and mammals; these, along with *AGAMOUS*, an *Arabidopsis* C function gene (Yanofsky *et al.*, 1990), formed the basis of the definition of the MADS-box genes (Schwarz-Sommer *et al.*, 1990). It rapidly became evident that nearly all genes responsible for the A, B and C functions belong to this large, transcription-factor gene family. The nucleotides encoding the DNA binding MADS domain are highly conserved, making it a relatively easy task to isolate MADS-box genes from any given plant species. Phylogenetic analysis of large numbers of these genes has shown that phylogenetic position and genetic function have been conserved to a surprising degree (Doyle, 1994; Purugganan et al., 1995; Tandre et al., 1995; Theißen et al., 1996; Yu et al., 1999). However, in addition to the four MADS-box genes directly implicated in the Arabidopsis A, B and C functions (the B function being fulfilled by two genes, APETALA3 and PISTILLATA: Krizek and Meyerowitz, 1996), all plant genomes may, like Arabidopsis, harbour several tens of MADS-box genes (Riechmann et al., 2000). While some of these genes are known to be important for aspects of vegetative development (Alvarez-Buylla et al., 2000), many are expressed specifically in flowers, and as discussed in this chapter, some seem to participate in the ABC model itself (Kotilainen et al., 2000; Pelaz et al., 2000).

11.2 Flower development in *Gerbera*

We are investigating flower development in a species with more complex inflorescences than those of *Arabidopsis* or *Antirrhinum*. As a model for the Asteraceae (Compositae), we have developed the ornamental plant *Gerbera hybrida* into a flexible experimental organism for molecular studies (Kotilainen et al., 1999a). We can genetically transform *Gerbera*, and work backwards from isolated genes to their mutations (Elomaa et al., 1993). Both ectopic/precocious gene expression and downregulation of gene expression through reintroduction of isolated *Gerbera* genes as sense or antisense constructs have proven to be extremely informative (Eckermann et al., 1998; Yu et al., 1999; Kotilainen et al., 1999b; Kotilainen et al., 2000). However, *Gerbera* cultivars are highly heterozygous, and due to the lack of pure lines and the size and relatively long generation time of the plant, mutant analysis is not possible.

As is typical of most Asteraceae (see Gillies et al., 2002), *Gerbera* bears several types of flowers on its inflorescence (capitulum). The marginal ray flowers are female and strongly zygomorphic, with three of the five petals fused together forming a large, bilabiate corolla. The central disc flowers are hermaphroditic, and while also zygomorphic, the five petals are of equal size. A third flower type can also be identified in *Gerbera*. The intermediate (or 'trans') flowers are located between the ray and disc flowers, and are female like the rays but range in size (from ray- to disc-like) in different *Gerbera* varieties. Flower development initiates similarly in all *Gerbera* flowers, with flower-type identity remaining obscure in young primordia. Accordingly, anther development in the female marginal flowers (ray and trans) initiates normally, but later aborts, resulting in small and withered staminodes in these flower types.

The flowers of Asteraceae are different in several aspects from the basic, symmetrical angiosperm flower. Whorl one, where sepals develop, is particularly peculiar. In most species, the petals are surrounded by hairy structures that later form a 'parachute' for the seed, but some species have scaly structures and others lack this whorl of organs altogether (Bremer, 1994). Whether the pappus hairs (or scales) are sepals at all has been debated (Lund, 1872, 1874), but our results with transgenic *Gerbera* plants show that they react to changes in MADS-box gene expression much like true sepals would (see Section 11.4).

11.3 *Gerbera* MADS-box genes

Using both the high sequence identity of the MADS-box region and random sequencing of cDNAs, we have identified eleven different MADS-box genes in *Gerbera* that are active in developing or mature flowers (Table 11.1; Yu *et al.*, 1999; Kotilainen *et al.*, 2000, unpubl. obs.). In a phylogenetic analysis, several of the *Gerbera* genes group in clades together with the *Arabidopsis* A, B and C encoding MADS-box genes, whereas several fall clearly outside (Figure 11.1). Based on experience with other plant species, the phylogenetic grouping of MADS-box genes can be used as a first prediction of their function. Thus, *GAGA1* and *GAGA2* are the *Gerbera* candidates for C function genes, *GDEF1*, *GDEF2* and *GGLO1* for B function genes, and *GSQUA1* for an A function gene. Expression data confirm that *GAGA1* and *GAGA2* are strongly expressed in whorls three and four, just like other C function genes (Purugganan *et al.*, 1995). *GDEF2* and *GGLO1* are likewise expressed in whorls two and three (as with other B function genes: Purugganan *et*

Table 11.1 Florally active MADS-box genes in *Gerbera hybrida*

Gene	Phylogenetic position	Expression pattern	Function
GSQUA1	SQUA group (with the A function gene AP1 from Arabidopsis)	In developing vascular bundles of the capitulum and flower primordia	Unknown, but not A function
GDEF1	DEF group (with B function genes)	Very low expression in flower primordia	Unknown, but not B function
GDEF2	DEF group (with B function genes)	Regions of flower primordia where petals and stamens develop, later in the ovule	B function
GGLO1	GLO group (with B function genes)	As with GDEF2, but not in the ovule	B function
GAGA1	AG group (with C function genes)	Central parts of flower primordia where stamens and carpels develop, later also in developing ovules	C function
GAGA2	AG group (with C function genes)	As with GAGA1	C function
GRCD1	AGL2 group (SEP3 branch)	First in all parts of flower primordia, but later stronger in stamen and carpel primordia and in the outer integument of the ovule	Necessary for the C function in whorl 3 (stamens)
GRCD2	AGL2 group (SEP3 branch)	First in all parts of flower primordia, but later concentrates to the central parts and in the ovule	Necessary for the C function in whorl 4 (carpels), but also for other processes of flower development (unpubl.)
GRCD3	AGL2 group (AGL6 branch)	Not determined	Unknown
G1005B02	AGL2 group	Not determined	Unknown
G1021A03	TM3 group	Not determined	Unknown

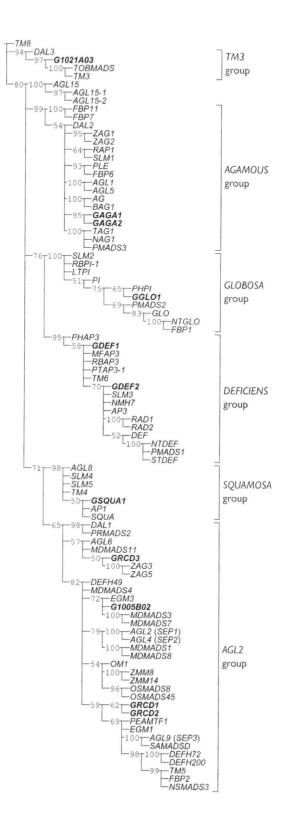

Figure 11.1 Phylogenetic relationships of selected seed-plant MADS-box genes, including eleven genes isolated from *Gerbera* (in bold). Parsimony jackknife analysis (Farris *et al.*, 1996) on aligned nucleotide sequences of the MADS and K boxes for ninety-six genes was performed exactly as in Yu *et al.* (1999). The consensus tree from the XAC application (J. S. Farris, unpubl.) provides estimates of support for individual tree branches. Jackknife values between 50 per cent and c.63 per cent reflect some robustness to extra steps, whereas those at or above c.63 per cent indicate stronger support, equivalent to at least one uncontradicted synapomorphy. The same data set was similarly analysed as translated amino acid residues, using another unpublished application written by J. S. Farris (not shown). The results of both analyses were highly inter-compatible, with the principal difference being that nucleotide data provided much higher resolution within the major MADS-box lineages. For example, the *GRCD1/GRCD2* group, which appears in 62 per cent of the nucleotide jackknife replicates, is not supported at or above the 50 per cent level in the amino acid tree. The sister-group relationship of *GRCD1* and *GRCD2* makes much sense from the perspective of the molecular functional findings reported in Kotilainen *et al.* (2000) and this chapter. The implication that these genes may be duplication products, highlighted in the nucleotide tree, would have been lost if more evolutionarily conservative information (i.e. amino acids) had been compared. An equivalent example is the *GAGA1/GAGA2* sister pair, which is strongly supported by nucleotide jackknifing (95 per cent of replicates) and molecular functional analysis (Yu *et al.*, 1999; Kotilainen *et al.*, 2000), but not by analysis of amino acid information, in which the pairing is lost entirely. Aside from the *Gerbera* genes already mentioned, *GGLO1* groups appropriately in the *GLOBOSA* lineage, and *GDEF1* and *GDEF2* (which confers the B function) group in the *DEFICIENS* lineage. As previously reported (Yu *et al.*, 1999), *GDEF2* and *GDEF1* appear at equivalent hierarchical levels with *AP3* and *TM6*, respectively (see Kramer *et al.*, 1998). *GSQUA1* groups with *APETALA1* and other *APETALA1* phylogenetic orthologues, despite its apparent lack of A function properties. *GRCD3*, not previously reported, groups closely with *AGL6* and related genes with the larger *AGL2* group. Likewise, *G1005B02* groups with another *AGL2*-like sub-clade that includes a gene isolated from pine. Lastly, *G1021A03* groups with a distinct lineage of genes characterised by predominantly vegetative expression (Pnueli *et al.*, 1991). Previously published plant MADS-box gene sequences were retrieved from the EMBL or GenBank databases (*DEF* [accession number X52023], *DEFH49* [X95467], *DEFH72* [X95468], *DEFH200* [X95469], *GLO* [X68831], *PLE* [S53900] and *SQUA* [X63701] from *Antirrhinum majus*; *AG* [X53579], *AGL1* [M55550], *AGL2* (*SEP1*) [M55551], *AGL4* (*SEP2*) [M55552], *AGL5* [M55553], *AGL6* [M55554], *AGL8* [U33473], *AGL9* (*SEP3*) [AF015552], *AGL15* [U22528], *AP1* [Z16421] *AP3* [D21125] and *PI* [D30807] from *Arabidopsis thaliana*; *OM1* [X69107] from *Aranda* x Deborah; *AGL15-1* [U22665], *AGL15-2* [U22681] and *BAG1* [M99415] from *Brassica napus*; *EGM1* [AF029975] and *EGM3* [AF029977]; *GAGA1* [AJ009722], *GAGA2* [AJ009723], *GDEF1* [AJ009724], *GDEF2* [AJ009725], *GGLO1* [AJ009726], *GSQUA1* [AJ009727], *GRCD1* [AJ400623], *GRCD2* [unpublished], *GRCD3* [unpublished], *G1021A03* [unpublished] and *G1005B02* [unpublished] from *Gerbera hybrida*; *MDMADS1* [U78947], *MDMADS3* [AF068722], *MDMADS4* [U78950], *MDMADS7* [AJ000761], *MDMADS8* [AJ001681] and *MDMADS11* [AJ000763] from *Malus domestica*; *LTP1* [AF052864] from *Liriodendron tulipifera*; *TAG1* [L26295], *TM3* [X60756], *TM4* [X60757], *TM5* [X60480] *TM6* [X60759] and *TM8* [X60760] from *Lycopersicon* (*Solanum*) *esculentum*; *NMH7* [L41727] from *Medicago sativa*; *MFAP3* [AF052877] from *Michelia figo*; *NSMADS3* [AF068722] from *Nicotiana sylvestris*; *NAG1* [L23925], *NTDEF* [X96428], *NTGLO* [X67959] and *TOBMADS* [X76188] from *Nicotiana tabacum*; *OSMADS8* [U78892] and *OSMADS45* [U31994] from *Oryza sativa*; *PTAP3-1* [AF052870] from *Pachysandra terminalis*; *PHAP3* [AF052879] and *PHPI* [AF052865] from *Peperomia hirta*; *PMADS1* [X69946], *PMADS2* [X69947], *PMADS3* [X72912], *FBP1* [M91190], *FBP2* [M91666], *FBP6* [X68675], *FBP7* [X81651] and *FBP11* [X81852] from *Petunia hybrida*; *DAL1* [X80902], *DAL2* [X79280] and *DAL3* [X79281] from *Picea abies*; *PRMADS2* [U42400] from *Pinus radiata*; *PEAMTF1* [AJ223318] from *Pisum sativum*; *RBAP3* [AF052876], *RAD1* [X89113], *RAD2* [X89108], *RAP1* [X89107] and *RBPI-1* [AF052859] from *Rumex acetosa*; *SLM1* [X80488], *SLM2* [X80489], *SLM3* [X80490], *SLM4* [X80491] and *SLM5* [X80492] from *Silene latifolia*; *SAMADSD* [Y08626] from *Sinapis alba*; *STDEF* [X67511] from *Solanum tuberosum*; *ZAG1* [L18924], *ZAG2* [L18925], *ZAG3* [L46397], *ZAG5* [L46398], *ZMM8* [Y09303] and *ZMM14* [AJ005338] from *Zea mays*).

al., 1995), but *GDEF1* is barely expressed in flowers at all and the expression pattern of *GSQUA1* is in what appear to be developing vascular bundles of the capitulum and floret primordia, clearly outside whorls one and two. Genetic transformation confirms the improved prediction based on expression patterns. *GDEF2* and *GGLO1* are indeed B function genes, and down-regulation of either of them causes transformation of petals to pappus hairs (the whorl one organs in *Gerbera*) and of stamens to carpel-like structures. Furthermore, ectopic expression of *GGLO1* (which also leads to ectopic expression of *GDEF2*) changes pappus hairs to petals and carpels to anther-like structures (Yu *et al.*, 1999).

Transformation studies also show that *GAGA1* and *GAGA2* are indeed C function genes, but unlike *GDEF1* and *GDEF2* they appear to be functionally identical. Down-regulation of their activity converts stamens to petals, and in the marginal flowers this homeotic change spares them from the developmental arrest typical of stamens; instead, both whorl two and four organs, when changed to stamens, develop but later wither in marginal flowers. Furthermore, in *GAGA1* or *GAGA2* antisense plants, carpels do not develop in whorl four, but instead the whorl is occupied by a greenish organ with pappus-like hairs on its surface, or in strongly down-regulated lines, by repeating pappus hairs and petal structures, exactly like the *AGAMOUS* mutation of *Arabidopsis* (Bowman *et al.*, 1989). Ectopic expression of either of the *GAGA* genes causes homeotic transformation of whorl two organs (petals, in wild type) into anther-like structures, but unlike in *Arabidopsis*, there is no change in the identity of whorl one organs (Yu *et al.*, 1999).

Therefore, in basic outline, the ABC model can be used to explain the determination of organ identity in *Gerbera* flowers; the deviation in ectopic expression of the C function genes is not sufficient to alter whorl one identity. Also, genes encoding at least the B and C functions do exist in *Gerbera*. *GSQUA1*, the *Gerbera* gene phylogenetically closest to *APETALA1*, an *Arabidopsis* A function gene (Mandel *et al.*, 1992), and *SQUAMOSA*, the *Antirrhinum* gene with an expression pattern like *APETALA1* but whose mutant is different from the mutated *APETALA1* (Huijser *et al.*, 1992), is not expressed on whorls one and two. It is possible that the 'true' *APETALA1* functional orthologue will be found later in *Gerbera*, but *GSQUA1* may also represent a deviation from the *Arabidopsis* ABC model, like *SQUAMOSA* itself.

11.4 *GRCD1* participates in the C function in *Gerbera*

In addition to the *Gerbera* genes that are phylogenetically related to ABC MADS-box genes (of which *GDEF1* and *GSQUA1* are not functionally ABC genes), we have isolated several others that group outside these better-characterised gene family lineages (Figure 11.1; Kotilainen *et al.*, 2000, and unpublished). One large clade, sister to the *APETALA1*-orthologous MADS-box genes, was first identified by cloning of the *Arabidopsis AGL2* cDNA (Ma *et al.*, 1991). Several other *Arabidopsis* genes fall into this clade, together with many genes isolated from a variety of angiosperm and gymnosperm species (Theissen *et al.*, 2000). As most of these genes are expressed specifically in flower primordia or flowers, we have been interested in their roles during *Gerbera* flower development.

GRCD1, one of the *AGL2*-clade MADS-box genes of *Gerbera*, is expressed early in flower development throughout the primordium, but later the expression is strongest in whorls three and four. Transgenic *Gerbera* plants in which *GRCD1* expression is down-regulated show a homeotic transformation in the flowers (Koti-lainen *et al.*, 2000). The transformation is most clear in the marginal (female) flower types, where the arrested staminodes are replaced by petals (Figure 11.2, Plate 2a). The anthers of disc flowers have minor petal-like characters (abaxial stomata) although they remain male-fertile. The observed homeotic change is distinctly similar to the homeotic change of the same whorl in transgenic plants with down-regulated *Gerbera* C function genes (*GAGA1* or *GAGA2*). However, the anti-*GRCD1* plants are not affected in whorl four, where normal and fertile carpels develop.

This type of mutation, in which a homeotic change (stamen to petal) occupies only whorl three, has not been described in *Arabidopsis*. Genetic redundancy may explain why *GRCD1* is not strictly necessary for anther development in the central disc flowers. Hypothetically, the marginal ray flowers may have lost expression of redundant genes, or these genes may have been recruited for the novel function of stamen abortion in ray and trans flowers.

The *Arabidopsis* model states that *AGAMOUS* is the only gene for the C func-tion, and that the C function is fulfilled through homodimerisation of the AGAMOUS protein (Riechmann *et al.*, 1996). It is not known which genes are regu-lated by the homodimer, and one hypothesis we had for *GRCD1* was that it may be a necessary downstream gene that mediates the fulfilment of C function (in stamens). Alternatively, we imagined that *GRCD1* could be regulating the C function genes in *Gerbera* (at least in whorl three). In order to clarify the functional relationship of these genes, we studied *GAGA1* and *GAGA2* expression in plants where *GRCD1* is down-regulated, and vice versa. Interestingly, expression of the *GAGA* genes is not

Figure 11.2 Down-regulation of *GRCD1*, a *Gerbera* gene grouping outside of the ABC function genes, causes a homeotic transformation in the marginal female flowers of the *Gerbera* capitu-lum. In the control plants (left) development of stamens initiates on whorl 3 but subse-quently arrests. In the *GRCD1* antisense transformant lines (right), the identity of whorl 3 organs is changed, and instead of withered staminodes, narrow petals emerge (arrow). In the central disc flowers, fertile stamens still develop in whorl 3.

Table 11.2 Interaction between Gerbera MADS-domain proteins in the Yeast Two-Hybrid Assay, shown as LacZ reporter activity in yeast cells. The background value in yeast cells was 6.9 (SD 3.3)

	GRCD1	GAGA1	GAGA2
GRCD1	4.13 (SD 1.2)	62.5 (SD 15.6)	37.8 (SD 12.5)
GAGA1		20.2 (SD 5.9)	16.0 (SD 7.9)
GAGA2			14.1 (SD 3.6)

affected by down-regulation of *GRCD1*, nor is *GRCD1* expression affected by *GAGA1* or *GAGA2*. Nevertheless, expression of both types of MADS-box gene is necessary for the identity of stamens in whorl three. Although the C function MADS-box genes were expected to make homodimers, an obvious possibility was that the GRCD1 protein might be a heterodimerising partner for either GAGA1 or GAGA2, or both. Indeed, yeast two-hybrid analysis showed that the GAGA1/GRCD1 and GAGA2/GRCD1 heterodimers are much more prevalent than the GAGA1 or GAGA2 homodimers, or the GAGA1/GAGA2 heterodimer (Table 11.2; Kotilainen *et al.*, 2000). *GRCD1* therefore appears to be a participant in the C function, and to act at the protein level through heterodimerisation with the classical C function MADS-domain proteins.

It is clear that if heterodimerisation is necessary for the C function in stamens of marginal *Gerbera* flowers, other MADS-domain proteins must be playing roles in the same process in different whorls and between different flower types. *GRCD1* may therefore not be required for the development of the fertile stamens in *Gerbera*, with the consequence that disc flowers must utilise other MADS-domain partners for GAGA1 and GAGA2. It is also possible that very low concentrations of GRCD1 are sufficient for stamen identity in disc flowers. However, carpels are able to develop in marginal flowers even when *GRCD1* is strongly down-regulated. The closest isolated relative to *GRCD1* in *Gerbera* is *GRCD2*, which is initially expressed at a low level in all four whorls of *Gerbera* flowers (Table 11.1; Anne Uimari, unpubl. obs.). It is tempting to speculate that *GRCD2* is the redundant gene responsible for stamen determination in disc flowers. Alternatively, *GRCD2* could play a role in carpel development, the other half of the C function. Preliminary analyses of transgenic plants favour the latter alternative. Down-regulation of *GRCD2* in transgenic plants phenocopies a C-type homeotic transformation limited to whorl four.

11.5 Implications for Asteraceae reproductive diversity

We have shown that *GRCD1* and perhaps *GRCD2* play important but different roles in sex expression within the *Gerbera* capitulum. For example, our findings strongly suggest that GRCD1 is more active in its control of the C function in marginal ray flowers than it is in central disc flowers. However, not all Asteraceae share the *Gerbera* sexual pattern of female outer flowers and hermaphroditic inner flowers. In species of *Barnadesia*, a phylogenetically primitive genus of the sunflower family (Bremer, 1994; Gustafsson *et al.*, 1999), the central flowers may be hermaphroditic, male, or even neuter, whereas the marginal flowers are always hermaphroditic. In *Huarpea*, a close relative of *Barnadesia*, there is only a single central flower, which is male. Moreover, in the same Asteraceae tribe that includes *Gerbera* (the Mutisieae), species of *Brachylaena* bear entirely male or female capitula. If our inferences about *GRCD1* function are correct, then the *Huarpea GRCD1* orthologue would be expected to have a primarily central rather than marginal effect, the converse of *Gerbera*. Likewise, some *Barnadesia* species might share this pattern, whereas others would be more like *Gerbera*. If such aspects of the C function can vary within a single, monophyletic genus of Asteraceae (as has been shown with molecular phylogenetics: Gustafsson *et al.*, 1999), then C-function modulation may be more common in angiosperms than previously thought. The *Brachylaena* case, in which the *GRCD1* orthologue might be differentially expressed in differently-sexed inflorescences, underscores that upstream regulation of the C function may be similarly variable. These potentially important observations highlight the utility of analysing flower development in plants that bear different kinds of flowers.

11.6 Expansion of the ABC model

How generally applicable are our findings to angiosperms, or have we observed only *Gerbera*- or Asteraceae-specific phenomena? The inference of genetic redundancy nearly masking down-regulation of *GRCD1* in the disc flowers of *Gerbera* stresses the possibility that many of the genes participating in fulfilment of the ABC model may be similarly masked by genetic redundancy in the model species. In *Antirrhinum*, the search for genes whose products might heterodimerise with *PLENA* (the *Antirrhinum* C function gene) in a yeast two-hybrid assay (Davies *et al.*, 1996) has yielded three genes, *DEF49*, *DEFH72* and *DEFH200*, which fall in the *AGL2* clade. Furthermore, recent analysis of three *Arabidopsis* AGL2 clade genes (*AGL2* itself, *AGL4* and *AGL9* – now renamed *SEPALLATA1*, *SEPALLATA2* and *SEPALLATA3*, respectively) in a construct in which all three genes are mutated, shows that these three genes are required for petal, stamen and carpel development in *Arabidopsis*; the triple mutant bears flowers with sepals only (Pelaz *et al.*, 2000).

It is becoming apparent that genes other than those encoding the classical ABC factors are needed where ABC function MADS-box genes were once thought to act alone. Here we show that the C function may be variously modulated by alternative heterodimerisation partners. The B function has long been known to involve heterodimerisation of two MADS-box genes in *Arabidopsis* and *Antirrhinum* (Schwarz-Sommer *et al.*, 1992; McGonigle *et al.*, 1996). However, although the heterodimer binds DNA, it suspiciously does not activate transcription (Honma and Goto, 2001). In *Arabidopsis*, alternate partners interact with the B gene products (APETALA1, SEPALLATA3 and AGAMOUS; Honma and Goto, 2001). It is possible that more alternative partners modulate the B function in other plants; for example, the GDEF1 protein of *Gerbera*, or the TM6 protein of tomato (see Kramer *et al.*, 1998; Kramer and Irish, 1999). It is becoming evident that a further complexity exists behind the classical ABC model as MADS-domain proteins are able to associate in higher level complexes than dimers, increasing the possibilities for modulation of activity by orders of magnitude (Egea-Cortines *et al.*, 1999; Honma and Goto, 2001; Theißen, 2001). The unique, multi-phenotype, yet single genotype of *Gerbera* has permitted an understanding of the C function beyond what could have been possible with *Arabidopsis* or *Antirrhinum*. At least one *SEPALLATA* orthologue, *GRCD1*, has different effects in different whorls of different flowers on the same plant. Clearly, the full story and details behind the ABC model are only beginning to emerge.

ACKNOWLEDGEMENTS

We thank Eija Takala, Marja Huovila, Anu Rokkanen and Sanna Peltola for excellent technical assistance and care of the plants in the greenhouse, and James S. Farris for providing access to parsimony jackknifing applications. This work was supported by the Academy of Finland (Grants 34533 and 44315). Additional support from the College of Arts and Sciences, University of Alabama, is also gratefully acknowledged.

REFERENCES

Alvarez-Buylla, E. R., Liljegren, S. J., Pelaz, S., Gold, S. E., Burgeff, C., Ditta, G. S., Vergara-Silva, F. and Yanofsky, M. F. (2000) MADS-box gene evolution beyond flowers: expression in pollen, endosperm, guard cells, roots and trichomes. *Plant Journal*, 24, 457–466.

Ambrose, B. A., Lerner, D. R., Ciceri, P., Padilla, C. M., Yanofsky, M. F. and Schmidt, R. J. (2000) Molecular and genetic analyses of the silky1 gene reveal conservation in floral organ specification between eudicots and monocots. *Molecular Cell*, 5, 569–579.

Bowman, J. L., Smyth, D. R. and Meyerowitz, E. M. (1989) Genes directing flower development in *Arabidopsis*. *Plant Cell*, 1, 37–52.

Bowman, J. L., Smyth, D. R. and Meyerowitz, E. M. (1991) Genetic interactions among floral homeotic genes in *Arabidopsis*. *Development*, 112, 1–20.

Bradley, D., Carpenter, R., Sommer, H., Hartley, N. and Coen, E. (1993) Complementary floral homeotic phenotypes result from opposite orientation of a transposon at the *plena* locus of *Antirrhinum*. *Cell*, 72, 85–95.

Bremer, K. (1994) *Asteraceae: Cladistics and Classification*. Timber Press, Portland.

Coen, E. S. and Meyerowitz, E. M. (1991) The war of the whorls: genetic interaction controlling flower development. *Nature*, 353, 31–37.

Davies, B., Egea-Cortines, M., de Andrade Silva, E., Saedler, H. and Sommer, H. (1996) Multiple interactions amongst floral homeotic MADS-box genes. *EMBO Journal*, 15, 4330–4343.

Doyle, J. J. (1994) Evolution of a plant homeotic multigene family: toward connecting molecular systematics and molecular developmental genetics. *Systematic Biology*, 43, 307–328.

Eckermann, S., Schröder, G., Schmidt, J., Strack, D., Edrada, R. A., Helariutta, Y., Elomaa, P., Mika Kotilainen, M., Kilpeläinen, I., Proksch, P., Teeri, T. H. and Schröder, J. (1998) New pathway to polyketides in plants. *Nature*, 396, 387–390.

Egea-Cortines, M., Saedler, H. and Sommer, H. (1999) Ternary complex formation between the MADS-box proteins SQUAMOSA, DEFICIENS and GLOBOSA is involved in the control of floral architecture in *Antirrhinum majus*. *EMBO Journal*, 18, 5370–5379.

Elomaa, P., Honkanen, J., Puska, R., Seppänen, P., Helariutta, Y., Mehto, M., Kotilainen, M., Nevalainen, L. and Teeri, T. H. (1993) Agrobacterium-mediated transfer of antisense chalcone synthase cDNA to *Gerbera hybrida* inhibits flower pigmentation. *Bio/Technology*, 11, 508–511.

Farris, J. S., Albert, V. A., Källersjö, M., Lipscomb, D. and Kluge, A. G. (1996) Parsimony jackknifing outperforms neighbor-joining. *Cladistics*, 12, 99–124.

Gillies, A. C. M., Cubas, P., Coen, E. S. and Abbott, R. J. (2002) Making rays in the Asteraceae: genetics and evolution of radiate versus discoid flower heads, in *Developmental Genetics and Plant Evolution* (eds Q. C. B. Cronk, R. M. Bateman and J. A. Hawkins), Taylor & Francis, London, pp. 233–246.

Gustafsson, M. H. G., Pepper, A. S.-R., Albert, V. A. and Källersjö, M. (1999) Molecular phylogeny and biogeography of the subfamily Barnadesioideae (Asteraceae). *Annals of the Missouri Botanical Garden*, 86, 57–117.

Honma, T. and Goto, K. (2001) Complexes of MADS-box proteins are sufficient to convert leaves into floral organs. *Nature*, 409, 525–529.

Huijser, P., Klein, J., Lönnig, W.-E., Meijer, H., Saedler, H. and Sommer, H. (1992) Bracteomania, an inflorescence anomaly, is caused by the loss of function of the MADS-box gene *squamosa* in *Antirrhinum majus*. *EMBO Journal*, 11, 1239–1249.

Kater, M. M., Colombo, L., Franken, J., Busscher, M., Masiero, S., Van Lookeren Campagne, M. M. and Angenent, G. C. (1998) Multiple AGAMOUS homologs from cucumber and petunia differ in their ability to induce reproductive organ fate. *Plant Cell*, 10, 171–182.

Kotilainen, M., Elomaa, P., Uimari, A., Albert, V. A., Yu, D. and Teeri, T. H. (2000) *GRCD1*, an *AGL2*-like MADS-box gene, participates in the C function during stamen development in *Gerbera hybrida*. *Plant Cell*, 12, 1893–1902.

Kotilainen, M., Albert, V. A., Elomaa, P., Helariutta, Y., Koskela, S., Mehto, M., Pöllänen, E., Uimari, A., Yu, D. and Teeri, T. H. (1999a) Flower development and secondary metabolism in *Gerbera hybrida*, an Asteraceae. *Flowering Newsletter*, 28, 20–31.

Kotilainen, M., Helariutta, Y., Mehto, M., Pöllänen, E., Albert, V. A., Elomaa, P. and Teeri, T. H. (1999b) *GEG* participates in the regulation of cell and organ shape during corolla and carpel development in *Gerbera hybrida*. *Plant Cell*, 11, 1093–1104.

Kramer, E. M. and Irish, V. F. (1999) Evolution of genetic mechanisms controlling petal development. *Nature*, 399, 144–148.

Kramer, E. M., Dorit, R. L. and Irish, V. F. (1998) Molecular evolution of genes controlling petal and stamen development: duplication and divergence within the *apetala3* and *pistillata* MADS-box gene lineages. *Genetics*, 149, 765–783.

Krizek, B. A. and Meyerowitz, E. M. (1996) The *Arabidopsis* homeotic genes *APETALA3* and *PISTILLATA* are sufficient to provide the B class organ identity function. *Development*, 122, 11–22.

Lund, S. (1872) Bægeret hos kurvblomsterne, et histologisk forsøg på at hævde udviklingens enhed i planteriget. *Botanisk Tidsskrift Ser. 2*, 2, 1–120.

Lund, S. (1874) Bemærkninger om bægeret hos kurvblomsterne, en antikritik. *Videnskabelige Meddelelser fra Dansk Naturhistorisk Forening i København*, 5, 10–37.

Ma, H., Yanofsky, M. F. and Meyerowitz, E. M. (1991) *AGL1-AGL6*, an *Arabidopsis* gene family with similarity to floral homeotic and transcription factor genes. *Genes and Development*, 5, 484–495.

Mandel, M. A., Gustafson-Brown, C., Savidge, B. and Yanofsky, M. F. (1992) Molecular characterization of the *Arabidopsis* floral homeotic gene *APETALA1*. *Nature*, 360, 273–277.

McGonigle, B., Bouhidel, K. and Irish, V. F. (1996) Nuclear localization of the *Arabidopsis* APETALA3 and PISTILLATA homeotic gene products depends on their simultaneous expression. *Genes and Development*, 10, 1812–1821.

Meyerowitz, E. M., Bowman, J. L., Brockman, L. L., Drews, G. N., Jack, T., Sieburth, L. E. and Weigel, D. (1991) A genetic and molecular model for flower development in *Arabidopsis thaliana*. *Development Supplement*, 1, 157–167.

Pelaz, S., Ditta, G. S., Baumann, E., Wisman, E. and Yanofsky, M. F. (2000) B and C floral organ identity functions require *SEPALLATA* MADS-box genes. *Nature*, 405, 200–203.

Pnueli, L., Abu-Abeid, M., Zamir, D., Nacken, W., Schwarz-Sommer, Z. and Lifschitz, E. (1991) The MADS-box gene family of tomato: temporal expression during floral development, conserved secondary structures and homology with homeotic genes from *Antirrhinum* and *Arabidopsis*. *Plant Journal*, 1, 255–266.

Purugganan, M. D., Rounsley, S. D., Schmidt, R. J. and Yanofsky, M. F. (1995). Molecular evolution of flower development: diversification of the plant MADS-box regulatory gene family. *Genetics*, 140, 345–356.

Riechmann, J. L., Heard, J., Martin, G., Reuber, L., Jiang, C.-Z., Keddie, J. *et al.* (2000) Arabidopsis transcription factors: genome-wide comparative analysis among eukaryotes. *Science*, 290, 2105–2110.

Riechmann, J. L., Krizek, B. A. and Meyerowitz, E. M. (1996) Dimerization specificity of *Arabidopsis* MADS domain homeotic proteins APETALA1, APETALA3, PISTILLATA, and AGAMOUS. *Proceedings of the National Academy of Sciences of the United States of America*, 93, 4793–4798.

Schwarz-Sommer, Z., Hue, I., Huijser, P., Flor, P. J., Hansen, R., Tetens, F., Lönnig, W.-E., Saedler, H. and Sommer, H. (1992) Characterization of the *Antirrhinum* floral homeotic MADS-box gene *deficiens*: evidence for DNA binding and autoregulation of its persistent expression throughout flower development. *EMBO Journal*, 11, 251–263.

Schwarz-Sommer, Z., Huijser, P., Nacken, W., Saedler, H. and Sommer, H. (1990) Genetic control of flower development by homeotic genes in *Antirrhinum majus*. *Science*, 250, 931–936.

Sommer, H., Beltrán, J.-P., Huijser, P., Pape, H., Lönnig, W.-E., Saedler, H. and Schwarz-Sommer, Z. (1990) *Deficiens*, a homeotic gene involved in the control of flower morphogenesis in *Antirrhinum majus*: the protein shows homology to transcription factors. *EMBO Journal*, 9, 605–613.

Tandre, K., Albert, V. A., Sundås, A. and Engström, P. (1995) Conifer homologues to genes that control floral development in angiosperms. *Plant Molecular Biology*, 27, 69–78.

Theißen, G., Becker, A., Kirchner, C., Munster, T., Saedler, H. and Winter, K.-U. (2002) How the plants on land learned the floral ABCs: the role of MADS-box genes in the evolutionary origin of flowers, in *Developmental Genetics and Plant Evolution* (eds Q. C. B. Cronk, R. M. Bateman and J. A. Hawkins), Taylor & Francis, London, pp. 173–205.

Theißen, G. (2001) Development of floral organ identity: stories from the MADS house. *Current Opinion in Plant Biology* 4, 75–85.

Theißen, G., Becker, A., Di Rosa, A., Kanno, A., Kim, J. T., Münster, T., Winter, K.-U. and Saedler, H. (2000) A short history of MADS-box genes in plants. *Plant Molecular Biology*, 42, 115–149.

Theißen, G., Kim, J. T. and Saedler, H. (1996) Classification and phylogeny of the MADS-box multigene family suggest defined roles of MADS-box gene subfamilies in the morphological evolution of eukaryotes. *Journal of Molecular Evolution*, 43, 484–516.

Yanofsky, M. F. (1995) Floral meristems to floral organs: genes controlling early events in *Arabidopsis* flower development. *Annual Review of Plant Physiology and Plant Molecular Biology*, 46, 167–188.

Yanofsky, M. F., Ma, H., Bowman, J. L., Drews, G. N., Feldman, K. A. and Meyerowitz, E. M. (1990) The protein encoded by the *Arabidopsis* homeotic gene *agamous* resembles transcription factors. *Nature*, 346, 35–39.

Yu, D., Kotilainen. M., Pöllänen, E., Mehto, M., Elomaa, P., Helariutta, Y., Albert, V. A. and Teeri, T. H. (1999) Organ identity genes and modified patterns of flower development in *Gerbera hybrida* (Asteraceae). *Plant Journal*, 17, 51–62.

Chapter 12

Making rays in the Asteraceae: genetics and evolution of radiate versus discoid flower heads

Amanda C. M. Gillies, Pilar Cubas, Enrico S. Coen and Richard J. Abbott

ABSTRACT

Variation in flower head form within the Asteraceae is largely dependent on the presence or absence, and position, of a small range of different floret types within the capitulum. We focus on variation for radiate versus discoid capitulum type, which is dependent on the replacement of an outer whorl of ray florets by a whorl of disk florets or vice versa. This variation has been shown to be mainly under the control of one or two major genes and an unidentified number of modifier genes. We propose a model for the molecular control of the development of the radiate capitulum, which is dependent on the expression of a *CYCLOIDEA* homologue in the production of zygomorphic ray florets. We review studies that have examined the establishment of a discoid capitulum variant within a self-incompatible, radiate species, and a radiate variant within a self-compatible, discoid species. These studies show that the factors determining the establishment of new capitulum variants in the wild are complex and difficult to resolve. A complete understanding of the phenomenon of establishment will be required before capitulum evolution in the Asteraceae is fully understood.

12.1 Introduction

A major challenge in plant evolutionary biology is to isolate the genes that control flower development. Mutations in these genes cause changes in flower structure, colour and shape which, in turn, may lead to alterations in the pollination biology and mating system of a species. Such alterations can establish prezygotic breeding barriers and be important, therefore, in the speciation process and the maintenance of species in sympatry (Hodges, 1997; Bradshaw *et al.*, 1998; Schemske and Bradshaw, 1999).

In recent years, an understanding has been gained of the molecular genetics of homeotic genes controlling the development of the main floral organs: sepals, petals, stamens and carpels (Coen and Meyerowitz, 1991; Weigel and Meyerowitz, 1994; Theißen *et al.*, 2000, 2002). More recently, a start has been made at isolating the genes that control the shape of flowers and examining their expression during development (Cubas *et al.*, 1999; Luo *et al.*, 1996; Luo *et al.*, 1999). This work on the molecular genetics of flower shape variation has been concerned with genes controlling the change from an asymmetrical, zygomorphic shape to a symmetrical,

In *Developmental Genetics and Plant Evolution* (2002) (eds Q. C. B. Cronk, R. M. Bateman and J. A. Hawkins), Taylor & Francis, London, pp. 233–246.

actinomorphic shape or vice versa. It has focused on species in the Scrophulariaceae and has shown that homologous genes are involved in the species examined. What is not known currently is whether the major genes that control floral shape in this family also have a role in determining similar changes of flower shape in more distantly related taxa (Donoghue *et al.*, 1998).

In this chapter, we focus on the family Asteraceae. We review the evidence for the genetic control of capitulum development and suggest a model for the molecular basis of its control and the candidate genes involved. In addition, we examine factors likely to influence the establishment of new capitulum variants in the wild. As emphasised by Cronk and Möller (1997), Donoghue *et al.* (1998) and Theißen (2000), to understand the evolution of floral form it is necessary to understand both the developmental genetics of the trait and how new floral variants become established. The latter requires 'studies on the population genetics of the system, its co-evolution with other organisms (e.g. pollinators) and its general ecology in the field' (Theißen, 2000).

12.2 Flower head variation in the Asteraceae

The Asteraceae is one of the largest families of flowering plants, comprising approximately 23,000 species and 1,500 genera (Bremer, 1994). Its most notable morphological feature is the flower head (capitulum), which is a highly compressed inflorescence that superficially resembles a solitary flower, but is composed of many florets.

Five main types of flower head are recognised: bilabiate, ligulate, radiate, discoid and disciform. Detailed descriptions of the types of floret found within each flower head are given by Jeffrey (1977) and Bremer (1994), and for a discussion of the role of MADS-box genes in patterning the Asteraceae flower see Teeri *et al.* (2002). Briefly, bilabiate capitula are composed entirely of bilabiate florets with corollas of equal size or with outer florets having larger corollas (Mabberley, 1997). The bilabiate floret has a corolla with either a 2-lobed inner (upper) and 3-lobed outer (lower) lip, or a 1-lobed inner and 4-lobed outer lip. Ligulate heads are composed of bisexual florets with zygomorphic corollas that are strap-shaped with five apical teeth. In contrast, radiate heads contain marginal ray or bilabiate florets that are female or sterile and a central disk of bisexual, male or sterile disk florets. Ray florets have zygomorphic corollas that are strap-shaped, but differ from ligulate florets by having three or fewer apical teeth, whereas disk florets have actinomorphic corollas, which are tubular and pentamerous or sometimes tetramerous. Discoid heads are composed entirely of disk florets, while disciform heads contain either central disk florets and marginal female florets with eligulate corollas or are entirely composed of the latter floret type (Mabberley, 1997).

The prevalence of particular types of flower head varies among the different subfamilies and tribes of the Asteraceae recognised by Bremer (1996) (Table 12.1). Members of the basal subfamily, Barnadesioideae, produce discoid, disciform or radiate capitula, while the Mutisieae, which is basal to the remaining subfamilies, produce discoid, disciform, bilabiate, radiate or ligulate capitula. In both of these basal groups, radiate capitula have marginal bilabiate florets, whereas the remainder of the Asteraceae produce radiate capitula with marginal ray florets. Jeffrey (1977)

Table 12.1 Distribution of capitulum and floret types among the subfamilies and tribes of Asteraceae (as recognised by Bremer, 1996) according to Jansen *et al.* (1991), Bremer (1994) and Mabberley (1997). Terms are explained in Section 12.2

Subfamily	Tribe	Capitulum type	Floret type
Barnadesioideae	Barnadesieae	discoid, disciform, radiate	disk, bilabiate, ligulate, eligulate
	Mutisieae*	discoid, disciform, bilabiate, radiate, ligulate	disk, bilabiate, ligulate, eligulate
Carduoideae	Cardueae	discoid	disk
Cichorioideae	Lactuceae	ligulate	ligulate
	Vernonieae	discoid (rarely ligulate)	disk, ligulate
	Liabeae	discoid and radiate	disk, ray
	Arctoteae	radiate (rarely discoid)	disk, ray
Asteroideae	Inuleae		
	Plucheeae		
	Gnaphalieae		
	Calenduleae		
	Astereae	radiate, disciform, discoid	disk, ray, eligulate
	Anthemideae		
	Senecioneae		
	Helenieae		
	Heliantheae		
	Eupatorieae	discoid	disk

Note:
*The Mutisieae is not classified with respect to subfamily because its phylogenetic position has not yet been resolved (Bremer, 1996).

suggested that the bilabiate corolla form is possibly a relictual feature from a previously elongated inflorescence and is not well suited to a capitate inflorescence. He argued that the inner lip of such a corolla may interfere spatially and functionally with the outer lip of the immediately inner floret on the same orthostichy within a capitulum. Subfamily Carduoideae, which is sister to the rest of the family and comprised entirely of the tribe Cardueae, is characterised by discoid capitula. Further changes of capitulum form are apparent in subfamily Cichorioideae, where the different tribes are characterised either by ligulate heads (Lactuceae), discoid (or rarely ligulate) heads (Vernonieae), radiate or discoid heads (Liabeae), or radiate and rarely discoid heads (Arctoteae). In the largest subfamily, the Asteroideae, comprising approximately two-thirds of the species in the family, the flower head is commonly radiate or disciform, although there are also many discoid taxa (Mabberley, 1997).

The evolutionary relationships of taxa within subfamilies remain poorly resolved (Bremer, 1996) and, consequently, it is not possible to comment with confidence on the evolution of capitulum and floret form within the family. However, all forms of capitula are present in the oldest Asteraceae, that is the Barnadesioideae and the Mutisieae (Bremer, 1996), except radiate capitula with marginal ray florets. It would seem, therefore, that the radiate capitulum with marginal ray florets evolved within the family after the splits of the Barnadesioideae and subsequently the Mutisieae from the rest of the family.

Of particular interest is the fact that major differences in capitulum type can occur between closely related species and even within species. For example, closely related taxa may produce either radiate or discoid capitula, due to the replacement of ray florets by disk florets in the capitulum. In these cases, the discoid form would seem to be derived from the radiate type (Bremer and Humphries, 1993), although there is evidence that the change can also occur in the opposite direction (Abbott *et al.*, 1992). The occurrence of such variation in closely related material allows an analysis of its genetic basis and an examination of its origin and maintenance in the wild. In the remainder of this chapter we focus on variation in capitulum form of the radiate versus discoid type. We discuss, in turn, its genetic basis, a possible mechanism of molecular control, and problems involved in the establishment of new capitulum variants in the wild.

12.3 Genetic control of variation for radiate versus discoid flower heads

12.3.1 Single gene control

The first formal genetic analysis of variation for radiate versus discoid capitula within a species was conducted by Trow (1912) on *Senecio vulgaris* L. (Common Groundsel). He showed this variation to be controlled by a single gene with radiate dominant to discoid. However, heterozygotes could be distinguished by possession of ray florets with rays approximately half the length of those produced by the radiate homozygote. Consequently, it was concluded that alleles at this locus (henceforth called *RAY*) exhibited codominance, and had an additive effect on ray length. Single gene control of radiate versus discoid capitula in *S. vulgaris* was confirmed by Abbott *et al.* (1992) and Comes (1998), and also in the related species, *Senecio squalidus* L. (Oxford Ragwort), by Ingram and Taylor (1982). Richards (1975) reported that ray length can vary considerably in the heterozygote of *S. vulgaris*, indicating that modifier genes affect the expression of dominance of alleles at the *RAY* locus. Ingram and Taylor (1982) also found that heterozygotes in *S. squalidus* exhibited wide variation in ray length and that mean ray length of F1 plants was inversely related to stamen, corolla tube and bilabiate floret development. The largely additive effect on ray length of alleles at the *RAY* locus in *Senecio* was confirmed by an analysis of the genetics of presence/absence of ray florets and ray length in *Senecio cambrensis* Rosser, the allopolyploid of *S. squalidus* and *S. vulgaris* (Ingram and Noltie, 1984). In *S. cambrensis* the *RAY* locus is replicated and ray length was largely determined by the number of radiate alleles present at each *RAY* locus. Thus, mean ray length was greatest in plants homozygous for the radiate allele at each of two *RAY* loci, and was reduced sequentially in individuals containing 3, 2 or 1 radiate alleles respectively.

Recently, Andersson (2001a) has completed a genetic analysis of variation for radiate versus discoid capitulum type in *Senecio jacobaea* (Ragwort), a species which is placed in a different section (Sect. *Jacobaea*) of the genus to *S. vulgaris* and *S. squalidus* (Sect. *Senecio*). In this species the radiate type is common, but a non-radiate variant forms populations at some coastal sites in northern Europe (Harper and Wood, 1957; Andersson, 2001b). Segregation of capitulum type in F2 and backcrossed (BC) families of crosses between the two morphs was consistent with a genetic model involving one major locus and an unknown number of modifiers. An excess of discoid offspring, and a deficiency of progeny expressing the intermediate radiate phenotype, indicated that many F2 and BC plants failed to produce rays (or occasionally produced a novel ray phenotype) despite being heterozygous at the major locus. Because certain offspring had ray florets with unusually large lobes or a bilabiate flower structure, it seemed that some parental genes interacted non-additively in these generations.

Single gene control of presence/absence of ray florets in radiate capitula has also been demonstrated in *Haplopappus phyllocephalus* (Jackson and Dimas, 1981) and in sunflowers (Fick, 1967).

12.3.2 Two gene control

In the genus *Layia*, variation for radiate versus discoid flower heads is predominantly under the control of two major genes (Clausen *et al.*, 1947; Ford and Gottlieb, 1990), based on breeding studies on the radiate species *L. glandulosa* and the closely related discoid species *L. discoidea*. Ford and Gottlieb (1990) showed that normal radiate plants were produced by genotypes *RRGG*, *RRGg*, *RrGG* and *RrGg*, and the discoid phenotype by genotypes *rrGG*, *rrGg* and *rrgg*. Their analysis revealed that material of *L. glandulosa* possessed genotype *RRGG*, whereas *L. discoidea* contained all three possible discoid genotypes, but was predominantly *rrgg*. Of particular interest was the finding that the 'recombinant' genotype, *RRgg*, produced novel peripheral florets designated 'gibbous florets'. These florets possessed some characteristics typical of ray florets and some typical of disc florets, as well as some unique traits. The 'gibbous' phenotype was also weakly expressed by *Rrgg*, although this genotype showed much-reduced penetrance and expressivity and therefore often produced a discoid phenotype. Evidence was also obtained that *L. discoidea* has retained additional genes that modify ray floret number, size, shape and colour despite the fact that it has lost the *R* allele controlling the production of ray florets.

All the genetic analyses conducted so far on variation for radiate versus discoid flower heads in the Asteraceae show that variation is under simple genetic control with one or two major genes involved. It would seem fruitful, therefore, to adopt a candidate gene approach when attempting to isolate the genes that control this variation. In the next section we focus on which candidate genes seem appropriate for investigation in future studies.

12.4 Molecular analysis of genetic control of capitulum development

Radiate capitula, where actinomorphic disk florets occur centrally and are surrounded by peripheral zygomorphic ray florets, resemble the *cen* phenotype of *Antirrhinum majus*. In this particular *Antirrhinum* mutant, the central terminal flower is radially symmetrical and the axillary flowers that surround it are zygomorphic (Figure 12.1). In both cases, the activation of genes controlling floral asymmetry seems to be restricted to a peripheral region around the inflorescence apex. In

Figure 12.1 (a) The cen mutant of *Antirrhinum majus*, with its actinomorphic terminal flower and zygomorphic peripheral flowers. (b) Radiate, (c) intermediate and (d) discoid forms of *Senecio vulgaris*.

(b)

(c)

(d)

cen this gives rise to all of the axillary flowers, whereas in radiate capitula it produces just the outermost florets (Coen *et al.*, 1995).

This similarity between *cen* and the radiate capitulum type suggests that similar genes may be involved in controlling development of asymmetrical flowers in both of them (Coen *et al.*, 1995). One of the genes known to control dorsiventral asymmetry in *Antirrhinum majus* is *CYCLOIDEA* (*CYC*: Luo *et al.*, 1996). *CYC* is a member of the TCP gene family and is thought to function as a DNA-binding protein and a transcription factor (Cubas *et al.*, 1999; Cubas, 2002). It has been proposed that production of zygomorphic wild-type *Antirrhinum* flowers is dependent on *CYC* activity establishing an axis of dorsiventral asymmetry. Activity is predicted to be greatest in the dorsal regions of the floral meristem and to decline towards the more ventral regions; this would account for the fact that flowers in *CYC* mutants are ventralised (Coen and Nugent, 1994; Luo *et al.*, 1996; Luo *et al.*, 1999).

Asymmetrical flower structure is thought to have evolved independently many times (Stebbins, 1974; Coen and Nugent, 1994; Donoghue *et al.*, 1998). Plants with actinomorphic instead of zygomorphic flowers, termed 'peloric mutants' (Bateman and DiMichele, 2002), have been described in several species in addition to *Antirrhinum*, and it is likely that *CYCLOIDEA* homologues are controlling dorsiventral asymmetry in all of them (but see Donoghue *et al.*, 1998). Genetic analysis of the naturally occurring peloric mutants in *Linaria vulgaris*, for example, has shown that the mutation is controlled by *CYC*-like genes (Cubas *et al.*, 1999). It is also speculated that *CYC*-like genes may be performing a similar function in *Saintpaulia velutina* and *Sinningia speciosa* (Citerne *et al.*, 2000). Therefore, it is reasonable to hypothesise that development of the ray floret in the Asteraceae is controlled in a similar manner (Coen and Nugent, 1994). We are currently using the polymorphism for radiate versus discoid capitula in *Senecio vulgaris* to test the hypothesis that the *RAY* gene in this species is a *CYCLOIDEA* homologue. *Senecio vulgaris* is a good species for investigating the molecular genetics of capitulum development, since it contains natural variation for radiate versus discoid capitulum type which is under single gene control. Moreover, it has a short generation time (approximately eight weeks under glasshouse conditions), is self-compatible so that inbred lines are readily produced, and is easy to cross, allowing large segregating families to be generated for analysis.

12.5 Establishment of new flower head variants in the wild

Examination of factors responsible for the spread of a new flower head variant in the wild requires consideration of the population genetics of the system and the conditions that lead to the replacement of one capitulum type by another. If capitulum type does not affect fitness, then the combined effects of genetic drift and gene flow will determine the fate of a new mutant form within a population, while founder effects, together with subsequent drift and gene flow, will be responsible for establishment in newly-formed populations. However, large changes in capitulum type are expected to directly affect fitness and, therefore, selection will be important in the spread of a new capitulum form. For example, there is evidence that ray florets increase a plant's attractiveness to pollinators (Burtt, 1977; Leppick, 1977; Lack,

1982; Stuessy *et al.*, 1986; Abbott and Irwin, 1988; Andersson, 1991), which may affect the amount of cross-pollination and, therefore, levels of seed-set in self-incompatible species (but see Section 12.5.1) and the outcrossing rate of self-compatible species (Sun and Ganders, 1990). In self-compatible species, a reduction in outcrossing (due to loss of rays) may result in a significant lowering of fitness caused by inbreeding depression (Charlesworth and Charlesworth, 1987).

Below we briefly review two case studies which have examined in some detail factors affecting the establishment of new capitulum variants in the wild. The first study investigated the success of a discoid capitulum variant in the self-incompatible and mainly radiate species, *Senecio jacobaea* (Andersson, 1996, 2001b), while the second study focused on the establishment of a radiate variant in the self-compatible and mainly discoid species, *S. vulgaris* (Abbott *et al.*, 1998).

12.5.1 Establishment of a discoid flower head variant in Senecio jacobaea

Although *S. jacobaea* normally produces radiate heads, populations monomorphic for discoid plants do occur in some coastal sites in north-west Europe (the Netherlands, Germany, Denmark, Poland, Norway, Sweden, Finland and the British Isles: Harper and Wood, 1957; Andersson, 2001b). Previous studies of other composites have shown that pollination success (measured as female fertility) is drastically reduced by removal of rays from radiate capitula (Lack, 1982; Stuessy *et al.*, 1986). In contrast, Andersson (1996) found no such effect in *S. jacobaea* and suggested that discoid plants of this species can maintain high visitation rates despite reduction in the individual attraction units, because such units are produced in dense clusters in terminal corymbs. In a subsequent study, Andersson (2001b) examined the fitness consequences of floral variation among segregating offspring within F2 and back-cross families of a cross between radiate and discoid *S. jacobaea*. Again, no effect was recorded in variation of ray length on pollination success (fruit set), nor on the proportion of heads infested by larvae of seed flies and the amount of resources retained for the next flowering season. However, a negative relationship was found between ray length and average germination rate of the maternal seed crop, due to a lower germination rate of achenes (seed) produced by ray florets. Andersson (2001b) hypothesised that the discoid morph may occupy a habitat that is 'a potentially more predictable and durable habitat than the anthropogenic sites occupied by the rayed phenotype', and that under these conditions there is 'selection for early seedling emergence – mediated through a conversion of ray achenes into disc achenes.' This hypothesis is not easy to accept, however, as radiate populations are often found in the same habitat as the discoid morph (duneland), but at different locations.

It is clear from the work conducted on *S. jacobaea* that the factors which determine the success of one flower head type relative to another in this species are likely to be complex and difficult to resolve. The genetics of flower head variation may be simple, in contrast to the ecological factors that determine establishment of a particular capitulum type in the wild.

12.5.2 Establishment of a radiate flower head variant in Senecio vulgaris

Throughout most of its native and introduced range, *S. vulgaris* is represented by a discoid form (var. *vulgaris*). However, in the British Isles a radiate form (var. *hibernicus*) is also found and frequently co-occurs with the discoid variant on open, disturbed ground in urban areas, such as waste-sites and gardens. Marshall and Abbott (1982, 1984a) showed that in wild polymorphic populations the radiate morph has a higher maternal outcrossing rate (6–36 per cent) than the discoid morph (1–15 per cent). This is because pistillate ray florets exhibit greater outcrossing than hermaphroditic disc florets (Marshall and Abbott, 1984b), and also because radiate plants are more attractive to pollinators (Abbott and Irwin, 1988). The higher outcrossing rate of the radiate morph places it at an immediate disadvantage to the highly selfing discoid morph in polymorphic populations, due to the 'cost of outcrossing' that results from the inherent transmission advantage of a selfing gene relative to an outcrossing gene (Fisher, 1941; Maynard-Smith, 1978).

The radiate morph originated recently in the British Isles following introgressive hybridisation between the discoid morph and the introduced, radiate *S. squalidus* (Ingram *et al.*, 1980; Abbott *et al.*, 1992). It was first recorded in 1832 in Oxford (Crisp, 1972) and, despite the disadvantage it suffers from the 'cost of outcrossing', subsequently spread to many parts of the British Isles (Abbott *et al.*, 1992; Lowe and Abbott, 1996). Studies of factors responsible for the spread of the radiate morph have shown that it frequently produces more seeds per plant relative to the discoid morph, both in the wild and under controlled conditions (Oxford and Andrews, 1977; Abbott, 1985). This difference is not due to inbreeding depression in the discoid morph (Abbott, 1985) and is normally of sufficient magnitude to cause the radiate morph to spread to fixation in a population (Marshall and Abbott, 1987). However, in certain populations there is no difference in seed production between the two morphs, or the difference is reversed (Abbott, 1985; Marshall and Abbott, 1987; Abbott and Horrill, 1991). Oxford *et al.* (1996) showed that associations in natural populations between flower head morphs and a suite of developmental, morphological and reproductive characters are largely due to genetic linkage in some populations, and linkage disequilibrium between the *RAY* locus and unlinked loci in other populations. They argued that linkage disequilibrium would be established at the time of the introgressive origin of the radiate morph and will be maintained over a number of generations due to low levels of outcrossing between the two morphs in the wild. However, ultimately, genetic linkages will be broken, linkage equilibrium generated, and only associations due to pleiotropy maintained.

Abbott *et al.* (1998) reported that the radiate and discoid morphs of *S. vulgaris* consistently differ in their germination behaviour, with the radiate morph producing a greater proportion of seed showing delayed germination. It is not known whether this consistent association results from tight genetic linkage or pleiotropy. In contrast to *S. jacobaea* (see Section 12.5.1), seed produced by ray and disc florets of radiate plants exhibit similar germination behaviour. The difference in germination behaviour between the radiate and discoid morphs greatly affects their relative fitness under different conditions in the wild. When mortality of early germinated seedlings is high, the radiate morph holds advantage, whereas the discoid morph has a much higher relative fitness under conditions of low mortality of early germinated plants (Abbott *et al.*, 1998).

Detailed investigation of the establishment of the radiate morph of *S. vulgaris* in the British Isles suggests that complex genetic and ecological factors are involved. It is feasible that the establishment and maintenance of this morph is largely dependent on its recurrent formation by introgressive hybridisation, the type of associations between flower head type and fitness characters generated by this process and maintained over numerous generations, and the opposing force of the 'cost of outcrossing'.

12.6 Conclusions

Variation for flower head form within the Asteraceae is considerable, but largely dependent on the presence/absence of a small range of different floret types within the capitulum. In this chapter we have focused attention on variation for radiate versus discoid capitulum type. The few studies which have reported a formal genetic analysis of this variation, either within or between species, have shown it to be largely under the control of one or two major genes plus an unidentified number of modifier genes. Such simple genetic control is important for two reasons. First, it should make it relatively easy to isolate and examine the expression of the key genes that control the development of the capitulum. We are currently testing the hypothesis that a *CYCLOIDEA* homologue controls development of the zygomorphic ray floret in *Senecio vulgaris* and that a mutation in this gene is responsible for the replacement of ray florets by symmetrical disk florets to produce a discoid head. Second, the effects of major genes would seem to be important in the evolution of what is one of the most important morphological features of the Asteraceae. Of course, genes other than those focused on in this chapter are likely to be involved in 'fine-tuning' the form of particular capitula; for example, by determining the size, colour, sex and intricate shape of florets, pubescence of different organs and presence/absence of pappus. However, it is likely that major genes are also important in controlling these characters (Ford and Gottlieb, 1990); the challenge now is to isolate all of these genes and determine how they act to control capitulum development.

Finally, it is clear from the little work done so far on the establishment and spread of a new capitulum variant of a species in the wild that the factors controlling this phenomenon are extremely complex and difficult to resolve. However, if we are to obtain a clear and detailed understanding of the evolution of capitulum form in the Asteraceae, it is vital that a proper grasp is obtained of the conditions that favour the establishment and spread of a particular capitulum variant. This may represent the greatest challenge of all to a full understanding of this topic.

ACKNOWLEDGEMENTS

Research discussed in this review was supported by grants from the NERC to RJA, from the BBSRC to RJA, ESC and ACMG, and from the EMBO and EC to PC and ESC. We are grateful to Stefan Andersson for providing us with the results of his recent research prior to its publication.

REFERENCES

Abbott, R. J. (1985) Maintenance of a polymorphism for outcrossing frequency in a predominantly selfing plant, in *Structure and Functioning of Plant Populations, 2* (eds J. Haeck and J. Woldendorp), North-Holland Publishing, Amsterdam, The Netherlands, pp. 227–286.

Abbott, R. J. and Horrill, J. C. (1991) Survivorship and fecundity of the radiate and non-radiate morphs of groundsel, *Senecio vulgaris* L., raised in pure stand and mixture. *Journal of Evolutionary Biology*, 4, 241–257.

Abbott, R. J. and Irwin, J. A. (1988) Pollinator movements and the polymorphism for outcrossing rate at the ray floret locus in groundsel, *Senecio vulgaris. Heredity*, 60, 295–299.

Abbott, R. J., Ashton, P. A. and Forbes, D. G. (1992) Introgressive origin of the radiate groundsel, *Senecio vulgaris* L. var. *hibernicus* Syme: *Aat-3* evidence. *Heredity*, 68, 425–435.

Abbott, R. J., Bretagnolle, F. C. and Thébaud, C. (1998) Evolution of a polymorphism for outcrossing rate in *Senecio vulgaris*: influence of germination behavior. *Evolution*, 52, 1593–1601.

Andersson, S. (1991) Floral display and pollination success in *Achillea ptarmica* (Asteraceae). *Holarctic Ecology*, 14, 186–191.

Andersson, S. (1996) Floral display and pollination success in *Senecio jacobaea* (Asteraceae): interactive effects of head and corymb size. *American Journal of Botany*, 83, 71–75.

Andersson, S. (2001a) The genetic basis of floral variation in *Senecio jacobaea* (Asteraceae). *Journal of Heredity*, 92, 409–414.

Andersson, S. (2001b) Fitness consequences of floral variation in *Senecio jacobaea* (Asteraceae): evidence from a segregating hybrid population and a resource manipulation experiment. *Biological Journal of the Linnean Society*, 74, 17–24.

Bateman, R. M. and DiMichele, W. A. (2002) Generating and filtering major phenotypic novelties: neoGoldschmidtian saltation revisited, in *Developmental Genetics and Plant Evolution* (eds Q. C. B. Cronk, R. M. Bateman and J. A. Hawkins), Taylor & Francis, London, pp. 109–159.

Bradshaw, H. D., Otto, K. G., Frewen, B. E., McKay, J. K. and Schemske, D. W. (1998) Quantitative trait loci affecting differences in floral morphology between two species of monkeyflower (*Mimulus*). *Genetics*, 149, 367–382.

Bremer, K. (1994) *Asteraceae: Cladistics and Classification*. Timber Press, Portland, Oregon.

Bremer, K. (1996) Major clades and grades of the Asteraceae, in *Compositae: Systematics* (eds D. J. N. Hind and H. J. Beentje), Royal Botanic Gardens, Kew, pp. 1–7.

Bremer, K. and Humphries, C. J. (1993) Generic monograph of the Asteraceae-Anthemideae. *Bulletin of the Natural History Museum London (Botany)*, 23, 71–177.

Burtt, B. L. (1977) Aspects of diversification of the capitulum, in *The Biology and Chemistry of the Compositae, Vol. 1* (eds V. H. Heywood, J. B. Harborne and B. L. Turner), Academic Press, London, pp. 41–59.

Charlesworth, D. and Charlesworth, B. (1987) Inbreeding depression and its evolutionary consequences. *Annual Review of Ecology and Systematics*, 18, 237–268.

Citerne, H. L., Möller, M. and Cronk, Q. C. B. (2000) Diversity of *cycloidea*-like genes in Gesneriaceae in relation to floral symmetry. *Annals of Botany*, 86, 167–176.

Clausen, J., Keck, D. D. and Hiesey, W. M. (1947) Heredity of geographically and ecologically isolated races. *American Naturalist*, 81, 114–133.

Coen, E. S. and Meyerowitz, E. M. (1991) The war of the whorls: genetic interactions controlling flower development. *Nature*, 353, 31–37.

Coen, E. S. and Nugent, J. M. (1994) Evolution of flowers and inflorescences. *Development*, 1994 (Supplement), 107–116.

Coen, E. S., Nugent, J. M., Luo, D., Bradley, D., Cubas, P., Chadwick, M., Copsey, L. and Carpenter, R. (1995) Evolution of floral symmetry. *Philosophical Transactions of the Royal Society of London B*, 350, 35–38.

Comes, H. P. (1998) Major gene effects during weed evolution: phenotypic characters coseg-regate with alleles at the ray floret locus in *Senecio vulgaris* L. (Asteraceae). *Journal of Heredity*, 89, 54–61.

Crisp, P. (1972) *Cytotaxonomic Studies in the Section* Annui *of* Senecio. PhD thesis, University of London.

Cronk, Q. C. B. and Möller, M. M. (1997) Genetics of floral asymmetry revealed. *Trends in Ecology and Evolution*, 12, 85–86.

Cubas, P. (2002) Role of TCP genes in the evolution of key morphological characters in angiosperms, in *Developmental Genetics and Plant Evolution* (eds Q. C. B. Cronk, R. M. Bateman and J. A. Hawkins), Taylor & Francis, London, pp. 247–266.

Cubas, P., Lauter, N. and Coen, E. (1999) The TCP domain: a motif found in proteins regu-lating plant growth and development. *Plant Journal*, 18, 215–222.

Cubas, P., Vincent, C. and Coen, E. (1999) An epigenetic mutation responsible for variation in floral symmetry. *Nature*, 401, 157–161.

Donoghue, M. J., Ree, R. H. and Baum, D. A. (1998) Phylogeny and the evolution of flower symmetry in the Asteridae. *Trends in Plant Science*, 3, 311–317.

Fick, G. N. (1967) Genetics of floral colour and morphology of sunflowers. *Journal of Hered-ity*, 67, 227–230.

Fisher, R. A. (1941) Average excess and average effect of a gene substitution. *Annals of Eugenics*, 11, 53–63.

Ford, V. S. and Gottlieb, L. D. (1990) Genetic studies of floral evolution in *Layia*. *Heredity*, 64, 29–44.

Harper, J. L. and Wood, W. A. (1957) Biological flora of the British Isles. *Senecio jacobaea*. *Journal of Ecology*, 45, 617–637.

Hodges, S. A. (1997) Rapid radiation due to a key innovation in columbines (Ranunculaceae: *Aquilegia*), in *Molecular Evolution and Adaptive Radiation* (eds T. J. Givnish and K. J. Sytsma), Cambridge University Press, Cambridge, pp. 391–406.

Ingram, R. and Noltie, H. J. (1984) Ray floret morphology and the origin of variability in *Senecio cambrensis* Rosser, a recently established allopolyploid species. *New Phytologist*, 96, 601–607.

Ingram, R. and Taylor, L. (1982) The genetic control of a non-radiate condition in *Senecio squalidus* L. and some observations on the role of ray florets in the Compositae. *New Phytologist*, 91, 749–756.

Ingram, R., Weir, J. A. and Abbott, R. J. (1980) New evidence concerning the origin of radiate groundsel, *Senecio vulgaris* L. var. *hibernicus*. *New Phytologist*, 84, 543–546.

Jackson, R. C. and Dimas, C. T. (1981) Experimental evidence for systematic placement of the *Haplopappus phyllocephalus* complex (Compositae). *Systematic Botany*, 6, 8–14.

Jansen, R. K., Michaels, H. J. and Palmer, J. D. (1991) Phylogeny and character evolution in the Asteraceae based on chloroplast DNA restriction site mapping. *Systematic Botany*, 16, 98–115.

Jeffrey, C. (1977) Corolla forms in Compositae – some evolutionary and taxonomic specula-tions, in *The Biology and Chemistry of the Compositae, Vol. 1* (eds V. H. Heywood, J. B. Harborne and B. L. Turner), Academic Press, London, pp. 111–118.

Lack, A. J. (1982) Competition for pollinators in the ecology of *Centaurea scabiosa* L. and *Centaurea nigra* L. *New Phytologist*, 91, 321–339.

Leppick, E. E. (1977) The evolution of capitulum types of the Compositae in the light of insect–flower interaction, in *The Biology and Chemistry of the Compositae, Vol. 1* (eds V. H. Heywood, J. B. Harborne and B. L. Turner), Academic Press, London, pp. 61–89.

Lowe, A. J. and Abbott, R. J. (1996) Origins of the new allopolyploid species *Senecio cam-brensis* (Asteraceae) and its relationship to the Canary Islands endemic *Senecio teneriffae*. *American Journal of Botany*, 83, 1365–1372.

Luo, D., Carpenter, R., Copsey, L., Vincent, C., Clark, J. and Coen, E. (1999) Control of organ asymmetry in flowers of *Antirrhinum*. *Cell*, 99, 367–376.

Luo, D., Carpenter, R., Vincent, C., Copsey, L. and Coen, E. (1996) Origin of floral asymmetry in *Antirrhinum*. *Nature*, 383, 794–799.

Mabberley, D. J. (1997) *The Plant Book* (2nd edn). Cambridge University Press, Cambridge.

Marshall, D. F. and Abbott, R. J. (1982) Polymorphism for outcrossing frequency at the ray floret locus in *Senecio vulgaris*. I. Evidence. *Heredity*, 48, 227–235.

Marshall, D. F. and Abbott, R. J. (1984a) Polymorphism for outcrossing frequency at the ray floret locus in *Senecio vulgaris*. II. Confirmation. *Heredity*, 52, 331–336.

Marshall, D. F. and Abbott, R. J. (1984b) Polymorphism for outcrossing frequency at the ray floret locus in *Senecio vulgaris*. III. Causes. *Heredity*, 53, 145–149.

Marshall, D. F. and Abbott, R. J. (1987) Morph differences in seed output and the maintenance of the polymorphism for capitulum type and outcrossing rate in *Senecio vulgaris* L. *Transactions Botanical Society of Edinburgh*, 45, 107–119.

Maynard-Smith, J. (1978) The ecology of sex, in *Behavioural Ecology: An Evolutionary Approach* (eds J. R. Krebs and N. B. Davies), Blackwell, Oxford, pp. 159–179.

Oxford, G. S. and Andrews, T. (1977) Variation in characters affecting fitness between radiate and non-radiate morphs in natural populations of groundsel (*Senecio vulgaris* L.). *Heredity*, 38, 367–371.

Oxford, G. S., Crawford, T. J. and Pernyes, K. (1996) Why are capitulum morphs associated with other characters in natural populations of *Senecio vulgaris* (groundsel)? *Heredity*, 76, 192–197.

Richards, A. J. (1975) The inheritance and behaviour of the rayed gene complex in *Senecio vulgaris*. *Heredity*, 34, 95–104.

Schemske, D. W. and Bradshaw, H. D. (1999) Pollinator preference and the evolution of floral traits in monkeyflowers (*Mimulus*). *Proceedings of the National Academy of Sciences USA*, 96, 11910–11915.

Stebbins, G. L. (1974) *Flowering Plants: Evolution Above the Species Level*. Harvard University Press, Cambridge, MA.

Stuessy, T. F., Spooner, D. M. and Evans, K. A. (1986) Adaptive significance of ray corollas in *Helianthus grossesserratus* (Compositae). *American Midland Naturalist*, 115, 191–197.

Sun, M. and Ganders, F. R. (1990) Outcrossing rates and allozyme variation in rayed and rayless morphs of *Bidens pilosa*. *Heredity*, 64, 139–143.

Teeri, T. H., Albert, V. A., Elomaa, P., Hämäläinen, J., Kotilainen, M., Pöllänen, E. and Uimari, A. (2002) Involvement of non-ABC MADS-box genes in determining stamen and carpel identity in *Gerbera hybrida* (Asteraceae), in *Developmental Genetics and Plant Evolution* (eds Q. C. B. Cronk, R. M. Bateman and J. A. Hawkins), Taylor & Francis, London, pp. 220–232.

Theißen, G. (2000) Evolutionary developmental genetics of floral symmetry: the revealing power of Linnaeus' monstrous flower. *Bioessays*, 22, 209–213.

Theißen, G., Becker, A., Di Rosa, A., Kanno, A., Kim, J. T., Munster, T., Winter, K. U. and Saedler, H. (2000) A short history of MADS-box genes in plants. *Plant Molecular Biology*, 42, 115–149.

Theißen, G., Becker, A., Winter, K., Münster, T., Kirchner, C. and Saedler, H. (2002) How the land plants learned their floral ABCs: the role of MADS-box genes in the evolutionary origin of flowers, in *Developmental Genetics and Plant Evolution* (eds Q. C. B. Cronk, R. M. Bateman and J. A. Hawkins), Taylor & Francis, London, pp. 173–205.

Trow, A. H. (1912) On the inheritance of certain characters in the common groundsel, *Senecio vulgaris*, and its segregates. *Journal of Genetics*, 2, 239–276.

Weigel, D. and Meyerowitz, E. M. (1994) The ABCs of floral homeotic genes. *Cell*, 78, 203–209.

Chapter 13

Role of TCP genes in the evolution of morphological characters in angiosperms

Pilar Cubas

ABSTRACT

The genetic basis of morphological diversity is still poorly understood. One approach to investigate this biological question is to identify genes responsible for the generation of particular traits in model species. The function of these genes is then compared to that of their counterparts in species with divergent morphologies to try to relate differences in function to changes in morphology. Such comparative studies are now providing evidence that alterations in the regulation and activity of developmental genes play key roles in the evolution of novel traits. Here I review our current knowledge of the function of one such gene, *CYCLOIDEA* (CYC), involved, in different species, in the generation of plant morphological characters as apparently unrelated as floral asymmetry and apical dominance. CYC codes for a member of the TCP family of putative transcription factors and may directly or indirectly affect cell proliferation and growth. The expression of CYC is spatially and temporally regulated during development. Comparison of CYC orthologs from various species may help us to describe the evolutionary history of CYC, how its ancestral function was modified to carry out new roles, and how this relates to the generation of new plant morphologies.

13.1 Introduction

One of the central questions in evolutionary biology is how changes in DNA have led to the diversity of biological forms. It is becoming clear that most of the morphological diversity must have been generated through modifications in the developmental mechanisms that control form (Gerhart and Kirschner, 1997; Purugganan, 1998; Carroll, 2000). Developmental genes have tightly regulated expression patterns and functions, so that changes in either aspect may result in subtle or dramatic alterations in pattern formation. Which DNA changes gave rise to those alterations and how were they tolerated and selected? In recent years, an interdisciplinary approach, 'evolutionary developmental biology', has emerged, aiming to address the question of how genes controlling development have evolved to generate new morphologies.

A common approach to study this problem is to compare the structure and function of similar developmental genes in different clades and then try to correlate genetic conservation and change to morphological similarities and differences. This analysis requires, first, the identification of the main players in model species

In *Developmental Genetics and Plant Evolution* (2002) (eds Q. C. B. Cronk, R. M. Bateman and J. A. Hawkins), Taylor & Francis, London, pp. 247–266.

amenable to molecular genetics such as the fruit fly (*Drosophila melanogaster*) or the thale cress (*Arabidopsis thaliana*). Once key genes have been identified and cloned, similar genes can be isolated from other species and their functions compared.

Evidence emerging from these studies suggests that many developmental genes involved in the evolution of characters may be transcriptional regulators. Transcription factors could act as evolutionary switches by providing for the coordinated expression of their targets in new spatial or temporal patterns as required to generate functional and well-integrated, yet novel, phenotypes (Doebley and Lukens, 1998). The new activities of these transcription factors could be acquired by changes in the function of their proteins (Eizinger *et al.*, 1999) or, more importantly, by changes in the cis-regulatory elements that control their own expression patterns (Doebley and Lukens, 1998; Carroll, 2000). In many instances these changes may have been tolerated because gene divergence has been preceded by gene duplication (Ohno, 1970; Ohta, 1989). This may allow one gene to retain the ancestral role and another to evolve and take on new functions. Indeed, most key developmental genes belong to large families that result from repeated duplications and divergences occurring during evolution. Well-known examples are the family of homeodomain HOM/Hox genes that determine regional body identity in animals (Lewis, 1978; McGinnis and Krumlauf, 1992; Ruddle *et al.*, 1994) and the MADS-box gene family of transcriptional regulators that control floral organ identity and many other processes in plants (Lawton-Rauh *et al.*, 2000; Theißen *et al.*, 2000).

When the gene studied belongs to a large family, it is important to identify and compare *ortholog* genes, with phylogenetic (and often functional) correspondence among clades, and not *paralog* genes, originated by duplication within a clade. Furthermore, in some cases, orthologs from very distantly related species may have diverged so much that their functions are no longer comparable (Wray and Abouheif, 1998). This does not imply that functional comparisons over long phylogenetic distances are not meaningful or should not be attempted. Indeed, many cases have already been described in which functional conservation among distantly related clades is striking (reviewed by Gerhart and Kirschner, 1997; Scott, 2000).

CYC is a plant developmental gene in which comparative evolutionary developmental studies are beginning to be performed. CYC was originally identified as a gene responsible for the control of floral asymmetry in the snapdragon (*Antirrhinum majus*). Independently, a related gene, *TEOSINTE BRANCHED1* (*TB1*), was isolated in maize, where it is involved in the control of apical dominance. These two genes code for putative transcription factors that affect cell proliferation and growth, and their expression is spatially and temporally regulated during development. The characteristics of these genes and the traits that they control make them excellent systems to address some important questions regarding the evolution of angiosperm morphology. Information about *CYC/TB1*-like genes from various species is becoming available, allowing a comparison of their structure and function across a wide range of clades. Our aim is to gain some insight into the evolutionary history of *CYC/TB1* and to describe how an ancestral *CYC/TB1* gene, of as yet unknown function, evolved to generate genes involved in the development of apparently unrelated traits such as floral shape and plant architecture.

CYC/TB1 belongs to a gene family, the TCP family, named after their founder members, *TB1*, CYC and *PROLIFERATING CELL FACTOR* (*PCF*) (Luo *et al.*,

1996; Doebley *et al.*, 1997; Kosugi and Ohashi, 1997; Cubas *et al.*, 1999). This family is relatively large in some species. For instance, sequence databases suggest that at least twenty-four TCP genes are present in *Arabidopsis*, though no function has yet been assigned to any of them. The evolutionary history of *CYC/TB1* therefore has to be placed in the context of the evolution of this gene family. Phylogenetic and functional analysis of the other TCP genes in different species will help to clarify the picture of the evolution and biological function of *CYC* and *TB1* in angiosperms.

13.2 *CYCLOIDEA* and the control of floral asymmetry

According to their symmetry, flowers are either actinomorphic (also called regular, polysymmetric or radially symmetric: Figure 13.1a) if they have several planes of reflexional symmetry, or zygomorphic (irregular, monosymmetric, bilaterally symmetric: Figure 13.1b) if they have a single plane of symmetry. In actinomorphic flowers, all the organs of one particular identity (sepals, petals, stamens and carpels) are morphologically identical. In zygomorphic flowers, organs of the same identity have different morphologies along the dorsiventral (or adaxial–abaxial) axis. The ancestral angiosperm flowers are thought to have been actinomorphic (Crane *et al.*, 1995; Friis *et al.*, 1999), zygomorphy being a derived condition associated with the evolution of specialised pollinators (Crepet, 1996; Dilcher, 2000). Looking at the phylogenetic distribution of floral zygomorphy among extant species reveals that clades mainly formed by species with zygomorphic flowers are distantly related and nested within groups with actinomorphic flowers (Figure 13.2a). This suggests that floral zygomorphy evolved independently several times (Stebbins, 1974; Donoghue *et al.*, 1998). Although zygomorphy is a recently evolved trait in the history of angiosperms, it is widespread; some of the largest families of angiosperms

(a)   (b)

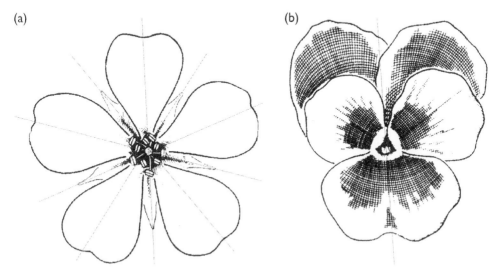

Figure 13.1 Radial symmetry versus bilateral symmetry. (a) Radially symmetrical flower of *Geranium pratense*. (b) Bilaterally symmetrical flower of *Viola tricolor*. Lines indicate planes of reflexional symmetry that divide the flower in two identical parts. (Modified from Weberling, 1989.)

(e.g. Orchidaceae, Zingiberaceae, Fabaceae, Lamiaceae, Scrophulariaceae, Asteraceae) are formed almost exclusively of species with zygomorphic flowers. This success has been attributed to the efficient promotion and control of the pollination achieved by these flowers (reviewed by Cronk and Möller, 1997; Endress, 1999).

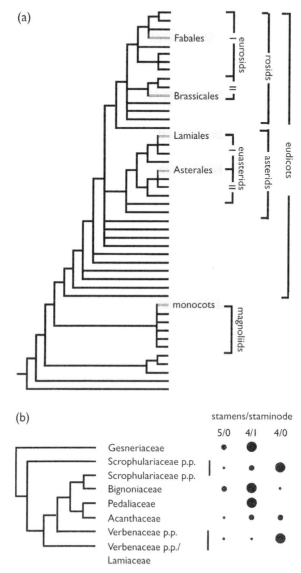

Figure 13.2 Angiosperm phylogeny and the evolution of floral zygomorphy. (a) Simplified angiosperm phylogenetic tree based on combined analysis of *rbcL*, *atpB* and 18S rDNA sequences. Names of the major clades in which floral zygomorphy is widely represented are included and boxed. (b) Phylogenetic tree for Lamiales *s.l.* Dots indicate the types of stamen pattern found in the different families. The size of the dots represents the degree of predominance of the corresponding pattern. 5/0, five developed stamens; 4/1, four stamens and one staminode; 4/0 four stamens and no staminode. ((a) after Soltis *et al.*, 1999; (b) based on Olmstead *et al.*, 1995, diagram reproduced from Endress, 1999.)

13.2.1 Snapdragon, a model system for the study of floral symmetry

Floral symmetry is a fascinating subject for evolutionary and developmental studies and raises many questions (as outlined by Coen and Nugent, 1994). Which genes control the floral symmetry of extant species? Do all species use the same genetic mechanisms to make zygomorphic flowers? If so, were these genes recruited once or several times independently? Finally, what was the ancestral role of these genes in plants with radially symmetric flowers before they were recruited to make zygomorphic flowers? To address all of these questions it was first necessary to identify and isolate the genes responsible for the generation of floral zygomorphy in a model species. These genes could then be used as a reference for comparative studies in other groups. Carpenter and Coen (1990) chose *Antirrhinum majus* as a model system for the genetic analysis of floral zygomorphy (reviewed by Coen, 1996). *Antirrhinum* (Scrophulariaceae) is ideal for this purpose as it has markedly bilaterally symmetric flowers and it is amenable to molecular genetics. The *Antirrhinum* flower (Figure 13.3a) has five petals with different shapes and sizes along the dorsiventral axis. The two dorsal petals are large and free of hairs, whereas the lateral and ventral petals are smaller, with yellow areas and hairs in the tube. Dorsal and

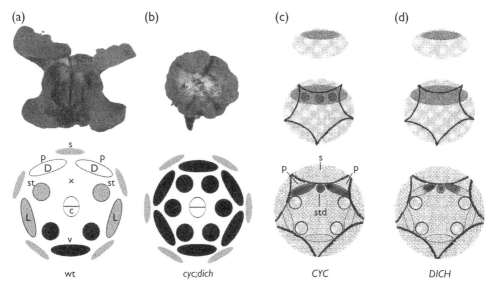

Figure 13.3 Control of floral asymmetry in *Antirrhinum majus*. (a) Top: front view of a wild-type *Antirrhinum* flower showing the dorsiventral asymmetry of its corolla. Bottom: floral diagram of the same flower. Each flower has five sepals (s), five petals (p), four stamens (st) and two carpels (c). Dorsal (D), lateral (L) and ventral (V) petals are indicated. Shades of grey represent different organ morphologies along the dorsiventral axis. (b) Top: flower of a double mutant *cyc;dich*. All petals and stamens resemble the ventral ones of the wild-type flower. Bottom: floral diagram of the mutant. (c) and (d). Expression patterns of *CYC* (C) and *DICH* (D) during normal floral development. Three stages of development of the floral meristem are represented. Dark grey indicates *CYC* and *DICH* mRNA distribution. Note that *DICH* expression becomes restricted to the dorsal-most part of dorsal petals at late stages of floral development. (Modified from Coen and Nugent, 1994; Luo *et al.*, 1999.)

lateral petals are individually asymmetric and the ventral petal is bilaterally symmetric. The stamens are also different along the dorsiventral axis: the dorsal-most is arrested to form a staminode, and the two lateral stamens are shorter than the ventral ones. To isolate the genes responsible for this bilateral symmetry, Coen and Carpenter generated, by transposon mutagenesis, a collection of mutants in which this dorsiventral asymmetry was reduced (Carpenter and Coen, 1990). One mutant they found with reduced zygomorphy was *cycloidea* (*cyc*). The *cyc* mutants have partially ventralised flowers in which lateral petals have ventral identity and dorsal petals have a combination of dorsal and lateral characteristics. This shows that *CYCLOIDEA* plays an important role in the generation of *Antirrhinum* floral zygomorphy. In addition, *cyc* mutants often have six petals and stamens instead of five. However, *cyc* mutant flowers are still somewhat bilaterally symmetric, which suggests that there are other genes involved in the generation of this trait.

DICHOTOMA (*DICH*) has been identified as another gene involved in the generation of *Antirrhinum* floral zygomorphy. In *dich* mutants the floral phenotype is subtler than in *cyc* mutants, with only dorsal petals having an altered shape. However, the phenotype of *cyc;dich* mutants is very dramatic: they produce radially symmetrical flowers in which all the petals and stamens resemble the ventral ones of the wild type (Figure 13.3b). This suggested that, without the activity of CYC and DICH, the flower could not establish its dorsiventral axis of symmetry. These genes, cloned by transposon tagging, encode two related proteins with features of transcription factors (Luo *et al.*, 1996, 1999; Cubas *et al.*, 1999). Expression pattern studies showed that they are expressed from very early stages of floral development in the dorsal part of the meristem. Expression continues through to later stages in the dorsal petal primordia and dorsal staminode. In the case of *DICH*, expression becomes restricted to the most dorsal half of each dorsal petal – the areas affected in *dich* mutants (Figure 13.3c–d). What is the function of CYC and DICH in the development of the flower?

When development of wild-type flowers was studied, it was found that, from very early in development, the dorsal part of the meristem (the region where CYC and DICH are expressed) is retarded in growth with respect to the ventral part. In *cyc;dich* mutants this retardation does not take place. As a consequence of having a larger meristem, *cyc;dich* mutants form more floral organ primordia. At later stages of development, wild-type flowers develop dorsal petals, different in shape and size from the lateral and ventral petals, and the dorsal stamen becomes arrested. In contrast, in *cyc;dich* mutants all petals and stamens (six instead of five) develop similarly to the ventral equivalents of the wild type.

In summary, the proteins CYC and DICH are related transcription factors that act at the dorsal part of the developing flower. At early stages they affect growth rate and primordium initiation. Later, they continue to act in dorsal primordia, controlling the bilateral symmetry, size, shape and cell types of dorsal petals and dorsal-most stamen (Luo *et al.*, 1996, 1999). The fact that CYC and DICH code for related proteins and are expressed in overlapping regions could explain the partial redundancy of their functions. However, CYC seems to play a more important role than DICH in the generation of floral zygomorphy and the control of floral organ number, as loss of CYC function more dramatically affects these two traits.

13.2.2 CYCLOIDEA and the evolution of floral zygomorphy

The cloning of *CYC* and *DICH* from *Antirrhinum* makes it possible to study the role of these genes in determining floral morphology in other groups. *CYC*-like genes can be isolated in other species with irregular flowers and their function explored using genetic, molecular and transgenic techniques (Coen and Nugent, 1994). Furthermore, species only distantly related to *Antirrhinum* and possessing radially symmetrical flowers might help understand the nature of the ancestral role of *CYC*.

13.2.2.1 Lamiales s.l.

Comparative studies among closely related species are more likely to give meaningful results than analysis over long phylogenetic distances. In close relatives, the probability that genes have undergone duplication, divergence or extinction is lower, therefore orthologs should be easier to identify (Reeves and Olmstead, 1998). The clade Lamiales *s.l.* constitutes a good compromise between the degree of closeness and morphological variation for the study of floral zygomorphy (Figure 13.2b). Lamiales *s.l.* is a monophyletic group containing the *Antirrhinum* family (Scrophulariaceae) and is believed to be ancestrally zygomorphic (Downie and Palmer, 1992; Olmstead *et al.*, 1992; Endress, 1997).

Zygomorphic flowers of the common ancestor are likely to have depended on *CYC*-like gene activity (Coen and Nugent, 1994). Moreover, different degrees of floral zygomorphy are found within Lamiales *s.l.*, which may be associated with variable levels of *CYC* function. The corollas range from strongly bilaterally symmetrical (e.g. *Antirrhinum*) to radially symmetrical (e.g. *Ramonda*, Gesneriaceae). The uppermost stamen primordium can be completely absent from the earliest stages of flower development (e.g. *Salvia*, Lamiaceae), become arrested during development giving a staminode (e.g. *Antirrhinum*), or give a fully fertile stamen (e.g. *Verbascum*, Scrophulariaceae). In some cases, not just the usual one but all three upper stamen primordia become arrested (e.g. *Mohavea*, Scrophulariaceae; *Saintpaulia*, Gesneriaceae). Organ number, which is controlled by *CYC* in *Antirrhinum*, also shows variation within the clade (i.e. *Sibthorpia*, Scrophulariaceae, has petal and stamen numbers that range from four to eight: Cronk and Möller, 1997; Endress, 1997, 1999; Donoghue *et al.*, 1998; Walker-Larsen and Harder, 2000). How does the function of *CYC*-like genes relate to these variations?

Toadflax (*Linaria vulgaris*, Scrophulariaceae) was the first species to be compared to *Antirrhinum* with respect to *CYC* function. *Linaria* has zygomorphic flowers that superficially resemble those of *Antirrhinum* except that they are smaller and have a long nectar spur in the ventral petal (Figure 13.4a, left; Plate 1c). It would not have been surprising to find that a *CYC/DICH* system similar to that of *Antirrhinum* controls floral asymmetry in this plant. This could be genetically tested because radially symmetrical mutants resembling *cyc;dich* double mutants exist in *Linaria* (Figure 13.4a, right: Coen and Nugent, 1994; Cubas *et al.*, 1999). These mutants, first described by Linnaeus more than 250 years ago (Linnaeus, 1749), can be found in small populations in the wild, where they reproduce vegetatively, as pollinators cannot enter the mutant flowers.

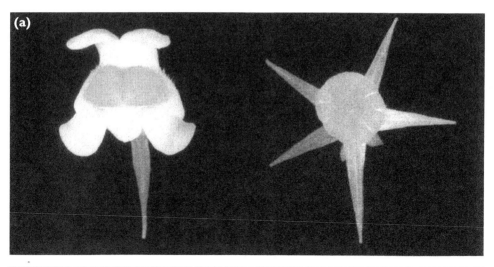

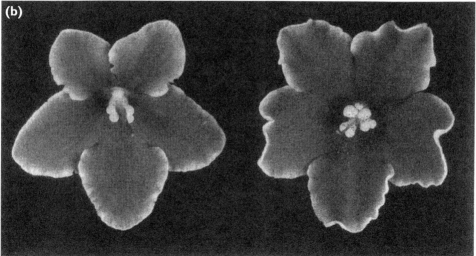

Figure 13.4 Photographs of wild-type (left) and radially symmetrical (right) flowers of *Linaria vulgaris* (a) and *Saintpaulia* (b). (Reproduced from Coen and Nugent, 1994.)

A gene very similar to *CYC* was isolated from wild-type *Linaria* (*LCYC*: Cubas *et al.*, 1999). Like *CYC* in *Antirrhinum*, *LCYC* is expressed in the dorsal part of floral meristems of *Linaria*. Moreover, its function is required in order to make zygomorphic flowers: in families of plants segregating for the mutant phenotype, all plants with radially symmetrical flowers have an *LCYC* gene heavily methylated and transcriptionally silenced. Furthermore, in plants in which *LCYC* is partially demethylated and its transcription has recovered, the flowers partially revert to wild type (Cubas *et al.*, 1999). All these data confirm that *LCYC* is involved in the generation of floral zygomorphy in *Linaria*. However, some functional differences with respect to the *Antirrhinum* system were also found. Genetic analysis suggested that a single locus, *LCYC*, rather than the two present in *Antirrhinum*, was sufficient to generate the radially symmetrical phenotype. Therefore, in *Linaria* there seems to be less genetic redundancy in the control of this trait. On the other hand, *Lcyc* epimutants have a wild-type floral organ number (Figure 13.4a, right), in contrast to the *Antirrhinum cyc* single mutants that have an excess of petals and stamens. This suggests that either *LCYC* did not control organ number or its activity in this respect was fully redundant. *CYC* and *LCYC* had similar expression patterns, therefore their functional differences should rely on other factors such as the activity of their proteins or their downstream genes, or in different trans-acting genes involved in the process in the two species. The activity of the proteins may be tested by introducing *LCYC* into *cyc;dich* and *cyc Antirrhinum* mutants and checking to what extent radial symmetry and altered organ number are rescued. As for trans-acting genes, it is interesting that no *DICH*-like genes were found in *Linaria*, but a second *LCYC*-like gene without a known counterpart in *Antirrhinum* was isolated (Cubas and Coen, unpubl. obs.). *DICH* and this second *LCYC* gene may be paralogs of *CYC* and *LCYC* respectively, evolved after the phylogenetic divergence of *Antirrhinum* and *Linaria*, and they may complement *CYC* and *LCYC* functions in different ways. These results show that even closely related species have developed slightly divergent strategies, involving *CYC*-like genes, to elaborate their floral asymmetry.

Other species of Scrophulariaceae are currently under study. Hileman and Baum have isolated *CYC*-like genes from several species of Antirrhineae and from one close relative, *Digitalis*. In most taxa studied they have found one likely *CYC* ortholog and (in contrast to our findings in *Linaria*) one likely *DICH* ortholog, suggesting that the *CYC/DICH* duplication occurred early in, or before, the radiation of Antirrhineae (Hileman and Baum, unpubl. data). On the other hand, Vieira *et al.* (1999) analysed several species from this same clade and found more than two *CYC*-like genes in most of the species studied. To date, the function of these genes has not been determined. Genetic and expression analysis will help identify which genes participate in the elaboration of floral asymmetry, and will allow comparison of their functions with those of *CYC* and *DICH*.

Gesneriaceae (Figure 13.2b; Plate 1d) is the basal-most family of Lamiales *s.l.* and an interesting family from the point of view of the evolution of floral symmetry. In this group, the corollas tend to be weakly bilateral and the uppermost stamen less reduced than in other Lamiales. Also, a large proportion of genera have regular flowers (Endress, 1997, 1999). It is therefore possible that this family represents an ancient and not too elaborated form of floral bilateral symmetry within the clade.

Several *CYC*-like genes, *GCYC*, have been isolated from species of Gesneriaceae (Möller *et al.*, 1999; Citerne *et al.*, 2000). Their phylogeny suggests that they evolved from a single ancestral *GCYC* gene present before Gesneriaceae split from Scrophulariaceae. After the separation, a number of duplications and gene losses seem to have taken place. This further supports the view that duplication of *CYC*-like genes is common (Citerne *et al.*, 2000).

The function of the *GCYC* genes is being tested with the help of radially symmetrical mutants. In *Siningia speciosa*, for instance, a mutant with regular flowers and altered floral organs number is available (Coen and Nugent, 1994; Citerne and Cronk, 1999). The *GCYC* gene found in this species has been analysed in the mutants. Interestingly, in the mutant plants, *GCYC* carries a frame-shift that yields a truncated protein (Citerne *et al.*, 2000). If genetic analysis confirms the linkage between the frame-shift and the mutant phenotype, it would suggest that a single *GCYC* gene might be controlling floral asymmetry and organ number in *Sinningia*.

The flowers of another Gesneriaceae, *Saintpaulia ionantha*, have a weakly zygomorphic corolla, and their three uppermost stamens are aborted (Figure 13.4b, left). A radially symmetric commercial cultivar of this species is available (Coen and Nugent, 1994; Figure 13.4b, right). Two related *GCYC* genes have been isolated from *Saintpaulia* in order to test their functions (Möller *et al.*, 1999; Citerne *et al.*, 2000; Cronk, unpubl. obs.). Mutant and expression analysis of *GCYC* genes during floral development will help determine the role of these genes. Also, it would determine whether the floral morphology of *Saintpaulia* correlates with a 'shift' of the *GCYC* expression to more lateral regions; a transient expression in dorsal petal primordia would give a just weakly zygomorphic corolla and a lasting expression in the uppermost and dorso-lateral stamen primordia would cause them to become aborted. A similar study of the role of *CYC* in stamen abortion is being carried out by Hileman and Baum in *Mohavea* (Scrophulariaceae: Baum, 2002).

Another interesting point is related to the reversals from zygomorphy to actinomorphy within Lamiales *s.l.*: some species with radially symmetrical flowers are found deeply nested within bilaterally symmetrical clades, suggesting that their radial symmetry is a derived condition (Cronk and Möller, 1997; Donoghue *et al.*, 1998). It has been proposed that reversions may be a result of an adaptation to generalist pollinators in areas where specialists are scarce (Cronk and Möller, 1997). Examples of this type of reversions are found, for instance, in Gesneriaceae. *Ramonda myconi*, *R. nathaliae* and *Conandron ramondioides* are radially symmetrical species growing in extreme mountainous habitats. In these habitats, an excess of generalist over specialist pollinators has been noted (Cronk and Möller, 1997), which perhaps favours reversals to less specialised and more accessible floral morphologies. Are those reversals due to loss-of-function mutations in *CYC*-like genes?

So far there is no evidence to support this conjecture. The *GCYC* genes of *Ramonda myconi*, *R. nathaliae* and *Conandron ramondioides* have been isolated and screened for putative mutations that would render their proteins non-functional. Möller *et al.* (1999) and Citerne *et al.* (2000) showed that these genes code for apparently functional GCYC proteins and their coding regions seem to be subjected to selection pressure. However, mutations in the regulatory region that might alter their expression patterns have not yet been ruled out. It is conceivable that *GCYC* expression is transient in those species and thus unable to generate floral zygomorphy.

It has been pointed out that mutations in *CYC*-like genes alone may not always confer the selective advantage necessary for a mutation to reach fixation (Coen and Nugent, 1994; Cronk and Möller, 1997; Donoghue *et al.*, 1998; Theißen, 2000). In *Antirrhinum* and *Linaria*, for example, *cyc* mutations would not be adaptive, as the bees would be unable to enter the narrow tube of the mutant flowers to pollinate them. Indeed, as we mentioned above, natural populations of radially symmetric *Linaria* propagate vegetatively. However, if other developmental changes took place that made the flower more open, or new visitors co-evolved that could efficiently pollinate the novel flowers, the mutation might become adaptive. In *Ramonda*, the corolla tube is almost missing, whereas its close zygomorphic relatives have tubular flowers. This makes the flower more accessible to generalist pollinators (Cronk and Möller, 1997).

It is likely that reversions to radial symmetry may not be easy to describe in terms of alterations in single genes; rather, they are likely to be the result of coordinated developmental and ecological changes that, to be properly understood, will require an interdisciplinary approach. However, genes such as *CYC* may, in some cases, play an important part in these processes; at least we now have the tools to investigate this point.

The picture of floral evolution and *CYC* function in Lamiales is still very incomplete. The results reviewed here are just the first steps in the reconstruction of its history, which will require further studies in species representative of other families from this order.

13.2.2.2 Outside Lamiales s.l: other Asteridae and beyond

An even more challenging question is whether floral asymmetry in species lying outside Lamiales *s.l.* is also *CYC*-dependent. It has been proposed that in asterids (Figure 13.2a) actinomorphy is ancestral and zygomorphic flowers originated at least eight times independently (Donoghue *et al.*, 1998). As we have seen, in one of the clades, Lamiales *s.l.*, the transition from regular to irregular flowers involved *CYC*-like genes. What happened to other groups such as Asterales? One possibility is that new molecular mechanisms were 'invented' when zygomorphic flowers evolved in this group. Another possibility is that *CYC* was recruited independently in Asterales. This would raise the question as to what makes *CYC* so suitable to perform this new function. Yet another possibility is that floral dorsiventral asymmetry was, in fact, more ancient than has been assumed and that *CYC* was already controlling this trait in the common ancestor of Lamiales and Asterales (Coen and Nugent, 1994), with reversion to radial symmetry occurring later in some derived groups. Asteraceae is an ideal family to investigate whether *CYC* is controlling floral shape in Asterales. The inflorescence of the Asteraceae, the capitulum, consists of numerous small flowers, called florets. Very often the shape and degree of zygomorphy of those florets varies within the capitulum. For instance, in the case of *radiate* capitula, the outer *ray* florets are markedly zygomorphic, whereas the inner *disc* florets are actinomorphic. Genetic analysis of the presence/absence of ray florets has been carried out in some species (Trow, 1912; Samata, 1959; Ford and Gottlieb, 1990; Abbott *et al.*, 1992; reviewed by Gillies *et al.*, 2002). Current work is being carried out to investigate whether *CYC*-like genes are involved in the generation of zygomorphic ray florets in several Asteraceae (Cubas, unpubl. obs.; Gillies *et al.*, 2002).

To complete the picture of the evolution of flower zygomorphy in angiosperms, future work should also involve the isolation and testing of *CYC*-like genes in even more distantly related clades (Figure 13.2a) with zygomorphic flowers. The pea family (Fabaceae, rosid dicots: Hawkins, 2002) or the orchid family (Orchidaceae, monocots: Johansen and Frederiksen, 2002; Bateman and DiMichele, 2002), both of which have yielded radially symmetrical mutants, would be interesting case studies.

13.3 *TEOSINTE BRANCHED I* and the control of apical dominance in maize

CYC-like genes have already been characterised in monocots, but their most obvious roles are not related to the control of floral asymmetry. The first *CYC*-like gene isolated from this group was the maize *TEOSINTE BRANCHED 1* (*TB1*) gene, which controls the branching pattern of the plant.

Maize was domesticated about 6,000 years ago from a wild Mexican grass called teosinte. One of the most striking differences between these two plants is the relative degree of apical dominance, reflecting the hormone-mediated control that the apical bud exerts on lateral buds (Cline, 1991). With weak apical dominance the lateral buds grow out to produce primary branches. With strong apical dominance, growth concentrates in the main axis and the lateral buds do not develop. In teosinte, apical dominance is weak, therefore the plant is extremely branched and resembles a candelabra (Figure 13.5a). In maize, in contrast, apical dominance is strong, lateral buds do not develop, and the plant looks more like a pole (Figure 13.5b). Another difference between teosinte and maize is in the position where female and male inflo-

(a)　　　　　　　　　　　　　　　　　(b)

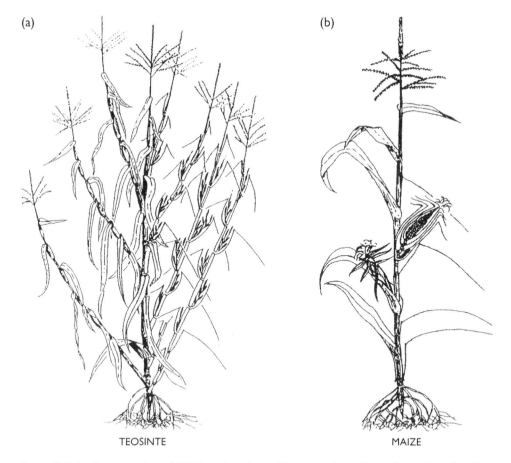

TEOSINTE　　　　　　　　　　　　　　MAIZE

Figure 13.5 (a) Teosinte plant. (b) Maize plant derived from teosinte. (Reproduced from Doebley, 1992.)

rescences develop. In teosinte, female inflorescences are borne on secondary branches, and male inflorescences are borne at the tips of primary branches. In maize, the female inflorescences ('ears') occur at the tip of short primary branches formed in upper nodes, and male inflorescences are at the apex of the plant (Iltis, 1983).

Doebley and colleagues have been working on the identification of genetic regions responsible for the morphological differences between teosinte and maize (reviewed in Doebley, 1992; White and Doebley, 1998). Quantitative trait locus mapping and molecular genetics of maize mutants has led to the isolation of *TB1*, the gene responsible for the differences in apical dominance and ear development (Doebley *et al.*, 1995, 1997) that turned out to be a CYC-like gene. Mutants in *TB1* resemble teosinte plants: in lower nodes they have lateral branches and in upper nodes, additional lateral branches replace ears. This suggests that, in maize, the wild type *TB1* controlled the development of axillary buds, preventing their outgrowth at lower nodes and promoting ear development at upper nodes.

Which were the DNA changes in the wild-type *TB1* gene of teosinte that altered its function and resulted in the radical morphological alterations found in maize? First, it was observed that the levels of *TB1* mRNA maize were double that of teosinte (Doebley *et al.*, 1997). Subsequently, several teosinte and maize *TB1* alleles were sequenced and compared, showing a much higher degree of polymorphism in the protein-coding region than in the gene regulatory region, suggesting that selection had mainly affected the latter (Wang *et al.*, 1999). This is a clear case in which a change in the regulation of a gene gives rise to a dramatic morphological novelty. In teosinte, *TB1* may be expressed in secondary axillary meristems where it controls their conversion into ear shoots. During domestication, humans have selected an allelic variant of *TB1* that is expressed at high levels in primary axillary meristems such that they form ears instead of branches (Doebley *et al.*, 1997).

How does *TB1* function relate to *CYC* function? In principle, the considerable phylogenetic distance between *Antirrhinum* and maize (Figure 13.2a) implies that their functions may not be comparable. However, common themes may be noted in their expression patterns and developmental effects (Doebley *et al.*, 1997), suggesting that there is an underlying functional conservation, which may reflect the function of their common ancestor. Both are expressed in axillary structures: flower meristems in the case of *Antirrhinum*, lateral buds in the case of maize. Also, both are expressed in petals (lodicules, in maize) and stamens. With respect to their activity, both genes affect the growth of the structures where they are expressed: *Antirrhinum* CYC retards the growth of the young floral meristem and is involved in the growth arrest of the uppermost stamen, where its activity has been related to downregulation of several cell-cycle genes (Gaudin *et al.*, 2000). Maize *TB1* prevents the growth of axillary meristems in lower nodes, causes the shortening of ear internodes, and is expressed in the stamen primordia of female florets that become arrested during development. Other functions of *CYC* and *TB1* as selector genes – specification of dorsal floral organ fate and promotion of ear development, respectively – seem to be specific to each gene.

We can speculate that the common ancestor of *CYC* and *TB1* may have been expressed in axillary meristems (flower and/or shoot meristems), and that it was involved in modulating the growth rate of some regions of that meristem. It is not difficult to imagine how one such gene may be recruited to generate new morphologies by altering the control of its spatial and temporal expression patterns. Floral symmetry, stamen abortion or inhibition of lateral bud outgrowth are just a few examples of morphological changes that can be generated by changing the growth patterns of a plant. It would not be surprising to find other evolutionary novelties generated during angiosperm evolution by recruitment of CYC-like genes.

13.4 The family grows: PCF1 and PCF2 and the control of *PCNA* transcription in rice

Comparison of the protein sequences predicted for the available CYC-like genes has revealed that only two regions of the CYC proteins are strongly conserved (Figure 13.6). This suggests that they are important for the biochemical function of the proteins. On the other hand, sequences outside those regions are highly variable, even among close relatives. The first region is predicted to form a non-canonical basic-Helix–Loop–Helix (bHLH) and contains a basic stretch, a putative nuclear localisation signal and two amphipathic helical regions that may promote protein–protein interactions. The second domain is predicted to form a hydrophilic α-helix rich in arginine (R) residues and hence was called the 'R domain' (Cubas *et al.*, 1999).

Interestingly, the bHLH domain was found in two other plant proteins with known function: PCF1 and PCF2, both DNA-binding factors from rice. PCF1 and PCF2 specifically bind to promoter elements of the *PROLIFERATING CELL NUCLEAR ANTIGEN (PCNA)* gene of rice. *PCNA* is transcriptionally regulated and its protein accumulates in proliferating cells of yeast, animals and plants, participating in a variety of processes such as DNA replication, DNA repair and cell cycle control (Jónsson and Hübscher, 1998; Warbrick, 2000). The bHLH domain of the PCFs is sufficient to bind DNA and necessary for homo- and hetero-dimerisation of these transcription factors (Kosugi and Ohashi, 1997). This suggests that CYC-like genes may also be DNA-binding proteins and may interact with other proteins through their bHLH domain. This domain was called the 'TCP domain', after the initial letters of the first characterised members, *TB1*, *CYC* and *PCFs* (Cubas *et al.*, 1999). According to the evidence presented for the PCFs, the TCP domain is probably involved in both DNA binding and protein–protein interactions.

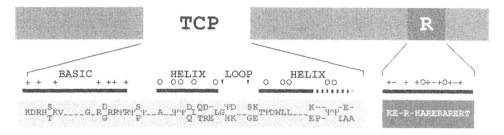

Figure 13.6 Schematic representation of a TCP gene. Consensus sequence of the TCP domain and the R domain are indicated. In the TCP consensus sequence, the top sequence corresponds to residues found in *CYC*-like genes and the bottom sequence to residues found in *PCF*-like gene. In the consensus sequence, ψ represents any hydrophobic residue; +, basic residues; O, hydrophobic residues; arrowheads, residues that prevent α-helix formation: proline, glicine, serine. *PCF*-like genes do not have R domain.

Although closely related, the TCP domain of CYC-like genes is clearly distinct from that of the PCFs. Phylogenetic trees comparing bHLH domains shows that CYC-like genes and PCFs fall into two distinct subclasses (Figure 13.7). Furthermore, the PCFs do not have R domains present in most of the CYC-like proteins.

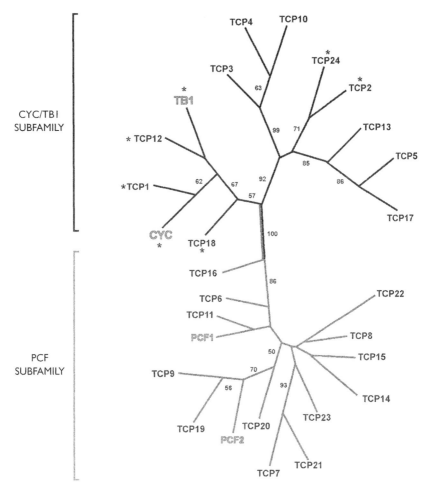

Figure 13.7 Unrooted phylogenetic tree showing relationships among the *Arabidopsis* TCP genes. *CYC*, *TB1* and *PCFs* were included for reference. TCP domains were aligned with CLUSTALW, 100 jackknife data sets were obtained with SEQBOOT, distance matrices calculated with PROTDIST (Dayhoff PAM matrix algorithm), trees constructed with NEIGHBOR and a consensus tree obtained with CONSENSE. SEQBOOT, PROTDIST, NEIGHBOR and CONSENSE are from the PHYLIP package (Felsenstein, 1988). Asterisks indicate genes containing an R domain. Branches with support ≥50 per cent are indicated. (Accessions: TCP1, AAG00255; TCP2, CAB78841, TCP3, AAC24010; TCP4 BAA97066; TCP5, BAB10646; TCP6 BAB09705; TCP7, BAB11183; TCP8, AAG27774; TCP9, AAC06168; TCP10, AAC63845; TCP11, AAD31585; TCP12, AAF15066; TCP13, AAF02122; TCP14, CAB61988; TCP15, AAF22916; TCP16, CAB72153; TCP17, CAB93708; TCP18, BAB02213; TCP19, BAA97226; TCP20, BAB01082; TCP21, CAC08333; TCP22, AAG27787; TCP23, AAF79358; TCP24, AAG30957.)

It is likely that the two groups of proteins have a similar biochemical function, namely transcriptional regulation, but they probably have different protein–protein interactions and downstream gene targets. The TCP domain allows dimerisation between PCF1 and PCF2 (Kosugi and Ohashi, 1997); it is also possible that different combinations of CYC-like proteins switch on different sets of downstream genes.

Furthermore, we can speculate that interactions between these two categories of proteins, PCF-like and CYC-like, may take place through the TCP domain at some stages of development. PCFs are likely to promote cell proliferation through their transcriptional control of *PCNA*. Proteins like CYC or TB1, when accumulating at high levels in the same regions as the PCFs, could interact with them through their TCP domain and interfere or modulate their proliferation-promoting activity. The inhibition of growth observed in areas where *CYC*-like genes are expressed might therefore be a result of these interactions. To date, there is no evidence *in vivo* or *in vitro* that either confirms or refutes this hypothesis.

13.5 *Arabidopsis* TCP genes: a family snapshot from a radially symmetrical species

The finding of *CYC*-like genes in all species studied, including monocots, suggests that they are present in a wide range of angiosperm species with very different floral symmetries and overall plant architectures. What are the functions of these *CYC*-like genes? Could they give us information about the ancestral role of *CYC* before it was recruited to make zygomorphic flowers? We can investigate this in a species with radially symmetrical flowers such as *Arabidopsis thaliana*. *Arabidopsis* provides an excellent system to clarify the role that *CYC*-like genes play in plant development. *Arabidopsis* (Brassicaceae, rosid dicots: Figure 13.2a) is distantly related to both *Antirrhinum* and maize. The complete *Arabidopsis* genome sequence is now available, so all TCP genes can be isolated and related to *CYC*, *TB1*, *PCFs* or novel subclasses. Moreover, genetic and molecular tools are available in *Arabidopsis* that allow functional analysis of these genes.

Our search in *Arabidopsis* for predicted TCP proteins has revealed twenty-four members of the family that map on all five chromosomes. Phylogenetic trees comparing the TCP domain of these proteins reveal two subclasses, the *CYC/TB1* subfamily and the PCF subfamily (Figure 13.7). Within the *CYC/TB1* subfamily two smaller groups are found, one group containing *CYC/TB1* and other genes with an R domain plus a sister clade with genes that may or may not have the R domain.

Sequence similarity indicates only a single putative *CYC* ortholog, *TCP1*. The adult *Arabidopsis* flower does not show any bilateral symmetry though, during development, the dorsal sepal shows a delay in growth relative to the ventral sepal. Therefore, *TCP1* may retain the role of the ancestral *CYC* gene before it was recruited to generate floral asymmetry in some groups. Mutant analysis and expression pattern studies are helping us to determine the function of this gene (Cubas, Coen and Martínez-Zapater, 2001). Two additional genes, *TCP18* and *TCP12*, are also related to *TB1* and *CYC*. Counterparts for *PCF1* and *PCF2* have also been found, along with a number of additional TCP genes unrelated to any of the founder members of the family.

The study of TCP genes in *Arabidopsis* has only just begun, but even the preliminary data are exciting. The number of experimental tools available in this species will surely help us to gain rapidly a more accurate picture of the biological role of TCP genes in *Arabidopsis* development and in angiosperm evolution.

ACKNOWLEDGEMENTS

I thank Enrico Coen, Desmond Bradley, Quentin Cronk, David Baum and Lena Hileman for constructive comments on the manuscript, and David Baum, Lena Hileman and Quentin Cronk for communication of unpublished results.

REFERENCES

Abbott, R. J. (1992) Plant invasions, interspecific hybridisation and the evolution of new plant taxa. *Trends in Ecology and Evolution*, 7, 401–404.

Abbott, R. J., Ashton, P. A. and Forbes, D. G. (1992) Introgressive origin of the radiate groundsel, *Senecio vulgaris* L. var. *hibernicus* Syme: *Aat-3* evidence. *Heredity*, 68, 425–435.

Averof, M. and Patel N. H. (1997) Crustacean appendage evolution associated with changes in Hox gene expression. *Nature*, 388, 682–6866.

Bateman, R. M. and DiMichele, W. A. (2002) Generating and filtering major phenotypic novelties: neoGoldschmidtian saltation revisited, in *Developmental Genetics and Plant Evolution* (eds Q. C. B. Cronk, R. M. Bateman and J. A. Hawkins), Taylor & Francis, London, pp. 109–159.

Baum, D. A. (2002) Identifying the genetic causes of phenotypic evolution: a review of experimental strategies. *Developmental Genetics and Plant Evolution*, 25, 493–507.

Carpenter, R. and Coen, E. (1990) Floral homeotic mutations produced by transposon mutagenesis in *Antirrhinum majus*. *Genes and Development*, 4, 1483–1493.

Carroll, S. (2000) Endless forms: the evolution of gene regulation and morphological diversity. *Cell*, 101, 577–580.

Citerne, H. and Cronk, Q. (1999) The origin of the peloric *Sinningia*. *New Plantsman*, 6, 219–222.

Citerne, H., Möller, M. and Cronk, Q. (2000) Diversity of *cycloidea*-like genes in Gesneriaceae in relation to floral symmetry. *Annals of Botany*, 86, 167–176.

Cline, M. G. (1991) Apical dominance. *Botanical Review*, 57, 318–358.

Coen, E. (1996) Floral symmetry. *EMBO Journal*, 15, 6777–6788.

Coen, E. and Nugent, J. (1994) Evolution of flowers and inflorescences. *Development*, Suppl. 107–116.

Crane, P. R., Friis, E. M. and Pedersen, K. R. (1995) The origin and early diversification of angiosperms. *Nature*, 374, 27–33.

Crepet, W. L. (1996) Timing in the evolution of derived floral characters: Upper Cretaceous (Turonian) taxa with tricolpate-derived pollen. *Review of Paleobotany and Palynology*, 90, 339–359.

Cronk, Q. and Möller, M. (1997) Genetics of floral asymmetry revealed. *Trends in Ecology and Evolution*, 12, 85–86.

Cubas, P., Coen, E. and Martínez-Zapater, J. M. (2001) Ancient asymmetries in the evolution of floral symmetry. *Current Biology*, 11, 1050–1052.

Cubas, P., Lauter, N., Doebley, J. and Coen, E. (1999) The TCP domain: a motif found in proteins regulating plant growth and development. *Plant Journal*, 18, 215–222.

Cubas, P., Vincent, C. and Coen, E. (1999) Epigenetic mutation responsible for natural variation in floral symmetry. *Nature*, 401, 157–161.

Dilcher, D. (2000) Towards a new synthesis: major evolutionary trends in the angiosperm fossil record. *Proceedings of the National Academy of Science of the USA*, 97, 7030–7036.

Doebley, J. (1992) Mapping the genes that made maize. *Trends in Genetics*, 8, 302–307.

Doebley, J. and Lukens, L. (1998) Transcriptional regulators and the evolution of plant form. *Plant Cell*, 10, 1075–1082.

Doebley, J., Stec, A. and Gustus, C. (1995) Teosinte branched 1 and the origin of maize: evidence for epistasis and the evolution of dominance. *Genetics*, 141, 333–346.

Doebley, J., Stec, A. and Hubbard, L. (1997) The evolution of apical dominance in maize. *Nature*, 386, 485–488.

Donoghue, M. J., Ree, R. and Baum, D. A. (1998) Phylogeny and the evolution of flower symmetry in Asteridae. *Trends in Plant Science*, 3, 311–317.

Downie, S. R. and Palmer, J. D. (1992) Restriction site mapping of the chloroplast inverted repeat: a molecular phylogeny of the Asteridae. *Annals of the Missouri Botanical Garden*, 79, 266–283.

Eizinger, A., Jungblut, B. and Sommer, R. (1999) Evolutionary change in the functional specificity of genes. *Trends in Genetics*, 15, 197–202.

Endress, P. K. (1997) *Antirrhinum* and Asteridae – evolutionary changes of floral symmetry. *Symposia of the Society for Experimental Biology*, 51, 133–140.

Endress, P. K. (1999) Symmetry in flowers: diversity and evolution. *International Journal of Plant Science*, 160 (6 Suppl.), S3–S23.

Felsenstein, J. (1989) PHYLIP – Phylogeny Inference Package (Version 3.2). *Cladistics*, 5, 164–166.

Ford, V. S. and Gottlieb, L. D. (1990) Genetic studies of floral evolution in *Layia*. *Heredity*, 64, 29–44.

Friis, E. M., Pedersen, K. R. and Crane, P. R. (1999) Early angiosperm diversification: the diversity of pollen associated with angiosperm reproductive structures in Early Cretaceous floras from Portugal. *Annals of the Missouri Botanical Garden*, 86, 256–296.

Gaudin, V., Lunnes, P. A., Fobert, P. R., Towers, M., Riou Khamlichi, C., Murray, J. A., Coen, E. and Doonan, J. H. (2000) The expression of D-cyclin genes defines distinct developmental zones in snapdragon apical meristems and is locally regulated by the *cycloidea* gene. *Plant Physiology*, 122, 1137–1148.

Gerhart, J. and Kirschner, M. (1997) *Cells, Embryos and Evolution*. Blackwell Science, Oxford.

Gillies, A. C. M., Cubas, P., Coen, E. S. and Abbott, R. J. (2002) Making rays in the Asteraceae: genetics and evolution of radiate versus discoid flower heads, in *Developmental Genetics and Plant Evolution* (eds Q. C. B. Cronk, R. M. Bateman and J. A. Hawkins), Taylor & Francis, London, pp. 233–246.

Hawkins, J. A. (2002) Evolutionary developmental biology: impact on systematic theory and practice, and the contribution of systematics. *Developmental Genetics and Plant Evolution*, 3, 32–51.

Iltis, H. H. (1983) From teosinte to maize: the catastrophic sexual transmutation. *Science*, 222, 886–894.

Johansen, B. and Frederiksen, S. (2002) Orchid flowers: evolution and molecular development. *Developmental Genetics and Plant Evolution*, 10, 206–219.

Jónsson, Z. O. and Hübscher, U. (1998) Proliferating cell nuclear antigen: more than a clamp for DNA polymerases. *Bioessays*, 19, 967–975.

Kosugi, S. and Ohashi, Y. (1997) PCF1 and PCF2 specifically bind to cis elements in a rice proliferating cell nuclear antigen gene. *Plant Cell*, 9, 1607–1619.

Lawton-Rauh, A. L., Alvarez-Buylla, E. R. and Purugganan, M. (2000) Molecular evolution of flower development. *Trends in Ecology and Evolution*, 15, 144–149.

Lewis, E. (1978) A gene complex controlling segmentation in *Drosophila*. *Nature*, 276, 565–570.

Linnaeus, C. (1749) *De Peloria*. Amoenitates Academia, Uppsala.

Luo, D., Carpenter R., Copsey, L., Vincent, C., Clark, J. and Coen, E. (1999) Control of organ asymmetry in flowers of *Antirrhinum*. *Cell*, 99, 367–376.

Luo, D., Carpenter, R., Vincent, C., Copsey, L. and Coen, E. (1996) Origin of floral asymmetry. *Nature*, 383, 794–799.

McGinnis, W. and Krumlauf, R. (1992) Homeobox genes and axial patterning. *Cell*, 68, 283–302.

Möller, M., Clokie, M., Cubas, P. and Cronk, Q. (1999) Integrating molecular and developmental genetics: a Gesneriaceae case study, in *Molecular Systematics and Plant Evolution* (eds P. M. Hollingsworth, R. M. Bateman and R. J. Gornall), Taylor and Francis, London, pp. 375–402.

Ohno, S. (1970) *Evolution by Gene Duplication*. Springer Verlag, Berlin, Heidelberg, New York.

Ohta, T. (1989) Role of gene duplication in evolution. *Genome*, 31, 304–310.

Olmstead, R. G., Bremer, B., Scott, K. M. and Palmer, J. D. (1993) A parsimony analysis of the Asteridae *sensu lato* based on *rbcL* sequences. *Annals of the Missouri Botanical Garden*, 80, 700–722.

Olmstead, R. G., Michaels, H. J., Scott, K. M. and Palmer, J. D. (1992) Monophyly of the Asteridae and identification of their major lineages inferred from DNA sequences of *rbcL*. *Annals of the Missouri Botanical Garden*, 79, 249–265.

Purugganan, M. (1998) The molecular evolution of development. *Bioessays*, 20, 700–711.

Reeves, P. A. and Olmstead, R. G. (1998) Evolution of novel morphological and reproductive traits in a clade containing *Antirrhinum majus* (Scrophulariaceae). *American Journal of Botany*, 85, 1047–1056.

Ruddle, F. H., Bartels, J. L., Bentley, K. L., Kappen, C., Murtha, M. T. and Pendleton, J. W. (1994) Evolution of Hox genes. *Annual Review of Genetics*, 28, 423–442.

Samata, Y. (1959) Genetic studies on *Cosmos bipinnatus*. *Japanese Journal of Breeding*, 8, 53–60.

Scott, M. (2000) Development: the natural history of genes. *Cell*, 100, 27–40.

Soltis, P., Soltis, D. and Chase, M. (1999) Angiosperm phylogeny inferred from multiple genes as a tool for comparative biology. *Nature*, 402, 402–403.

Stebbins, G. L. (1974) *Flowering Plants: Evolution Above the Species Level*. Harvard University Press, Harvard, Mass.

Theißen, G. (2000) Evolutionary developmental genetics of floral symmetry: the revealing power of Linnaeus' monstrous flower. *Bioessays*, 22, 209–213.

Theißen, G., Becker, A., Di Rosa, A., Kanno, A., Kim, J. T., Munster, T., Winter K.-U. and Saedler, H. (2000) A short history of MADS-box genes in plants. *Plant Molecular Biology*, 42, 115–149.

Trow, A. H. (1912) On the inheritance of certain characters in the common groundsel, *Senecio vulgaris*, and its segregates. *Journal of Genetics*, 2, 239–276.

Vieira, C. P., Vieira, J. and Charlesworth, D. (1999) Evolution of the *cycloidea* gene family in *Antirrhinum* and *Misopates*. *Molecular Biology and Evolution*, 16, 1474–1483.

Walker-Larsen J. and Harder L. D. (2000) The evolution of staminodes in angiosperms: patterns of reduction, loss and functional reinvention. *American Journal of Botany*, 87, 1367–1384.

Wang, R., Stec, A., Hey, J., Lukens, L. and Doebley, J. (1999) The limits of selection during maize domestication. *Nature*, 398, 236–239.

Warbrick, E. (2000) The puzzle of PCNA's many partners. *Bioessays*, 22, 997–1006.

Weberling, F. (1989) *Morphology of Flowers and Inflorescences*. Cambridge University Press, Cambridge.

White, S. and Doebley, J. (1998) Of genes and genomes and the origin of maize. *Trends in Genetics*, 14, 327–332.

Wray, G. and Abouheif, E. (1998) When homology is not homology. *Current Opinion in Genetics and Development*, 8, 675–680.

Chapter 14

Floral diversity and evolution in the Solanaceae

Sandra Knapp

ABSTRACT

The flowers of members of the Solanaceae are usually viewed as radially symmetric, and those genera with bilaterally symmetric flowers as the exception. In fact, a wide range of floral form occurs in the Solanaceae, from some members of the genus *Solanum* with completely radially symmetric flowers, through genera with zygomorphic androecia such as *Schultesianthus*, to species of *Schizanthus* with markedly zygomorphic flowers. Zygomorphy can occur in from one to all of the floral whorls, and members of the Solanaceae exhibit a bewildering variety of floral forms. The relationship between floral form and pollination syndrome has been assumed by many botanists to be driving the evolution of floral symmetry in the Solanaceae, but it is clear that a phylogenetic component to the distribution of flower shape diversity also exists. A close relationship with the Convolvulaceae has led botanists to assume that zygomorphy is derived in the Solanaceae. Recent phylogenetic analyses, however, have shown that this is not the case; genera with markedly zygomorphic flowers occupy basal positions in the tree. In addition to variation in symmetry between genera, floral zygomorphy has clearly evolved several times in several of the larger genera of the family (e.g. *Solanum*, *Cestrum*, *Nicotiana*). The scope of floral form diversity in the Solanaceae is reviewed in light of current phylogenetic treatments of the family and of selected genera in the family. I also examine the potential for future developmental work on the evolution of floral form in the family, particularly in relation to taxa with potential as model organisms such as *Petunia* and *Nicotiana*, which are well understood at the genetic level. Finally, I examine the relationship between pollination and floral form, and propose ways of linking the study of genetics, development and the study of plants in their native habitats.

14.1 Introduction

The Solanaceae are an economically important, cosmopolitan family with over 2500 species in some ninety genera. The family has members occurring in all habitats, from the driest deserts on the western coasts of South America to the dense wet tropical rainforests of the Amazon and South East Asia. Life forms in the family range from canopy trees to minute ephemeral herbs. Members of the family include globally important food crops such as potatoes and tomatoes (*Solanum tuberosum* L. and *S. lycopersicum* L., respectively) and widely-used drug plants such as *Nicotiana* L., the

In *Developmental Genetics and Plant Evolution* (2002) (eds Q. C. B. Cronk, R. M. Bateman and J. A. Hawkins), Taylor & Francis, London, pp. 267–297.

Table 14.1 Largest genera of the Solanaceae, in rank order

Genus	Number of species
Solanum L.	1000–1500
Lycianthes Bitter	c. 200
Cestrum L.	c. 175
Nicotiana L.	75–80
Physalis L.	60–70
Brunfelsia L.	45
Nierembergia Ruiz & Pavón	36

source of nicotine, and *Atropa* L., the source of atropine. The genomes of tomato and tobacco are among the most thoroughly studied of angiosperms, ranking alongside those of many cereal grasses of economic importance, now rivalled amongst dicotyledons only by *Arabidopsis* Heynh. in the Brassicaceae (Arabidopsis Genome Initiative, 2000). Approximately half of the species in the family are contained in five genera, the largest and most diverse of which is *Solanum* L. (Table 14.1).

14.1.1 Classification

The Solanaceae has traditionally been divided into two subfamilies. The traditional Cestroidae, including petunias, tobaccos, cestrums and their relatives, have non-compressed, often prismatic seeds and tropane alkaloids. The traditional Solanoidae, containing the large majority of species in the family, includes *Solanum* and its relatives which have compressed and mainly steroidal alkaloids. Several morphologically distinctive genera have traditionally (Hunziker, 1979; Thorne, 1992) been maintained as satellite families to the Solanaceae, based largely on their possession of autapomorphies such as a drupaceous (*Duckeodendron* Kuhlm.) or schizocarpic (*Nolana* L.f.) fruit or embryos with straight, fleshy cotyledons (*Henoonia* Griseb., *Goetzea* Miers, *Espadaea* A. Rich, *Coeloneurum* Radkl., and all members of the Goetzeaceae *sensu* Hunziker, 1979). This traditional classification, which has held sway since the early twentieth century (see Cronquist, 1981), has recently been challenged by cladistic analysis using chloroplast and nuclear datasets, and the family can now be divided into approximately seven monophyletic groups (see Olmstead *et al.*, 1999). Current phylogenetic opinion includes the Amazonian canopy tree *Duckeodendron cestroides* Kuhlm. in the family (traditionally treated as the monotypic family Duckeodendraceae: see Fay *et al.*, 1998), as well as the families Goetzeaceae and Nolanaceae (Olmstead *et al.*, 1999). All of these groups are nested well within the Solanaceae *sensu stricto*.

14.1.2 Floral variation

In addition to the huge variation in life form, members of the Solanaceae also exhibit a wide range of floral morphologies. The family is traditionally (Cronquist, 1981; Thorne, 1992) considered to have 'primitively' actinomorphic flowers, with those genera with zygomorphy of varying kinds assumed to be 'advanced'. In this chapter I will use the terms 'bilaterally symmetric' or 'zygomorphic' to characterise those flowers with a single plane of symmetry (synonym = monosymmetric, including those more specifically designated asymmetric: Endress, 1999; Bateman and DiMichele, 2002; Cubas, 2002) and radially symmetric or actinomorphic to characterise those flowers with multiple planes of symmetry (synonym = polysymmetric: Endress, 1999). Floral zygomorphy (monosymmetry *sensu* Endress, 1999) occurs in many genera of Solanaceae, and has been the cause of much debate as to taxonomic placement of genera and of species in the family. Within the traditional subfamily Cestroideae, several genera have long been subject to confusion and contention over their placement, both in relation to their affinities in the family Solanaceae itself and as to whether or not they belong in the family at all. This group of genera, often characterised as the tribe Salpiglossidae (Hunziker, 1979), consists of taxa with marked flower zygomorphy and a reduced number of stamens (generally to two or four). The tribe is named for the genus *Salpiglossis* Ruiz & Pav., which was placed in the Bignoniaceae by Hooker (1827) and Bartling (1830) in their natural systems. Sweet (1830) and Don (1837) placed *Salpiglossis* in the family Solanaceae. George Bentham (1835) regarded it as a member of the Scrophulariaceae, and erected the tribe Salpiglossidae to encompass *Salpiglossis*, *Brunfelsia* L., *Schizanthus* Ruiz & Pav., *Browallia* L., *Franciscea* Pohl (=*Brunfelsia*) and *Anthocercis* Labill. Many authors followed Bentham's placement of these genera in the Scrophulariaceae (Endlicher, 1839; Brogniart, 1843; Clos, 1849; Baillon, 1887, 1888), despite the work of Miers (1849). He studied in detail most of the known genera of the Solanaceae and proposed that floral zygomorphy (or 'irregularity') was quite common in the family, thus necessitating a series of new tribes in a new 'intermediate' family, the Atropaceae. His system was largely followed by Bentham and Hooker in their monumental revision of the natural system (1876), where Miers' Atropaceae was subsumed into a larger Solanaceae, but his tribal divisions were left more or less intact.

Wydler (1866) and Eichler (1869) found that the members of the Solanaceae were characterised by a unique type of floral zygomorphy, with the gynoecium having an oblique orientation (Figure 14.1b). This oblique orientation is found in all genera of

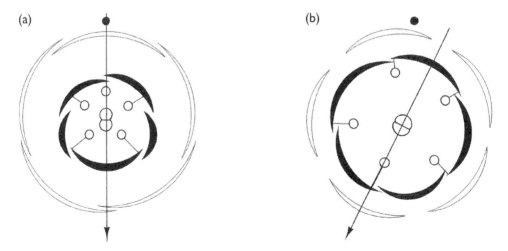

Figure 14.1 (a) Generalised floral diagram of a member of the Scrophulariaceae. (b) Generalised floral diagram of a member of the Solanaceae (see text; open circles in the stamen position indicate fertile stamens, black indicates reduced stamens, or staminodes).

the family (except *Nicandra* Adans.: see Robyns, 1931), and along with internal phloem (Cronquist, 1981) it constitutes a good synapomorphy for the family (Robyns, 1931; Coccuci, 1989b; but see Huber, 1980; Ampornan and Armstrong, 1988, 1989, 1991). This orientation means that in the Solanaceae, the adaxial petal is on the plane of perianth symmetry, rather than the abaxial petal as occurs in the Scrophulariaceae and Lamiaceae (Figure 14.1a). Essentially, all members of the Solanaceae are zygomorphic to at least some extent. Robyns (1931), in his review of floral zygomorphy in the family, characterised three basic floral forms:

1   actinomorphic (only *Nicandra*, which does not possess the oblique orientation of the gynoecium);
2   oblique (many genera with actinomorphic corollas); and
3   bilateral (several floral whorls exhibiting bilateral symmetry).

He further defined two types of zygomorphy in the family: (a) functional zygomorphy, which primarily involved the order of anther maturation and dehiscence, and (b) morphological zygomorphy, affecting the form of floral organs.

Recent phylogenetic studies in the Solanaceae (Olmstead *et al.*, 1999; Olmstead, pers. comm. 2001) allow us to examine the patterns of floral form in relation to monophyletic clades rather than in isolation. This examination of the range of floral morphologies in the family should in no way be seen as a substitute for the careful reconstruction of phylogeny using these same morphological characters, and is only a first glance at what is a complex problem.

In this chapter I review floral zygomorphy in the Solanaceae clade by clade, using the recently published cpDNA phylogeny of Olmstead *et al.* (1999) and personal observations of floral morphology of living plants and herbarium specimens. I assess these patterns and how robust the phylogeny is for each clade, examining genera throughout the family. The contention that pollination syndromes have driven the evolution of the diverse array of floral forms in the family is reviewed in the light of phylogeny and the distribution of zygomorphy amongst genera. I finally relate floral form and architecture in the Solanaceae to recent molecular genetic studies on floral development and variation that may have relevance to the future study of the diversity of floral symmetry types in the Solanaceae.

14.2 Types of floral zygomorphy

14.2.1 Functional

Robyns (1931) defined functional zygomorphy primarily in terms of the order of dehiscence of anthers during anthesis. In many members of the Solanaceae, all five anthers dehisce simultaneously and filament growth post-anthesis is limited. However, in some genera such as *Lycianthes* Bitter (Dean, pers. comm. July 2000) or *Jaltomata* Schltdl. (Figure 14.2), post-anthesis growth, particularly of the filaments, can markedly alter floral form. In his studies of functional zygomorphy, Robyns (1931) found two basic patterns of dehiscence order (Figure 14.3). In *Physalis* L. (the ground cherries and their relatives) anthers are all of the same length and mature in the order lateral > adaxial > abaxial (Figures 14.3 and 14.4a). Many

Figure 14.2 *Jaltomata procumbens* (Cav.) J. L. Gentry. (a) Early anthesis, filaments not elongate. (b) Late anthesis, filaments elongate.

species of the traditional genus *Physalis* (not including *Margaranthus* Schltdl. and *Quincula* Raf.) exhibit this pattern, but other genera in the clade (Physalinae *sensu* Estrada and Martinez, 1999, including *Larnax* Miers, *Deprea* Raf., *Chamaesaracha* (A. Gray) Benth. and *Tzeltalia* Estrada & Ma. Martínez) have not been examined.

However, the genus *Hyoscyamus* L. (the henbanes, represented in Robyns' studies by *H. niger* L.) has anthers that are markedly different in filament length, and dehisce in the order adaxial > lateral > abaxial (Figures 14.3 and 14.4b). All members of the genus *Hyoscyamus* I have seen exhibit this pattern of anther dehiscence, as do other members of the Hyoscyameae such as *Physochlaina* G. Don (see Hoare and Knapp, 1997). Other genera included in the *Hyoscyamus* clade (Hyoscyameae *sensu* Hoare and Knapp, 1997; Olmstead *et al.*, 1999), including *Anisodus* Link, the sister genus of *Hyoscyamus* itself, appear to have simultaneous anther dehiscence.

14.2.2 Morphological

Robyns (1931) examined ten genera of the Solanaceae in his survey of floral zygomorphy. Members of the family exhibit a 3/2 arrangement of petals (with three adaxial and two abaxial) rather than the 2/3 arrangement commonly found in the Scrophulariaceae (Figure 14.1: Robyns, 1931; Donoghue *et al.*, 1998). The development of bilateral symmetry in flowers of the Solanaceae apparently follows what I will term 'Robyns' rule', where the abaxial stamen is the first to be lost or modified, then the adaxial ones, and finally the lateral ones. This is in marked contrast to the pattern of organ suppression seen in members of the Lamiales (*sensu* Savolainen *et al.*, 2000; see Endress, 1999) such as the Scrophulariaceae or Gesneriaceae, where in general the adaxial stamen is modified, then the abaxial pair. Robyns (1931) postulated a sequence of increasing specialisation in floral form from the actinomorphic

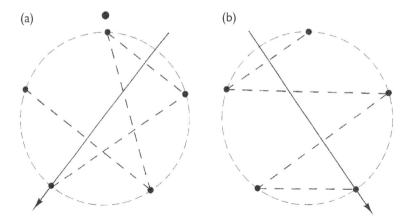

Figure 14.3 Anther maturation order. (a) *Physalis*. (b) *Hyoscyamus* (modified from Robyns, 1939).

taxa to the highly zygomorphic species of *Schizanthus*, and fitted the scheme to the existing classification of the family.

14.2.3 Solanaceae phylogeny and zygomorphy

Using a combination of restriction site maps and DNA sequences from two chloroplast genes, *rbcL* and *ndhF*, Olmstead *et al.* (1999) examined thirty genera and thirty-six species in the Solanaceae and assessed relationships on a variety of hierarchical levels. A new intrafamilial classification was proposed, in which seven monophyletic subfamilies are recognised, as compared to the traditionally recognised Cestroidae and Solanoidae (see Section 14.1.1). Major differences in the new classification and the traditional one are few, but focus on the relationships of *Nicotiana* L.

Figure 14.4 (a) *Physalis crassifolia* Benth. (b) *Hyoscyamus albus* L.

(in the new classification forming its own tribe, with the rest of the traditional Nicotianae plus *Brunfelsia* moving to the Petunieae), and some rearrangements in the relationships amongst actinomorphic genera. Details of the classification can be found in Olmstead *et al.* (1999), and further sampling of more genera and species of the family has largely supported this new hypothesis of generic relationships (analyses in Fay *et al.*, 1998 were performed subsequent to those in Olmstead *et al.*, 1999; Olmstead, pers. comm. July 2000). A major clade based on the base chromosome number 12, identified by Olmstead and Palmer (1992) as the 'X = 12' clade, is strongly supported in these analyses (Olmstead *et al.*, 1999). This clade includes all of *Solanum*, *Lycianthes* and the traditional Solanoidae, but also includes some genera more traditionally placed in the Cestroideae such as *Nicotiana* and members of the strictly Australian tribe Anthocercidae. Unfortunately, the subfamilial and tribal names proposed by Olmstead *et al.* (1999) were not validly published in accordance with the International Code of Botanical Nomenclature (Greuter *et al.*, 2000) and so have no formal standing. In this chapter, the names for groups proposed by Olmstead *et al.* (1999) are placed in quotation marks; this indicates that I am using their concept of the group in question rather than delimiting the group in a more traditional sense.

Since Olmstead *et al.* (1999) performed several analyses, most of which included different sets of taxa, for the purposes of discussion of floral diversity in the Solanaceae I have slightly adapted their published trees. In no way should the abstract tree presented here be taken as a phylogeny of the family: readers are instead referred to the original cladograms. I have used the most generically robust analysis and added (using dotted lines) those genera that appear in other analyses or were not included in the DNA sampling, but were included in the proposed classification. In order to better visualise the distribution of floral zygomorphy in the Solanaceae, only taxa with morphological zygomorphy in one or more floral whorls (sepals, petals or stamens) were retained. Taxa with oblique or actinomorphic flowers (*sensu* Robyns, 1931) were deleted, so that their terminals bear no names. If we 'prune' all taxa not exhibiting morphological floral zygomorphy from the 'modified' strict consensus tree based on the second analysis of Olmstead *et al.* (1999), including a broad sampling of genera, it is apparent that zygomorphy is not evenly distributed across all clades in the family (Figure 14.5). In fact, genera with marked bilateral symmetry are clustered in basal clades of the family. Another glaringly obvious pattern is that floral zygomorphy occurs multiple times in the family. This 'character', however, is not homologous across the genera of the Solanaceae; the zygomorphy of *Schizanthus*, with its 'papilionaceous' flowers, is not at all similar to that of *Solandra* Sw., with a declinate androecium. Details of the morphology of symmetry in each of the clades where it occurs is discussed below. Subfamilial and tribal names are those of Olmstead *et al.* (1999) except where otherwise noted, and a synopsis of the genera with floral zygomorphy in each of the clades discussed here can be found in Table 14.2.

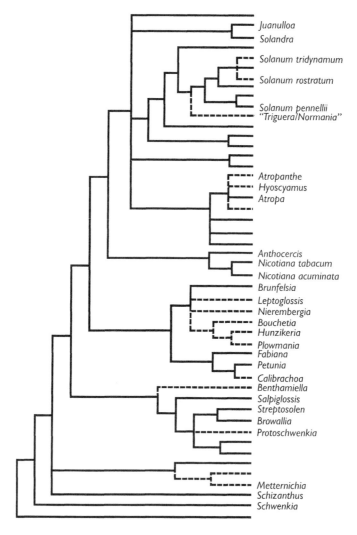

Figure 14.5 Simplified summary phylogeny of the Solanaceae based on chloroplast DNA data (modified from Fay *et al.*, 1998; Olmstead *et al.*, 1999). Names of all taxa with radial symmetry
(polysymmetric) have been deleted, so that only those taxa with floral zygomorphy are
indicated. Most species names have been excluded.

14.3 The commonly zygomorphic clades of Solanaceae

14.3.1 Hyoscyamus clade (tribe 'Hyoscyameae')

The genus *Hyoscyamus* has clearly zygomorphic flowers, with bilateral symmetry in
all floral whorls (Figure 14.6a, b, Plate 3c). Other genera in the tribe exhibit varying
degrees of bilateral symmetry in the sepals (*Scopolia* Jacq. and *Anisodus*: Figure
14.6d), petals (*Hyoscyamus* and *Atropanthe* Pascher) and androecium (*Atropa*:
Figure 14.6c). Calyx zygomorphy in *Hyoscyamus* is largely in the tube, whereas in

Table 14.2 Synopsis of Solanaceae (adapted from Olmstead et al., 1999)

Subfamily	Tribe	Genera
'Cestroideae'	'Browallieae'	Browallia, Streptosolen
	'Cestreae'	Metternichia
	'Salpiglossideae'	Salpiglossis, Reyesia
'Petunoideae'		Benthamiella[2], Bouchetia, Brunfelsia, Calibrachoa, Combera[2], Fabiana, Hunzikeria, Leptoglossis, Nierembergia, Pantacantha[2], Petunia, Plowmania
'Schizanthoideae'		Schizanthus
'Schwenkoideae'		Schwenkia, Protoschwenkia
'Nicotianoideae'	'Anthocercideae'	Anthocercis, Crenidium, Cyphanthera, Duboisia, Grammosolen, Symonanthus
	'Nicotianeae'	Nicotiana
'Solanoideae'	'Capsiceae'	Lycianthes[1]
	'Daturaeae'	Brugmansia[1]
	'Hyoscyameae'	Anisodus, Atropa, Atropanthe, Hyoscyamus, Scopolia
	'Solandreae'	Juanulloa, Markea[1], Schultesianthus[1], Solandra
	'Solaneae'	Solanum[1], Normania, Triguera

Notes:
Only genera in which zygomorphy has been confirmed are listed.
    Subgeneric and tribal names proposed by Olmstead et al. (1999) are not validly published and so are used in quotation marks, to indicate delimitation sensu Olmstead et al. (1999) rather than in their traditional sense.
1 Those genera in which only some species are zygomorphic.
2 Genera placed in the 'Petunoideae' by Olmstead et al. (1999) but perhaps more closely related to members of the 'Salpiglossideae' (fide R. Olmstead, pers. comm. 2001).

Scopolia and Anisodus it is manifested in the size of the calyx lobes, with the adaxial and abaxial lobes larger and more elongate (Hoare and Knapp, 1997). In addition, Robyns (1931) found that the axis of symmetry in Hyoscyamus goes from left to right (Figure 14.7), rather than from right to left as in the rest of the genera that he examined (see Section 14.1.2 and Figure 14.1). None of the other species in the tribe were examined by Robyns (1931), so it is unclear whether this 'handed-ness' is found only in Hyoscyamus or is more generally distributed within the clade (but see discussion of Schizanthus in Section 14.3.5).

14.3.2 The genus Nicotiana and relatives (subfamily 'Nicotianoideae')

The large and diverse genus Nicotiana has traditionally been treated as related to Petunia Juss. and members of the Cestroideae (Hunziker, 1979), due in part to their superficially similar flowers and the presence of sticky glandular pubescence on leaves and stems. Olmstead et al. (1999) have shown that Nicotiana is most closely related to the members of the group traditionally known as the Anthocercideae, whose members (Table 14.2) are all Australian endemics (Haegi, 1981; Knapp et al., 2000). With the exception of Anthotroche Endl. and some specimens of the dioecious Symonanthus L. Haegi, which have radially symmetric flowers, all of the

Figure 14.6 (a) *Hyoscyamus niger* L. (b) *Hyoscyamus muticus* L. (c) *Atropa belladonna* L. (d) *Anisodus luridus* Link.

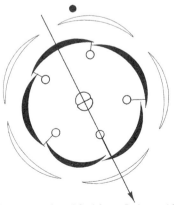

Figure 14.7 Floral diagram of *Hyoscyamus* (modified from Robyns, 1939).

Australian genera have bilateral symmetry only in the androecium, where the stamens are generally didynamous. In some genera (e.g. *Anthocercis*), a staminode persists in the abaxial position, but in others this is absent. All species of *Nicotiana*, in contrast, possess five stamens, but these vary in insertion position and length (Figure 14.8). Some sections of *Nicotiana* have markedly zygomorphic flowers in all floral whorls (e.g. Tomentosae, *N. glutinosa* L.: Figure 14.9a), while other sections, most notably the Alatae, have distinct bilateral symmetry in the corolla and androecium but little in the calyx (*N. bonariensis* Lehm.: Figure 14.9b).

Patterns of zygomorphy across the genus have not been analysed in detail, but it is clear that zygomorphy in the corolla has arisen at least twice. The symmetry pattern in the androecium of *Nicotiana* is generally similar to that found in the rest of the family, with the abaxial stamen being different from the lateral or adaxial ones (see floral diagrams and Figure 14.8). In general, the abaxial stamen is shorter than the other four, and those four occur either in two differentiated pairs (*N. glutinosa*, *N. tabacum* L.) or are all equal (*N. alata* Link & Otto, *N. sylvestris* Speg. & Comes). Species of alloploid hybrid origin in *Nicotiana* (e.g. *N. tabacum*, *N. arentsii* Goodsp., *N. rustica* L.) are generally intermediate in floral zygomorphy compared with their 'parents' (Goodspeed, 1954; Chase *et al.*, unpubl. obs.).

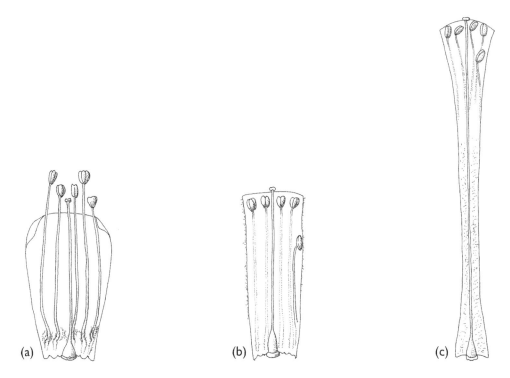

Figure 14.8 Longitudinal sections of representative *Nicotiana* flowers, showing differences in filament insertion and length. (a) *Nicotiana benavidesii* Goodsp. (b) *Nicotiana rotundifolia* Lindl. (c) *Nicotiana longiflora* Cav.

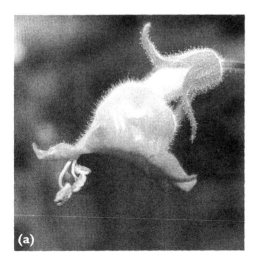 

Figure 14.9 (a) *Nicotiana glutinosa* L. (b) *Nicotiana bonariensis* Lehm.

14.3.3 Salpiglossis *clade (tribe 'Salpiglossideae', tribe 'Browallieae')*

This clade contains some of the members of the traditional tribe Salpiglossidae (Table 14.2): *Salpiglossis*, *Streptosolen* Miers and *Browallia* L. Here zygomorphy is found in both the corolla and androecium, and is particularly pronounced in the genus *Browallia*, where the anthers are highly modified and packed into the corolla throat (Cocucci, 1995). Olmstead *et al.* (1999) divided the genera here into two tribes (Table 14.2), but more recent analyses including more genera suggest that they form a single monophyletic group (Olmstead, pers. comm. 2001), with the addition of some genera placed by Olmstead *et al.* (1999) in the 'Petunoideae' (*Benthamiella* Speg., *Pantacantha* Speg., *Combera* Sandwith: Olmstead, pers. comm.).

Corolla zygomorphy in these taxa varies from slight (in *Salpiglossis*: Figure 14.10a) to almost two-lipped (in *Browallia*: Figure 14.10b). Stamen number also varies, with *Salpiglossis* having five stamens, one of which is reduced, and the rest of the genera having only four and occasionally a staminode in the abaxial position (Figure 14.11a). Cocucci (1995) demonstrated that the stamens of *Browallia* were versatile and could be shown to deposit pollen on artificial butterfly tongues inserted in flowers. The genus *Streptosolen* has always been considered the sister group of *Browallia*, differing from it mainly in orange flower colour and more open flower tube (Figure 14.10c). *Streptosolen* also has resupinate flowers (considered not to occur in the family by Robyns, 1931; however, he did not examine *Streptosolen*), where the floral tube is twisted through 180° (Cocucci, 1995) and the adaxial petal thus appears to be abaxial. This has been related to a shift to hummingbird pollination (Cocucci, 1995, 1999), although why this should be so is unclear.

Figure 14.10 (a) *Salpiglossis sinuata* Ruiz & Pav. (b) *Browallia americana* L. (c) *Streptosolen jamesonii* (Benth.) Miers.

14.3.4 Petunia clade (subfamily 'Petunoideae')

Several of the genera included in the 'Petunoideae' (*sensu* Olmstead *et al.*, 1999), such as *Brunfelsia, Nierembergia* Ruiz & Pav., *Hunzikeria* D'Arcy and *Leptoglossis* Benth., have traditionally been treated as members of a tribe Salpiglossideae together with the above clade. Zygomorphy in members of the *Petunia* clade occurs in all floral whorls; Robyns (1931) showed that, despite having a more or less regular corolla and five stamens (the abaxial of which is reduced: Figures 14.11b

(a)　　　　　　　　　　　(b)

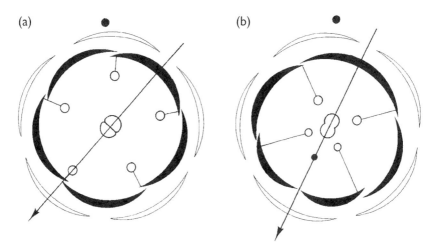

Figure 14.11 Floral diagrams of members of the Salpiglossidae (as traditionally circumscribed). (a) *Salpiglossis*. (b) *Petunia* (open circles in the stamen position indicate fertile stamens, black indicates reduced stamens or staminodes).

and 14.12a, b), *Petunia axillaris* (Lam.) Britton, Stern & Poggenb. showed marked calyx epitrophy. *Petunia* and its segregate genus *Calibrachoa* La Llave & Lex. are differentiated in part due to their symmetry differences in bud, species of *Calibra-choa* have strongly bilaterally symmetrical buds, while aestivation in *Petunia* is whorled or irregular (Wijsman and de Jong, 1985; Wijsman, 1990; Ando *et al.*, 1992). Patterns of corolla and calyx aestivation have been considered important in Solanaceae systematics (Baehni, 1915), but have rarely been used in phylogenetic studies, in part due to difficulties in character assessment from herbarium specimens (Knapp *et al.*, 1997).

The rest of the genera of this tribe with bilaterally symmetric flowers show marked zygomorphy in both the corolla and androecium (Table 14.2, Figures 14.12c, d, 14.13a). In addition, the stigma in several of these genera (e.g. *Bouchetia* Dunal, *Leptoglossis*, *Hunzikeria*) is bilabiate or strongly elaborated. D'Arcy (1978) hypothesised a series of evolutionary stages in floral morphology from *Petunia* (stage I: regular corolla with open mouth, five stamens, small stigma) to *Leptoglossis* and *Hunzikeria* (incl. *Browallia*: stage II: zygomorphic corolla with narrow mouth, 4(5) didynamous stamens, two-armed stigma) to *Nierembergia* and *Bouchetia* (stage III: slightly zygomorphic corolla with open mouth, 4(5) stamens, stigma elaborated). He pointed out that it was artificial to assume relatedness of genera with reduction in stamen number; this is supported by the cpDNA phylogeny (Olmstead *et al.*, 1999). *Nierembergia* has the most 'regular' flowers in this group, with an actino-morphic, cup-shaped corolla; however, the five stamens differ in anther size and fila-ment length, with the abaxial filament the shortest, followed by the adaxial filaments (D'Arcy, 1978; Cocucci, 1991). This 'primitive' floral morphology is almost certainly derived in this group, and related to pollination by oil-collecting bees (Cocucci, 1991).

Reduction of stamen number in this group in general follows the pattern

Figure 14.12 (a) *Petunia axillaris* (Lam.) Britton, Stern & Poggenb. (b) *Petunia hybrida* Hort ex Vilm., showing packing of stamens and stigma in corolla mouth. (c) *Brunfelsia grandiflora* (D.) Don. (d) *Brunfelsia densifolia* Krug & Urb., showing tight packing of reproductive organs at opening of corolla tube.

described by Robyns (1931), with the abaxial stamen being modified or absent (Figure 14.13). In *Leptoglossis*, however, several species have only two stamens, but rather than the adaxial pair being modified (as was assumed to be the rule in the family by Robyns, 1931), the lateral stamen pair is reduced (see illustrations in Hunziker and Subils, 1979). The genus *Reyesia* Gay (placed by Olmstead *et al.*, 1999, with *Salpiglossis*, but not sequenced) also exhibits this pattern (Hunziker and Subils, 1979; Cocucci, 1995). The fertile adaxial stamens of *Leptoglossis linifolia* Benth. & Hook.f. have a necrotic distal filament tip, so the anther is versatile, like that of *Browallia*, and can be moved by artificial butterfly tongues (Cocucci, 1995). The flowers of all species of *Leptoglossis* are very small (*c.* 3–10 mm long) and the

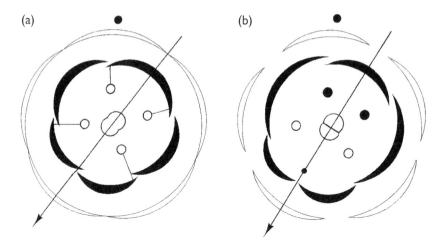

Figure 14.13 Floral diagrams: (a) *Brunfelsia*. (b) *Schizanthus* (open circles in the stamen position indicate fertile stamens, black indicates reduced stamens or staminodes).

corolla throat is very narrow and completely blocked by the anthers and expanded stigma.

Brunfelsia flowers also have a very narrow corolla mouth completely blocked by the anthers and stigma (Figure 14.12c, d). The corolla lobes are strongly zygomorphic in all *Brunfelsia* species, and in some taxa the adaxial lobe can be extremely large. The anthers are unusual in that they have fusion of sister thecae, very hard pollen sac walls, and are versatile due to a thin specialised filament attachment (Cocucci, 1995). Considerable variation in corolla form exists in the genus, from the long-tubed Puerto Rican endemic *B. densifolia* Krug & Urb. with a straight, narrow tube (Figure 14.12d) to the more open-mouthed *B. amazonica* C. V. Morton, to *B. grandiflora* D. Don (Figure 14.12c) with a narrow and strongly bent floral tube (Robyns, 1931; Plowman, 1998).

14.3.5 Schizanthus (subfamily 'Schizanthiodeae') and Schwenkia (subfamily 'Schwenkiodeae')

Members of the tribe Schwenkieae, like *Leptoglossis*, have very small flowers. Species of *Schwenkia* L. have either two or four stamens, but the corolla is regular or only very slightly zygomorphic (Carvahlo, 1978; Hunziker, 1979). D'Arcy and Benítez (1991) suggested that reduction to two stamens had occurred at least twice in the genus *Schwenkia*, but they did not perform a cladistic analysis. Those taxa with two stamens have reductions in the abaxial and adaxial stamens, as predicted by Robyns (1931). The two fertile stamens exhibit precocious development in early flower primordia (Ampornan and Armstrong, 1990). *Protoschwenkia* Solereder and *Melananthus* Walp. possess four stamens and also have regular corollas. Recent evidence from cpDNA sequences suggests that the tribe as currently circumscribed may not be monophyletic (Olmstead, pers. comm. July 2000).

Schizanthus has long been the most problematic of the bilaterally symmetric

Solanaceae, due to its extremely zygomorphic flowers (Figures 14.13b and 14.14). Flowers of *Schizanthus* exhibit zygomorphy in all floral whorls and the corolla in particular has a unique 'papilionaceous' shape. The adaxial petal is enlarged and the abaxial two petals often (but not always: Figure 14.14b, Plate 3b) form a keel in which the two fertile lateral stamens are held under pressure (for a description of the floral mechanism in *Schizanthus* see Cocucci, 1989a). The fertile stamens are always the lateral pair (Figure 14.13b); the others are completely sterile, although the abaxial stamen occasionally bears a minute, empty anther. As in *Schwenkia*, the fertile stamens exhibit precocious growth in early primordial stages (Ampornan and Armstrong, 1988). Although both Wydler (1866) and Robyns (1931) demonstrated the oblique symmetry of the gynoecium in *Schizanthus*, there has been some debate as to the validity of their observations (Ampornan and Armstrong, 1988, 1991). Coccuci (1989b), studying the development of *Schizanthus grahamii* Gillies, confirmed the oblique orientation of primordia, and also showed that pedicel twisting after anthesis brought the flowers to a position where their symmetry appeared medial. Intriguingly, in the monochasial inflorescence of *Schizanthus*, flowers alternate in being left- or right-handed with respect to the axis (Cocucci, 1989b). This type of oscillating architecture has been termed 'pendulum symmetry' (Charlton, 1998) and is also found in the enantiostylous flowers of *Solanum* section *Androceras* and similar species (see below). The degree to which this oscillation is found across Solanaceae is not known.

14.4 Scattered occurrences of bilateral symmetry in the rest of the family

Scattered occurrences of floral zygomorphy are found in the 'X = 12' (*sensu* Olmstead *et al.*, 1999) clade, especially in the tribe 'Solandreae' and in the genus *Solanum s.l.* These range from very subtle curvature of the androecium (*Solanum pennellii* Correll) or declination of stamens (*Solandra*) to marked floral zygomorphy, occurring in several floral whorls (e.g. *Solanum tridynamum* Dunal). Differences in

Figure 14.14 (a) *Schizanthus pinnatus* Ruiz & Pav. (b) *Schizanthus gilliesii* Phil.

anther size, thus affecting floral symmetry, also occur in the genus *Lycianthes* (Bitter, 1919; Dean, pers. comm.).

The tribe 'Solandreae' contains seven genera of epiphytic shrubs (Knapp *et al.*, 1997), four of which exhibit floral zygomorphy. *Juanulloa* Ruiz & Pav., *Markea* Rich. and some species of *Schultesianthus* Hunz. have a bilaterally symmetric corolla, with a somewhat expanded adaxial petal (Figure 14.15a). *Solandra* and the remaining species in *Schultesianthus* show marked stamen displacement, with the stamens declinate, usually abaxially (positional monosymmetry *sensu* Endress, 1999; also Knapp *et al.*, 1997: Figure 14.15b). It has been suggested that this displacement is due to gravity (Endress, 1999), but in some members of this group the stamens are displaced adaxially. This displacement of stamens to one side of the flower may function in pollination, but neither the developmental nor the phylogenetic components are clear (Knapp *et al.*, 1997).

In the large and vegetatively diverse genus *Solanum*, floral zygomorphy has arisen multiple times (Whalen, 1984; Lester *et al.*, 1999; Knapp, 2000), and is particularly prevalent in the androecium. The strongly down-turned anther cone of *Solanum pennellii* is apparent even in small buds, but the corolla of this species is completely regular. Species in several unrelated sections show stamen differentiation, with one or more stamens being longer, or larger, than the others in the whorl (Knapp, 2000, 2001). In these species, both the filament and the anther are modified. In *Solanum thelopodium* Sendtn. and its relatives (subgenus *Solanum*), the abaxial anther is modified and the corolla somewhat zygomorphic, but in *Solanum wendlandii* Hook.f. (subgenus *Leptostemonum*), the adaxial pair of anthers is longer than the others and the corolla is regular. In *S. turneroides* Chodat (subgenus *Solanum*), the abaxial anther exhibits post-anthesis filament growth and the filament elongates such that this anther appears two to three times longer than the others. Floral zygomorphy of this less obvious type is common in *Solanum*, but often overlooked by herbarium botanists.

Figure 14.15 (a) *Juanulloa mexicana* (Schltdl.) Miers. (b) *Solandra grandiflora* Sw. (photograph courtesy of J. Mallet).

Marked development of zygomorphy is found in both subgenera of *Solanum*: in three putatively unrelated groups of spiny solanums and in the species of section *Normania* of subgenus *Solanum* (Bohs and Olmstead, 2001; Knapp, 2001). All of these species have corolla zygomorphy, with the abaxial petals enlarged (Figures 14.16 and 14.17). The heterandrous spiny solanums occur in Mexico, the Canary Islands and in Africa. They all share a common morphology, with one anther (the abaxial) extremely enlarged and curved (Whalen, 1979); most of these taxa are andromonoecious. In section *Androceras* (Figure 14.16a), right- and left-handed flowers alternate along the inflorescence axis (see also *Schizanthus* above). Heterandry and enantiostyly have been interpreted as an adaptation to promote out-crossing in these self-compatible taxa (Bowers, 1975; Whalen, 1979; Lester *et al.*, 1999). *Solanum tridynamum* Dunal is a member of Whalen's (1984) *S. vespertilio*

Figure 14.16 (a) *Solanum rostratum* Dunal. (b) *Solanum tridynamum* Dunal, showing inflorescence with functionally male and hermaphroditic flowers. (c) *S. tridynamum*; functionally hermaph-roditic flower. (d) *S. tridynamum*; functionally staminate flower.

Figure 14.17 *Solanum trisectum* Dunal. (a) Front view of slightly zygomorphic corolla. (b) Lateral view of horned and dimorphic anthers.

group, containing species from both the Canary Islands and Mexico. It is the most strongly andromonoecious member of the group, bearing only a single hermaphroditic flower at the base of the inflorescence (Figure 14.16b). In these flowers, the abaxial anther is not at all enlarged, all five anthers are equal in size, and the style is elongate and curved (Figure 14.16c). In more distal, functionally male flowers, the style is minute and hidden within the anther cone, while three anthers (the abaxial and two laterals) are enlarged and elongate (Figure 14.16d). In other andromonoecious solanums sex expression is regulated by a combination of genetic and environmental effects (Whalen *et al.*, 1981; Diggle, 1993). Differences in the production of female sterile (male) flowers can be only partly attributed to pollination and fruit set of hermaphrodite flowers, and proportions of flower types are under some genetic control (Diggle, 1993, 1994). Developmental trajectories in these morphologically very different hermaphroditic and functionally male flowers have not been investigated, nor have manipulative experiments that remove hermaphroditic flowers to affect the sex expression of more distal flowers (Whalen *et al.*, 1981) been done in any of the andromonoecious zygomorphic solanums.

The genera *Normania* Lowe, two species endemic to Macaronesia, and *Triguera* Cav., from coastal habitats in the Mediterranean (both now considered to be part of *Solanum s.l.*: Bohs and Olmstead, 2001) exhibit a completely different form of anther zygomorphy. In *Triguera*, the five stamens are subequal, but the lateral pair of anthers has small apical horns. The corolla is tubular, and slightly curved with an expanded adaxial lobe. *Normania* has an only slightly zygomorphic, rotate corolla (Figure 14.17a), but the stamens are markedly bilaterally symmetric, largely in anther morphology (Figure 14.17b). The abaxial stamen has a very small anther (*c*. 3–4 mm), the lateral pair have the largest anthers (to 11 mm), while the adaxial pair are intermediate in size. Both the adaxial and lateral pairs of anthers bear a pronounced horn near the middle of each anther (Figure 14.17b). The Madeiran species

(*Normania triphylla = Solanum trisectum* Dunal) was at one time placed in the genus *Nycterium* Vent. along with *Solanum vespertilio* Aiton (see above), recognising its unusual anther structure.

14.5 Discussion

It is clear that floral zygomorphy in the Solanaceae is very labile and has been lost and gained independently several times. It should not therefore be thought of as an 'all or nothing' state, but instead actinomorphy and zygomorphy are endpoints of a developmental continuum. Not only are occurrences of bilateral symmetry scattered amongst genera through the phylogenetic tree based on cpDNA, but forms of zygomorphy found in the family are remarkably varied, certainly reflecting differences in development and genetic control. Botanists have tended to link the evolution of floral symmetry with pollination syndromes (Coccuci, 1999), but it is clear that there is a strong phylogenetic component to the evolution of floral form. This is not to negate the importance of pollination as a selective force in natural populations, but it is important to seek integrated, rather than single character, explanations.

14.5.1 Zygomorphy and pollination

Flower symmetry has long played an important role in the study of pollination ecology. Regularity has been presumed to predominate unless irregularity (i.e. zygomorphy) proves adaptive (Guirfa *et al.*, 1999), and pollination ecologists continue to search for the advantages of irregularity. The concept of pollination syndromes (Faegri and van der Pijl, 1979) has dominated pollination ecology, with pollinators being predicted on the basis of floral form, opening time or reward (Cocucci, 1999; Ollerton and Watts, 2000). Many workers attempt to categorise flowers according to their perceived syndrome, often in the absence of information on actual pollinators. For example, the discussion of butterfly 'pollination' in *Browallia* (Cocucci, 1995) was framed entirely on the basis of experiments with artificial butterfly tongues rather than on observations on animals and plants in the field. Field observations often reveal that the commonest or most obvious visitors to a plant are not the actual pollinators, and data of this sort are crucial to the eventual understanding of how pollinators affect the evolution of floral form.

It is evident that if flower symmetry is influenced by pollinators, pollinators must be able to perceive it. This has only been shown to be the case recently (Gould, 1985; Horridge, 1996), and most work has involved a single species, the honeybee (*Apis mellifera*). Bees can clearly distinguish between radially and bilaterally symmetric patterns (Horridge, 1996) and can be trained to respond to a particular symmetry type (Guirfa *et al.*, 1999; Glover and Martin, 2002). Bumblebees prefer symmetric (including bilaterally symmetric) flowers over asymmetric ones (Møller, 1995). If pollinators discriminate against asymmetric flowers, selection for pollination efficiency will have an element of frequency dependence (Møller and Eriksson, 1994). Many groups of insects have been shown to have learning abilities; honeybees can be trained to visit certain shapes (Guirfa *et al.*, 1999) and can remember generalised flower shapes (Gould, 1985), butterflies that visit the same plants day after day learn to avoid localities where they have been caught (Mallet *et al.*, 1987), and hawkmoths develop a repertoire of flower preferences based on innate preferences for certain floral shapes (Kelber, 1997).

Vertebrate pollinators, such as hummingbirds, are known to 'trapline' flowers on a regular basis (Stiles, 1985), and bats re-visit resource-rich sites (Heithaus *et al.*, 1974; Dinerstein, 1986). Flower symmetry differences are probably also perceived by vertebrate pollinators, but it is unclear whether shape or colour is the more important factor (Stiles, 1985). The mechanisms by which insects perceive form (Dafni *et al.*, 1997; Kelber, 1997) and symmetry (Horridge, 1996) will clearly have selective effects on flowers through differences in pollination efficiency. The phylogenetic basis for patterns in pollination systems has only been investigated in a few groups; a common pattern appears to be that pollination systems are quite labile and probably evolve quite rapidly (Ollerton, 1996). Different flower visitors may each contribute to selection on floral traits, and floral features that seem to be an evolutionary response to one pollinator (e.g. butterflies or bees) may in fact reflect a more complex and diverse pollination history (Waser *et al.*, 1996).

14.5.2 Zygomorphy and molecular developmental genetics

The isolation of numerous genes controlling development and morphology from model organisms such as *Arabidopsis* and *Antirrhinum* L. has revolutionised our understanding of plant development. Both *Arabidopsis* and *Antirrhinum*, though apparently rather different on a superficial morphological level, are monopodial herbaceous plants with a single terminal, indeterminate inflorescence. Members of the Solanaceae all have sympodial growth, with terminal, determinate inflorescences and continued shoot growth from axillary meristems, giving branches and shoots a characteristic zig-zag appearance (Danert, 1958; Silvy, 1974; Child, 1979). From work done using the tomato (*Solanum lycopersicum* = *Lycopersicon esculentum* Mill.), tobacco (*Nicotiana tabacum*) and the garden petunia (*Petunia hybrida* (Hook.) Vilm.), it is clear that genes such as *FALSIFLORA* (Coen and Nugent, 1994: Allen and Sussex, 1996; Molinero-Rosales *et al.*, 1999), *SELFPRUNING* (Pnueli *et al.*, 1998) and *CET* (Amaya *et al.*, 1999) function similarly to genes known from *Arabidopsis* (*TERMINALFLOWER1* and *LEAFY*) and *Antirrhinum* (*CENTRORADIALIS* and *FLORICAULA*) in the transition of vegetative to reproductive shoots (Table 14.3).

Table 14.3 Some key developmental genes identified in members of the Solanaceae that control flower and inflorescence development

Gene	Genus	Reference	Antirrhinum, Arabidopsis 'equivalent'
FALSIFLORA	Solanum (Lycopersicon)	Allen and Sussex, 1996; Molinero-Rosales et al., 1999	FLORICAULA, LEAFY
SP (SELFPRUNING)	Solanum (Lycopersicon)	Pnueli et al., 1998	CEN, TFL1
NFL	Nicotiana	Kelly et al., 1995	FLORICAULA, LEAFY
CET (CEN-like in tobacco)	Nicotiana	Amaya et al., 1999	CEN, TFL1

Flower primordia in both *Solanum* and *Petunia* are produced from inflorescence meristems by bifurcation (Allen and Sussex, 1996; Souer *et al.*, 1998), rather than initiated as lateral primordia on the flanks of the meristem as in both *Arabidopsis* and *Antirrhinum*. This is a reflection of the cymose rather than racemose nature of the solanaceous inflorescence, where the axis of the inflorescence, like that of the main shoot, is sympodial (Weberling, 1989). The simple differentiation between racemes, with persistent inflorescence meristems giving rise to the entire inflorescence, and cymes, with axillary meristems responsible for continued inflorescence growth, breaks down somewhat with detailed developmental study (Allen and Sussex, 1996; Souer *et al.*, 1998). The gene required for the bifurcation of the inflorescence meristem in *Petunia*, *EXP* (*EXTRAPETALS*: Souer *et al.*, 1998), does not affect primordium identity, while another gene, *ALF* (*ABERRANT LEAF AND FLOWER*, a homolog of the *Arabidopsis* gene *LEAFY*: Souer *et al.*, 1998), is required for the determination of meristem identity. Although the development of inflorescences and inflorescence types may not appear to be related to floral zygomorphy, it has been suggested that there is a correlation between inflorescence type and flower symmetry (Coen and Nugent, 1994), with taxa bearing indeterminate, racemose inflorescences also having irregular flowers. Mutants of *Antirrhinum* provide strong evidence for developmental constraints to the linking of inflorescence and flower morphology (Coen and Nugent, 1994).

The fundamental symmetry of Solanaceae flowers, reflecting the oblique orientation of the gynoecium, must be established early in floral meristem development. The degree to which this is structurally or genetically determined has not been investigated, nor have any of the studies in tomatoes, tobacco or petunias specifically mentioned this important characteristic of solanaceous flowers. It may be that the bifurcation of inflorescence meristems to produce a flower primordium and a continuation meristem may introduce structural constraints that force the oblique orientation of flowers. Investigation of this may not necessarily require genetic studies, but rather simple developmental analysis (Tucker, 1999). Comparative studies of *Nicandra*, which lacks the oblique gynoecium (Robyns, 1931), and related genera such as *Solanum* (Olmstead *et al.*, 1999) will be particularly interesting.

14.5.3 Floral symmetry genes

The ultimate shape and size of flowers is determined by differential growth of organs in the various floral whorls. Organs are either all the same within a whorl, generating radial symmetry, or differ in size or shape, generating zygomorphy or asymmetry. Dorsiventral asymmetry is thought to have arisen many times in angiosperms from a radially symmetrical ancestral condition (Stebbins, 1974). Genes involved in the development of dorsiventral asymmetry have been extensively studied in the Scrophulariaceae, especially in *Antirrhinum majus* L. (Coen and Nugent, 1994; Luo *et al.*, 1995, 1999; Almeida *et al.*, 1997). The existence of peloric mutants, with radially symmetrical flowers, has facilitated the study of the genetic control of floral symmetry (Cubas, 2002). These peloric mutants lack genetic functions normally active in the lateral and adaxial regions of the flower and have a completely ventralised phenotype (Luo *et al.*, 1995). Peloric mutants are known in

Scrophulariaceae, Gesneriaceae (Coen and Nugent, 1994; Harrison *et al.*, 1999; Citerne *et al.*, 2000) and Orchidaceae (Bateman and DiMichele, 2002).

In *Antirrhinum*, several genes controlling floral symmetry have been isolated, all of these affecting growth rate and primordium initiation. Organ asymmetry is progressively built upon through flower development through the interaction of genes active in subdomains (Luo *et al.*, 1999). The principal genes involved in the establishment of floral symmetry in *Antirrhinum* are the TCP genes *CYCLOIDEA* (*CYC*) and *DICHOTOMA* (*DICH*), working in the adaxial regions of the flower (Luo *et al.*, 1995, 1999; Cubas, 2002), and *DIVARICATA* (*DIV*), which acts in ventral regions and in a dosage dependent manner (Almeida *et al.*, 1997). Related TCP genes have been found in other families such as the Asteraceae (*Senecio* L., Gillies *et al.*, 2002) and the Gesneriaceae (Citerne *et al.*, 2000). TCP genes have also been found in non-zygomorphic families such as Brassicaceae (*Arabidopsis*: Cubas, 2002). The degree to which dorsiventral asymmetry is derived in angiosperms can be investigated using *CYC* homologues in distantly related taxa (Coen and Nugent, 1994). It is clear that TCP genes exist in families outside the Lamiales, thus providing support for their ancient role in vegetative and floral development. Whether asymmetric expression of TCP genes was present in a common ancestor of the eudicots (*sensu* Savolainen *et al.*, 2000) or independently evolved several times remains to be tested.

14.5.4 Phylogenetic perspectives on floral zygomorphy in Solanaceae

The Scrophulariaceae, Gesneriaceae and Solanaceae all belong to the large and diverse clade Asteridae, where Donoghue *et al.* (1998) have suggested that zygomorphy has evolved independently at least eight times, assuming equal probability of transformation from one state to another. Members of the Lamiales (including Scrophulariaceae) are thought to be ancestrally zygomorphic, with actinomorphic taxa evolving secondarily (Endress, 1999), especially in Gesneriaceae, the basal family in the clade. Authors have assumed that the Solanaceae have actinomorphic flowers, with *Schizanthus* the notable exception (Coen and Nugent, 1994; Endress, 1999; Tucker, 1999). To a certain extent, this is an unconscious application of the 'common is primitive' concept, as the large and extremely diverse genus *Solanum* (Table 14.1) is largely (but not always!) actinomorphic. *Solanum* has extensively and recently radiated (Hawkes and Smith, 1965; Olmstead and Palmer, 1997; Bohs, in press) and the assumption that the majority of the rest of the family has flowers like most solanums has no basis in fact.

The 2/3 ground-plan common to members of the Scrophulariaceae and Gesneriaceae (Lamiales) is considered by Donoghue *et al.* (1998) to be a nearly universal feature of the Asteridae, and exceptions to this are rare and confined to a few genera such as *Rhododendron* L. (Ericaceae), which has a 3/2 pattern like that found in the Solanaceae. More important than the numbers of adaxial or abaxial petals to the pattern of floral development is whether or not the axis of the flower passes through the adaxial (Scrophulariaceae, Gesneriaceae, Lamiaceae) or abaxial (Solanaceae) stamen.

The production of *CYC*-dependent asymmetry seems to depend upon the positional relationship between the inflorescence meristem and the lateral floral meristem

(Coen and Nugent, 1994). In flowers of Solanaceae, the dorsiventral asymmetry is quite different from that normally found in the Scrophulariaceae (see Figure 14.1 and discussion above; also Reeves and Olmstead, 1998). Rather than exhibiting suppression of the adaxial anther, as in *Antirrhinum*, species of Solanaceae such as *Leptoglossis*, *Browallia* and *Schizanthus* have the abaxial anther suppressed (see Section 14.3) and the abaxial petals usually enlarged (Figure 14.14). In *Browallia speciosa* Hook.f. (Robyns, 1931), the abaxial stamen is suppressed about midway through bud development, at about the same time as the differentiation of the two carpels of the gynoecium, when the bud is approximately 2 mm in diameter (Robyns, 1931, plate 6). If dorsiventral asymmetry is *CYC*-dependent in these taxa, it is clearly controlled differently. In *Antirrhinum*, *CYC* expression in adaxial regions is thought to be influenced by the proximity of the inflorescence meristem, but in Solanaceae the area morphologically similar to the *CYC*-controlled region in the snapdragon flower is abaxial. However, if the inflorescence meristem in all Solanaceae is bifurcated at the initiation of each floral primordium, the spatial arrangement of primordia may be involved in a distinctly different way. In leaves, the dorsal–ventral (or abaxial–adaxial) polarity is established very early in the development, and it has been suggested that cells of the apical meristem promote adaxial cell fate (Bowman, 2000).

14.6 Conclusions

The family Solanaceae has clear potential for investigation of some of these important phenomena. The sister group of the Solanaceae is the family Convolvulaceae (Savolainen *et al.*, 2000), whose members in general have radially symmetric flowers. Genera with zygomorphic flowers occur in clusters at basal nodes in the cpDNA phylogeny of the Solanaceae, however, suggesting that there may be a clear developmental genetic component to flower morphology in the family, rather than a purely selective set of constraints on morphology. Reeves and Olmstead (1998) have suggested that it is not necessary to work at higher taxonomic levels in order to study some of the major types of morphological changes that characterise angiosperm evolution. Studies conducted in a modern phylogenetic context using closely related taxa differing in key morphological traits of interest will suffer fewer problems with the identification of gene homologies, and may allow us to link more effectively the recent advances in flower development with what we know about the amazing diversity of flower form.

Molecular work on *Schizanthus*, the most zygomorphic genus in the family, has revealed the presence of a member of the *CYCLOIDEA* gene family (TCP genes: Coen and Cronk, pers. comm. 2001), but other genera in the family have not been assayed, and studies of the pattern of expression in primordia of *Schizanthus* are ongoing. Botanists have presumed that *Schizanthus* was unusual in the family in having zygomorphic flowers (Coen and Nugent, 1994; Tucker, 1999). From the above review, it is clear that *Schizanthus*, although extreme, is clearly not unusual. Genera such as *Leptoglossis*, where great variation in the expression and form of floral zygomorphy occurs, will be interesting candidates for future study. It may be that simple switching of TCP genes can explain the variation in floral symmetry such as that found in the traditional Salpiglossideae (see Section 14.3.3). The scattered occurrences of zygomorphy in the rest of the family, especially in *Solanum*, tend to involve mainly the androecium. Genes affecting the androecium, such as *CYC* and its homologues, clearly could play a role in these taxa. Enantiostyly, like that found in *Solanum*, occurs in *Saintpaulia* Wendland (Gesneriaceae: Harrison *et al.*, 1999), where *CYC*-like genes have been found but expression patterns have not yet been tested (Citerne *et al.*, 2000). It will be important to effectively assess the scope and morphology of zygomorphy in *Solanum* and to then carefully choose taxa in which to initiate genetic studies. Tomatoes and their relatives, already well known genetically (e.g. Pnueli *et al.*, 1998), may be a good place to begin to unravel floral development in the remarkably diverse genus *Solanum*.

Except for a few well-studied species, symmetry in flowers is largely unexplored territory. Future studies will need to link people and disciplines, and should involve truly collaborative partnerships. Molecular developmental genetics, with its ability to unravel the detailed workings of how organisms make themselves (Coen, 1999), detailed studies of phylogeny in specific groups, where relationships can be postulated and hypotheses proposed, and studies of plants in natural habitats, where selection and evolution really happen, are all critically important in order to truly understand the diversity of floral variation and the evolution of plant form.

ACKNOWLEDGEMENTS

I thank Quentin Cronk and Richard Bateman for inviting me to the DGPE meeting, and also for encouraging me to link my morphological knowledge of Solanaceae to recent developments in developmental genetics; several helpful reviewers for critical comments on the emerging manuscript; the Photographic Unit at the Natural History Museum, especially Derek Adams, for help in preparing the photographic plates; James Mallet for preparing Figure 14.5; Phil Rye for preparing the floral diagrams; and Jim Mallet, Jeff Ollerton, Paula Rudall, Michael Nee and Lynn Bohs for helpful discussions about Solanaceae. Michael Whalen and Tim Plowman inspired a generation of solanologists to think broadly about evolution in the family; they are sorely missed.

REFERENCES

Allen, K. D. and Sussex, I. M. (1996) *Falsiflora* and *anantha* control early stages of floral meristem development in tomato (*Lycopersicon esculentum* Mill.). *Planta*, 200, 254–264.
Almeida, J., Rocheta, M. and Galego, L. (1997) Genetic control of flower shape in *Antirrhinum majus*. *Development*, 124, 1387–1392.
Amaya, I., Ratcliffe, O. J. and Bradley, D. J. (1999) Expression of CENTRORADIALIS (CEN) and CEN-like genes in tobacco reveals a conserved mechanism controlling phase change in diverse species. *Plant Cell*, 11, 1405–1417.
Ampornan, L. and Armstrong, J. E. (1988) The floral ontogeny of *Schizanthus*, a zygomorphic member of the Solanaceae. *American Journal of Botany*, 75, suppl., 54.
Ampornan, L. and Armstrong, J. E. (1989) The floral ontogeny of *Salpiglossis*, a zygomorphic member of the Solanaceae. *American Journal of Botany*, 76, suppl., 64.

Ampornan, L. and Armstrong, J. E. (1990) The floral ontogeny of *Schwenkia* (Solanaceae). *American Journal of Botany*, 77, suppl., 168.

Ampornan, L. and Armstrong, J. E. (1991) In quest of the oblique ovary in Solanaceae, an adventure in floral development. *American Journal of Botany*, 78, suppl., 164.

Ando, T., Ueda, Y. and Hashimoto, G. (1992) Historical survey and present status of systematics in the genus *Petunia* Jussieu (Solanaceae). *Technical Bulletin of the Faculty of Horticulture, Chiba University*, 45, 17–26.

Arabidopsis Genome Initiative (AGI) (2000) Analysis of the genome sequence of the flowering plant *Arabidopsis thaliana*. *Nature*, 408, 796–815.

Baehni, C. (1915) L'ouverture du bouton chez les fleurs de Solanées. *Candollea*, 10, 399–492.

Baillon, H. E. (1887) Sur quelques types de groups intermediares aux Solanacées et aux Scrofulariacées. *Bulletin Mensual de la Société Linnéene de Paris*, 83, 660–663.

Baillon, H. E. (1888) *Histoire des plantes 9*, 281–359. L. Hachette & Cie., Paris.

Bartling, F. T. (1830) *Ordines Naturales Plantarum Eorumque Characteres et Affinates Adjecta Generum Enumeratione*. Dieterich, Göttingen.

Bateman, R. M. and DiMichele, W. A. (2002) Generating and filtering major phenotypic novelties: neoGoldschmidtian saltation revisited, in *Developmental Genetics and Plant Evolution* (eds Q. C. B. Cronk, R. M. Bateman and J. A. Hawkins), Taylor & Francis, London, pp. 109–159.

Bentham, G. (1835) Revision of Scrophulariaceae: trib. Salpiglossidae. *Botanical Register*, 21, t. 1770.

Bentham, G. and Hooker, J. D. (1876) *Genera Plantarum 2*, 882–980, 1244–1245.

Bitter, G. (1919) Die Gattung *Lycianthes*. *Abhandlungen herausgeben von Naturwissenschaften Verein zu Bremen*, 24, 292–520.

Bohs, L. (in press) Major clades in *Solanum* based on *ndhF* sequence analyses. *Monographs in Systematic Botany from the Missouri Botanical Garden*.

Bohs, L. and Olmstead, R. G. (2001) A reassessment of *Normania* and *Triguera* (Solanaceae). *Plant Systematics and Evolution*, 228, 33–48.

Bowers, K. A. W. (1975) The pollination ecology of *Solanum rostratum*. *American Journal of Botany*, 62, 633–638.

Bowman, J. L. (2000) Axial patterning in leaves and other lateral organs. *Current Opinion in Genetics and Development*, 10, 399–404.

Brogniart, A. (1843) *Enumeration des Genres des Plantes Cultivés au Muséum d'Histoire Naturelle de Paris*, 1–136. Paris.

Carvahlo, L. d'A Freire de (1978) O gênero *Schwenkia* D. Van Rooyen ex Linnaeus no Brasil – Solanaceae. *Rodriguésia*, 44, 307–524.

Charlton, W. A. (1998) Pendulum symmetry, in *Symmetry in Plants* (eds R. V. Jean and D. Barabè), World Scientific, River Edge, New Jersey, pp. 61–88.

Chase, M. W., Cox, A. V., Knapp, S., Komarnitsky, I., Komarnitsky, S., Butsko, Y., Marshall, J. A., Joseph, J., Savolainen, V. and Parokonny, A. S. (subm.) Molecular systematics, GISH and the origin of hybrid taxa in *Nicotiana* (Solanaceae).

Child, A. (1979) A review of branching patterns in the Solanaceae, in *The Biology and Taxonomy of the Solanaceae* (eds J. G. Hawkes, R. N. Lester and A. D. Skelding), Academic Press, London, pp. 345–356.

Citerne, H. L., Moller, M. and Cronk, Q. C. B. (2000) Diversity of *cycloidea*-like genes in Gesneriaceae in relation to floral symmetry. *Annals of Botany*, 86, 1667–1676.

Clos, D. (1849) Escrofularíneas, in *Historia Fisica y Politica de Chile, Bot. 5*, 100–188 (ed. C. Gay). Privately published by C. Gay, Paris.

Cocucci, A. (1989a) El mechanismo floral en *Schizanthus* (Solanaceae). *Kurtziana*, 20, 113–132.

Cocucci, A. (1989b) Sobre el diagrama floral de *Schizanthus* (Solanaceae) y su interpretación. *Kurtziana*, 20, 133–137.

Cocucci, A. (1991) The floral biology of *Nierembergia* (Solanaceae). *Plant Systematics and Evolution*, 174, 17–35.

Cocucci, A. (1995) Floral mechanisms in the tribe Salpiglossidae (Solanaceae). *Plant Systematics and Evolution*, 194, 207–230.

Cocucci, A. (1999) Evolutionary radiation in Neotropical Solanaceae, in *Solanaceae IV: Advances in Biology and Utilization* (eds M. Nee, D. E. Symon, R. N. Lester and J. P. Jessop), Royal Botanic Gardens Kew, Richmond, Surrey, pp. 9–22.

Coen, E. S. (1999) *The Art of Genes: How Organisms Make Themselves.* Oxford University Press, Oxford.

Coen, E. S. and Nugent, J. M. (1994) Evolution of flowers and inflorescences. *Development Supplement*, 1994, 107–116.

Cronquist, A. (1981) *An Integrated System of Classification of Flowering Plants.* Columbia University Press, New York.

Cubas, P. (2002) The role of TCP genes in the evolution of key morphological characters in angiosperms, in *Developmental Genetics and Plant Evolution* (eds Q. C. B. Cronk, R. M. Bateman and J. A. Hawkins), Taylor & Francis, London, pp. 247–266.

Dafni, A., Lehrer, M. and Kevan, P. G. (1997) Spatial flower parameters and insect spatial vision. *Biological Reviews*, 72, 239–282.

Danert, S. (1958) Die Verzweigung der Solanaceen im Reproduktiven Bereich. *Abhandlungen der Deutschen Akademie der Wissenschaften zu Berlin, Klasse fur Chemie, Geologie und Biologie*, 6, 1–183.

D'Arcy, W. G. (1978) A preliminary synopsis of *Salpiglossis* and other Cestreae (Solanaceae). *Annals of the Missouri Botanical Garden*, 65, 698–724.

D'Arcy, W. G. and Benítez, C. E. (1991) Biogeographical mapping: the Schwenkiae example, in *Solanaceae III: Taxonomy, Chemistry, Evolution* (eds J. G. Hawkes, R. N. Lester, M. Nee and E. Estrada), Royal Botanic Gardens Kew, London, pp. 169–179.

Diggle, P. K. (1991) Labile sex expression in andromonoecious *Solanum hirtum*: floral development and sex determination. *American Journal of Botany*, 78, 377–393.

Diggle, P. K. (1993) Developmental plasticity, genetic variation, and the evolution of andromonoecy in *Solanum hirtum* (Solanaceae). *American Journal of Botany*, 80, 967–973.

Diggle, P. K. (1994) The expression of andromonoecy in *Solanum hirtum* (Solanaceae): phenotypic plasticity and ontogenetic contingency. *American Journal of Botany*, 81, 1354–1365.

Dinerstein, E. (1986) Reproductive ecology of fruit bats and the seasonality of fruit production in a Costa Rican cloud forest. *Biotropica*, 18, 307–318.

Don, G. (1837) *A General System of Gardening and Botany*, vol. 4. J. G. & F. Rivington *et al.*, London.

Donoghue, M. J., Ree, R. H. and Baum, D. A. (1998) Phylogeny and evolution of flower symmetry in the Asteridae. *Trends in Plant Science*, 3, 311–317.

Eichler, A. W. (1869) Einige Bemerkungen über de Bau Cruciferen blüthe und das Dédoublement. *Flora*, 52, 102–109.

Endlicher, S. L. (1839) *Genera Plantarum.* Fr. Beck, Vienna.

Endress, P. K. (1999) Symmetry in flowers: diversity and evolution. *International Journal of Plant Science*, 160, S3–S23.

Estrada, E. and Martínez, M. (1999) *Physalis* L. (Solanoideae: Solaneae) and allied genera: I. A morphology-based cladistic analysis, in *Solanaceae IV: Advances in Biology and Utilization* (eds M. Nee, D. E. Symon, R. N. Lester and J. P. Jessop), Royal Botanic Gardens Kew, London, pp. 139–159.

Faegri, K. and van der Pijl, L. (1979) *The Principles of Pollination Ecology* (3rd edn). Cambridge University Press, Cambridge.

Fay, M. F., Olmstead, R. G., Richardson, J. E., Santiago, J. E., Prance, G. T. and Chase, M. W. (1998) Molecular data support the inclusion of *Duckeodendron cestroides* in Solanaceae. *Kew Bulletin*, 53, 203–212.

Gillies, A. C. M., Cubas, P., Coen, E. S. and Abbott, R. J. (2002) Making rays in the Asteraceae: genetics and evolution of radiate versus discoid flower heads, in *Developmental Genetics and Plant Evolution* (eds Q. C. B. Cronk, R. M. Bateman and J. A. Hawkins), Taylor & Francis, London, pp. 233–246.

Glover, B. J. and Martin, C. (2002) Evolution of adaptive petal cell morphology, in *Developmental Genetics and Plant Evolution* (eds Q. C. B. Cronk, R. M. Bateman and J. A. Hawkins), Taylor & Francis, London, pp. 160–172.

Goodspeed, T. H. (1954) The genus *Nicotiana*. *Chronica Botanica*, 16, 1–536.

Gould, J. L. (1985) How bees remember flower shapes. *Science*, 227, 1492–1494.

Greuter, W., McNeill, J., Barrie, F. R., Burdet, H. M., Demoulin, V., Filgueiras, T. S., Nicolson, D. H., Silva, P. C., Skog, J. E., Trehane, P., Turland, N. J. and Hawksworth, D. L. (members of the editorial committee). (2000) International Code of Botanical Nomenclature. *Regnum vegetabile* 138. Koeltz Scientific Books, Königstein.

Guirfa, M., Dafni, A. and Neal, P. R. (1999) Floral symmetry and its role in plant–pollinator systems. *International Journal of Plant Science*, 160, S41–S50.

Haegi, L. (1981) A conspectus of Solanaceae tribe Anthocercideae. *Telopea*, 2, 173–180.

Harrison, J., Möller, M. and Cronk, Q. C. B. (1999) Evolution and development of floral diversity in *Streptocarpus* and *Saintpaulia*. *Annals of Botany*, 84, 49–60.

Hawkes, J. G. and Smith, P. (1965) Continental drift and the age of angiosperm genera. *Nature*, 207, 48–50.

Heithaus, E. P., Opler, P. A. and Baker, H. G. (1974) Bat activity and the pollination of *Bauhinia pauletia*: plant–pollinator coevolution. *Ecology*, 55, 412–419.

Hoare, A. L. and Knapp, S. (1997) A phylogenetic conspectus of the tribe Hyoscyameae: Solanaceae. *Bulletin of the Natural History Museum, London (Botany)*, 27, 11–29.

Hooker, W. J. (1827) *Exotic Flora*, vol. 3. T. Cadell, London.

Horridge, G. A. (1996) The honeybee (*Apis mellifera*) detects bilateral symmetry and discriminates its axis. *Journal of Insect Physiology*, 42, 755–764.

Huber, K. A. (1980) Morphologische und entwicklungsgeschichtliche Untersuchungen an Blüten und Blütenständen von Solanaceen und von *Nolana paradoxa* Lindl. (Nolanaceae). *Dissertationes Botanicae*, 55, 1–252, 696 figs.

Hunziker, A. T. (1979) South American Solanaceae: a synoptic survey, in *The Biology and Taxonomy of the Solanaceae* (eds J. G. Hawkes, R. N. Lester and A. D. Skelding), Academic Press, London, pp. 49–86.

Hunziker, A. T. and Subils, R. (1979) *Salpiglossis*, *Leptoglossis* and *Reyesia* (Solanaceae), a synoptical survey. *Botanical Museum Leaflets*, 27, 1–43.

Kelber, A. (1997) Innate preferences for flower features in the hawkmoth *Macroglossum stellatarum*. *Journal of Experimental Biology*, 200, 827–836.

Knapp, S. (2000) Revision of the *Solanum thelopodium* species group (section *Anthoresis sensu* Seithe, *pro parte*): Solanaceae. *Bulletin of the Natural History Museum, London (Botany)*, 30, 13–30.

Knapp, S. (2001) Is morphology dead in *Solanum* taxonomy?, in *Solanaceae V* (eds G. van der Werden, G. Barendse and R. van den Berg), University of Nijmegen, Nijmegen, 23–38.

Knapp, S., Persson, V. and Blackmore, S. (1997) A phylogenetic conspectus of the tribe Juanulloeae (Solanaceae). *Annals of the Missouri Botanical Garden*, 84, 67–89.

Knapp, S., Stafford, P. and Persson, V. (2000) Pollen morphology in the Anthocercideae (Solanaceae). *Kurtziana*, 28, 7–18.

Lester, R. N., Francisco-Ortega, J. and Al-Ani, M. (1999) Convergent evolution of heterandry (unequal stamens) in *Solanum*, proved by spermoderm SEM, in *Solanaceae IV: Advances in Biology and Utilization* (eds M. Nee, D. E. Symon, R. N. Lester and J. P. Jessop), Royal Botanic Gardens Kew, London, pp. 51–69.

Luo, D., Carpenter, R., Copsey, L., Vincent, C., Clark, J. and Coen, E. (1999) Control of organ asymmetry in flowers of *Antirrhinum*. *Cell*, 99, 367–376.

Luo, D., Carpenter, R., Vincent, C., Copsey, L. and Coen, E. (1995) Origin of floral asymmetry in *Antirrhinum*. *Nature*, 383, 794–799.

Mallet, J., Longino, J. T., Murawski, D., Murawski, A. and Simpson de Gamboa, A. (1987) Handling effects in *Heliconius*: where do all the butterflies go? *Journal of Animal Ecology*, 56, 377–386.

Miers, J. (1849) Observations upon several genera hitherto placed in Solanaceae, and upon others intermediate between that family and the Scrophulariaceae. *Annals and Magazine of Natural History*, series 2, 15, 161–183.

Molinero-Rosales, N., Jamilena, M., Zurita, S., Gómez, P., Capel, J. and Lozano, R. (1999) FALSIFLORA, the tomato orthologue of FLORICAULA and LEAFY, controls flowering time and floral meristem identity. *Plant Journal*, 20, 685–693.

Møller, A. P. (1995) Bumblebee preference for symmetric flowers. *Proceedings of the National Academy of Sciences, USA*, 92, 2288–2292.

Møller, A. P. and Eriksson, M. (1994) Patterns of fluctuating asymmetry in flowers: implications for sexual selection in plants. *Journal of Evolutionary Biology*, 7, 93–113.

Ollerton, J. (1996) Reconciling ecological processes with phylogenetic patterns: the apparent paradox of plant–pollinator systems. *Journal of Ecology*, 84, 767–769.

Ollerton, J. and Watts, S. (2000) Phenotype space and floral typology: towards an objective assessment of pollination syndromes. [Scandinavian Association for Pollination Ecology honour Knut Faegri]. *Det Norske Videnskaps-Akademi, I. Matematisk-Naturvidenskapelige Klasse, Avhandlinger, Ny Serie*, 39, 149–159.

Olmstead, R. G. and Palmer, J. D. (1992) A chloroplast DNA phylogeny of the Solanaceae. *Annals of the Missouri Botanical Garden*, 79, 346–360.

Olmstead, R. G. and Palmer, J. D. (1997) Implications for the phylogeny, classification, and biogeography of *Solanum* from cpDNA restriction site variation. *Systematic Botany*, 22, 19–29.

Olmstead, R. G., Sweere, J. A., Spangler, R. A., Bohs, L. and Palmer, J. D. (1999) Phylogeny and provisional classification of the Solanaceae based on chloroplast DNA, in *Solanaceae IV: Advances in Biology and Utilization* (eds M. Nee, D. E. Symon, R. N. Lester and J. P. Jessop), Royal Botanic Gardens Kew, London, pp. 111–137.

Plowman, T. [Knapp, S. and Press, J. R., eds.] (1998) A revision of the South American species of *Brunfelsia* (Solanaceae). *Fieldiana, Botany n.s.*, 39, 1–135.

Pnueli, L., Carmel-Goren, L., Haraven, D., Gutfinger, T., Alvarez, J., Ganal, M., Zamir, D. and Lifschitz, E. (1998) The *SELF-PRUNING* gene of tomato regulates vegetative to reproductive switching of sympodial meristems and is the ortholog of *CEN* and *TFL1*. *Development*, 125, 1979–1989.

Reeves, P. A. and Olmstead, R. G. (1998) Evolution of novel morphological and reproductive traits in a clade containing *Antirrhinum majus* (Scrophulariaceae). *American Journal of Botany*, 85, 1047–1056.

Robyns, W. (1931) L'organisation florale des Solanacées zygomorphes. *Memoires de l'Academie Royale de Belgique, Classe des Sciences*, 11, 1–84.

Savolainen, V., Fay, M. F., Albach, D. C., Backlund, A., van den Bank, M., Cameron, K. M., Johnson, S. A., Lledo, M. D., Pintaud, J.-C., Powell, M., Sheahan, M. C., Soltis, D. E., Soltis, P. S., Weston, P., Whitten, W. M., Wurdack, K. J. and Chase, M. W. (2000) Phylogeny of the eudicots: a nearly complete familial analysis based on *rbcL* gene sequences. *Kew Bulletin*, 55, 257–309.

Silvy, A. (1974) Etude des modes de ramification sympodial chez *Lycopersicon esculentum* et *L. pimpinellifolium*. *Canadian Journal of Botany*, 52, 2207–2218.

Souer, E., van der Krol, A., Kloos, D., Spelt, C., Bleik, M., Mol, J. and Koes, R. (1998) Genetic control of branching pattern and floral identity during *Petunia* inflorescence development. *Development*, 125, 733–742.

Stebbins, G. L. (1974) *Flowering Plants: Evolution Above the Species Level*. Belknap, Boston.

Stiles, F. G. (1985) Seasonal patterns and coevolution in the hummingbird-flower community of a Costa Rican subtropical forest, in *Neotropical Ornithology* (eds P. A. Buckley, M. S. Foster, E. S. Morton, R. S. Ridgely and F. G. Buckley), Ornithological Monographs 36. American Ornithologists' Union, Washington, DC, pp. 757–787.

Sweet, R. (1830) *Hortus Britannicus* (2nd edn). James Ridgway, London.

Thorne, R. T. (1992) Classification and geography of the flowering plants. *Botanical Review*, 58, 225–348.

Tucker, S. C. (1999) Evolutionary lability of symmetry in early floral development. *International Journal of Plant Sciences*, 160, S25–S39.

Waser, N. M., Chittka, L., Price, M. V., Williams, N. M. and Ollerton, J. (1996) Generalization in pollination systems, and why it matters. *Ecology*, 77, 1043–1060.

Weberling, F. (1989) *Morphology of Flowers and Inflorescences* (English version translated by R. J. Pankhurst, first published in German as *Morphologie der Blüten und der Blütenstände* in 1981). Cambridge University Press, Cambridge.

Whalen, M. D. (1979) Taxonomy of *Solanum* section *Androceras*. *Gentes Herbarum*, 11, 359–426.

Whalen, M. D. (1984) Conspectus of species groups in *Solanum* subgenus *Leptostemonum*. *Gentes Herbarum*, 12, 179–282.

Whalen, M. D., Costich, D. E. and Heiser, C. B. (1981) Taxonomy of *Solanum* section *Lasiocarpa*. *Gentes Herbarum*, 12, 41–129.

Wijsman, H. J. W. (1990) On the interrelationships of certain species of *Petunia* VI. New names for the species of *Calibrachoa* formerly included in *Petunia* (Solanaceae). *Acta Botanica Neerlandica*, 39, 101–102.

Wijsman, H. J. W. and de Jong, J. H. (1985) On the interrelationships of certain species of *Petunia* IV. Hybridization between *P. linearis* and *P. calycina* and nomenclature consequences in the *Petunia* group. *Acta Botanica Neerlandica*, 34, 337–349.

Wydler, H. (1866) Morphologische Mitteilungen. *Flora*, 49, 513–525.

Chapter 15

Integrating phylogeny, developmental morphology and genetics: a case study of inflorescence evolution in the 'bristle grass' clade (Panicoideae: Poaceae)

Andrew N. Doust and Elizabeth A. Kellogg

ABSTRACT

Our studies in the grass family (Poaceae) have concentrated on understanding diversity in inflorescence morphology. We have focused on analysis of inflorescence evolution in the panicoid 'bristle grass' clade (including *Setaria*, *Pennisetum* and *Cenchrus*) as an example of the way in which molecular phylogenetic hypotheses can be combined with developmental and genetic data to understand morphological evolution. Analyses of developmental morphology with phylogenies derived from molecular data sets have enabled us to identify a small number of parameters that can control morphological diversification. These include numbers of orders of branching, numbers of primordia produced on each branch, and the timing and amount of branch axis elongation. We predict that this small number of parameters will prove to be controlled by a correspondingly small number of genes, a prediction that we are testing by genetic studies of two species of *Setaria*.

15.1 Introduction

Our interest as evolutionary biologists is to understand and explain the evolution of morphological diversity in large clades. Methods of analysis have changed rapidly over the last decade as phylogenies have shifted from being primarily based on morphological data to being increasingly based on DNA sequence data (Soltis *et al.*, 1998). With this shift has come the opportunity to map morphological characters onto phylogenies constructed from independently derived molecular data sets. This has eliminated possible circularity arising from the use of morphological data both to construct trees and to analyse character evolution on the basis of those trees. A further advance is underway as detailed knowledge of the developmental genetics of model plant systems such as *Arabidopsis* and maize becomes available. Knowing how genes control morphological development can potentially help to understand how evolution of gene pathways can result in morphological differentiation. A crucial link between genetic studies in model systems and evolution of morphological diversity in large clades lies in the study of comparative developmental morphology. This is because developmental morphology is the direct outcome of spatial and

In *Developmental Genetics and Plant Evolution* (2002) (eds Q. C. B. Cronk, R. M. Bateman and J. A. Hawkins), Taylor & Francis, London, pp. 298–314.

temporal patterns of gene expression, and the evolution of developmental morphology in related taxa results in morphological diversification. Analysing changes in developmental morphology with molecular phylogenies adds a temporal component to the study of morphological evolution that is not available when only mature structures are analysed.

15.1.1 The grass inflorescence as a model system

We are particularly interested in the evolution of inflorescence morphology in the grass family (Poaceae) both because its members exhibit great diversity in inflorescence form and because inflorescence morphology is central to delimitation of grass taxa at all levels (Clayton and Renvoize, 1986; Watson and Dallwitz, 1992). The agronomic importance of crop species such as maize, rice and wheat has also meant that enormous efforts have been made to understand the genetic control of morphology in these species. Thus, information on genetic control of inflorescence morphology available for crop systems such as rice and maize may be useful in analysing the evolution of inflorescence morphology in other grass groups. In this chapter we stress how detailed comparative developmental morphology is critical in any attempt to understand how evolution of genetic pathways results in morphological diversification. We will also attempt to show how molecular phylogenies, developmental morphology and information on genetic control of morphology might be integrated into a more complete understanding of morphological evolution. As an example of such an approach we present preliminary data on the evolution of inflorescence development in the panicoid bristle grass ('bristle') clade, which includes *Setaria*, *Pennisetum* and *Cenchrus* (Gómez-Martinez and Culham, 2000; Zuloaga *et al.*, 2000).

One of us (Kellogg, 2000a) analysed inflorescence morphology in the panicoid tribe Andropogoneae and showed that inflorescence evolution was complex, and that most inflorescence characters exhibited numerous parallelisms and reversals. In this study and in subsequent work we have found it difficult to apply definitions of inflorescence characters consistently to such a large group of taxa (86 genera, *c.* 1100 species). Therefore we have focused on a smaller group, the bristle clade (in the tribe Paniceae, the sister tribe to the Andropogoneae), with 25 genera and 310 species, 110 of which are in the genus *Setaria*. This clade has inflorescences that range from elongate and diffuse to highly branched and condensed (Figure 15.1). The clade has been found to be monophyletic in the morphological analysis of Zuloaga *et al.* (2000), as well as in all molecular studies to date (Gómez-Martinez and Culham, 2000; Giussani *et al.*, 2001; Doust and Kellogg, unpubl. obs.), and includes all grass species in which some inflorescence branch meristems become sterile and are converted to setae or bristles. Somewhat surprisingly, the molecular phylogenies also include *Panicum bulbosum* in the bristle clade, a species that does not obviously have bristles in its inflorescence. The morphological synapomorphy of the inflorescence bristles was not recognised by previous workers, who placed the members of the clade in different subtribes (Clayton and Renvoize, 1986). The clade contains the crop species *Setaria italica* (foxtail millet) and *Pennisetum glaucum* (pearl millet), for which there are known mutants and genome maps (Koduru and Krishna Rao, 1983; Anand Kumar and Andrews, 1993; Devos *et al.*, 1998, 2000;

Figure 15.1 Inflorescence diversity in *Setaria*. (a) *S. palmifolia*, showing distal half of inflorescence with long secondary branches and elongated primary axis internodes. (b) *S. grisebachii*, with contracted secondary branches and a small amount of internode elongation on the primary axis. (c) *S. verticillata*, with contracted secondary branches and a condensed primary axis (sbc = secondary branch complex). Scale: (a)–(c) = 10 mm.

Wang *et al.*, 1998). We intend to use knowledge gained from analysis of the bristle clade to help understand inflorescence diversification in the grasses as a whole.

15.1.2 Methodological approach

Our methodology integrates (a) construction of molecular phylogenies, (b) detailed comparative studies of developmental morphology, (c) mapping of developmental characters onto the molecular phylogeny, and (d) genetic analysis of variation in interspecific crosses between morphologically diverse parents. In general, such an integrated approach has been uncommon, in part because of the difficulties in getting sufficient detailed information on a phylogenetically interesting array of taxa. However, the integration of genetic and morphological data in a phylogenetic context has been attempted by several authors, including Doebley and Stec (1991, 1993), who examined the evolution of maize from its teosinte ancestors, and Harrison *et al.* (1999), who examined the phylogenetic relationship between *Saintpaulia* and *Streptocarpus* (Gesneriaceae). Phylogenies have been used to analyse trends in developmental character evolution in a variety of contexts, such as centrarchid fishes (Mabee, 1993, 2000), floral evolution in *Besseya* (Scrophulariaceae) (Hufford, 1995), evolution of floral morphology in Proteaceae (Douglas, 1997), floral ontogeny in the Loasaceae (Moody and Hufford, 2000), endosperm development in basal flowering plants (Floyd and Friedman, 2000), and growth patterns in *Strepto-*

carpus (Gesneriaceae) (Möller and Cronk, 2001). Construction of molecular phylogenies and optimisation of morphological characters on those phylogenies have been discussed in detail elsewhere (Harvey and Pagel, 1991; Losos and Miles, 1994; Maddison, 1994; Swofford *et al.*, 1996; Cunningham *et al.*, 1998), and will not be discussed here.

In both developmental and morphological analysis, the method and criteria of character delimitation need to be explicitly stated (Stevens, 1991). An important initial step in developmental analysis is delimitation of the developmental pathway from which characters are to be extracted. Mabee (1993) observed that multiple ontogenetic pathways could be analysed in organisms, ranging from development of a single structure to that of the whole organism, and that comparisons should only be made between homologous pathways. However, recognition of homologous pathways in divergent taxa may be difficult because of heterochrony (evolutionary changes in the timing of developmental processes) or heterotopy (evolutionary changes in the position at which developmental processes occur) (Zelditch and Fink, 1996). In grasses, the beginning of the inflorescence developmental pathway is unequivocal as it is marked by the transformation of the vegetative meristem into an inflorescence meristem (Evans and Grover, 1940). In every grass we have examined, this change is accompanied by elongation of the meristem and diminution in the size of the initiated primordia. The end of inflorescence development is similarly clear, and can be considered as either the point at which all inflorescence axes have ceased meristematic activity, anthesis, or dispersal of the seed. Within the overall pattern of inflorescence development other developmental pathways can be distinguished, including inflorescence ramification and spikelet development.

Potential characters useful in understanding the evolution of development can be found at points in the developmental continuum where differences between taxa emerge. Character states are then those variants that can be consistently distinguished from each other; that is, where there is a discontinuity in character variation, such as when a structure is produced consistently earlier or later in one species as compared to others. In the case of the bristle clade, variation in character states was analysed both within the clade as well as in outgroup taxa, to make sure that discontinuities among character states were consistent.

Our ultimate aim is to identify the genes involved in morphological diversity and to elucidate how these genes and the pathways to which they belong have changed over evolutionary time. In the grass family the great number of morphological mutants found in maize allows the use of a 'candidate gene' approach, where mutant phenotypes can be compared to the morphology of other grass species. For example, the *tasselseed 4* mutant of maize produces an extensively ramified inflorescence strikingly similar to the inflorescence of *Setaria italica*. Where the phenotypes are similar it is a reasonable hypothesis that genes controlling the mutant phenotype may also control similar morphology in other grasses. Such an approach can be made more powerful by identifying major quantitative trait loci (QTLs) controlling inflorescence form, as we have done for a cross between *Setaria italica* and *S. viridis* (unpubl. obs.). Because of the high level of synteny between grass genomes, it may be possible to map these QTLs to the almost completely sequenced rice genome (Sasaki and Burr, 2000; Wolfe, 2000). It will then be possible to search these regions of the rice genome for candidate genes known to control inflorescence morphology.

Such genes could be responsible for patterns of morphological diversification in *Setaria* and perhaps in the bristle clade as a whole.

15.2 Materials and methods

15.2.1 Phylogenetic analysis

A preliminary analysis of the bristle clade included 29 taxa. DNA was extracted using the CTAB method as described in Giussani *et al.* (2001). The *trnL* intron was PCR-amplified using primer pairs C–F, designed by Taberlet *et al.* (1991). The *ndhF* gene was PCR-amplified in two overlapping fragments: 5F/1318R, 972F/2110R. Ten sequencing primers were used, nine as designed by Olmstead and Sweere (1994) and one (1821R) designed by Clark *et al.* (1995). PCR products were cleaned with the QIAquick PCR purification kit (Qiagen Inc., Valencia, CA), quantified by comparison with DNA of a known concentration (pGEM 10 and 25 ng, Applied Biosystems) and fluorescence-labelled using the 'Big Dye' (Applied Biosystems) cycle sequencing protocol. Both forward and reverse strands were sequenced on an ABI 377 automated sequencer. Contig assembly and editing of sequences used Sequencher, version 3.1 (Gene Codes Corporation, Ann Arbor, MI). All sequences were aligned in Clustal W (Thompson *et al.*, 1994) followed by manual adjustments in Se-Al version 1 (available via FTP from Andrew Rambaut, University of Oxford, at ftp://evolve.zo.ox.ac.uk/packages/Se-Al/). Sequence data are available on request from the authors.

Phylogenetic analyses were performed with a maximum parsimony algorithm, treating character states as unordered. Analyses were conducted using PAUP* version 4.0b4a (Swofford, 1999), with heuristic searches, TBR branch swapping, 100 random addition sequence replicates, and gaps treated as missing data. Full heuristic bootstrap analyses were conducted using 250 replicates, five random addition sequence replicates, and MAXTREES set to 10,000, with remaining parameters identical to those used in the parsimony analysis (Felsenstein, 1985). Incongruence between the two chloroplast data sets was tested using the incongruence length difference (ILD) test with a significance level of $p < 0.01$ (Farris *et al.*, 1994; Cunningham, 1997), implemented in PAUP*.

Optimisation of developmental characters on one of the most parsimonious trees used MacClade 4.0 (Maddison and Maddison, 2000). *Stenotaphrum* was omitted from the tree because the homology of its unusual inflorescence was unclear (see Figure 15.2). In those cases where character states were equivocal at a node all equally parsimonious states were analysed.

15.2.2 Morphological and developmental analysis

Most species of *Setaria*, *Pennisetum* and *Cenchrus* used in this study were obtained as seed from United States Department of Agriculture and grown in a glasshouse; inflorescences were harvested at different stages of development. Inflorescences of *Setaria palmifolia* were obtained from plants in the Missouri Botanical Garden. Inflorescences were dissected while fresh, and fixed either in PFA/Glutaraldehyde (phosphate buffered 4 per cent paraformaldehyde for twenty-four hours followed by phosphate buffered 4 per cent glutaraldehyde for two days) or FAA (formalin–acetic acid–70 per cent ethanol, 10:5:85 v/v). A proportion of the material was dehydrated using an ethanol series, critical-point dried in an SPI Jumbo critical point drier and sputter-coated with gold in a Polaron E5000 sputter coater. Specimens prepared in this manner showed significant charging of bristles in the electron beam and therefore the rest of the specimens were re-hydrated and post-fixed with osmium tetroxide, using the OTOTO method of Murphy (1978). This treatment ensured that all of the tissue of each specimen was fully electron-conductive. The OTOTO specimens also underwent dehydration and critical point drying, but were not sputter-coated with gold. Both types of specimens were imaged in either an Hitachi S450 or an Amray AMR1000 SEM at 20 kV. Developmental sequences were constructed for each taxon and characters delimited.

15.3 Results

15.3.1 Phylogeny

A preliminary phylogeny, based on a combination of data from *ndhF* and *trnL*, shows that the bristle clade is well supported (Figure 15.2). *Setaria* appears monophyletic in some of the most parsimonious trees, including the tree presented, but in the strict consensus there is no support for monophyly. However, two sub-clades of *Setaria* are monophyletic with high bootstrap support. Two species of *Setaria*, *S. grisebachii* and *S. lachnea*, form a clade in the tree presented but their position in the strict consensus tree is uncertain. *Pennisetum* is paraphyletic and forms a clade with *Cenchrus*. *Paspalidium*, *Stenotaphrum* and *Panicum bulbosum* are in an uncertain position relative to *Setaria*, *Pennisetum* and *Cenchrus*.

15.3.2 Development

Many aspects of development are the same for all species of the bristle clade investigated so far (mostly species of *Setaria*, *Pennisetum* and *Cenchrus*). In the transition to reproductive growth, the apical meristem elongates and may become broader in transverse section. This becomes the primary (main) axis of the inflorescence. The secondary branch primordia (secondary axes of the inflorescence) are initiated in a polystichous arrangement, with species differing in the number of secondary branches produced. Basal secondary branches grow and initiate tertiary branch primordia while the inflorescence meristem is still initiating secondary branch primordia distally. In most species the tertiary branch primordia are arranged distichously along the axis of each secondary branch, and quaternary branch primordia distichously along the axis of each tertiary branch (Figure 15.3a). The number of orders of branching varies, as does the number of primordia per order of branching. Branch primordia become either spikelets or bristles. The secondary branch, the higher orders of branching that it bears, and the spikelets and bristles that terminate those branches form a unit, the secondary branch complex (Figures 15.1a–c, 15.3b), whose structure is repeated throughout the inflorescence. Late in development, the internodes of the primary axis of the inflorescence elongate in some species, producing a lax inflorescence.

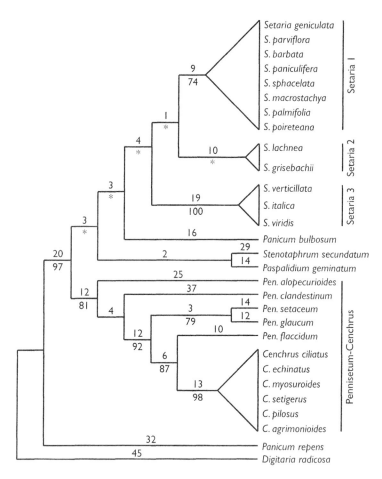

Figure 15.2 One of twenty-one most parsimonious trees (asterisks indicate nodes that collapse in the strict consensus tree) for a combined analysis of *ndhF* and *trnL* (data sets congruent at 0.01 significance level by ILD test). Length = 306, CI = 0.582, RI = 0.754, with uninformative characters excluded. Numbers above branches are branch lengths, those below branches are bootstrap values.

We have found so far only two exceptions to this developmental description. *Stenotaphrum* has distichously arranged secondary branches, and thus its primary axis appears similar to a secondary branch axis in other species. Tertiary branches may be arranged polystichously on basal secondary branches of *Setaria palmifolia*, similar to the arrangement of secondary branches on the primary axis of other species.

In most species of *Setaria*, bristles and spikelets are generally paired in early development, with terminal axes becoming bristles and the first lateral axis to any terminal axis becoming a spikelet. Each secondary branch complex produces several spikelets. In some species of *Setaria* only spikelets on the lower orders of branching develop to maturity while spikelets on higher orders of branching are arrested at

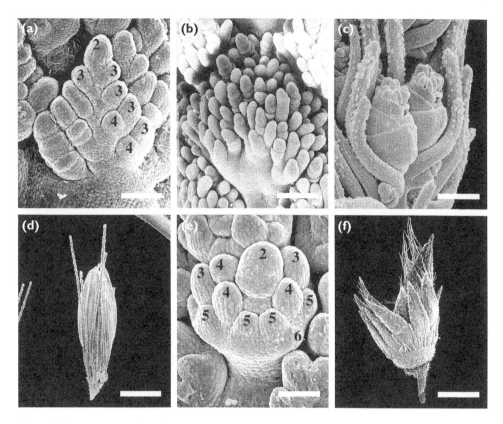

Figure 15.3 Inflorescence development in the bristle clade; secondary branch complexes. (a) *Setaria palmifolia*, showing distichous tertiary and quaternary branches. (b) *S. italica*, showing multiple higher orders of branching. (c) Developing spikelets and bristles in *S. verticillata*, showing suppressed spikelets. (d) Mature secondary branch complex in *S. viridis*, showing suppressed spikelets, and multiple bristles per mature spikelet. (e) *Pennisetum glaucum*, showing terminal spikelet and higher-order branches which grow out as bristles. (f) Mature secondary branch complex of *Cenchrus echinatus*, showing flattened bristles and terminal spikelet. Scale: (a) = 75 μm, (b), (c) = 200 μm, (d) = 1 mm, (e) = 50 μm, (f) = 750 μm.

early stages of development (Figure 15.3c–d). In other species, however, all of the spikelets mature. In *Pennisetum* and *Cenchrus* there are only one or two spikelets in each secondary branch complex. The spikelets terminate each secondary branch and, in some cases, the last initiated tertiary branches (Figure 15.3e). All other higher order axes terminate in bristles. All initiated spikelets in *Pennisetum* and *Cenchrus* appear to develop to maturity.

The *Pennisetum/Cenchrus* clade shares a unique form of inflorescence branching where the internodes of the secondary branch complex do not elongate. Instead, the branch axes become enlarged in the region where the branch primordia have initiated (Figure 15.3e). These enlarged nodal regions coalesce to form a disc, from the centre of which the secondary branch axis protrudes, and towards the edge of which higher order branch axes appear to be arranged in concentric rings. This is in contrast to all other genera of the bristle clade, where the branch internodes elongate to produce a ramified secondary branch complex (Figure 15.3b, d). The secondary branch complexes in both *Pennisetum* and *Cenchrus* disarticulate below the disc bearing the branches, so that the spikelet and branches fall as a unit. This is unlike the other members of the bristle clade, where the bristles remain on the inflorescence when the spikelet falls. The differences between *Pennisetum* and *Cenchrus* at maturity derive from changes that occur very late in development; in *Pennisetum* the higher order branches elongate distally, whereas in most species of *Cenchrus* they also flatten laterally (Figure 15.3f), and may become fused with each other.

In general, most differences in inflorescence morphology can be described by a small number of parameters, including numbers of orders of branching, numbers of primordia produced in each order, and the timing and amount of branch axis elongation. Although these characters are quantitative, they vary discretely, rather than continuously, among the taxa we have sampled.

15.3.3 Optimisation of developmental characters onto the phylogeny

A simplified phylogeny was used for character optimisation. *Stenotaphrum* was excluded from this phylogeny because the distichously (instead of polystichously) arranged branches on the single inflorescence axis may be either a novel phyllotaxis or equivalent to a single secondary branch with the primary axis suppressed. Three developmental characters important in the overall appearance of the inflorescence were optimised onto the phylogeny. These were (a) degree of elongation of the primary inflorescence axis internodes, (b) degree of elongation of the secondary branch internodes, and (c) number of tertiary branches (Figure 15.4a–c). The optimisations show that condensation of the primary axis evolved twice within the bristle clade, in the *Pennisetum/Cenchrus* clade and in the *Setaria* clade (Figure 15.4a). Reversal to elongate inflorescences has occurred in *Setaria* clade 1, and this is coded as polymorphic pending a more resolved phylogeny. Condensed primary axis internodes are also present in *Digitaria radicosa*, one of the outgroups. Elongation of the secondary branch evolved once within the bristle clade, in *Panicum bulbosum*, and is also present in *Panicum repens*, one of the outgroups (Figure 15.4b). Higher numbers of tertiary branches on each secondary branch (>10) evolved twice in the bristle clade: once in *Paspalidium* and once in some species of *Setaria* clade 1 (Figure 15.4c). High numbers of tertiary branches are also present in *Digitaria radicosa*, one of the outgroups.

Figure 15.4 Optimisation of three inflorescence characters onto one of the twenty-one most parsimonious trees. (a) Internode elongation on the primary axis of the inflorescence (condensed, elongate). (b) Internode elongation on the secondary branches (condensed, elongate). (c) Number of tertiary branches initiated on each secondary branch (<5 versus >10). Shading of branches indicate optimised character states along those branches.

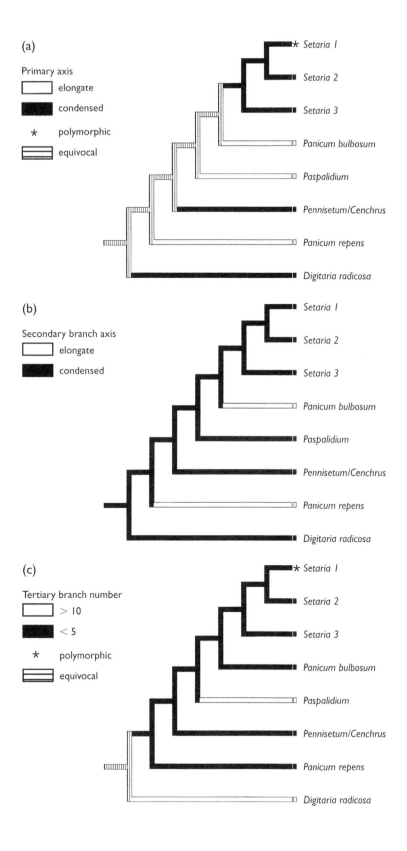

(a)

Primary axis

☐ elongate

■ condensed

∗ polymorphic

▤ equivocal

∗ Setaria 1
Setaria 2
Setaria 3
Panicum bulbosum
Paspalidium
Pennisetum/Cenchrus
Panicum repens
Digitaria radicosa

(b)

Secondary branch axis

☐ elongate

■ condensed

Setaria 1
Setaria 2
Setaria 3
Panicum bulbosum
Paspalidium
Pennisetum/Cenchrus
Panicum repens
Digitaria radicosa

(c)

Tertiary branch number

☐ > 10

■ < 5

∗ polymorphic

▤ equivocal

∗ Setaria 1
Setaria 2
Setaria 3
Panicum bulbosum
Paspalidium
Pennisetum/Cenchrus
Panicum repens
Digitaria radicosa

15.4 Discussion

The phylogenetic analysis shows the bristle clade to be well supported, with three main groups distinguishable. The monophyly of *Setaria* is not supported in the strict consensus tree, although the general lack of resolution of the branches at the base of the bristle clade does not preclude monophyly for the genus. Two clades of *Setaria* species are well supported, but the third grouping of *S. grisebachii* and *S. lachnea* has poor support, collapsing in the strict consensus tree. *Pennisetum* and *Cenchrus* form a well-supported clade, with *Cenchrus* nested within *Pennisetum*. The place-ment of *Pennisetum* and *Cenchrus* is not a surprise as they have always been closely allied in morphological classifications and some authors (e.g. Correll and Johnston, 1970) merged the genera. This is in part because bristles of some species of *Cenchrus* are transitional between the flattened 'normal' *Cenchrus* bristle and the elongate bristle of *Pennisetum*. *Stenotaphrum*, *Paspalidium* and *Panicum bulbosum* are uncertainly placed. The placement in the bristle clade of *Panicum bulbosum*, the only species without obvious bristles, may indicate that bristles have been lost in this species or that they occur only in early development. Because the large genus *Panicum* is polyphyletic (Zuloaga *et al.*, 2000; Giussani *et al.*, 2001; Aliscioni, unpubl. obs.), it is not clear what other *Panicum* species might also be expected to group with *Panicum bulbosum*. A nuclear gene phylogeny based on granule-bound starch synthase I (*waxy*: Mason-Gamer *et al.*, 1998) is being constructed to examine the relationships between the groups within the bristle clade.

15.4.1 Development in a phylogenetic context

Comparative analysis of inflorescence development across a range of species shows that only a few changes in developmental pathways can account for the considerable range of variation seen at maturity. These changes include variation in numbers of orders of branching, numbers of primordia produced in each order, and the timing and amount of internode elongation. It is possible to identify all of these differences at maturity, yet they occur at very different times during development. Differences in number of orders of branching and numbers of primordia produced by each order of branching are the result of initiation of primordia by branch meristems early in development, whereas differences in elongation of internodes occur late in develop-ment, just before the inflorescence emerges from the sheath. Thus, the analysis of development adds a temporal dimension to differences observable at maturity.

Our approach has been to use phylogenies to understand development, rather than the more common use of developmental characters for generating phylogenies (e.g. Nelson, 1973, 1978, 1985; de Queiroz, 1985; Weston, 1988; Mabee, 1989, 2000; Wheeler, 1990; Mabee and Humphries, 1993). Mabee (1993) and Hufford (1995) have both suggested that ontogeny could be described as a linear series of transformations that could be mapped onto a phylogeny. From this it is possible to determine the relative frequency of terminal deletions or additions, or of novel or reciprocal substitutions. One example in the bristle clade of a linear transformation series is the number of orders of branching, where change in the number of branches can be seen as terminal additions or deletions. However, other variation does not readily fit into a linear transformation series; for example, the plesiomorphic state for secondary branches is to be arranged polystichously, but in taxa such as *Steno-taphrum* in which the secondary branches are arranged distichously, either the underlying phyllotaxy has changed or there has been a non-terminal deletion of the primary inflorescence axis stage. A non-terminal addition to the developmental pathway has apparently occurred in *Setaria palmifolia* where the arrangement of the tertiary branches on basal secondary branches may be polystichous, rather than distichous, as in all other species examined. In this case the apical meristem of the secondary branch that is producing tertiary branch primordia polystichously is appreciably larger than secondary branch meristems that produce tertiary branch primordia distichously. Such changes to the inflorescence development pathway do not obviously involve a slowing down or speeding up of the ancestral developmental pathway and are therefore not heterochronic (Guerrant, 1988; Klingenberg, 1998; Zelditch *et al.*, 2000). In general, the variation we see is difficult to fit into a linear framework, and may be better described by combinatorial factors (Kellogg, 2000b).

The combinatorial nature of inflorescence variation can be seen in the three characters that were mapped onto one of the most parsimonious trees (Figure 15.4; without *Stenotaphrum*, because of the uncertain interpretation of its inflorescence). The ancestral condition for the internodes of the primary axis of the inflorescence is equivocal, but gains or losses of this character have occurred twice in the bristle clade (if the elongate state is derived, there have been three gains in the bristle clade as some species of *Setaria* clade 1 have elongate internodes). The reversal from a condensed to an elongate primary axis in some species of *Setaria* clade 1 contradicts the conclusions of Rominger (1962), who regarded the elongate axis in *Setaria* as the ancestral state. The ancestral condition for the second character, internode length between tertiary branches on each secondary branch, is condensed, with a change to elongate secondary branch internodes in *Panicum bulbosum*. The inflorescence of this species looks identical at maturity to other species of *Panicum*, such as the outgroup taxon *Panicum repens*, echoing results of earlier studies that recognised that morphological inflorescence types are not necessarily homologous (Kellogg, 2000a, b). The ancestral condition for the third character, numbers of tertiary branches, in the bristle clade is less than five. Branch numbers greater than ten have evolved twice, once in *Paspalidium* and once in some species of *Setaria* clade 1. High branch numbers was regarded by Rominger (1962) as ancestral but is shown as derived when mapped onto the molecular phylogeny. Tertiary branch number greater than ten is also present in the outgroup *Digitaria radicosa*. When the distribution of states of the three mapped characters are compared, it is clear that the characters can potentially (and actually) vary independently of each other. Various combinations of these characters produce the different mature inflorescence forms. Not all combinations of these characters are present in the taxa examined and it remains to be seen whether all combinations are possible.

Another interesting set of developmental characters pertains to the fate of individual branch meristems. In most species of *Setaria*, bristles and spikelets appear to be paired, with spikelets lateral to bristles. However, at maturity, species such as *Setaria viridis* appear to have a number of bristles subtending each spikelet in the primary branch complex. The discrepancy between the appearance of the inflorescence in these species of *Setaria* in early development and at maturity stems from the failure of the spikelets on higher order branch meristems to mature, while the bristles that are initiated around them grow on to maturity. Unlike *Setaria*, in *Pennisetum* and *Cenchrus* there are many bristles but only one or two spikelets, one of which appears to terminate the secondary branch axis. Our observations indicate that all spikelets develop to maturity in these two genera.

A diagnostic character often used in identifying *Setaria* species is the number of bristles subtending the spikelet. However, based on developmental analysis, this 'character' is in fact a composite of at least three characters, being determined by the numbers of orders of branching, the numbers of primordia per order of branching, and the number of spikelets whose development is suppressed. At maturity, spikelets in *Setaria* may have similar numbers of subtending bristles as spikelets in *Pennisetum* even though the development of these superficially similar structures is very different. Here, developmental data give not just an added temporal dimension to the analysis of morphological form, but also information that is otherwise lost at maturity.

Clayton and Renvoize (1986) regarded the bristles in *Pennisetum* and *Cenchrus* to be non-homologous with those of *Setaria*, and they assigned *Setaria* to a separate tribe. Our observations have shown that the bristles appear to be homologous and that the most striking difference between the *Pennisetum* clade and the rest of the bristle clade is instead the broadening of the base of the secondary branch axis. The phylogeny enables us to hypothesise that elongate bristles are plesiomorphic within the bristle clade, and focuses our attention on the evolution of flattened branch axes in *Cenchrus*.

15.4.2 Genetic architecture of developmental changes

On the basis of our developmental observations, we hypothesise that most inflorescence variation in the bristle clade is the result of changes in only a handful of parameters, which may correspond to single genes. To test this theory we have begun QTL mapping studies of a cross between *Setaria italica* and *S. viridis*. Preliminary data suggest that a relatively small number of QTL of major effect control inflorescence differences between the two species, as expected. This is similar to the results of Doebley and Stec (1991, 1993), who found that only a few major QTL could explain most of the morphological differences between maize and teosinte. We hypothesise that each QTL may consist of only one or a few genes and that these genes may control differences among all species in the genus, and even among all species in the bristle clade. The developmental analysis has enabled us to identify the parameters that need to be changed to create these inflorescence differences, and we are now searching for candidate genes that might control these developmental parameters. Some possible genes identified from maize mutants with phenotypes similar to those seen in the bristle clade include *ramosa 1* and *2*, which convert spikelet pairs to branches, and *tasselseed 4*, which controls numbers of orders of branching (Neuffer *et al.*, 1997). We also hope that the co-linearity of grass genomes will allow us to use the position of the QTL to identify other candidate genes in rice and maize that may control phenotypic variation in the bristle clade. With a robust phylogeny and with precise description of phenotypes, we can then extrapolate to deep branches of the phylogeny to hypothesise the underlying genetic changes involved.

The grasses are a particularly good group to investigate by the integrated approach outlined above because of the high level of synteny between genomes and the detailed genetic information available for a number of crop species. However, such an approach to the understanding of morphological diversification across large clades may be more difficult in groups where genetic information is less readily available. Difficulties of inter-breeding morphologically interesting species may prevent the analysis of morphology through QTL mapping populations. Likewise, investigating patterns of candidate gene expression throughout development can be a laborious and time-consuming task. However, the use of robust molecular phylogenies to investigate trends in evolution of developmental morphology will provide a better understanding of morphological diversification and will lay the ground-work for future analysis as genetic techniques become easier to apply. In groups such as the grasses, the integration through phylogeny of developmental morphology and genetic information from well-studied model systems is allowing a better understanding of the evolution of morphological diversity in large clades.

ACKNOWLEDGEMENTS

We wish to thank Richard Bateman, Quentin Cronk and Julie Hawkins for organising this conference. We also wish to thank the United States Department of Agriculture for seeds for developmental analysis, Katrien Devos and Mike Gale of the John Innes Institute for providing seed of the *Setaria italica* × *S. viridis* F2 populations for QTL analysis, and the Missouri Botanical Garden for fresh plant material. Thanks to Lili Giussani and Hugo Cota for several of the *ndhF* and *trnL* sequences used in this analysis. Thanks also to Jan Barber, Lili Giussani, Ken Hiser, Vicki Mackenzie, Simon Malcomber, Tony Verboom, Quentin Cronk, and two anonymous reviewers for their helpful comments on this chapter. Funding was provided by NSF grant DEB-9815392 to E. A. Kellogg and the E. Desmond Lee Endowment at the University of Missouri–St Louis.

REFERENCES

Anand Kumar, K. and Andrews, D. J. (1993) Genetics of qualitative traits in pearl millet: a review. *Crop Science*, 33,1–20.

Clark, L. G., Zhang, W. and Wendel, J. F. (1995) A phylogeny of the grass family (Poaceae) based on *ndhF* sequence data. *Systematic Botany*, 20, 436–460.

Clayton, W. D. and Renvoize, S. A. (1986) *Genera Graminum*. HMSO, London.

Correll, D. S. and Johnston, M. C. (1970) *Manual of the Vascular Plants of Texas*. Texas Research Foundation, Renner.

Cunningham, C. W. (1997) Can three incongruence tests predict when data should be combined? *Molecular Biology and Evolution*, 14, 733–740.

Cunningham, C. W., Omland, K. E. and Oakley, T. H. (1998) Reconstructing ancestral character states: a critical appraisal. *Trends in Ecology and Evolution*, 13, 361–366.

de Queiroz, K. (1985) The ontogenetic method for determining character polarity and its relevance to phylogenetic systematics. *Systematic Zoology*, 34, 280–299.

Devos, K. M., Pittaway, T. S., Reynolds, A. and Gale, M. D. (2000) Comparative mapping reveals a complex relationship between the pearl millet genome and those of foxtail millet and rice. *Theoretical and Applied Genetics*, 100, 190–198.

Devos, K. M., Wang, Z. M., Beales, J., Sasaki, T. and Gale, M. D. (1998) Comparative genetic maps of foxtail millet (*Setaria italica*) and rice (*Oryza sativa*). *Theoretical and Applied Genetics*, 96, 63–68.

Doebley, J. and Stec, A. (1991) Genetic analysis of the morphological differences between maize and teosinte. *Genetics*, 129, 285–295.

Doebley, J. and Stec, A. (1993) Inheritance of the morphological differences between maize and teosinte: comparison of results for two F2 populations. *Genetics*, 134, 559–570.

Douglas, A. W. (1997) The developmental basis of morphological diversification and synorganization in flowers of Conospermeae (*Stirlingia* and Conosperminae, Proteaceae). *International Journal of Plant Science*, 158 (6 Suppl.), S13–S48.

Evans, M. W. and Grover, F. O. (1940) Developmental morphology of the growing point of the shoot and the inflorescence in grasses. *Journal of Agricultural Research*, 61, 481–520.

Farris, J. S., Källersjö, M., Kluge, A. G. and Bult, C. (1994) Testing the significance of incongruence. *Cladistics*, 10, 315–319.

Felsenstein, J. (1985) Confidence limits of phylogenies: an approach using the bootstrap. *Evolution*, 39, 783–791.

Floyd, S. K. and Friedman, W. E. (2000) Evolution of endosperm developmental patterns among basal flowering plants. *International Journal of Plant Science*, 161 (6 Suppl.), S57–S81.

Giussani, L. M., Cota-Sánchez, J. H., Zuloaga, F. O. and Kellogg, E. A. (2001) A molecular phylogeny of the grass subfamily Panicoideae (Poaceae) shows multiple origins of C_4 photosynthesis. *American Journal of Botany*, 88, 1993–2012.

Gómez-Martinez, R. and Culham, A. (2000) Phylogeny of the subfamily Panicoideae with emphasis on the tribe Paniceae: evidence from the *trnL-F* cpDNA region, in *Grasses: Systematics and Evolution* (eds S. W. L. Jacobs and J. Everett), CSIRO, Melbourne, pp. 136–140.

Guerrant, E. O. (1988) Heterochrony in plants: the intersection of evolution, ecology, and ontogeny, in *Heterochrony in Evolution: a Multidisciplinary Approach* (ed. M. L. McKinney), Plenum Press, New York and London.

Harrison, C. J., Moller, M. and Cronk, Q. C. B. (1999) Evolution and development of floral diversity in *Streptocarpus* and *Saintpaulia*. *Annals of Botany*, 84, 49–60.

Harvey, P. H. and Pagel, M. D. (1991) *The Comparative Method in Evolutionary Biology*. Oxford University Press, Oxford.

Hufford, L. (1995) Patterns of ontogenetic evolution in perianth diversification of *Besseya* (Scrophulariaceae). *American Journal of Botany*, 82, 655–680.

Kellogg, E. A. (2000a) Molecular and morphological evolution in the Andropogoneae, in *Grasses: Systematics and Evolution* (eds S. W. L. Jacobs and J. Everett), CSIRO, Melbourne, pp. 149–158.

Kellogg, E. A. (2000b) A model of inflorescence development, in *Monocots: Systematics and Evolution* (eds K. L. Wilson and D. A. Morrison), CSIRO, Melbourne, pp. 84–88.

Klingenberg, C. P. (1998) Heterochrony and allometry – the analysis of evolutionary change in ontogeny. *Biological Reviews of the Cambridge Philosophical Society*, 73, 79–123.

Koduru, P. R. K. and Krishna Rao, M. (1983) Genetics of qualitative traits and linkage studies in pearl millet. *Zeitschrift für Pflanzenzüchtung*, 90, 1–22.

Losos, J. B. and Miles, D. B. (1994) Adaptation, constraint, and the comparative method: phylogenetic issues and methods, in *Ecological Morphology: Integrative Organismal Biology* (eds P. C. Wainwright and S. M. Reilly), University of Chicago Press, Chicago, pp. 60–98.

Mabee, P. M. (1989) An empirical rejection of the ontogenetic polarity criterion. *Cladistics*, 5, 409–416.

Mabee, P. M. (1993) Phylogenetic interpretation of ontogenetic change: sorting out the actual and artefactual in an empirical case study of cetrarchid fishes. *Zoological Journal of the Linnean Society*, 107, 175–291.

Mabee, P. M. (2000) The usefulness of ontogeny in interpreting morphological characters, in *Phylogenetic Analysis of Morphological Data* (ed. J. J. Wiens), Smithsonian Institution, Washington, DC.

Mabee, P. M. and Humphries, J. (1993) Coding polymorphic data: examples from allozymes and ontogeny. *Systematic Biology*, 42, 166–181.

Maddison, D. R. (1994) Phylogenetic methods for inferring the evolutionary history and processes of change in discretely valued characters. *Annual Review of Entomology*, 39, 267–292.

Maddison, D. R. and Maddison, W. P. (2000) *MacClade 4: Analysis of Phylogeny and Character Evolution. Version 4.0.* Sinauer, Sunderland, MA.

Mason-Gamer, R. J., Weil, C. F. and Kellogg, E. A. (1998) Granule-bound starch synthase: structure, function, and phylogenetic utility. *Molecular Biology and Evolution*, 15, 1658–1673.

Möller, M. and Cronk, Q. C. B. (2001) Evolution of morphological novelty: a phylogenetic analysis of growth patterns in *Streptocarpus* (Gesneriaceae). *Evolution*, 55, 918–929.

Moody, M. L. and Hufford, L. (2000) Floral ontogeny and morphology of *Cevallia*, *Fuertesia*, and *Gronovia* (Loasaceae subfamily Gronovioideae). *International Journal of Plant Science*, 161, 869–883.

Murphy, J. A. (1978) Non-coating techniques to render biological specimens conductive. *Scanning Electron Microscopy*, 2, 115–194.

Nelson, G. J. (1973) The higher-level phylogeny of vertebrates. *Systematic Zoology*, 22, 87–91.

Nelson, G. J. (1978) Ontogeny, phylogeny, paleontology, and the biogenetic law. *Systematic Zoology*, 27, 324–345.

Nelson, G. J. (1985) Outgroups and ontogeny. *Cladistics*, 1, 29–45.

Neuffer, M. G., Coe, E. H. and Wessler, S. R. (1997) *Mutants of Maize.* Cold Spring Harbor Laboratory Press, Plainview, New York.

Olmstead, R. G. and Sweere, J. A. (1994) Combining data in phylogenetic systematics: an empirical approach using three molecular data sets in the Solanaceae. *Sytematic Botany*, 43, 467–481.

Rominger, J. M. (1962) Taxonomy of *Setaria* (Gramineae) in North America. *Biological Monographs*, 29, 1–127.

Sasaki, T. and Burr, B. (2000) International rice genome sequencing project: the effort to completely sequence the rice genome. *Current Opinions in Plant Biology*, 3, 138–141.

Soltis, D. E., Soltis, P. S. and Doyle, J. J. (eds) (1998) *Molecular Systematics of Plants II: DNA Sequencing.* Kluwer, Boston, Dordrecht, London.

Stevens, P. F. (1991) Character states, morphological variation, and phylogenetic analysis: a review. *Systematic Botany*, 16, 553–583.

Swofford, D. L. (1999) PAUP*. *Phylogenetic Analysis Using Parsimony.* Version 4. Sinauer Associates, Sunderland, MA.

Swofford, D. L., Olsen, G. J., Waddell, P. J. and Hillis, D. M. (1996) Phylogenetic inference, in *Molecular Systematics* (eds D. M. Hillis, C. Moritz and B. K. Mable), Sinauer, Sunderland, MA., pp. 407–514.

Taberlet, P. L., Gielly, L., Pautou, G. and Bouvet, J. (1991) Universal primers for amplification of three non-coding regions of chloroplast DNA. *Plant Molecular Biology*, 17, 1105–1109.

Thompson, J., Gibson, T. and Higgins, D. (1994) Clustal W, version 1.7: improving the sensitivity of progressive multiple sequence alignment through sequence weighting, position-specific gap penalties and weight matrix choice. *Nucleic Acids Research*, 22, 4673–4680.

Wang, Z. M., Devos, K. M., Liu, C. J., Wang, R. Q. and Gale, M. D. (1998) Construction of RFLP-based maps of foxtail millet, *Setaria italica* (L.) P. Beauv. *Theoretical and Applied Genetics*, 96, 31–36.

Watson, L. D. and Dallwitz, M. J. (1992) *The Grass Genera of the World*. CAB International, Wallingford, UK.

Weston, P. W. (1988) Indirect and direct methods in systematics. In *Ontogeny and Systematics* (ed. C. J. Humphries), Columbia University Press, New York, pp. 25–56.

Wheeler, Q. D. (1990) Ontogeny and character phylogeny. *Cladistics*, 6, 225–268.

Wolfe, K. (2000) The rice genome. *Nature Reviews Genetics*, 1, 7.

Zelditch, M. L. and Fink, W. L. (1996) Heterochrony and heterotopy – stability and innovation in the evolution of form. *Paleobiology*, 22, 241–254.

Zelditch, M. L., Sheets, H. D. and Fink, W. L. (2000) Spatiotemporal reorganization of growth rates in the evolution of ontogeny. *Evolution*, 54, 1363–1371.

Zuloaga, F. O., Morrone, O. and Giussani, L. M. (2000) A cladistic analysis of the Paniceae: a preliminary approach, in *Grasses: Systematics and Evolution* (eds S. W. L. Jacobs and J. Everett), CSIRO, Melbourne, pp. 123–135.

Chapter 16

Identification of genes involved in evolutionary diversification of leaf morphology

Tracy McLellan, Helen L. Shephard and Charles Ainsworth

ABSTRACT

Comparisons of expression of candidate genes during development have revealed that the structure and function of many genes has been conserved throughout evolution. Gene expression can be an indication of homology, but a gene may also be expressed in a wide variety of non-homologous structures. Therefore, correlation between gene expression and evolutionary change does not necessarily indicate that the candidate gene examined is primarily responsible for diversification. The primary genes in evolutionary diversification will show inherited differences, while secondary genes may not be different in structure, although they may be expressed differentially. The estimated number of primary genes that differ between plant species, based on QTL studies, is far higher than the number of candidate genes known as mutations in model species. Therefore, there is a need to identify more genes and to determine whether they are primary determinants of evolutionary variation. We present an approach for the identification of primary genes responsible for naturally occurring variation in leaf morphology using a combination of differential gene expression and QTL mapping. Preliminary results show great variation in expressed genes between plants within the same species, *Begonia dregei*. Linkage analysis will be used to determine which of these transcripts correspond to primary genes. This approach takes advantage of the many useful properties of plants for the study of microevolution and therefore has an important contribution to make in the field of evolution and development.

16.1 Introduction

Resurgence of interest in evolution and development over the past decade has been facilitated by advances in the study of the molecular biology of development. Most of the research in evolution and development, as with that in developmental biology, has been carried out on animals rather than plants. Precedents from research on animals might be used to some advantage in designing approaches that will provide insight into both evolutionary and developmental processes in plants. This chapter critically appraises recent findings in evolution and development and describes a novel approach to finding the genetic basis for evolutionary change in plants.

In *Developmental Genetics and Plant Evolution* (2002) (eds Q. C. B. Cronk, R. M. Bateman and J. A. Hawkins), Taylor & Francis, London, pp. 315–329.

16.1.1 Evolution and development in animals and plants

Genes that were originally identified as mutations, primarily in *Drosophila*, have been found to be highly conserved in structure and function in a wide variety of animals. Initially, mutants in *Drosophila* and mice with similar phenotypes were found to be based on genes of similar sequence, such as *eyeless* and *Pax-6* in the development of eyes, and *Distalless/Dlx* in the determination of the proximal–distal axis of appendages (Halder *et al.*, 1995; Panganiban *et al.*, 1997). In plant development, there are also mutants with similar phenotypes based on homologous genes, such as those for floral organ identity (Irish and Yamamoto, 1995).

The conservation of gene sequences has made it possible to examine gene expression during development in many species with diverse morphology and which are not otherwise easily amenable to genetic investigations. For example, broad specificity antibodies to the gene products of both *Distalless* (*Dll*) and *engrailed* have been used in comparative studies of development in many types of animals (Patel *et al.*, 1989; Panganiban *et al.*, 1997). Expression of *Dll*, or its lack of expression, has been used as an indicator of homology of structures with implications for phylogenetic relationships within arthropods (Popadic *et al.*, 1996). Similarly, the lack of expression of orthologues of *APETALA3* and *PISTILLATA* (B-function MADS box genes expressed in developing stamens and petals in *Arabidopsis* and *Antirrhinum*) in the petals of the Ranunculidae suggests that lower and higher eudicot petals are not homologous (Kramer and Irish, 1999).

There are some limitations to the use of gene expression as an indication of homology in more distantly related organisms. Extensive examination of *Dll* expression throughout development and in a wider variety of animals revealed that homology between appendages was not so easily demonstrated (Popadic *et al.*, 1998). *Dll* is expressed in the distal ends of developing appendages that cannot be homologous with each other, in the sense that they were derived from the same structure in a common ancestor, such as the legs of flies and of mice (Tabin *et al.*, 1999; Wray and Lowe, 2000). Wide comparisons of *Dll* expression in animals, some without appendages and some with highly modified body plans, showed that this gene is also expressed in the central nervous system, in developing teeth, and in the eyespot colour pattern elements on butterfly wings (Brakefield *et al.*, 1996; Panganiban *et al.*, 1997). Thus, the expression of *Dll* alone cannot identify the distal ends of appendages, as *Dll* is expressed elsewhere, nor can its presence in developing limbs indicate that those limbs are homologous. The broader comparisons have been used to infer the ancestral function of the gene. A role in the evolutionary origins of appendages as outgrowths from the body wall, which were perhaps originally sensory organs, is consistent with observations of expression in the nervous system (Panganiban *et al.*, 1997; Tabin *et al.*, 1999). Thus, comparisons of expression of *Dll* have demonstrated more about the conservation of the gene than about the evolutionary diversification of appendages.

Fewer comparative studies of gene expression have been made between plants, but work is currently in progress on several genes, such as the MADS box floral organ identity genes, and *CYCLOIDEA*, *LEAFY* and *KNOTTED* (e.g. Baum, 1998; Frohlich and Parker, 2000; Shu *et al.*, 2000; Gleissberg, 2002; Theißen *et al.*, 2002). Although phylogenetic coverage is sparse, we know that the structures of many homeotic plant genes are highly conserved. One such gene, *LEAFY*, is known from all extant seed plant groups, as well as from ferns. *LEAFY*'s function in reproductive specification of the apical meristem also appears to be conserved. However, in the case of pea, *LEAFY* has been recruited to a role in specifying compound leaves (Hofer *et al.*, 1997; Frohlich and Parker, 2000) whilst retaining a reproductive function. We might expect other plant genes to have more than one role during development. In plants as well as animals, gene expression may not always be a good indicator of homology of structures.

Not only have individual genes been conserved in evolution, but several steps in developmental signalling pathways have been found to be remarkably similar between diverse animal groups. In the case of the *WNT* signalling pathway in *Hydra*, the function in determining the anterior–posterior axis is conserved (Hobmayer *et al.*, 2000), while the *hedgehog* pathway has been recruited from its original role in determining wing pattern to determine the position of butterfly eyespots (Keys *et al.*, 1999). The pathways for the three-dimensional genetic patterning of anterior–posterior, dorsal–ventral and proximal–distal axes in animals appear to be determined in the same ways across the animal kingdom (Hobmayer *et al.*, 2000; Wray and Lowe, 2000). There are fewer developmental pathways known in plants than in animals, but it seems likely that pathways will be found to have been conserved, and sometimes to have been recruited into new roles in development. A number of signal transduction pathways appear to be conserved between animals and plants (Cox *et al.*, 1998; Fletcher and Meyerowitz, 2000), and it will be interesting to see how these mechanisms for communication between cells function in plant development.

Genes may have the same structure and function at the molecular level and protein sequences have been highly conserved, but they do not always have the same role in morphogenesis. It seems that *cis* regulatory elements of transcriptional regulators have varied in evolution and their variation may be responsible for diversification (Doebley and Lukens, 1998; Stern, 2000; Wray and Lowe, 2000). Molecular studies have provided information on the evolution of developmental mechanisms; knowledge of the conservation of gene structure and function should facilitate framing questions and approaches to understand the developmental and genetic mechanisms underlying evolutionary diversification.

16.1.2 Primary and secondary genes: a key distinction

Molecular methods make it possible to examine gene expression during development in a wide variety of species. When differences in gene expression are correlated with evolutionary modifications of structures, it is tempting to infer that the gene examined is the one that is responsible for evolutionary change. However, there are several reasons why caution should be exercised in the interpretation of the results of studies that compare gene expression. A gene that is expressed differently in two species could be a downstream target of another gene modified in evolution, its differential expression resulting from a role that is secondary to the inheritance of the trait. For example, MADS-box genes are differentially expressed in male and female flowers of the dioecious species *Rumex acetosa* (Ainsworth *et al.*, 1995). Expression patterns of putative B-function and C-function homologues are similar to those in *Arabidopsis* and *Antirrhinum* early in flower development. However, expression ceases in the stamen primordia of female flowers at the same time that the growth of stamen primordia is arrested. Although the cessation of gene expression and organ arrest are correlated, it has not been determined whether gene expression changes are a cause or simply a consequence of organ arrest (Ainsworth *et al.*, 1995). Furthermore, the interesting questions in the evolution of dioecy concern the origin of a species with separate male and female plants from an ancestor with hermaphroditic flowers, and therefore the genetic basis for determination of separate sexes. Differential expression of MADS-box genes is correlated with the structure of male and female flowers, but MADS-box genes are unlikely to be the primary sex-determining genes (Ainsworth, 2000). The genes that are responsible for sex determination are not currently known in any dioecious plant species, although many genes have been found to be differentially expressed in male and female flowers in several species (Ainsworth, 2000). Differences in gene expression in male and female flowers are thus part of the phenotype, as is the difference in the timing of flowering between male and female plants, expressed as a consequence of sex determination, but not directly by sex-determining genes.

Evolutionary change depends on variation that is inherited. The finding of differences in gene expression between two species in a way that is correlated with morphological difference is not necessarily an indication that the gene in question is inherited in a different form. Genes involved in the control of development occur in pathways, or networks, in which the expression of one gene is regulated by other genes. It is therefore possible for a gene to be expressed differently in two species as a consequence of regulation by another gene that occurs in heritably different forms between the species (Palopoli and Patel, 1996). We will use the term 'primary gene' to refer to genes whose inherited modifications result in alterations to developmental pathways and their morphological results. Those genes whose expression is modified as a consequence of the heritable change in primary genes as their downstream targets will be referred to as 'secondary genes'. Secondary genes are not necessarily different between species and may be of critical importance in producing a phenotype on which natural selection acts, but it is only the inherited variation in primary genes that actually produces evolutionary change. For example, in *R. acetosa*, sex determination is inherited as a primary genetic difference. MADS-box gene expression is then influenced somehow by the sex determining genes, and thus MADS-box genes are secondary genes with respect to sex determination. A complete understanding of the genetic basis of dioecy should involve both the primary genes and the developmental pathways in which the expression of many genes has been modified. Correlations between developmental events and gene expression do not provide strong evidence that these single genes have had a role in determining evolutionary diversification. (This key distinction between primary and secondary genes has been voluntarily adopted by several other authors in this volume – eds.)

16.2 Number of genes

There has long been a debate about the number and magnitude of effects of mutations that become established during evolution (Orr and Coyne, 1992; Stern, 2000). Plant morphological evolution might be attributed to a small number of genes (Hilu, 1983; Gottlieb, 1984), a suggestion based on analyses of cultivated plants and early Mendelian genetic work on wild plants. Some resolution of this issue is now possible because the number and relative effect of genes in polygenic traits can be estimated via quantitative trait locus (QTL) mapping with molecular markers (Lander and Botstein, 1989; Tanksley, 1993). In QTL mapping, numerous molecular markers are used to detect linkage with the genes that underlie a trait by examining both the traits and markers in crosses between different inbred strains or species (Lander and Botstein, 1989).

16.2.1 Estimates of numbers of genes determining morphological differences between plant species

Several studies in plants show that the number of QTL, a minimum estimate of the number of genes that determine morphological differences between closely related species, ranges from eighteen to seventy-four, and that the variance of any one trait that is explained by each QTL ranges from 2 to 84 per cent (Table 16.1). QTL will represent primary genes (as defined in Section 16.1.2), since they represent inherited differences. Each individual trait may be based on a small number of QTL, but differences between species when all traits are considered add up to reasonably large numbers. Some QTL appear to be involved in more than one trait, in part because traits that are correlated, such as various measurements of leaves that are related to overall size, will be related to at least some of the same QTL. The study of leaf morphology in *Gossypium* (cotton) showed that one 'major gene' corresponded to a known mutant locus in one of the parents. Four genes for leaf shape are known in

Table 16.1 Differences between species in number of quantitative trait loci

Species	Traits	No. of Traits	No. of QTL	% PVE*	Reference
Gossypium hirsutum G. barbadense	Leaf shape	14	62	6–49	Jiang et al., 2000
Helianthus annuus H. debilis	Morphology, pollen sterility	15	56	6–36	Kim and Rieseberg, 1999
Mimulus lewisii M. cardinalis	Flower shape, colour, nectar	12	66	18–84	Bradshaw et al., 1998
Populus trichocarpa P. deltoides	Leaf shape	9	50	18–55	Wu et al., 1997
Lycopersicon esculentum L. pennellii	Size of organs	11	74	3–34	DeVincente and Tanksley, 1993
Zea mays subsp. mays Z. mays subsp. mexicana	Inflorescence form, plant architecture	12	61	5–78	Doebley and Stec, 1991
Arabidopsis thaliana A. thaliana	Flower morphology	8	18	2–77	Juenger et al., 2000
Drosophila mauritiana D. simulans	Genital lobe size and shape	1	19	1–15	Zeng et al., 2000
Ambystoma trigrinum A. mexicanum	Paedomorphosis (body form at day 450)	1	1	93	Voss and Schaffer, 1997

Note:
*Phenotypic Variance Explained.

cultivated cottons, and the development of leaves in plants with those mutations has been described quantitatively (Hammond, 1941a, b; Stephens, 1944). One might predict, then, that variation in leaf shape between species could be determined by those four loci, but the QTL study shows that many more genes affect leaf shape (Jiang et al., 2000). Similarly, variation in leaf shape between two species of Populus is based on 50 QTL (Wu et al., 1997). A study of pollinator attractiveness of the flowers of two species of Mimulus showed that QTL for pigment concentrations had large effects on the phenotype, whereas those for the shape of petals and size of stamens had much smaller effects, perhaps indicating that the modification of colour is based on a smaller number of loci than changes in morphology (Bradshaw et al., 1998).

Transitions between cultivated crop plants and their wild progenitors have provided another type of comparison between 'species', where both the genetics of the cultivated species is well known and the ancestral species has been identified. Maize and teosinte differ greatly in plant architecture (Doebley and Stec, 1991), and all differences between them are related to 61 QTL. The major difference in inflorescence architecture was related to five QTL, of which some of the corresponding genes have been identified (Dorweiler et al., 1993; Doebley et al., 1995). Similarly, the differences between cultivated tomatoes and their wild relatives were attributed to 74 QTL (DeVincente and Tanksley, 1993). Therefore, the genetic differences between cultivated species and their wild ancestors are generally similar to the differences between closely related wild species.

Also included in this survey is a QTL study of variation in floral morphology within Arabidopsis thaliana, which showed eighteen QTL associated with quantitative variation in flowers between two inbred strains (Juenger et al., 2000). This study is of interest because it was found that seven QTL did not correspond to any of the currently known 'candidate' genes.

16.2.2 Estimates of numbers of genes determining morphological differences between species of animals

There are two interspecific QTL studies of animals included in Table 16.1, compared to seven of plants, an indication of greater difficulty of crossing between species in animals than in plants. The lower numbers of QTL found in the animal studies suggest that there might be fewer genetic differences between species in animals than in plants, but this may not be generally true. The study of paedomorphosis in axolotls considered only one trait, not all the differences between species, and that trait was measured as presence or absence, not as a continuous variable. The failure to metamorphose in salamanders is known to be due to the disruption of the thyroxine pathway (Frieden, 1981; Voss et al., 2000), and so the observation that one QTL explains the difference between species for this trait is not surprising and may, or may not, be indicative of evolution of life history traits in animals generally. The study of the genital arch in Drosophila deals with size and shape of the one organ that differs between species. These Drosophila species, D. mauritiana and D. simulans, together with D. melanogaster and D. sechellia, have been used extensively for studying the genetic differences between species because they are amenable to some interspecific hybridisation (Coyne, 1983).

Considering that the studies listed in Table 16.1 differed in sample size, crossing design, and analytical methods, the estimates of numbers of QTL that differ between plant species are surprisingly similar, and provide support for an oligogenic model rather than either the infinitesimal model or a single-gene model as the basis for morphological differences between species.

There are several biases inherent in QTL analysis. Small sample sizes provide underestimates of the number of QTL, as well as overestimates of the proportion of variance explained by each QTL detected. If anything, larger sample sizes should yield estimates for larger numbers of QTL, each with a smaller effect on the trait (Bradshaw *et al.*, 1998). Another problem with QTL studies between species is that segregation distortion is common in crosses between species, and some combinations of parental genes do not occur in the hybrid offspring (Kim and Rieseberg, 1999).

16.2.3 The need to identify additional genes

QTL that account for 25 per cent of the variance of a trait are considered to be 'major genes' in QTL studies (Bradshaw *et al.*, 1998). However, the 'major genes' in quantitative genetics are not the same as the single candidate genes that are usually the subject of investigation in evolution and development. The resemblance of the mutant phenotypes of candidate genes to evolutionary transitions has led to conclusions that a gene such as *CYCLOIDEA* is responsible for the origin of floral asymmetry (Luo *et al.*, 1996), or that *KNOTTED* is responsible for transitions between compound and simple leaves (Sinha, 1999). Variation in phenotypes that have occurred through evolution is much greater than can be accounted for by mutations in candidate genes; many more genes may be involved in evolutionary diversification than the few that are currently being considered (Donoghue *et al.*, 1998). Redundancy in developmental pathways may mean that some genes are difficult to detect as mutations because another gene has a similar function. Currently available data are consistent in indicating a fairly large number of genes determining differences between species of plants, far more than the candidate genes that are known in plant development. It is critical to learn not only the number of genes but also their identity and function in order to understand the evolution of morphology.

It is difficult to identify specific genes in QTL studies at present. A QTL is often located within 20 cM of a marker, corresponding to hundreds of thousands of base pairs of DNA and hundreds of genes. Finer scale mapping may localise a gene to within a smaller recombination distance making possible positional cloning of the gene. Alternatively, candidate genes that map to the same region can be examined to determine whether they differ between species (Falconer and Mackay, 1996). This approach is feasible in *Drosophila* and *Arabidopsis*, given the large number of characterised mutants and the completely sequenced genomes (Nuzhdin *et al.*, 1998; Juenger *et al.*, 2000), but will be more problematic in other species. However, even in these model species, it is still difficult to locate the precise genetic difference that underlies a trait. QTL analysis can, with some certainty, exclude a candidate gene from involvement in a trait (Voss *et al.*, 2000). Given the need to identify primary genes and the large number of genes that appear to be involved in evolutionary change, we propose here an additional way to identify genes using QTL mapping of expressed genes and naturally occurring variation, that should be capable of detecting genes in addition to those originally found as mutations in model systems.

16.3 Properties of plants

The study of naturally occurring variation, where numerous genes are undoubtedly involved and phenotypes show a greater range than can be accounted for by known single gene mutants (Donoghue *et al.*, 1998), will lead to the identification of genes that have produced evolutionary change. Fortunately, plants possess a number of properties that make them amenable to genetic studies of the investigation of development and evolution, especially in the study of microevolution. Many of the major shifts in evolution, such as simple to compound leaves and symmetrical to asymmetrical flowers, occur in closely related species. Crosses between divergent species, or even between genera in the Orchidaceae, can yield fertile offspring in plants, whereas genetic studies between species of animals are far more limited by incompatibility. Many plant organs, such as leaves and flowers, are produced repeatedly on the same individual, facilitating the examination of organs throughout development without having to construct inbred lines. The ease of clonal propagation of plants also facilitates genetic and developmental investigations of naturally occurring variation.

Although most botanists have concentrated on evolutionary studies of reproductive not vegetative structures, vegetative structures are also highly diverse and worthy of investigation. The size, shape and complexity of simple leaves varies greatly within many dicotyledonous plant families (e.g. Aceraceae, Asteraceae, Cucurbitaceae, Fagaceae, Geraniaceae, Malvaceae, Solanaceae and Papaveraceae). Leaf shape and size are often used for species identification, and some aspects of leaf shape can be characteristic of an entire plant family, as with the asymmetrical leaf bases in Begoniaceae. The adaptive significance of leaf shape and size has been examined in many groups of plants (Givnish, 1978, 1987). Leaf shape is a good trait with which to study quantitative genetics, as most leaves can be pressed into two dimensions and the shape then quantified with various morphometric methods (Kincaid and Schneider, 1983; McLellan and Endler, 1998). Although there is variation in leaf shape in the Brassicaceae, *Arabidopsis thaliana*, the favoured model system in plants, has little naturally occurring leaf shape variation. Many mutants for leaf morphology have been identified in *Arabidopsis* (Berna *et al.*, 1999), but for most of them, the genes have not yet been characterised. Genes that have been extensively characterised in *Arabidopsis* and *Antirrhinum* leaf development appear to be involved in leaf initiation and dorsal–ventral determination, and it is not known whether any of them is involved in determining naturally occurring variation in leaf shape (Tsukaya, 1995; Hudson, 1999; Scanlon, 2000).

16.4 QTL mapping of expressed genes

We have taken a novel approach to identify genes that determine evolutionary variation in morphology which combines methods from molecular biology and quantitative genetics and uses naturally occurring variation in leaf morphology. This approach is intended to bridge the gap between comparative studies of gene expression in distantly related taxa and quantitative and population genetics studies of variation within and between species by providing a way of identifying specific genes that determine differences between species. Expressed genes in developing leaves have been examined in two forms of *Begonia dregei* that differ in leaf morphology and which have been crossed with each other to produce an F_2 generation for QTL mapping (Figure 16.1). *Begonia dregei* are small, perennial herbaceous plants native to forests on the eastern coast of South Africa. The plants exhibit much variation in the shape of leaves, as well as in the size and distribution of foliar trichomes (Irmscher, 1961; Hilliard, 1976; McLellan, 2000). Leaf morphology is fairly constant regardless of growing conditions (McLellan, 2000), and each population is small and highly uniform (Matolweni *et al.*, 2000), permitting the application of genetic methods for inbred strains. Minimum estimates of the number of genes causing differences in leaf shape range from six to ten, based on the Castle–Wright method (Lande, 1981), and it is likely that there are about ten to twenty genes involved.

16.4.1 Method

We have isolated RNA from three developmental stages of leaves. The first stage included the shoot apical meristem, developing stipules and a leaf primordium that is less than 300 µm long and was determined by the lack of visible trichomes. At this stage leaves already differ in shape (McLellan, 1990). The second stage contained leaf primordia from the point when trichomes first appear, at a length of about 300 µm,

Figure 16.1 Silhouettes of leaves from *Begonia dregei*. Top row left shows a leaf from a plant from Ntumeni Forest near Eshowe in Kwazulu-Natal (T. McLellan 338). Top row right shows a leaf from Ntlopeni Forest, Port St Johns, Eastern Cape (T. McLellan 339). These two plants were crossed; the three leaves in the second row were collected from F_1 hybrid plants. The leaves in rows 3, 4 and 5 were collected from twelve plants in the F_2 generation, produced by selfing F_1 plants. They vary in the degree of incision between lobes and in the size of teeth on the margins. Leaves of *B. dregei* are asymmetrical and are produced in two mirror-image orientations. All leaves here are shown in the same orientation to facilitate comparison of leaf shape. The leaf on the top left is 40 mm long. (For collection numbers see McLellan, 2000.)

through to primordia that are still enclosed within overlapping stipules and are about 1 cm in length. During this stage, chlorophyll becomes visible in the primordia. The third stage was defined by leaf length being longer than the stipules, and the end of this stage included leaves that were not yet completely flat, as are mature leaves. It is during this third stage that the region of the lamina between the petiole and the sinuses between lobes in deeply incised leaves grows rapidly (McLellan, 1990). Complementary DNA (cDNA) was synthesised from each of the developmental stages of leaves in the two plants that have been used as parents of an F_2 mapping population, and variation between plants and stages was examined using Amplified Fragment Length Polymorphism (AFLP). AFLP is a method for detecting variation in DNA sequence that is used extensively in gene mapping in plants and for detecting variation within species in population studies (Vos *et al.*, 1995; Mueller and Wolfenbarger, 1999). The cDNA-AFLP technique is a method of differential display of expressed genes that has a potential for screening the entire population of transcripts in an organ (Bachem *et al.*, 1996, 1998; Money *et al.*, 1996). The cDNA-AFLP method used here involved digestion with the restriction enzymes Apo1 and Mse1, followed by ligation of adapters to each restriction site, and amplification by PCR with non-specific primers. Another round of PCR was then performed using primers with from one to five additional selective nucleotides, in addition to the sequence that matches the adapters and restriction sites. One of these primers was labelled with ^{33}P and variation in the presence and absence of cDNA fragments was visualised after electrophoretic separation of the DNA fragments on polyacrylamide gels using autoradiography.

16.4.2 Preliminary results

There exists a variety of patterns of differential expression of transcripts between the two plants and between stages of leaf development within a plant (Figure 16.2). Between 11 per cent and 19 per cent of the AFLP bands varied between the two plants in the first stage of leaf development, depending on the selective primers used. This represents from 550 to 950 of the estimated 5000 transcripts in leaf primordia and the shoot apical meristem. There are several sources for this variation:

1   the ten to twenty primary genes that are probably responsible for differences;
2   the differential expression of the downstream targets of the primary genes (the 'secondary' genes); and
3   sequence polymorphism unrelated to gene expression.

Assuming that many of the primary genes will have transcripts that occur in low numbers of copies, we have optimised the AFLP technique for the detection of rare fragments by using primers with more selective nucleotides than are commonly used in cDNA-AFLP studies. The strategy for identification of primary genes is to test for linkage between morphological traits and cDNA-AFLPs in the F_2 mapping population, concentrating on ten to twenty plants out of 600 that most resemble each of the original parents to minimise the number of samples examined (Lander and Botstein, 1989). Primary genes should show no recombination with traits, but at least some secondary genes will also appear tightly linked with their cDNA, as they are downstream targets of the primary genes and their expression is likely to be correlated with that of

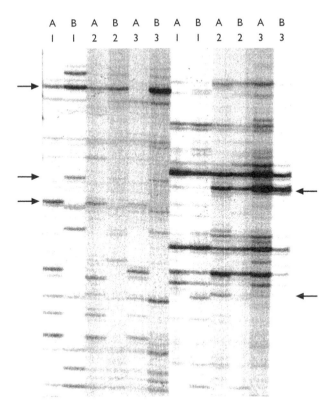

Figure 16.2 Autoradiograph of an AFLP gel for three developmental stages (1, 2 and 3) of two plants (A: McLellan 338, Ntumeni Forest; B: McLellan 339, Ntlopeni Forest) using two combinations of selective primers. Arrows indicate bands that differ between samples. Left, from top to bottom, band absent in plant A, stage 3; band present in plant B, all stages; band present in plant A, all stages. Right, top to bottom, band at low concentration in stage 1 and highest concentration in stage 3 in both plants; band present in plant B, stage 1 only.

some of the primary genes. QTL mapping of cDNA-AFLP differentials should distinguish genomic polymorphisms that are unrelated to differential expression, which are unlikely to show linkage with traits, from the primary and secondary genes which show linkage. Primary and secondary genes can then be distinguishable in a further screen for variation in the genomic DNA corresponding to the cDNA fragments, since primary genes should show polymorphism correlated with morphological variation, but secondary genes will not. Primary genes can then be characterised for structure and function at the molecular level by examining their expression during development in both parental forms. Together with sequence similarity to other genes, this information may provide clues to gene function. One advantage of working with naturally occurring variation is that genes that have been found to be involved in determining morphological variation can also be examined for their variation and distribution in natural populations, and related to phylogeny and the wide variety of leaf forms that are found in *B. dregei* and other species of *Begonia*.

This approach is not without its limitations. It might be possible that some primary genes are not expressed at levels that are sufficiently different to be detected as differential bands by cDNA-AFLP. The large number of genes that might be involved in a trait such as leaf shape is an obstacle to their thorough characterisation at the molecular level. Focusing on a trait such as trichome number, which might be simple genetically, or on events that occur during a short period during development (Stern, 2000) would make both identification and characterisation of genes more tractable. Since the determination of primary genes depends in part on assessing inheritance, it will be difficult to identify them with certainty at higher taxonomic levels where divergent species cannot be crossed.

Generalisations about the developmental mechanisms that underlie morphological diversification in evolution will be based on accumulated data on many traits and many plants sampled over a range of phylogenetic levels, and employing a variety of approaches. The approach to microevolution and development presented here has an important contribution to make to the understanding of plant development and evolution.

ACKNOWLEDGEMENTS

We thank D. Baum and M. Ntwasa for their comments. This work was made possible by financial support from the Foundation for Research Development (South Africa) and the Anderson–Capelli Fund of the University of the Witwatersrand.

REFERENCES

Ainsworth, C. (2000) Boys and girls come out to play: the molecular biology of dioecious plants. *Annals of Botany*, 86, 211–221.

Ainsworth, C., Crossley, S., Buchanan-Wollaston, V., Thangavelu, M. and Parker, J. (1995) Male and female flowers of the dioecious plant sorrel show different patterns of MADS box gene expression. *Plant Cell*, 7, 1583–1598.

Bachem, C. W. B., Oomen, R. J. F. J. and Visser, R. G. F. (1998) Transcript imaging with cDNA-AFLP: a step-by-step protocol. *Plant Molecular Biology Reporter*, 16, 157–173.

Bachem, C. W. B., van der Hoeven, R. S., de Bruijn, S. M., Vreugdenhil, D., Zabeau, M. and Visser, R. G. F. (1996) Visualization of differential gene expression using a novel method of RNA fingerprinting based on AFLP: analysis of gene expression during potato tuber development. *Plant Journal*, 9, 745–753.

Baum, D. A. (1998) The evolution of plant development. *Current Opinion in Plant Biology*, 1, 79–86.

Berna, G., Robles, P. and Micol, J. L. (1999) A mutational analysis of leaf morphogenesis in *Arabidopsis thaliana. Genetics*, 152, 729–742.

Bradshaw, D., Otto, K. G., Frewen, B. E., McKay, J. K. and Schemske, D. W. (1998) Quantitative trait loci affecting differences in floral morphology between two species of monkeyflower (*Mimulus*). *Genetics*, 149, 367–382.

Brakefield, P., Gates, J., Keys, D., Kesbeke, F., Wijngaarden, P. J., Monteiro, A., French, V. and Carroll, S. B. (1996) Development, plasticity and evolution of butterfly eyespot patterns. *Nature*, 384, 236–242.

Cox, D. N., Chao, A., Baker, J., Chang, L., Qiao, D. and Lin, H. (1998) A novel class of evolutionarily conserved genes defined by *piwi* are essential for stem cell self-renewal. *Genes and Development*, 12, 3715–3727.

Coyne, J. A. (1983) Genetic basis of differences in genital morphology among three sibling species of *Drosophila*. *Evolution*, 37, 1101–1118.

DeVincente, M. C. and Tanksley, S. D. (1993) QTL analysis of transgressive segregation in an interspecific tomato cross. *Genetics*, 134, 585–596.

Doebley, J. and Lukens, L. (1998) Transcriptional regulators and the evolution of plant form. *Plant Cell*, 10, 1075–1082.

Doebley, J. and Stec, A. (1991) Genetic analysis of the morphological differences between maize and teosinte. *Genetics*, 129, 285–295.

Doebley, J., Stec, A. and Gustus, C. (1995) *Teosinte branched1* and the origin of maize: evidence for epistasis and the evolution of dominance. *Genetics*, 141, 333–346.

Donoghue, M., Ree, R. H. and Baum, D. A. (1998) Phylogeny and the evolution of flower symmetry in the Asteridae. *Trends in Plant Sciences*, 3, 311–317.

Dorweiler, J., Stec, A., Kermicle, J. and Doebley, J. (1993) *Teosinte glume architecture1*: a genetic locus controlling a key step in maize evolution. *Science*, 262, 233–235.

Falconer, D. and Mackay, T. F. C. (1996) *Introduction to Quantitative Genetics* (4th edn). Longman, Harlow.

Fletcher, J. C. and Meyerowitz, E. M. (2000) Cell signalling with the shoot meristem. *Current Opinion in Plant Biology*, 3, 23–30.

Frieden, E. (1981) The dual role of thyroid hormones in vertebrate development and calorigenesis, in *Metamorphosis, a Problem in Developmental Biology* (eds L. I. Gilbert and E. Frieden), Plenum Press, New York, pp. 545–563.

Frohlich, M. W. and Parker, D. S. (2000) The mostly male theory of flower evolutionary origins: from genes to fossils. *Systematic Botany*, 25, 155–170.

Givnish, T. J. (1978) On the adaptive significance of leaf form, in *Topics in Plant Population Biology* (eds O. T. Solbrig, S. Jain, G. B. Johnson and P. H. Raven), Columbia University Press, New York, pp. 375–407.

Givnish, T. J. (1987) Comparative studies of leaf form: assessing the relative roles of selective pressures and phylogenetic constraints. *New Phytologist*, 106, 131–160.

Gleissberg, S. (2002) Comparative developmental and molecular genetic aspects of leaf dissection, in *Developmental Genetics and Plant Evolution* (eds Q. B. C. Cronk, R. M. Bateman and J. A. Hawkins), Taylor and Francis, London, pp. 404–417.

Gottlieb, L. D. (1984) Genetics and morphological evolution in plants. *American Naturalist*, 123, 681–709.

Halder, G., Callaerts P. and Gehring, W. J. (1995) Induction of ectopic eyes by targeted gene expression of the *eyeless* gene in *Drosophila*. *Science*, 267, 1788–1792.

Hammond, D. (1941a) The expression of genes for leaf shape in *Gossypium hirsutum* L. and *Gossypium arboreum* L. I. The expression of genes for leaf shape in *Gossypium hirsutum* L. *American Journal of Botany*, 28, 124–138.

Hammond, D. (1941b) The expression of genes for leaf shape in *Gossypium hirsutum* L. and *Gossypium arboreum* L. II. The expression of genes for leaf shape in *Gossypium arboreum* L. *American Journal of Botany*, 28, 139–150.

Hilliard, O. M. (1976) Begoniaceae, in *Flora of Southern Africa*, 22, 136–144.

Hilu, K. W. (1983) The role of single-gene mutations in the evolution of flowering plants. *Evolutionary Biology*, 16, 97–128.

Hobmayer B., Rentzch, F., Kuhn, K., Happel, C. M., von Laue, C. C., Snyder, P., Rothbacher, U. and Holstein, T. W. (2000) WNT signalling molecules act in axis formation in the diploblastic metazoan *Hydra*. *Nature*, 407, 186–189.

Hofer, J., Turner, L., Hellens, R., Ambrose, M., Matthews, P. and Ellis, N. (1997) Unifoliata regulates leaf and flower morphogensis in pea. *Current Biology*, 7, 581–587.

Hudson, A. (1999) Axioms and axes in leaf formation? *Current Opinion in Plant Biology*, 2, 56–60.

Irish, V. F. and Yamamoto, Y. T. (1995) Conservation of floral homeotic gene function between *Arabidopsis* and *Antirrhinum. Plant Cell*, 7, 1635–1644.

Irmscher, E. (1961) Monographische Revision de Begoniaceen Afrikas. I. Sekt. Augustia und Rostrobegonia sowie einige neue Sippen aus anderen Sektionen. *Botanische Jahrbücher für Systematik*, 81, 106–188.

Jiang, C., Wright, R. J., Woo, S. S., Delmonte, T. A. and Paterson, A. H. (2000) QTL analysis of leaf morphology in tetraploid *Gossypium* (cotton). *Theoretical and Applied Genetics*, 100, 409–418.

Juenger, T., Purugganan, M. and Mackay, T. F. C. (2000) Quantitative trait loci for floral morphology in *Arabidopsis thaliana. Genetics*, 156, 1379–1392.

Keys, D. N., Lewis, D. L., Selegue, J. E., Pearson, B. J., Goodrich, L. V., Johnson, R. L., Gates, J., Scott, M. P. and Carroll, S. B. (1999) Recruitment of a *hedgehog* regulatory circuit in butterfly eyespot evolution. *Science*, 283, 532–534.

Kim, S. C. and Rieseberg, L. H. (1999) Genetic architecture of species differences in annual sunflowers: implications for adaptive trait introgression. *Genetics*, 153, 965–977.

Kincaid, D. T. and Schneider, R. B. (1983) Quantification of leaf shape with a microcomputer and Fourier transformation. *Canadian Journal of Botany*, 61, 2333–2342.

Kramer, E. M. and Irish, V. F. (1999) Evolution of genetic mechanisms controlling petal development. *Nature*, 399, 144–148.

Lande, R. (1981) The minimum number of genes contributing to quantitative variation between and within populations. *Genetics*, 99, 541–553.

Lander, E. S. and Botstein, D. (1989) Mapping Mendelian factors underlying quantitative traits using RFLP linkage maps. *Genetics*, 121, 185–199.

Luo, D., Carpenter, R., Vincent, C., Copsey, L. and Coen, E. (1996) Origin of floral asymmetry in *Antirrhinum. Nature*, 383, 794–799.

McLellan, T. (1990) Development of differences in leaf shape in *Begonia dregei* (Begoniaceae). *American Journal of Botany*, 77, 796–804.

McLellan, T. (2000) Geographic variation and plasticity in the size and shape of leaves of *Begonia dregei* and *B. homonyma* (Begoniaceae). *Botanical Journal of the Linnean Society*, 132, 79–95.

McLellan, T. and Endler, J. A. (1998) The relative success of some methods for measuring and describing the shape of complex objects. *Systematic Biology*, 47, 264–281.

Matolweni, L. O., Balkwill, K. and McLellan, T. (2000) Genetic diversity and gene flow in the morphologically variable, rare endemics *Begonia dregei* and *Begonia homonyma* (Begoniaceae). *American Journal of Botany*, 87, 431–439.

Money, T., Reader, S., Qu, L. J., Dunford, R. P. and Moore, G. (1996) AFLP-based mRNA fingerprinting. *Nucleic Acids Research*, 24, 2616–2617.

Mueller, U. G. and Wolfenbarger L. L. (1999) AFLP genotyping and fingerprinting. *Trends in Ecology and Evolution*, 14, 389–394.

Nuzhdin, S. V., Keightley, P. D., Paryukova, E. G. and Morozova, E. A. (1998) Mapping quantitative trait loci affecting sternopleural bristle number in *Drosophila melanogaster* using changes in marker allele frequencies in divergently selected lines. *Genetical Research*, 72, 79–91.

Orr, H. A. and Coyne, J. A. (1992) The genetics of adaptation: a reassessment. *American Naturalist*, 140, 725–742.

Palopoli, M. F. and Patel, N. H. (1996) Neo-Darwinian developmental evolution: can we bridge the gap between pattern and process? *Current Opinion in Genetics and Development*, 6, 502–508.

Panganiban, G., Irvine, S. M., Lowe, C., Roehl, H., Corley, L. S., Sherbon, B., Grenier, J. K.,

Fallon, J. F., Kimble, J., Walker, M., Wray, G. A., Swalla, B. J., Martindale, M. Q. and Carroll, S. B. (1997) The origin and evolution of animal appendages. *Proceedings of the National Academy of Sciences USA*, 94, 5162–5166.

Patel, N., Martin-Bianco, E., Coleman, K. G., Poole, S. J., Ellis, M. C., Kornberg, T. B. and Goodman, C. S. (1989) Expression of *engrailed* proteins in arthropods, annelids, and chordates. *Cell*, 58, 955–968.

Popadic, A., Panganiban, G., Rusch, D., Shear, W. A. and Kaufman, T. C. (1998) Molecular evidence for the gnathoblastic derivation of arthropod mandibles and for the appendicular origin of the labrum and other structures. *Development, Genes and Evolution*, 208, 142–150.

Popadic, A., Rusch, D., Peterson, M., Rogers, B. T. and Kaufman, T. C. (1996) Origin of the arthropod mandible. *Nature*, 380, 395.

Scanlon, M. J. (2000) Developmental complexities of simple leaves. *Current Opinion in Plant Biology*, 3, 31–36.

Shu, G., Amaral, W., Hileman, L. C. and Baum, D. A. (2000) *Leafy* and the evolution of rosette flowering in violet cress (*Jonopsidium acaule*, Brassicaceae). *American Journal of Botany*, 87, 634–641.

Sinha, N. (1999) Leaf development in angiosperms. *Annual Review of Plant Physiology and Plant Molecular Biology*, 50, 419–446.

Stephens, S. G. (1944) The genetic organization of leaf-shape development in the genus *Gossypium*. *Journal of Genetics*, 46, 28–51.

Stern, D. L. (2000) Evolutionary developmental biology and the problem of variation. *Evolution*, 54, 1079–1091.

Tabin, C. J., Carroll, S. B. and Panganiban, G. (1999) Out on a limb: parallels in vertebrate and invertebrate limb patterning and the origin of appendages. *American Zoologist*, 39, 650–663.

Tanksley, S. D. (1993) Mapping polygenes. *Annual Review of Genetics*, 27, 205–233.

Theißen, G., Becker, A., Winter, K-U., Münster, T., Kirchner, C. and Saedler, H. (2002) How the land plants learned their floral ABCs: the role of MADS-box genes in the evolutionary origin of flowers, in *Developmental Genetics and Plant Evolution* (eds Q. C. B. Cronk, R. M. Bateman and J. A. Hawkins), Taylor & Francis, London, pp. 173–205.

Tsukaya, H. (1995) Developmental genetics of leaf morphogenesis in dicotyledonous plants. *Journal of Plant Biology*, 108, 407–416.

Vos, P., Hogers, R., Bleeker, M., Reijans, M., Van De Lee, T., Hornes, M., Frijters, A., Pot, J., Peleman, J., Kuiper, M. and Zabeau, M. (1995) AFLP: a new concept in DNA fingerprinting. *Nucleic Acids Research*, 23, 4407–4414.

Voss, S. R. and Shaffer, H. B. (1997) Adaptive evolution via a major effect: paedomorphosis in the Mexican axolotl. *Proceedings of the National Academy of Sciences USA*, 94, 14185–14189.

Voss, S. R., Shaffer, H. B., Taylor, J., Safi, R. and Laudet, V. (2000) Candidate gene analysis of thyroid hormone receptors in metamorphosing vs. nonmetamorphosing salamanders. *Heredity*, 85, 107–114.

Wray, G. A. and Lowe, C. J. (2000) Developmental regulatory genes and echinoderm evolution. *Systematic Biology*, 49, 28–51.

Wu, R., Bradshaw, H. D. and Stettler, R. F. (1997) Molecular genetics of growth and development in *Populus* (Salicaceae). 5. Mapping quantitative trait loci affecting leaf variation. *American Journal of Botany*, 84, 143–153.

Zeng, Z.-B., Liu, J., Stam, L. F., Kao, C.-H., Mercer, J. and Laurie, C. C. (2000) Genetic architecture of a morphological shape difference between two *Drosophila* species. *Genetics*, 154, 299–310.

Chapter 17

Evolution of vascular plant body plans: a phylogenetic perspective

Harald Schneider, Kathleen M. Pryer, Raymond Cranfill,
Alan R. Smith and Paul G. Wolf

ABSTRACT

Extant vascular plants comprise three major lineages: Lycophytina, Moniliformopses and Spermatophytata. We have investigated the evolution of body plans of vascular plants using a phylogenetic framework to reconstruct morphological character state changes. Our phylogenetic definition of body plans is based on synapomorphies of the lineages of extant vascular plants. Fundamental body plan features considered include the structure of meristems, the position of sporangia, spore/pollen wall development, and life cycle changes. Phylogenetic evidence supports the presence of roots in the common ancestor of extant vascular plants and a single origin of euphylls prior to the divergence of extant euphyllophytes. Heterochronic and heterotopic mutations and morphological simplification have each played major roles in the evolution of vascular plants. Phylogenetic evidence and the fossil record are integrated to reflect our current understanding of the evolution of vascular plants since their origin in the late Palaeozoic. The phylogenetic position of model organisms commonly used in developmental gene studies illustrates the importance of improving and diversifying taxon selection in future evolutionary studies that use developmental genes.

17.1 Introduction

Current studies in plant evolution focus on three themes: (1) phylogenetic relationships among extant and/or extinct lineages of vascular plants; (2) evolution of plant structures and shapes; and (3) the evolution of genes that control plant development. All of these themes are pertinent to what has become known as evolutionary developmental genetics. In the past, and to some extent still today, studies focusing on phylogenetic relationships and the evolution of form interpreted the anatomy and morphology of extant plant taxa using *ad hoc* statements to identify primitive or derived characters (Goebel, 1933; Bower, 1935; Troll, 1937, 1939; Campbell, 1940; Wardlaw, 1952, 1965; Kaplan, 1977; Kaplan and Groff, 1995; Kato and Imaichi, 1997; Hagemann, 1999; Niklas, 2000a, b). Several studies have also used the fossil record to reconstruct the first appearance of taxa and characters that were then used as empirical data to interpret plant relationships and the evolution of plant morphology (Zimmermann, 1959, 1965; Gensel, 1977, 1992; Gensel *et al.*, 2001). Investigations of developmental and functional aspects of plant structures, such as

In *Developmental Genetics and Plant Evolution* (2002) (eds Q. C. B. Cronk, R. M. Bateman and J. A. Hawkins), Taylor & Francis, London, pp. 330–364.

biomechanical properties in the reconstruction of fossils, have also been popular approaches to infer the evolution of vascular plants (Wardlaw, 1952, 1965; Speck and Rowe, 1999; Niklas, 2000a, b; DiMichele *et al.*, 2001).

Recently, two new approaches have made possible remarkable advancements in our understanding of plant evolution. In the first approach, plant phylogeny is inferred from the application of stringent analytical methods (e.g. maximum parsimony and maximum likelihood optimisation criteria) to both DNA sequence data and morphological data. These studies have allowed new insights into plant relationships (Donoghue and Doyle, 2000; Soltis and Soltis, 2000; Pryer *et al.*, 2001) and the interpretation of morphological character evolution (Crane and Kenrick, 1997; Kenrick and Crane, 1997; Bateman *et al.*, 1998; Doyle and Endress, 2000; Graham *et al.*, 2000; Renzaglia *et al.*, 2000). The second approach is based on the growing understanding of the role of dedicated genes (e.g. transcription factors) in controlling plant development, which has inspired new studies that focus on plant development in an evolutionary context (Doyle, 1994; Kramer and Irish, 1999, 2000; Frohlich and Parker, 2000; Lawton-Rauh *et al.*, 2000; Riechmann *et al.*, 2000; Vergara-Silva *et al.*, 2000). Varied terms have been used for the genetic factors involved in the regulation of plant development, such as receptors, transducers and transcription factors (Doebley and Lukens, 1998). Here, we use the term 'developmental genes' in a broad sense (Arthur, 1997; Gilbert, 2000; Morange, 2000). In plants, MADS-box genes are the most commonly studied developmental genes used to infer the evolution of key features of seed plants, such as the evolution of flowers (Hasebe, 1999; Hasebe and Ito, 1999; Shindo *et al.*, 1999; Winter *et al.*, 1999; Alvarez-Buylla *et al.*, 2000a, b; Becker *et al.*, 2000; Krogan and Ashton, 2000; Smyth, 2000; Svensson *et al.*, 2000; Theißen, 2000; Theißen *et al.*, 2000; Vergara-Silva *et al.*, 2000). Other kinds of plant developmental genes, such as homeodomain genes (Bharathan *et al.*, 1997, 1999; Aso *et al.*, 1999; Richards *et al.*, 2000; Sakakibara *et al.*, 2001), MYB genes (Kranz *et al.*, 2000) and phytochrome genes (Schneider-Poetsch *et al.*, 1998; Basu *et al.*, 2000), have been utilised in only a few evolutionary studies. Other studies have explored the evolution of actin genes, which encode a major component of the cytoskeleton, because duplication and modification of these genes is involved in the evolution of morphological complexity at the cellular level (Bhattacharya *et al.*, 2000).

The potential of these new sources of data to answer long-standing questions about plant evolution is staggering. Developmental genes, such as HOX-box genes, have already provided critical insights into the genetic basis of the developmental evolution of animals (Hall, 1996; Raff, 1996; Arthur, 1997; Gellon and McGinnis, 1998; Graham, 2000; Grbic, 2000; Jenner, 2000; Kappen, 2000; Peterson and Davidson, 2000; Wray and Lowe, 2000), prompting the application of similar approaches to plants (Hasebe, 1999; Kramer and Irish, 1999, 2000; Theißen, 2000; Vergara-Silva *et al.*, 2000). A future challenge will be to integrate phylogenetic reconstruction, morphological studies and developmental genetic data (Bateman, 1999; Valentine *et al.*, 1999; Kellogg, 2000a; Mabee, 2000). A series of nested studies might be envisaged to meet this challenge: (1) nucleotide sequence data of coding and/or non-coding DNA regions can be used to reconstruct the phylogeny; (2) extensive data sets comprising anatomical, biochemical, cytological and morphological characters can be used to infer character evolution on the resultant phylogeny; and (3) the phylogeny, with its explicit character transformation state-ments, can be compared to gene trees based on sequence data of developmental genes to further our understanding of the evolution of plant development. Researchers favouring a total evidence approach (de Queiroz, 2000; Hillis and Wiens, 2000) could combine steps 1 and 2 to construct a phylogeny based on both morphological and molecular data. Whatever the approach used, it seems advisable to maintain step 3 as an independent exercise.

A recent phylogenetic study by Pryer *et al.* (2001) utilising five data sets comprising three chloroplast genes (*atpB*, *rbcL*, *rps4*), nuclear small subunit (SSU) ribosomal DNA, and an extensive morphological matrix resulted in a new understanding of the relationships among major lineages of extant vascular plants. They refuted previous hypotheses of spore-bearing vascular plants as transitional evolutionary grades between bryophytes and seed plants. In particular, the hypothesis that *Psilotum* is a 'living fossil' with a close relationship to Lower Devonian psilophytes (Kaplan, 1977; Wagner, 1977; Rothwell, 1999) no longer appears tenable. These results call for a reinterpretation of the evolution of plant morphology. Reconstruction of phylogeny and morphological character state changes allows us to infer the relationship between ontogeny and phylogeny (Rieppel, 1993; Bang *et al.*, 2000; Collazo, 2000), mechan-isms of evolution (Hall, 1996; Raff, 1996, 1999; Arthur, 1997, 2000a; Budd, 1999; Donoghue and Ree, 2000; Gibson and Wagner, 2000; Wagner and Schwenk, 2000), and the acquisition of 'key innovations' and body plans in the evolution of organisms (Arthur, 2000b; Graham *et al.*, 2000; Wagner *et al.*, 2000).

17.2 Methodology

17.2.1 Reconstruction of phylogenetic relationships

A phylogeny of vascular plants comprising representatives from all major extant clades was reconstructed using maximum likelihood analysis of nucleotide sequences from three chloroplast genes (*atpB*, *rbcL*, *rps4*) and nuclear SSU rDNA (Pryer *et al.*, 2001). This phylogeny is referred to subsequently as 'Phylogeny 2001'. Relation-ships among the bryophyte outgroups are the subject of current controversy (Lewis *et al.*, 1997; Nickrent *et al.*, 2000; Qiu and Lee, 2000). Because Phylogeny 2001 exhibited a polytomy among the outgroups, we follow here a most recent hypothesis of the relationships among the four lineages of land plants (Lewis *et al.*, 1997; Nick-rent *et al.*, 2000; Qiu and Lee, 2000) in order to optimise character state reconstruc-tion (Maddison and Maddison, 1992). Three alternative topologies, in which the sister group to tracheophytes is either (1) hornworts, (2) mosses or (3) a clade com-prising liverworts and mosses, were considered initially. Reconstruction of character states within the vascular plants was not affected by these outgroup choices; there-fore, character state changes were reconstructed using liverworts (Marchantiomor-pha) as outgroup.

17.2.2 Reconstruction of character evolution

Characters taken from an extensive morphological data set were mapped onto Phy-logeny 2001 (see Section 17.2.1). This morphological data set consists of 136 characters, including features of general morphology, anatomy, cytology, biochemistry and some structural DNA data. The data set was especially designed for an independent analysis of phylogenetic relationships among vascular plants and does not generally include characters that are of interest only for terminal groups such as flowering plants, horsetails, derived ferns, or for relationships among the outgroups. The morphological data matrix is available from the senior author. Character evolution was reconstructed for this data set using both accelerated transformation (Acctran) and delayed transformation (Deltran) optimisation as implemented in MacClade 3.0 (Maddison and Maddison, 1992). Character state changes were treated as ambiguous if the application of the two optimisation criteria resulted in different reconstructions. All characters shown in Figures 17.2–17.8 are treated as unordered. The reconstructed phylogeny (Section 17.3) and optimised character state changes (Section 17.4) formed the basis of an investigation into the nature of evolutionary transformations among vascular plants (Section 17.5).

17.3 Phylogeny of vascular plants (phylogenetic statements)

Extant vascular plants comprise three major lineages (Phylogeny 2001, Figure 17.1): lycophytes, seed plants and non-lycophyte pteridophytes. The third lineage comprises leptosporangiate ferns (Polypodiidae), two extant lineages of eusporangiate ferns (Marattiidae, Ophioglossidae), whisk ferns (Psilotidae) and horsetails (Equisetopsida). This clade is referred to throughout this chapter as Moniliformopses (or moniliforms), reflecting a classification first introduced by Kenrick and Crane (1997). The horsetails (Equisetopsida) and Marattiidae form a clade that is, in turn, sister to the leptosporangiate ferns (Polypodiidae). The basalmost branch of the Moniliformopses is a clade that includes Psilotidae and Ophioglossidae. Within seed plants, angiosperms are shown as sister to a monophyletic gymnosperm clade. This is not the first time that the anthophyte hypothesis has been refuted. Other phylogenetic analyses using DNA sequence data and denser taxonomic sampling also show that Gnetaceae is not sister to angiosperms (Doyle, 1996, 1998; Donoghue and Doyle, 2000; Sanderson et al., 2000). Here, *Gnetum* is sister to the conifer *Pinus*; similar topologies are reported in recent studies focused on seed plant phylogeny (Doyle, 1998; Barkman et al., 2000; Bowe et al., 2000; Chaw et al., 2000; Donoghue and Doyle, 2000).

17.4 Character evolution of vascular plants

17.4.1 Character state changes and vascular plant lineages

The total number of character state changes, as well as the number of unambiguous character state changes, are reported for each branch in Figure 17.1 for the 136 morphological characters that were mapped onto Phylogeny 2001. The morphological data set did not include characters that are informative only for terminal groups (e.g. floral characters). The number of character state changes, therefore, is relatively low for several derived branches.

The branches supporting the main lineages of vascular plants have relatively high

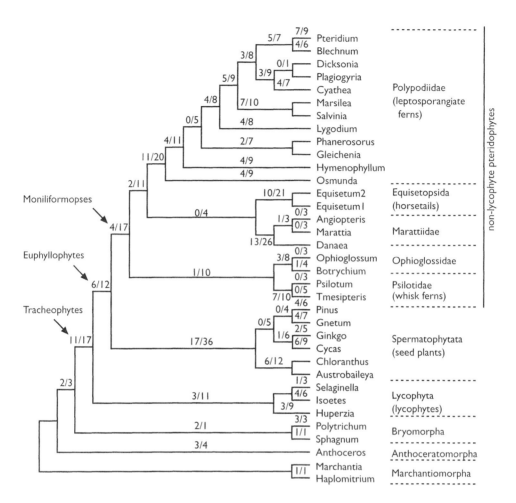

Figure 17.1 Phylogeny of vascular plants (referred to as Phylogeny 2001 throughout the text) as shown in Pryer *et al.* (2001), except for outgroup relationships, which have been redrawn (see Section 17.2.1). Character state evolution was reconstructed for 136 morphological characters using this phylogeny. The number of unambiguous morphological character state changes is given for each branch; total number of morphological character states changes (ambiguous + unambiguous) is shown after the slash. Taxonomy follows Kenrick and Crane (1997).

numbers of unambiguous and total character state changes: Lycophyta (3/11), Euphyllophytes (6/12), Spermatophytata (17/36), Moniliformopses (4/17) (Figure 17.1). The number of character state changes for the seed plant clade is high, reflecting the remarkable evolutionary transformation of major morphological features within this lineage after its divergence from other vascular plants. The majority of these spermatophyte character state changes are associated with the evolution of seeds. Within Moniliformopses, each of the five principal lineages shows a relatively high number of character state changes: Equisetopsida (10/21), Marattiidae (13/26),

Ophioglossidae (3/8), Polypodiidae (11/20), Psilotidae (7/10). In striking contrast, the deeper branches that support clades such as Ophioglossidae + Psilotidae (1/10) and Equisetopsida + Marattiidae (0/4) show one and no unambiguous character state changes, respectively. This imbalance between support for these deeper clades and clades one step higher in the phylogenetic hierarchy is remarkable.

17.4.2 Phylogenetic classification of body plans of vascular plants

The reconstruction of morphological character state changes on a robust phylogeny provides us with guidelines to define the body plans of vascular plants. This approach differs from previous attempts to define body plans of vascular plants, which were based on *ad hoc* interpretations combining historical and functional aspects (Rothwell, 1995; Niklas, 2000a). A brief definition of tracheophyte body plans, based primarily on characters of extant taxa, is given here.

All vascular plants share such character states as an independent sporophyte and multiple sporangia. Nearly all vascular plants, except some genera of Lemnaceae (Cook, 1999), possess differentiated vascular tissues and endodermal sheaths. Most vascular plants also produce lignin and mechanical tissues such as sclerenchyma and collenchyma. Tracheophytes exhibit two major body plans: (1) shoots with exarch protoxylem poles, dichopodial roots with endarch protoxylem poles, and lycophylls (=microphylls), which are characteristic of lycophytes; (2) shoots with endarch to mesarch protoxylem poles, monopodial roots with exarch protoxylem poles, and euphylls (=megaphylls), which are characteristic of euphyllophytes. Roots are absent from few groups of euphyllophytes.

Extant lycophytes include two major kinds of body plans: (1) the ligulate type and (2) the non-ligulate type. The ligulate type possesses a ligule on the adaxial surface of the microphylls and has differentiated structures (rhizophores or rhizomorphs) that bear roots. Extant euphyllophytes also include two distinct body plan types: (1) the seed plant type with eusteles, general occurrence of secondary growth, lateral roots borne from pericycle/pericambium cells, extreme heterospory and seeds; and (2) the moniliform type with solenosteles (or dictyosteles), generally lacking secondary growth, lateral roots borne from endodermis cells, periplasmodial tapetum, pseudoendospore and spore wall development that is exclusively centrifugal.

Extant moniliforms include five main body plans that correspond to each of the main lineages: (1) the psilotoid-type is defined by the absence of roots, reduced euphylls, and differentiation of the shoot into an erect photosynthetic portion and a creeping non-photosynthetic portion; (2) the ophioglossoid-type is defined by a reduction in the number of euphylls to one per shoot produced at any given time, usually unbranched roots, and the absence of root hairs; (3) the marattioid-type is defined by shoots with polycyclic steles, roots with septate root hairs, and leaves with pulvini, scattered pneumathodes, and polycyclic vascular bundles; (4) the equisetoid-type is defined by reduced euphylls that are arranged in whorls, shoots differentiated into creeping and erect parts, presence of extensive lacunae systems in the ribbed shoots, and endogenous origin of lateral shoots; and (5) the polypodioid-type is defined by the occurrence of leptosporangiate sporangia formed from single epidermal cells, a reduced number of protoxylem poles per root (in general two), and the absence of a root pith.

Extant seed plants include two major body plans: (1) the gymnosperm-type with embryos that arise from a multinucleate zygote, phloem tissue with Strassburger cells, and secondary xylem cells of the coniferoid-type; and (2) the angiosperm-type with embryos that arise from a uninucleate zygote, phloem tissue with companion cells, a secondary endosperm, and flowers. Detailed definitions of the four body plan subtypes nested within the gymnosperm-type (cycadoid, ginkgoid, gnetoid, coniferoid) are not presented here because definitions need to be based on a phylogenetic analysis with a broader taxon sampling of seed plants.

17.4.3 Evolution of main features of vascular plants

The evolution of tracheophyte characters in comparison to other land plants and green algae has been examined in detail in previous studies (Kenrick and Crane, 1997; Edwards, 1999; Graham *et al.*, 2000; Renzaglia *et al.*, 2000). In the following text, we infer the evolution of a few selected characters (Figures 17.2–17.8) within vascular plants using Phylogeny 2001 (Pryer *et al.*, 2001).

A. Life cycle

Although the evolution of the life cycle of land plants has been explored in previous studies (Kenrick, 1994; Kenrick and Crane, 1997), differences in the life cycles among tracheophytes have yet to be examined in detail. It has been suggested that bryophytes and tracheophytes share a common ancestor possessing isomorphic gametophytic and sporophytic phases (Kenrick, 1994; Kenrick and Crane, 1997), whereas extant land plants have two phases that differ in form (heteromorphic) and duration. It is well known that vascular plants differ from bryophytes in having a dominant (or co-dominant) and independent sporophyte (Figure 17.2a, b). In general, tracheophytes have a gametophytic phase that is short-lived, although this is not the case in several basal lineages (Lycopodiales, Marattiidae, Ophioglossidae, Psilotidae). Nevertheless, it appears that the condition of having extremely short-lived gametophytes and long-lived sporophytes has evolved at least three times within vascular plants: heterosporous lycophytes (Isoëtales, Selaginellales), seed plants (Spermatophytata), and leptosporangiate ferns (Polypodiidae) (Figure 17.2a).

Another interesting aspect of the life cycle of land plants is the existence of a period of dormancy, which is intercalated between the sporophytic and gametophytic phases in bryophytes and pteridophytes (in the form of a haploid spore), but which occurs between the gametophytic and sporophytic phases in seed plants (in the form of a diploid embryo enclosed in a seed) (Figure 17.2b).

B. Meristems

Sporophytes of euphyllophytes possess at least three kinds of apical or marginal meristems that are involved in the formation of new organs: shoot meristems, root meristems, and leaf meristems. Intercalary meristems and cambia are ignored here because in general they are not involved in the formation of new organs. Fossil

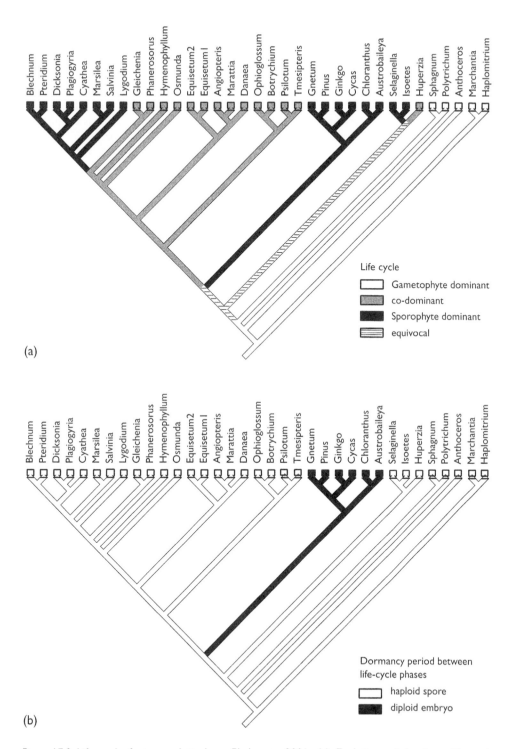

Figure 17.2 Life cycle features plotted on Phylogeny 2001. (a) Evolution of dominant life cycle phases in land plants. (b) Dormancy period between life cycle phases in land plants.

evidence suggests that the common ancestor of vascular plants possessed only one type of meristem per sporophyte and one type of meristem per gametophyte (Philipson, 1990; Kenrick and Crane, 1997). In bryophytes, the gametophytes possess only one meristem type, whereas the sporophytes have one apical shoot meristem or grow exclusively via an intercalary meristem (Kenrick and Crane, 1997).

Reconstruction of root evolution (Figure 17.3a) suggests a differentiation of shoot and root meristems in the common ancestor of all extant lineages of tracheophytes. Subsequent differentiation of meristem types in the common ancestor of the euphyllophytes resulted in a leaf meristem that produces euphylls (Figure 17.4b), whereas lycophylls grow exclusively with an intercalary meristem (Figure 17.4a). This scenario is consistent with the phylogeny but it needs to be confirmed with studies that address the genetic control of organogenesis and, in particular, organ identity. Genes controlling leaf identity of euphylls are assumed to be different from those controlling leaf identity of lycophylls (Kerstetter and Poethig, 1998; Foster and Veit, 2000; Frugis et al., 1999; Tsukaya, 2000), whereas root identity genes are assumed to be homologous among tracheophytes (Benfey, 1999; Bai et al., 2000; Costa and Dolan, 2000).

The three organs found in moniliforms (leaves, roots, shoots) have a meristem structure that has a single apical cell, similar to that found in bryophytes and the lycophyte lineage Selaginellales (Figure 17.5a). In contrast, the two other extant lineages of lycophytes (Lycopodiales and Isoëtales) and seed plants possess complex meristems. The meristems of lycophytes and seed plants differ substantially, and it is still unclear whether complex meristems are homologous, as proposed by Philipson (1990), or merely analogous.

C. Root–shoot differentiation

The root is one of the three basic organs of vascular plants, yet the phylogenetic origin of roots is rarely discussed (Zimmermann, 1965; Kutschera and Sobotik, 1997; Gensel et al., 2001; Raven and Edwards, 2001). Some authors (Goebel, 1933; Hagemann, 1992, 1997, 1999) suggest that roots originated as tuberous storage organs. However, some Lower Devonian vascular plant fossils suggest that roots evolved from creeping, elongate shoot-like structures (Remy et al., 1997; Gensel et al., 2001; Raven and Edwards, 2001). It has been argued that roots of lycophytes and euphyllophytes are not homologous because roots are unknown from many Lower Devonian trimerophytes and zosterophytes (Gensel, 1992; Stewart and Rothwell, 1993; Taylor and Taylor, 1993; Gensel et al., 2001; Raven and Edwards, 2001). However, fossil evidence for roots is often ambiguous (Kenrick and Crane, 1997; Gensel et al., 2001; Raven and Edwards, 2001) and root-like structures are known for some Lower Devonian taxa (Remy et al., 1997; Gensel et al., 2001; Raven and Edwards, 2001). Phylogenetic evidence indicates that roots of lycophytes and euphyllophytes are homologous (Figure 17.3a).

Roots of euphyllophytes and lycophytes share several structural features such as a calyptra, endogenous origin of the shoot-borne root and presence of root hairs, but they differ in two notable characters. First, shoot-borne roots of lycophytes show dichopodial branching, whereas shoot-borne roots of euphyllophytes show monopodial branching with lateral roots differentiated endogenously (Figure 17.3b). This

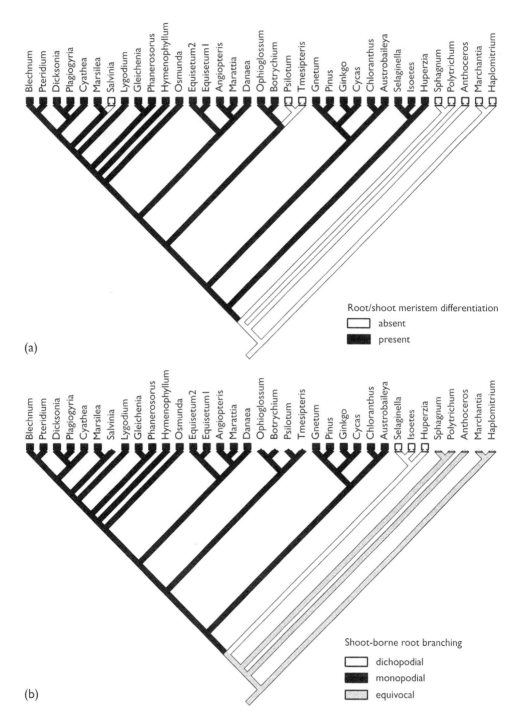

Figure 17.3 Root characters plotted on Phylogeny 2001. (a) Root/shoot meristem differentiation. (b) Branching of shoot-borne roots. The latter character is not applicable to taxa with unbranched shoot-borne roots (*Ophioglossum*, *Botrychium*) and rootless taxa such as bryophytes, whisk ferns (*Psilotum*, *Tmesipteris*) and *Salvinia*. Shoot-borne roots of *Ophioglossum* are unbranched, with the exception of *Ophioglossum palmatum*, which sometimes has dichotomously branched roots.

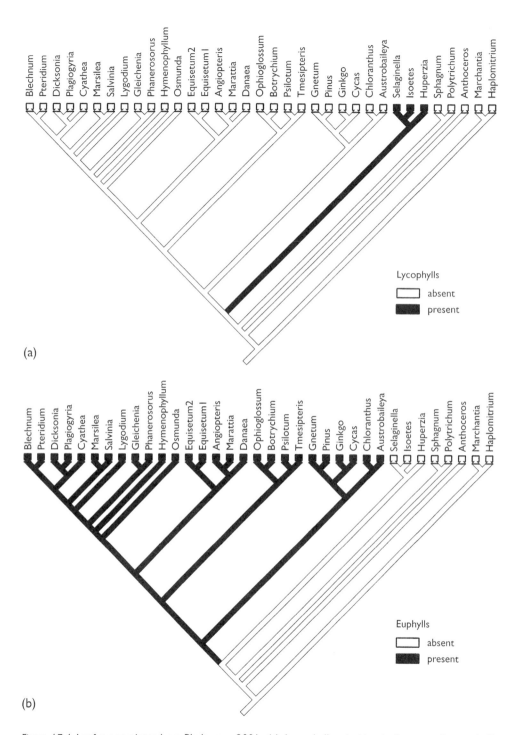

Figure 17.4 Leaf types plotted on Phylogeny 2001. (a) Lycophylls = lacking leaf gaps and an apical/marginal leaf meristem. (b) Euphylls = possessing leaf gaps and an apical/marginal leaf meristem.

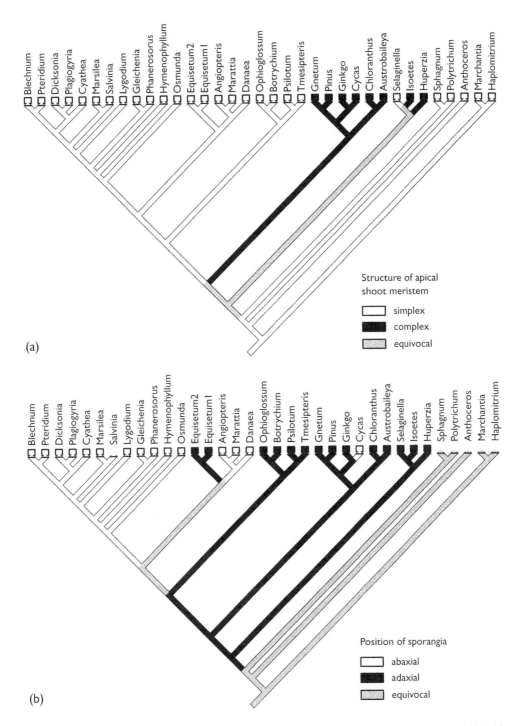

Figure 17.5 Shoot apical meristem structure and sporangial position plotted on Phylogeny 2001. (a) Shoot apical meristem organisation; simplex (with single apical cell) and complex (without a single apical cell), definitions according to Philipson (1990). (b) Position of sporangia relative to leaf-like organs; adaxial sporangia are attached to the shoot above the leaf or on the adaxial surface of the leaf, abaxial sporangia are attached to the abaxial surface of the leaf. *Salvinia* is scored as unknown because the interpretation of the highly modified, submerged, sporangia-bearing organ is unclear. Seed plants are scored according to Doyle (1996). Bryophyte sporophytes lack leaf-like structures and the character is therefore not applicable.

character is not applicable to rootless taxa such as Psilotidae and taxa with unbranched roots, such as the majority of Ophioglossidae. Second, protoxylem poles are located in an endarch position in lycophytes but exarch in euphyllophytes. This character has a reverse correlation with the position of protoxylem strands in the shoot stele, which are exarch in lycophytes and endarch or mesarch in euphyllophytes.

The main features of roots are conserved in the evolution of euphyllophytes except in the Ophioglossideae + Psilotidae clade, where root systems are reduced or absent (Figure 17.3a, b). Leptosporangiate ferns (Polypodiidae) are characterised by the absence of a root pith. This loss of a root pith may have occurred twice in closely related lineages, Polypodiidae and Equisetopsida, or it may be a synapomorphy of the clade including these two lineages and Marattiidae, with a reversal in Marattiidae.

Several other characters of root systems correlate with characters found in other organs; for example, roots with secondary growth are found only in taxa with secondary shoot growth, and homorhizy is correlated with the presence of seeds.

D. Leaf–shoot differentiation

The evolution of leaves is often discussed with reference to various 'leaf' characters, such as dorsiventral organisation, leaf gaps, and branched venation (Arber, 1950; Wagner, 1977; Wagner et al., 1982, Rutishauser, 1999; Dengler and Tsukaya, 2001). Recent phylogenetic studies support the independent origin of two leaf-like organs in vascular plants (Figure 17.4a, b): the lycophylls (=microphylls) of lycophytes, and the euphylls (=megaphylls) of euphyllophytes (Kenrick and Crane, 1997; Pryer et al., 2001). Crane and Kenrick (1997) proposed that lycophylls are transformed sporangia, whereas euphylls appear to be modified shoot systems (Zimmermann, 1959, 1965). Differences of opinion surrounding leaf origin in land plants (Niklas, 2000a, b) can be attributed, in part, to the use of different criteria to define leaves (Rutishauser, 1999). Only two features are consistently present in all leaves of euphyllophytes (with very few exceptions) but always absent from lycophytes: leaf gaps and development by an apical or marginal meristem. A further observation is the association of euphylls with lateral branches (Arber, 1950; Rutishauser, 1999), which are always axial only in extant seed plants. In moniliforms, lateral branches are generally located close to, but rarely within, the axils of leaves (Galtier, 1999). In addition, shoot branching patterns were more varied in Palaeozoic seed plants than they are in extant ones, and included non-axial and axial lateral branches (Galtier, 1999).

Other features used to define leaves often reflect functional specialisation and therefore are not useful for determining homology. For example, leaf-like structures of bryophytes and vascular plants share a planar shape, yet this is not an indicator of homology but is probably the result of functional constraints (Beerling et al., 2001; Raven and Edwards, 2001). Branched veins in leaves of a few species of *Selaginella* (Wagner et al., 1982) are also likely to be the result of independent evolutionary innovation and not evidence for their homology with euphyllophyte leaves. Similarly, several leaf characters, such as dorsiventral organisation, petiole-blade differentiation, marginal meristem, simple blades, anastomosing venation, and differentiation of palisade and spongy parenchyma, may have evolved or been lost independently in different lineages after the establishment of euphylls.

The homology of leaves of ferns and seed plants has been questioned (Wagner *et al.*, 1982; Rutishauser, 1999) even though they share the occurrence of leaf gaps and of apical and/or marginal meristems. Leaves of extant members of these lineages do differ substantially in their development (Hagemann, 1984), but similar foliage patterns observed in progymnosperms and ferns suggest that a shared developmental program of leaf formation existed in the common ancestor of moniliforms and seed plants. Angiosperms have a notable diversity of leaf development patterns (Tsukaya, 2000; Kaplan, 2001), but a comparative study including other seed plant lineages is lacking. Some features such as basipetal growth (Hagemann, 1984; Tsukaya, 2000) are likely to be restricted to flowering plants.

Current studies of genes that control leaf identity and formation have been carried out exclusively on derived angiosperms (Bowman, 2000; Foster and Veit, 2000; Tsukaya, 2000; Dengler and Tsukaya, 2001), and the results were rarely reported in a comparative framework and, unfortunately, never with a phylogenetic perspective. The remarkable diversity of leaf shapes and structures in early euphyllophytes (Taylor and Taylor, 1993; Galtier and Phillips, 1996) is evident in some features of the leaves of horsetails (Equisetopsida), whisk ferns (Psilotidae) and moonworts (Ophioglossidae). In horsetails and whisk ferns the leaves are extremely reduced and no leaf gap is present in *Psilotum*. However, the closely related genus *Tmesipteris* possesses larger leaves with leaf gaps. The leaves of Psilotidae and Ophioglossidae are not associated with lateral branches but with fertile structures called sporangiophores. The homology of these structures is unclear, but it is thought that they are reduced branches. In addition, the long and extensive fossil record of members of the horsetail lineage documents a reduction (simplification) of the leaves during their evolution (Zimmermann, 1965; Stewart and Rothwell, 1993; Taylor and Taylor, 1993).

E. Position of sporangia

Sporangia are found attached either to the adaxial or abaxial surface of a leaf-like structure (Figure 17.5b). Abaxial sporangia are found in Polypodiidae and Marattiidae, whereas adaxial sporangia are found in the most basal lineage of moniliforms comprising Ophioglossidae and Psilotidae. Traditional interpretations of the position of sporangia in Equisetopsida (Stewart and Rothwell, 1993; Taylor and Taylor, 1993) suggest an attachment of the sporangia to an adaxial sporangiophore, similar to the condition found in Psilotidae and Ophioglossidae. This hypothesis indicates either an independent origin of abaxial sporangia in Marattiidae and Polypodiidae or a reversal to the adaxial position in Equisetopsida (Figure 17.5b). According to Doyle (1996), most extant seed plants bear sporangia in an adaxial position.

F. Spore/pollen wall evolution

Spore and pollen wall formation is a highly conserved character within the major lineages of land plants. The exine of bryophytes, lycophytes and seed plants develops in two directions, centripetally and centrifugally, but the exine of moniliforms develops exclusively centrifugally (Figure 17.6a) (Rowley, 1996). Moniliforms share several unique features of spore development and structure, such as periplasmodial

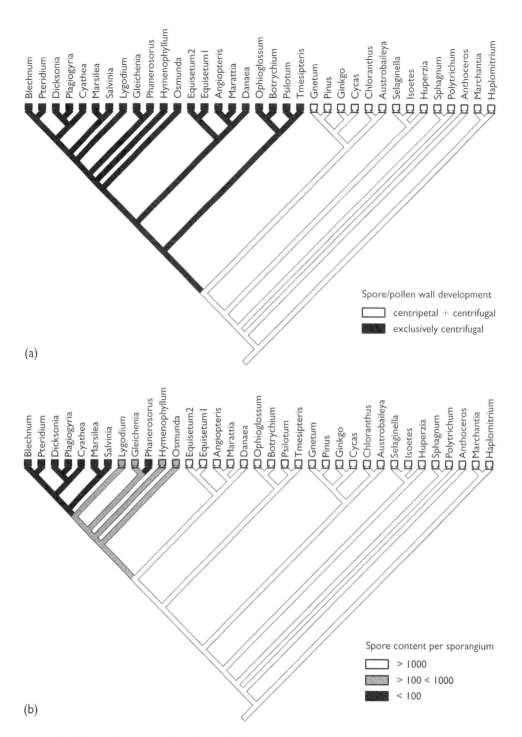

Figure 17.6 Spore characters plotted on Phylogeny 2001. (a) Spore/pollen wall develops in two directions, centripetally (inwardly) and centrifugally (outwardly), or in only one direction (centrifugally). (b) Spore content per sporangium.

tapetum and pseudoendospore. In turn, the seed plant pollen wall is very complex and differs in several ultrastructural aspects (e.g. differentiation of nexine and sexine) from spore walls found in bryophytes, lycophytes and moniliforms.

G. Other characters of vascular plants

Several other features of land plants are worthy of discussion in a phylogenetic context. The evolution of the stele is among the most critical of these. Lycophytes and euphyllophytes differ from each other in the position of the protoxylem poles in shoots and roots (Figure 17.7a, b). Lycophytes have endarch roots and exarch shoots, whereas euphyllophytes have exarch roots and endarch to mesarch shoots (except *Psilotum*). This indicates a substantially different organisation of tissue formation in the root and shoot, but very little is known about the positional control of vascular tissue development in plants (Bowman, 2000; Scheres, 2000). Further notable aspects of stele types are the mesarch position of protoxylem poles (Figure 17.7a) and the formation of siphonosteles in most moniliforms (Figure 17.8a). These character states distinguish the lineage from the seed plants, which always possess endarch protoxylem poles and a eustele. These differences have been noted as the primary distinctions between these sister lineages (Stein, 1993; Kenrick and Crane, 1997; Pryer *et al.*, 2001), but the conspicuous variability of the protoxylem pole position within the Ophioglossidae + Psilotidae clade introduces some ambiguity in the phylogenetic reconstruction of this character.

The presence of collateral vascular bundles in leaves of seed plants and ophioglossoids (Figure 17.8b) was used as evidence for a relationship between extinct progymnosperms and extant ophioglossoid ferns (Kato, 1988). Phylogenetic reconstruction indicates that this character is probably the plesiomorphic condition for euphyllophytes, in contrast to previous interpretations in which it was considered to be a synapomorphy for seed plants (Kato, 1988). Amphicribal vascular bundles appear to have evolved at least twice: in the Cycadatae and in the clade comprising horsetails, marattiaceous and leptosporangiate ferns. Reconstruction of this character using fossil evidence would be useful in corroborating this interpretation.

Gametophyte form and life-history are among the more interesting characters when considering the evolution of early land plants. Cylindrical, heterotrophic gametophytes are found in several basal lineages (Lycopodiales, Ophioglossidae, Psilotidae). However, some gametophytes of Lycopodiales are also autotrophic. In vascular plants, green thalloid gametophytes are restricted to leptosporangiate ferns and marattiaceous ferns, whereas the gametophytes of horsetails are also green but fimbriate. Reconstruction of the ancestral life-history of vascular plant gametophytes is obscured by heterosporous lineages such as ligulate lycophytes and seed plants that have extremely reduced, semiautotrophic to heterotrophic gametophytes. The reconstruction of the ancestral gametophyte condition is also complicated by the extinction of early Silurian vascular plant lineages (Kenrick and Crane, 1997) and the likelihood that extant bryophyte gametophytes have derived body plans (Proctor, 2000). So far, neither fossils (Kenrick, 1994; Kenrick and Crane, 1997) nor phylogenetic reconstructions have recovered unambiguous evidence for thalloid gametophytes as the plesiomorphic condition, as hypothesised in some interpretations of plant evolution (Hagemann, 1999). Upper Silurian and Lower Devonian

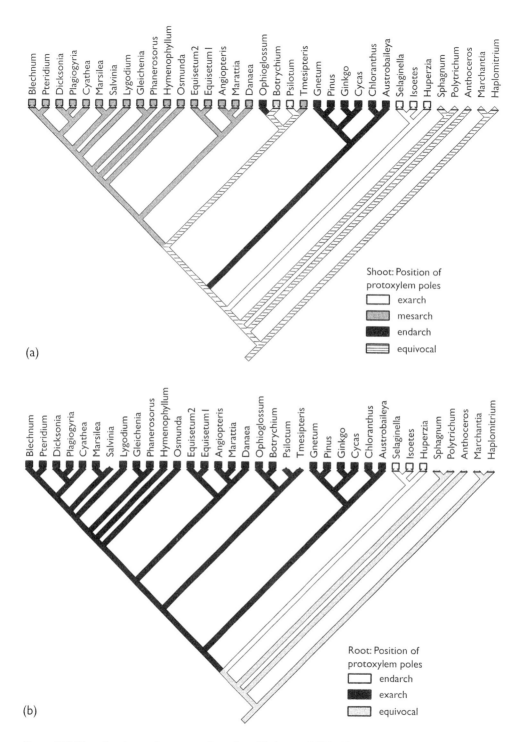

Figure 17.7 Vascular tissue characters plotted on Phylogeny 2001. (a) Position of protoxylem poles in shoot stele. (b) Position of protoxylem poles in root vascular bundle. The latter character is not applicable to rootless taxa such as *Salvinia* and Psilotaceae. Neither character is applicable to bryophytes.

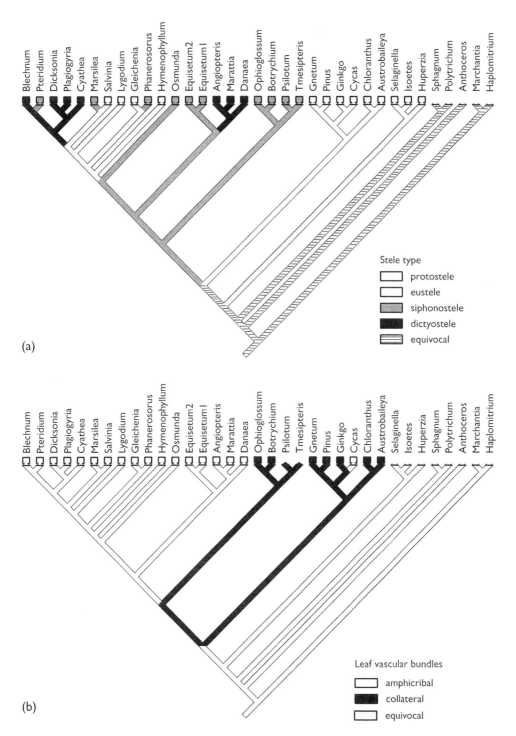

Figure 17.8 Vascular tissue characters plotted on Phylogeny 2001. (a) Structure of the stele in the mature shoot. (b) Structure of the vascular bundles in the leaf. Vascular tissue in leaves of Psilotidae is highly reduced and is scored as an unknown character state. Neither character is applicable to bryophytes.

fossils have suggested that early land plant gametophytes were cylindrical (Gensel, 1992; Kenrick, 1994; Kenrick and Crane, 1997; Edwards, 1999; Gensel et al., 2001).

17.5 Categories of transformations involved in the evolution of vascular plants

17.5.1 Iteration

Plants are modular organisms and in general each structure exists in several multiplications (Tomlinson, 1984). It is, therefore, difficult to identify duplication events of phylogenetic significance. Nevertheless, one example is the multiplication of vascular stele cycles resulting in polycyclic structures within shoots or petioles. Polycyclic steles are rare, occurring in only a few lineages of moniliforms, but they characterise the Marattiidae. A further example of multiplication without modification is the increased number of sperm cell flagellae in euphyllophytes (Renzaglia et al., 2000).

17.5.2 Modification of plant development

A. Heterochrony

Several character state changes may be caused by heterochronic mutations that result in an alteration in the sequence and timing of developmental processes (Mosbrugger, 1995; Raff, 1996; Friedman and Carmichael, 1998; Klingenberg, 1998; Gould, 2000; Kellogg, 2000a; Li and Johnson, 2000). Changes in the length of the gametophytic or sporophytic phases, as discussed above in Section 17.4.3A, are likely to be the result of heterochronic events in the evolution of vascular plants (Figure 17.2a). For example, extremely short-lived gametophytes have arisen at least three times: ligulate lycophytes, seed plants and heterosporous leptosporangiate ferns. Another possible example of heterochrony is the shift of the dormancy period between life phases from the haploid spore to the diploid embryo enclosed in the seed (Figure 17.2b). This transformation is correlated with the evolution of seeds, and recent studies of the evolution of seed storage globulins have demonstrated that a vicilin-like protein is specifically expressed in fern spores (Shutov et al., 1998). In seed plants, members of this gene family are expressed exclusively in the seed (dormancy phase). Heterochronic transformations may also be responsible for the reduction in number of spores produced per sporangium (Section 17.5.3; Figure 17.6b).

B. Heterotopy

Several character state changes may be caused by heterotopic mutations that result in relocation of structures in the evolution of vascular plant body plans (Sattler, 1988, 1994; Sattler and Rutishauser, 1997; Kellogg, 2000a). Examples of heterotopic mutations are observed in anatomical characters, such as in the position of protoxylem poles or sclerenchymatous tissue. As discussed in Section 17.4.3G, the position of protoxylem poles in the root and the shoot is an important distinction between lycophytes and euphyllophytes (Figure 17.7a, b). The endarch or mesarch position of protoxylem in the shoot distinguishes the seed plants and moniliforms except for the Ophioglossidae + Psilotidae clade, which is distinct from all other vascular plant lineages in having taxa with endarch, exarch or mesarch protoxylem poles. Notably, it includes *Psilotum*, the only extant euphyllophyte with exarch protoxylem poles in the shoot stele, a character state otherwise restricted to lycophytes, whereas its sister genus *Tmesipteris* has mesarch protoxylem poles, which are typical of moniliforms. Phylogenetic changes in the localisation of tissues such as protoxylem are likely due to changes in the positional control of cell differentiation (Benfey, 1999; Dolan and Okada, 1999; Costa and Dolan, 2000). Sclerenchymatous tissue in the root cortex of leptosporangiate ferns (Polypodiidae) exemplifies relocation of cell types (Schneider, unpubl. obs.). Sclerenchymatous cells, if present, are differentiated either in the inner or the outer cortex. Sporangial position is also an example of structural relocation in the evolution of vascular plants (Figure 17.5b). The sporangia in moniliforms are located either on the abaxial side of the leaves or adaxially on sporangiophores (Section 17.4.3E). The phylogeny indicates one or perhaps two transitions of sporangia from an adaxial to abaxial position in moniliforms (Figure 17.5b).

C. Heterometry

Little evidence for heterometric mutations that result in changes in size of structures (Zelditch and Fink, 1996; Gould, 2000) was found with this data set because quantitative characters were excluded. They are of great interest in studies of the evolution of closely related species but less informative for studies of deep phylogenies.

17.5.3 Simplification is ubiquitous in plant evolution

Duplication and subsequent modification result in a general trend towards increasing the complexity of body plans of vascular plants (Valentine, 2000), but several derived lineages are characterised by the reduction or absence of structures (Bateman, 1996; Pryer *et al.*, 2001). Obvious examples of simplification are the deletion of organs during evolution. Psilotidae are rootless, but phylogenetic reconstructions indicate that their ancestors possessed roots (Figure 17.3a). Rootless plants are found also in other clades of vascular plants, such as the heterosporous fern *Salvinia* (Polypodiidae) and in flowering plants (e.g. *Ceratophyllum*, *Wolffia*). The absence of lateral roots and root hairs in the Ophioglossidae, the sister clade of Psilotidae, indicates that reduction of the root system is a shared trait of the Ophioglossidae + Psilotidae clade (Figure 17.3b), in which roots are either completely absent (Psilotidae) or develop only as unbranched, shoot-borne roots without root hairs (Ophioglossidae).

Other simplifications include the absence of mechanical tissue (collenchyma and sclerenchyma) in Ophioglossidae and some Marattiidae, the reduction of euphylls to scale-like structures in Psilotidae and Equisetopsida, and the absence of a root pith in all Polypodiidae. The reduction in spore wall thickness and the number of spores produced per sporangium in leptosporangiate ferns (Polypodiidae) are both examples of simplification that may be explained by heterochronic or heterometric mutations. The relatively gradual reduction in spore number per sporangium in leptosporangiate ferns is particularly notable (Figure 17.6b), proceeding sequentially from more than 1000, to less than 1000 but more than 100, and finally to less than 100 (usually exactly 64). Heterochronic mutations may also be responsible for the reduction observed in leaf production in Ophioglossidae; only a single leaf is produced each growing season.

17.6 Implications of a phylogeny for current studies

17.6.1 Time-scale and the evolution of vascular plant lineages

It is now generally accepted that the age of a given lineage of organisms can be inferred from a combination of phylogenetic reconstruction and dates of first appearance of the lineage in the fossil record (Norell and Novacek, 1992; Wagner, 1995; Kenrick and Crane, 1997). Data regarding first appearances of various vascular plant lineages are available in several recent studies (Stewart and Rothwell, 1993; Taylor and Taylor, 1993; Collinson, 1996; Kenrick and Crane, 1997; Crane, 1999; Miller, 1999; Liu *et al.*, 2000). This approach for dating lineages is limited by gaps in the fossil record and is dependent on differentiating and correctly identifying the relationships among early Palaeozoic tracheophytes (Gensel, 1992; Galtier and Philips, 1996; Miller, 1999; Berry and Stein, 2000; Liu *et al.*, 2000; Berry and Fairon-Demaret, 2001; Gensel *et al.*, 2001). As a general rule, phylogenetic evidence provides age estimates that considerably pre-date first appearances in the fossil record. Psilotidae and Ophioglossidae are among the most prominent examples (Figure 17.9). Both are known only from Cenozoic fossils (Stewart and Rothwell, 1993; Taylor and Taylor, 1993; Kenrick and Crane, 1997), whereas their phylogenetic placement necessitates an origin of the Ophioglossidae + Psilotidae clade no later than the Devonian. A similar conflict between phylogenetic topology and the fossil record exists among the seed plants, especially with regard to the origin of flowering plants (=Magnolidra). However, estimates for the age of the seed plant lineages based on DNA nucleotide sequences appear to be consistent with the topology in Figure 17.9 (Goremykin *et al.*, 1997; Magallón *et al.*, 2000). Further aspects of the phylogeny and fossil record were reviewed recently for Coniferidra (Miller, 1999), Marattiidae (Liu *et al.*, 2000) and Polypodiidae (Collinson, 1996; Schneider and Kenrick, 2001). With regard to reconstructing the evolution of plant development, it is important to note that the phylogeny (Figure 17.9) suggests that the major lineages of vascular plants, namely lycophytes, moniliforms and seed plants, have been evolving independently since the Devonian. In addition, extant representatives or close relatives of members of these three major lineages can be traced to a time before the Upper Devonian (lycophytes and moniliforms) or the Permian (seed plants). Such long separation times between the three major lineages may cause problems in studies attempting to infer the evolution of developmental genes (Becker *et al.*, 2000).

17.6.2 Model organisms reconsidered in a phylogenetic framework

A full understanding of the diversity and complexity of organisms is one of the major challenges in biology. Complex structures, such as genomes, can presently be

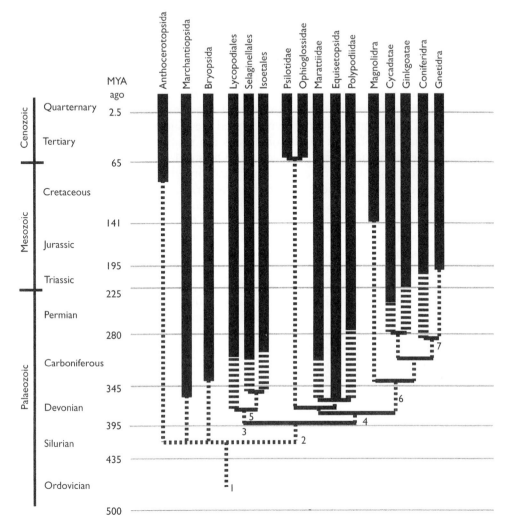

Figure 17.9 Estimate of the age of major lineages of land plants based on stratigraphic evidence (Stewart and Rothwell, 1993; Taylor and Taylor, 1993; Collinson, 1996; Kenrick and Crane, 1997; Doyle, 1998; Miller, 1999; Liu *et al.*, 2000) plotted on our best estimate of land plant phylogeny. Thick continuous lines indicate a fossil record for an extant lineage, thick dashed lines indicate a fossil record for close relatives of extant lineages (such as early conifers), thin dashed vertical lines indicate branches of ambiguous length caused by a conflict between the fossil record and phylogenetic evidence. Numbers indicate first fossil appearances used as calibration points: (1) first land plants, (2) first tracheophytes, (3) first lycophytes, (4) first Moniliformopses (e.g. *Ibyka*) and Radiatopses (e.g. *Crossia*), (5) first ligulate lycophytes (e.g. *Leclerqia*), (6) first seed plants (e.g. *Elkinsia*), (7) first conifers. The Ophioglossidae + Psilotidae clade is known only since the Tertiary; the age of the split between both lineages is ambiguous. The taxonomic classification is based on Kenrick and Crane (1997).

studied in detail only for a few species due to enormous cost considerations. Consequently, current scientific progress relies on the assumption that similar structures and processes are identical in various and often distantly related organisms, and plant developmental genetics focuses on only a few plant species as model organisms. To evaluate the current selection of plant model organisms used in developmental genetic studies, we identified their position in Phylogeny 2001 (Figure 17.1) and in more detailed phylogenetic studies for ferns (Hasebe *et al.*, 1995; Pryer *et al.*, 1995), flowering plants (Qiu *et al.*, 1999; Soltis *et al.*, 1999) and mosses (Goffinet and Cox, 2000). The phylogenetic positions of those model organisms for which developmental gene sequences have been reported are indicated in Figure 17.10.

The vast majority of plant model organisms are members of the more recently evolved lineages of angiosperms (Mandoli and Olmstead, 2000), and many of these are of noted economic importance, such as monocotyledons (Poaceae: *Zea, Oryza, Triticum*) and eudicots (Brassicaceae: *Arabidopsis, Brassica*; Scrophulariaceae: *Antirrhinum*; Solanaceae: *Lycopersicon* (=*Solanum*), *Nicotiana, Petunia*). A few studies have attempted to establish some gymnosperms (e.g. *Gnetum*, and the conifers *Picea* and *Pinus*) as additional model organisms, although their long generation times limit their usefulness for genetic studies (Lev-Yada and Sederoff, 2000). The fern *Ceratopteris* (Pteridacae, Polypodiidae) and the moss *Physcomitrella patens* (Funariaceae, Bryomorpha) have been widely used to represent pteridophytes and bryophytes, respectively (Chatterjee and Roux, 2000; Cove, 2000). All model organisms are members of the crown groups of their lineages, and most exhibit derived rather than ancestral features in their clade. In angiosperms, the herbaceous growth form typical of all model organisms is the derived condition (Qiu *et al.*, 1999; Soltis *et al.*, 1999; Doyle and Endress, 2000), and the fern *Ceratopteris* has an unusually rapid reproductive cycle (Hickok *et al.*, 1995; Banks, 1999; Chatterjee and Roux, 2000), which may indicate a fundamental modification in its reproductive biology from other leptosporangiate ferns. The reproductive biology of *Ceratopteris* is unlikely to be representative of the common ancestor of moniliforms and seed plants.

Several recent studies have inferred the evolution of various genes that were demonstrated to control the development of various plant structures (Riechmann *et al.*, 2000; Riechmann and Ratcliffe, 2000). These studies often include a broad taxon sample, but several critical taxa are usually lacking. MADS-box gene evolution is the best studied among these examples. These studies usually focus on seed plants (Hasebe, 1999; Shindo *et al.*, 1999; Winter *et al.*, 1999; Alvarez-Buylla *et al.*, 2000a, b; Becker *et al.*, 2000; Theißen *et al.*, 2000, 2002). *Ceratopteris*, and sometimes *Ophioglossum*, are included in some of these studies as the only non-seed plant representatives. Only recently have MADS-box genes been described for some other critical taxa, including the bryophyte *Physcomitrella patens* (Krogan and Ashton, 2000) and the lycophyte *Lycopodium annotinum* (Svensson *et al.*, 2000). The last two taxa have not yet been included in a phylogenetic study of MADS-box genes (but see Langdale *et al.*, 2002). In contrast, the sampling of seed plants for MADS-box genes has been much improved during the last few years, with sequences from Gnetatae (*Gnetum*), Coniferidra (*Picea, Pinus*), Ginkgoatae (*Ginkgo*) and Magnolidra (e.g. *Arabidopsis, Brassica, Oryza, Petunia, Solanum*) now available, although Cycadatae and representatives of basal lineages of angiosperms are still lacking.

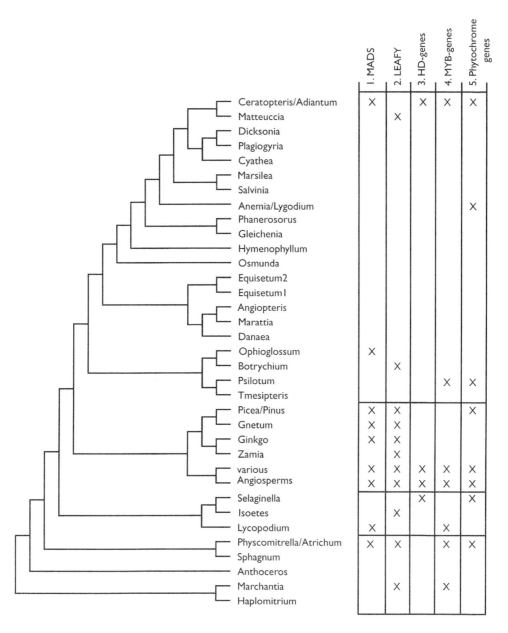

Figure 17.10 Phylogenetic position of taxa with reported sequence data for five developmental gene families. (1) MADS-box (Winter *et al.*, 1999; Becker *et al.*, 2000; Krogan and Ashton, 2000; Svensson *et al.*, 2000; Theißen, 2000; Theißen *et al.*, 2002), (2) *LEAFY* (Frohlich and Estabrook, 2000; Frohlich and Parker, 2000; Frohlich, 2002), (3) HD-genes (Bharathan *et al.*, 1997, 1999; Juarez and Banks, 1997; Aso *et al.*, 1999; Reiser *et al.*, 2000; Champagne and Ashton, 2001; Sakakibara *et al.*, 2001), (4) *MYB* genes (Kranz *et al.*, 2000), (5) Phytochrome genes (Schneider-Poetsch *et al.*, 1998; Basu *et al.*, 2000). Not all of the sequences available for MADS-box, *LEAFY*, HD-genes and *MYB*-genes have been included in single comprehensive studies. Unfortunately, some sequences are not accessible because they have not been submitted to public gene databases.

The phylogeny of another developmental gene, *LEAFY*, was inferred in recent studies including representatives of all five main lineages of seed plants and *Nymphaea* as representative of the basal lineage of angiosperms, but only two non-seed vascular plants were included (Frohlich and Estabrook, 2000; Frohlich and Parker, 2000; Frohlich, 2002). A third group of developmental genes, homeodomain proteins (HD genes), has been studied in the broad context of the evolution of this gene family in a clade including animals, fungi and plants (Bharathan *et al.*, 1997), but in plants they are known nearly exclusively from angiosperms. This is especially the case with one class of HD genes, the *KNOTTED* genes (Bharathan *et al.*, 1999). Although *KNOTTED* genes have been reported from the fern *Ceratopteris* (Juarez and Banks, 1997; Banks, 1999; Reiser *et al.*, 2000) and the bryophyte *Physcomitrella* (Champagne and Ashton, 2001), they have not been included in an extensive phylogenetic study. Several copies of homeodomain-leucine-zip genes (HD-zip genes) are known from the fern *Ceratopteris* and the bryophyte *Physcomitrella*, and have been included in a comprehensive phylogenetic analysis together with derived angiosperms (e.g. *Arabidopsis*, *Daucus*, *Oryza*) (Aso *et al.*, 1999; Sakakibara *et al.*, 2001). Other developmental genes, such as the *MYB* genes (Kranz *et al.*, 2000; Langdale *et al.*, 2002) and phytochromes (Schneider-Poetsch *et al.*, 1998; Basu *et al.*, 2000) have been studied with a better taxon sampling of bryophytes and pteridophytes than in MADS-box gene studies. For several developmental gene families, such as *YABBY* genes, which are involved in the control mechanisms of axial patterning (Bowman, 2000), no homologous sequences are known from bryophytes or pteridophytes. The actin gene family is a noteworthy exception because its evolution has been inferred in studies (Meagher *et al.*, 1999; Bhattacharya *et al.*, 2000) that included a wide sampling of algae, liverworts, lycophytes, moniliforms and seed plants.

The phylogenetic framework we discuss here underscores the importance of appropriate taxon selection when inferring the evolution of developmental genes, including the detection of gene duplication and functional shifts (Eizinger *et al.*, 1999; Ganfornina and Sanchez, 1999; Holland, 1999; Wray, 1999; Kellogg, 2000b). A denser and more diverse phylogenetic sampling is a critical issue in studies of the evolution of development (Browne *et al.*, 2000; Hughes and Kaufman, 2000; Wray, 2000) because it is essential to distinguish convergence, parallelism and reversal. There is an obvious positive trend to broaden taxon sampling, and several aspects need to be considered in selecting new 'model' organisms: phylogenetic position, developmental mode and experimental practicality (Hughes and Kaufman, 2000). Our phylogenetic framework, which includes statements about the relationships of taxa (phylogenetic statements) and the character state changes that support lineages (taxic statements), provides a sound basis for selecting additional taxa that are critical in studies of the evolution of plant development.

17.6.3 Significance of phylogenetic studies in evolutionary developmental biology

Phylogeny estimation is best approached by analysing DNA sequence data and/or morphological data (de Queiroz, 2000; Hillis and Wiens, 2000; Thornton and DeSalle, 2000). The evolution of development should be evaluated by comparing these independent data sources and their resultant trees with phylogenies generated from developmental genes. A step-by-step procedure that advances from an estimate of phylogenetic relationships, to the reconstruction of morphological character evolution, and finally to the identification of evolutionary changes in development is recommended for moving towards a synthesis of developmental and evolutionary biology.

ACKNOWLEDGEMENTS

This research was supported by grants from the National Science Foundation to K. M. P., A. R. S., P. G. W. and R. C. Thanks are due to the organisers of the symposium, J. A. Hawkins, Q. C. B. Cronk and R. M. Bateman, for their patience and support. Several colleagues provided stimulating discussions or comments on various aspects of this study and/or critical comments on earlier drafts: D. L. Des Marais, D. Kaplan, R. Lupia, S. Magallón and R. Rutishauser.

REFERENCES

Alvarez-Buylla, E. R., Liljegren, S. J., Pelaz, S., Gold, S. E., Burgeff, C., Ditta, G. S., Vergara-Silva, F. and Yanofsky, M. F. (2000b) MADS-box gene evolution beyond flowers: expression in pollen, endosperm, guard cells, roots and trichomes. *Plant Journal*, 24, 457–466.

Alvarez-Buylla, E. R., Pelaz, S., Liljegren, S. J., Gold, S. E., Burgeff, C., Ditta, G. S., de Pouplana, L. R., Martinez-Castilla, L. and Yanofsky, M. F. (2000a) An ancestral MADS-box duplication occurred before the divergence of plants and animals. *Proceedings of the National Academy of Sciences USA*, 97, 5328–5333.

Arber, A. (1950) *The Natural Philosophy of Plant Form*. Cambridge University Press, Cambridge.

Arthur, W. (1997) *The Origin of Animal Body Plans*. Cambridge University Press, Cambridge.

Arthur, W. (2000a) The concept of developmental reprogramming and the quest for an inclusive theory of evolutionary mechanisms. *Evolution and Development*, 21, 49–57.

Arthur, W. (2000b) Intraspecific variation in developmental characters: the origin of evolutionary novelties. *American Zoologist*, 40, 811–818.

Aso, K., Kato, M., Banks, J. A. and Hasebe, M. (1999) Characterization of homeodomain-leucine zipper genes in the fern *Ceratopteris richardii* and the evolution of the homeodomain-leucine zipper gene family in vascular plants. *Molecular Biology and Evolution*, 16, 544–552.

Bai, S., Chen, L., Yund, M.-A. and Sung, Z.-R. (2000) Mechanisms of plant embryo development. *Current Topics in Developmental Biology*, 50, 61–88.

Bang, R., DeSalle, R. and Wheeler, W. (2000) Transformalism, taxism and developmental biology in systematics. *Systematic Biology*, 49, 19–27.

Banks, J. A. (1999) Gametophyte development in ferns. *Annual Review of Plant Physiology and Molecular Biology*, 50, 163–186.

Barkman, T. J., Chenery, G., McNeal, J. R., Lyons-Weiler, L., Ellisens, W. J., Moore, G., Wolfe, A. D. and dePamphilis, C. W. (2000) Independent and combined analyses of sequences from all three genomic compartments converge on the root of flowering plant phylogeny. *Proceedings of the National Academy of Sciences USA*, 97, 13166–13171.

Basu, D., Debesh, K., Schneider-Poetsch, H.-J., Harringon, S. E., McCouch, S. R. and Quail, P. H. (2000) Rice *PHYC* gene: structure, expression, map position and evolution. *Plant Molecular Biology*, 44, 27–42.

Bateman, R. M. (1996) Non-floral homoplasy and evolutionary scenarios in living and fossil plants, in *Homoplasy and the Evolutionary Process* (eds M. J. Sanderson and L. Hufford), Academic Press, London, pp. 91–130.

Bateman, R. M. (1999) Architectural radiations cannot be optimally interpreted without morphological and molecular phylogenies, in *The Evolution of Plant Architecture* (eds M. H. Kurmann and A. R. Hemsley), Royal Botanic Gardens, Kew, London, pp. 221–250.

Bateman, R. M., Crane, P. R., DiMichele, W. A., Kenrick, P., Rowe, N. P., Speck, T. and Stein, W. E. (1998) Early evolution of land plants: phylogeny, physiology, and ecology of the primary terrestrial radiation. *Annual Review of Ecology and Systematics*, 29, 263–292.

Becker, A., Winter, K.-U., Meyer, B., Saedler, H. and Theißen, G. (2000) MADS-box gene diversity in seed plants 300 million years ago. *Molecular Biology and Evolution*, 17, 1425–1434.

Beerling, D. J., Osborne, C. P. and Chaloner, W. G. (2001) Evolution of leaf-form in land plants linked to atmospheric CO_2 decline in the Late Palaeozoic era. *Nature*, 410, 352–354.

Benfey, P. N. (1999) Is the shoot a root with a view? *Current Opinion in Plant Biology*, 2, 39–43.

Berry, C. M. and Fairon-Demaret, M. (2001) The Middle Devonian flora revisited, in *Plants Invade the Land: Evolutionary and Environmental Perspectives* (eds P. G. Gensel and D. Edwards), Columbia University Press, Cambridge, pp. 120–139.

Berry, C. M. and Stein, W. E. (2000) A new iridopteridalean from the Devonian of Venezuela. *International Journal of Plant Sciences*, 161, 807–827.

Bharathan, G., Janssen, B.-J., Kellogg, E. A. and Sinha, N. (1997) Did homeodomain proteins duplicate before the origin of angiosperms, fungi, and metazoa? *Proceedings of the National Academy of Sciences USA*, 94, 13749–13753.

Bharathan, G., Janssen, B.-J., Kellogg, E. A. and Sinha, N. (1999) Phylogenetic relationships and evolution of the *KNOTTED* class of plant homeodomain proteins. *Molecular Biology and Evolution*, 16, 553–563.

Bhattacharya, D., Aubry, J., Twait, E. C. and Jurk, S. (2000) Actin gene duplication and the evolution of morphological complexity in land plants. *Journal of Psychology*, 38, 813–820.

Bowe, L. M., Coat, G. and dePamphilis, C. W. (2000) Phylogeny of seed plants based on all three genomic compartments: extant gymnosperms are monophyletic and Gnetales' closest relatives are conifers. *Proceedings of the National Academy of Sciences USA*, 97, 4092–4097.

Bower, F. O. (1935) *Primitive Land Plants*. Macmillan, New York.

Bowman, J. L. (2000) Axial patterning in leaves and other lateral organs. *Current Opinion in Genetics and Development*, 10, 399–404.

Browne, W. E., Davis, G. K. and McClintock, J. M. (2000) Ancestors and variants: tales from the cryptic. *Evolution and Development*, 2, 130–132.

Budd, G. E. (1999) Does evolution in body patterning genes drive morphological change – or vice versa? *BioEssays*, 21, 325–332.

Campbell, D. H. (1940) *The Evolution of the Land Plants [Embryophyta]*. Stanford University Press, Stanford, CA.

Champagne, C. E. M. and Ashton, N. W. (2001) Ancestry of KNOX genes revealed by bryophyte (*Physcomitrella patens*) homologs. *New Phytologist*, 150, 23–36.

Chatterjee, A. and Roux, S. J. (2000) *Ceratopteris richardii*: a productive model for revealing secrets of signalling and development. *Journal of Plant Growth and Regulation*, 19, 284–289.

Chaw, S.-M., Parkinson, C. L., Cheng, Y., Vincent, T. M. and Palmer, J. D. (2000) Seed plant phylogeny inferred from all three plant genomes: monophyly of extant gymnosperms and origin of Gnetales from conifers. *Proceedings of the National Academy of Sciences USA*, 97, 4086–4091.

Collazo, A. (2000) Developmental variation, homology, and the pharyngula stage. *Systematic Biology*, 49, 3–18.

Collinson, M. E. (1996) 'What use are fossil ferns?' – 20 years on: with a review of the fossil history of extant pteridophyte families and genera, in *Pteridology in Perspective* (eds J. M. Camus, M. Gibby and R. J. Johns), Royal Botanic Gardens, Kew, London, pp. 349–394.

Cook, C. D. K. (1999) The number and kinds of embryo-bearing plants which have become aquatic: a survey. *Perspectives in Plant Ecology, Evolution and Systematics*, 2, 79–102.

Costa, S. and Dolan, L. (2000) Development of the root pole and cell patterning in *Arabidopsis* roots. *Current Opinion in Genetics and Development*, 10, 405–409.

Cove, D. (2000) The moss, *Physcomitrella patens. Journal of Plant Growth and Regulation*, 19, 275–283.

Crane, P. R. (1999) Major patterns in botanical diversity, in *Evolution: Investigating the Evidence* (eds J. Scotchmore and D. A. Springer), *Palaeontological Society Special Publication*, 9, 171–187.

Crane, P. R. and Kenrick, P. (1997) Diverted development of reproductive organs: a source of morphological innovation in land plants. *Plant Systematics and Evolution*, 206, 161–174.

Dengler, N. G. and Tsukaya, H. (2001) Leaf morphogenesis in dicotyledons: current issues. *International Journal of Plant Sciences*, 162, 459–464.

de Queiroz, K. (2000) Logical problems associated with including and excluding character during tree reconstruction and the implications for the study of morphological character evolution, in *Phylogenetic Analysis of Morphological Data* (ed. J. J. Wiens), Smithsonian Institution Press, Washington, pp. 192–212.

DiMichele, W. A., Stein, W. E. and Bateman, R. M. (2001) Ecological sorting of vascular plant classes during the Paleozoic evolutionary radiation, in *Evolutionary Paleoecology: The Ecological Context of Macroevolutionary Change* (eds W. D. Allmon and D. J. Bottjer), Columbia University Press, New York, pp. 285–335.

Doebley, J. and Lukens, L. (1998) Transcriptional regulators and the evolution of plant form. *The Plant Cell*, 10, 1075–1082.

Dolan, L. and Okada, K. (1999) Signalling in cell type specification. *Cell and Development*, 10, 149–156.

Donoghue, M. J. and Doyle, J. A. (2000) Seed plant phylogeny: demise of the anthophyte hypothesis? *Current Biology*, 3, R106–R109.

Donoghue, M. J. and Ree, R. H. (2000) Homoplasy and developmental constraint: a model and an example from plants. *American Zoologist*, 40, 759–769.

Doyle, J. A. (1996) Seed plant phylogeny and relationships of Gnetales. *International Journal of Plant Sciences*, 157 (Suppl. 6), S3–S39.

Doyle, J. A. (1998) Molecules, morphology, fossils, and the relationship of angiosperms and Gnetales. *Molecular Phylogenetics and Evolution*, 9, 448–462.

Doyle, J. A. and Endress, P. K. (2000) Morphological phylogenetic analysis of basal angiosperms: comparison and combination with molecular data. *International Journal of Plant Sciences*, 161 (Suppl. 6), S121–S153.

Doyle, J. J. (1994) Evolution of a plant homeotic multigene family: toward connecting molecular systematics and molecular developmental genetics. *Systematic Biology*, 43, 307–328.

Edwards, D. (1999) Origins of plant architecture: adapting to life in a brave new world, in *The Evolution of Plant Architecture* (eds M. H. Kurmann and A. R. Hemsley), Royal Botanic Gardens, Kew, London, pp. 3–21.

Eizinger, A., Jungblot, B. and Sommer, R. J. (1999) Evolutionary change in the functional specificity of genes. *Trends in Genetics*, 15, 197–202.

Foster, T. and Veit, B. (2000) Genetic analysis of leaf development and differentiation, in *Leaf Development and Canopy Growth* (eds B. Marshall and J. A. Roberts), Sheffield Academic Press, Sheffield, pp. 59–95.

Friedman, W. E. and Carmichael, J. S. (1998) Heterochrony and developmental innovation: evolution of female gametophyte ontogeny in *Gnetum*, a highly apomorphic seed plant. *Evolution*, 52, 1016–1030.

Frohlich, M. W. (2002) The Mostly Male theory of flower origins: summary and update regarding the Jurassic pteridosperm *Pteroma*, in *Developmental Genetics and Plant Evolution* (eds Q. C. B. Cronk, R. M. Bateman and J. A. Hawkins), Taylor & Francis, London, 85–108.

Frohlich, M. W. and Estabrook, G. F. (2000) Wilkinson Support calculated with exact probabilities: an example using *Floricaula/LEAFY* amino acid sequences that compares three hypotheses involving gene gain/loss in seed plants. *Molecular Biology and Evolution*, 17, 1914–1925.

Frohlich, M. W. and Parker, D. S. (2000) The mostly male theory of flower evolutionary origins: from genes to fossils. *Systematic Botany*, 25, 155–170.

Frugis, G., Giannino, D., Mele, G., Nicolodi, C., Innocenti, A. M., Chiappetta, A., Bitoni, M. B., Dewitte, W., van Onckelen, H. and Mariotti, D. (1999) Are homeobox *Knotted*-like genes and cytokinins the leaf architects? *Plant Physiology (Lancaster)*, 119, 371–373.

Galtier, J. (1999) Contrasting diversity of branching patterns in early ferns and early seed plants, in *The Evolution of Plant Architecture* (eds M. H. Kurmann and A. R. Hemsley), Royal Botanic Gardens, Kew, London, pp. 5–64.

Galtier, J. and Phillips, T. L. (1996) Structure and evolutionary significance of Palaeozoic ferns, in *Pteridology in Perspective* (eds J. M. Camus, M. Gibby and R. J. Johns), Royal Botanic Gardens, Kew, London, pp. 417–433.

Ganfornina, M. D. and Sanchez, D. (1999) Generation of evolutionary novelty by functional shift. *BioEssays*, 21, 432–439.

Gellon, G. and McGinnis, W. (1998) Shaping animal body plans in development and evolution by modulation of Hox expression patterns. *BioEssays*, 20, 116–125.

Gensel, P. G. (1977) Morphologic and taxonomic relationships of the Psilotaceae relative to evolutionary lines in early vascular plants. *Brittonia*, 29, 14–29.

Gensel, P. G. (1992) Phylogenetic relationships of zosterophylls and lycopsids: evidence from morphology, paleoecology, and cladistic methods of inference. *Annals of the Missouri Botanical Garden*, 79, 450–473.

Gensel, P. G., Kotyk, M. E. and Basinger, J. F. (2001) Morphology of above- and below-ground structures in Early Devonian (Pragian–Emsian) plants, in *Plants Invade the Land: Evolutionary and Environmental Perspectives* (eds P. G. Gensel and D. Edwards), Columbia University Press, Cambridge, MA, pp. 83–102.

Gibson, G. and Wagner, G. (2000) Canalization in evolutionary genetics: a stabilizing theory? *BioEssays*, 22, 372–380.

Gilbert, S. F. (2000) Genes classical and genes developmental. The different use of genes in evolutionary synthesis, in *The Concept of the Gene in Development and Evolution* (eds P. J. Beurton, R. Falk and H.-J. Rheinberger), Cambridge University Press, Cambridge, pp. 178–192.

Goebel, K. von (1933) *Organographie der Pflanzen*, 3rd edn. G. Fischer, Jena.

Goffinet, B. and Cox, C. J. (2000) Phylogenetic relationships among basal-most arthrodontous mosses with special emphasis on the evolutionary significance of the Funariineae. *The Bryologist*, 10, 212–223.

Goremykin, V., Hansmann, S. and Martin, W. F. (1997) Evolutionary analysis of 58 proteins encoded in six completely sequenced chloroplast genomes: revised molecular estimates of two seed plant divergence times. *Plant Systematics and Evolution*, 206, 337–351.

Gould, S. J. (2000) Of coiled oysters and big brains: how to rescue the terminology of heterochrony, now gone astray. *Evolution and Development*, 2, 241–248.

Graham, A. (2000) The evolution of the vertebrates – genes and development. *Current Opinion in Genetics and Development*, 10, 624–628.

Graham, L. E., Cook, M. E. and Busse, J. S. (2000) The origin of plants: body plan changes contributing to a major evolutionary radiation. *Proceedings of the National Academy of Sciences USA*, 97, 4535–4540.

Grbic, M. (2000) 'Alien' wasps and evolution of development. *BioEssays*, 22, 920–932.

Hagemann, W. (1984) Morphological aspects of leaf development in ferns and angiosperms, in *Contemporary Problems in Plant Anatomy* (eds R. A. White and W. C. Dickinson), Academic Press, Orlando, FL, pp. 301–349.

Hagemann, W. (1992) What is a root? in *Root Ecology and its Practical Application – A Contribution to the Investigation of the Whole Plant* (ed. L. Kutschera), Verein für Wurzelforschung, Klagenfurt, pp. 1–8.

Hagemann, W. (1997) Über die Knöllchenbildung an den gametophytes der Farngattung *Anogramma. Stapfia*, 50, 375–391.

Hagemann, W. (1999) Towards an organismic concept of land plants: the marginal blastozone and the development of the vegetative body of selected frondose gametophytes of liverworts and ferns. *Plant Systematics and Evolution*, 216, 81–133.

Hall, B. K. (1996) Baupläne, phylotypic stages, and constraint: why there are so few types of animals. *Evolutionary Biology*, 29, 215–255.

Hasebe, M. (1999) Evolution of reproductive organs in land plants. *Journal of Plant Research*, 112, 463–474.

Hasebe, M. and Ito, M. (1999). Evolution of reproductive organs in vascular plants, in *The Biology of Biodiversity* (ed. M. Kato), Springer, New York, pp. 243–254.

Hasebe, M., Wolf, P. G., Pryer, K. M., Ueda, K., Ito, M., Sano, R., Gastony, G. J., Yokoyama, J., Manhart, J. R., Murakami, N., Crane, E. H., Haufler, C. H. and Hauk, W. D. (1995) Fern phylogeny based on *rbcL* nucleotide sequences. *American Fern Journal*, 85, 134–181.

Hickok, L. G., Warne, T. R. and Fribourg, R. S. (1995) The biology of the fern *Ceratopteris* and its use as a model system. *International Journal of Plant Sciences*, 156, 332–345.

Hillis, D. M. and Wiens, J. J. (2000) Molecules versus morphology in systematics: conflicts, artefacts, and misconceptions, in *Phylogenetic Analysis of Morphological Data* (ed. J. J. Wiens), Smithsonian Institution Press, Washington, DC, pp. 1–19.

Holland, P. E. H. (1999) The effect of gene duplication on homology, in *Homology* (eds G. R. Bock and G. Cardew), Novarites Foundation Symposium 222. Wiley, Chichester, pp. 226–242.

Hughes, C. L. and Kaufman, T. C. (2000) A diverse approach to arthropod development. *Evolution and Development*, 2, 6–8.

Jenner, R. A. (2000) Evolution of animal body plans: the role of metazoan phylogeny at the interface between pattern and process. *Evolution and Development*, 2, 208–221.

Juarez, C. M. and Banks, J. A. (1997) Studies of sex determination and meristem development in the fern *Ceratopteris richardii*. *Annual Meeting of the American Society of Plant Physiologists 1997*. [Abstract]. (available at URL http://www.rycomusa.com/aspp1997/public/).

Kaplan, D. R. (1977) Morphological status of the shoot systems of Psilotaceae. *Brittonia*, 29, 30–53.

Kaplan, D. R. (2001) Fundamental concepts of leaf morphology and morphogenesis: a contribution to the interpretation of molecular genetic mutants. *International Journal of Plant Sciences*, 162, 465–474.

Kaplan, D. R. and Groff, P. A. (1995) Developmental themes in vascular plants: functional and evolutionary significance, in *Experimental and Molecular Approaches to Plant Biosystematics* (eds P. Hoch and A. G. Stephenson), Monographs in Systematic Botany from the Missouri Botanical Garden 53, Missouri Botanical Garden, St. Louis, MO, pp. 71–86.

Kappen, C. (2000) Analysis of a complete homeobox gene repertoire: applications for the evolution of diversity. *Proceedings of the National Academy of Sciences USA*, 97, 4481–4486.

Kato, M. (1988) The phylogenetic relationships of Ophioglossaceae. *Taxon*, 37, 381–386.

Kato, M. and Imaichi, R. (1997) Morphological diversity and evolution of vegetative organs in pteridophytes, in *Evolution and Diversification of Land Plants* (eds K. Iwatsuki and P. H. Raven), Springer Press, Berlin, pp. 27–43.

Kellogg, E. A. (2000a) The grasses: a case study in macroevolution. *Annual Review in Ecology and Systematics*, 31, 217–238.

Kellogg, E. A. (2000b) Genetics of character evolution. *American Journal of Botany*, 87, 104 [Abstract].

Kenrick, P. (1994) Alternation of generations in land plants: new phylogenetic and morphological evidence. *Biological Reviews*, 69, 293–330.

Kenrick, K. and Crane, P. R. (1997) *The Origin and Early Diversification of Land Plants: A Cladistic Study*. Smithsonian Institution Press, Washington, DC.

Kerstetter, R. A. and Poethig, R. S. (1998) The specification of leaf identity during shoot development. *Annual Review of Cell Development and Biology*, 14, 373–398.

Klingenberg, C. P. (1998) Heterochrony and allometry: the analysis of evolutionary change in ontogeny. *Biological Reviews*, 73, 79–123.

Kramer, E. M. and Irish, V. F. (1999) Evolution of genetic mechanisms controlling petal development. *Nature*, 399, 144–148.

Kramer, E. M. and Irish, V. F. (2000) Evolution of the petal and stamen developmental programs: evidence from comparative studies of the lower eudicots and basal angiosperms. *International Journal of Plant Sciences*, 161 (Suppl. 6), S29–S40.

Kranz, H., Scholz, K. and Weisshaar, B. (2000) c-MYB oncogene-like genes encoding three MYB repeats occur in all major plant lineages. *Plant Journal*, 21, 231–235.

Krogan, N. T. and Ashton, N. W. (2000) Ancestry of plant MADS-box genes revealed by bryophyte (*Physcomitrella patens*) homologues. *New Phytologist*, 147, 505–517.

Kutschera, L. and Sobotik, M. (1997) Bewurzelung von Pflanzen in den verschiedenen Lebensräumen. 5. Buch der Wurzelatlas-Reihe. Allgemeiner Teil. *Stapfia*, 49, 5–54.

Langdale, J. A., Scotland, R. W. and Corley, S. B. (2002) A developmental perspective on the evolution of leaves, in *Developmental Genetics and Plant Evolution* (eds Q. C. B. Cronk, R. M. Bateman and J. A. Hawkins), Taylor & Francis, London, pp. 388–394.

Lawton-Rauh, A. L., Alvarez-Buylla, E. R. and Purugganan, M. D. (2000) Molecular evolution of flower development. *Trends in Ecology and Evolution*, 15, 144–149.

Lev-Yada, S. and Sederoff, R. (2000) Pines as model gymnosperms to study evolution, wood formation and perennial growth. *Journal of Plant Growth and Regulation*, 19, 290–305.

Lewis, L. A., Mishler, B. D. and Vilgalys, R. (1997) Phylogenetic relationships of the liverworts (Hepaticae), a basal embryophyte lineage, inferred from nucleotide sequence data of the chloroplast gene *rbcL*. *Molecular Phylogenetics and Evolution*, 7, 377–393.

Li, P. and Johnston, M. O. (2000) Heterochrony in plant evolutionary studies through the twentieth century. *Botanical Review (Lancaster)*, 66, 57–88.

Liu, Z.-H., Hilton, J. and Li, C.-S. (2000) Review on the origin, evolution and phylogeny of Marattiales. *Chinese Bulletin of Botany*, 17, 39–52.

Mabee, P. M. (2000) Developmental data and phylogenetic systematics: evolution of the vertebrate limb. *American Zoologist*, 40, 789–800.

Maddison, W. P. and Maddison, D. R. (1992) *MacClade. Analysis of Phylogeny and Character Evolution, Version 3*. Sinauer, Sunderland, MA.

Magallón, S. A., Sanderson, M. J., Doyle, J. A. and Wojciechowski, M. F. (2000) Estimate of the age of the angiosperm crown group derived from integrated analysis of molecular and paleontological data. *American Journal of Botany*, 87, 141 [abstract].

Mandoli, D. F. and Olmstead, R. (2000) The importance of emerging model systems in plant biology. *Journal of Plant Growth and Regulation*, 19, 249–252.

Meagher, R. B., McKinnet, E. C. and Vitale, A. V. (1999) The evolution of new structures: clues from plant cytoskeletal genes. *Trends in Genetics*, 15, 278–283.

Miller, C. N. Jr. (1999) Implications of fossil conifers for the relationships of living families. *Botanical Review (Lancaster)*, 65, 239–277.

Morange, M. (2000) The developmental gene concept, in *The Concept of the Gene in Development and Evolution* (eds P. J. Beurton, R. Falk and H. J. Rheinberger), Cambridge University Press, Cambridge, pp. 193–215.

Mosbrugger, V. (1995) Heterochrony and the evolution of land plants, in *Evolutionary Change and Heterochrony* (ed. K. J. McNamara), Wiley, Chichester, pp. 93–105.

Nickrent, D. L., Parkinson, C. L., Palmer, J. D. and Duff, R. J. (2000) Multigene phylogeny of land plants with special reference to bryophytes and the earliest land plants. *Molecular Biology and Evolution*, 17, 1885–1895.

Niklas, K. J. (2000a) The evolution of plant body plans – a biomechanical perspective. *Annals of Botany (Oxford)*, 85, 411–438.

Niklas, K. J. (2000b) The evolution of leaf form and function, in *Leaf Development and Canopy Growth* (eds B. Marshall and J. A. Roberts), Sheffield Academic Press, Sheffield, pp. 1–35.

Norell, M. A. and Novacek, M. J. (1992) Congruence between superpositional and phylogenetic patterns: comparing cladistic patterns with fossil records. *Cladistics*, 8, 319–337.

Peterson, K. J. and Davidson, E. H. (2000) Regulatory evolution and the origin of the bilaterians. *Proceedings of the National Academy of Sciences USA*, 97, 4430–4433.

Philipson, W. R. (1990) The significance of apical meristems in the phylogeny of land plants. *Plant Systematics and Evolution*, 173, 17–38.

Proctor, M. C. F. (2000) Mosses and alternative adaptation to life on land. *New Phytologist*, 148, 1–3.

Pryer, K. M., Smith, A. R. and Skog, J. E. (1995) Phylogenetic relationships of extant ferns based on evidence from morphology and *rbcL* sequences. *American Fern Journal*, 85, 205–282.

Pryer, K. M., Schneider, H., Smith, A. R., Cranfill, R., Wolf, P. G., Hunt, J. S. and Sipes, S. D. (2001) Horsetails and ferns are a monophyletic group and the closest living relatives to seed plants. *Nature*, 409, 618–622.

Qiu, Y.-L., Bernasconi-Quadroni, F., Soltis, D. E., Soltis, P. S., Zanis, M., Zimmer, E. A., Chen, Z., Savolainen, V. and Chase, M. W. (1999) The earliest angiosperms: evidence from mitochondrial, plastid and nuclear genomes. *Nature*, 402, 404–407.

Qiu, Y.-L. and Lee, J. (2000) Transition to a land flora: a molecular phylogenetic perspective. *Journal of Phycology*, 36, 799–802.

Raff, R. A. (1996) *The Shape of Life: Genes, Development and the Evolution of Animal Form*. University of Chicago Press, Chicago, IL.

Raff, R. A. (1999) Larval homologies and radical evolutionary changes in early development, in *Homology* (eds G. R. Bock and G. Cardew), Novarites Foundation Symposium 222, Wiley, Chichester, pp. 110–121.

Raven, J. A. and Edwards, D. (2001) Roots: evolutionary origin and biogeochemical significance. *Journal of Experimental Botany*, 22, 381–401.

Reiser, L., Sánchez-Baracaldo, P. and Hake, S. (2000) Knots in the family tree: evolutionary relationships and function of *knox* homeobox genes. *Plant Molecular Biology*, 42, 151–166.

Remy, W., Remy, D. and Hass, H. (1997) Organisation, Wuchsformen und Lebensstrategien früher Landpflanzen des Unterdevons. *Botanische Jahrbücher für Systematik*, 119, 509–562.

Renzaglia, K. S., Duff, R. J., Nickrent, D. L. and Garbary, D. J. (2000) Vegetative and reproductive innovations of early land plants: implications for a unified phylogeny. *Philosophical Transactions of the Royal Society of London*, B355, 768–793.

Richards, D. E., Peng, J. and Harberd, N. (2000) Plant *GRAS* and metazoan *STATs*: one family? *BioEssays*, 22, 573–577.

Riechmann, J. L., Heard, J., Marti, G., Reuber, L., Jiang, C.-Z., Keddie, J., Adam, L., Pineda, O., Ratcliffe, O. J., Samaha, R. R., Creelman, R., Pilgrim, M., Broun, P., Zhang, J.-Z., Ghandelhari, D., Sherman, B. K. and Yu, G.-L. (2000) *Arabidopsis* transcription factors: genome-wide comparative analysis among eukaryotes. *Science*, 290, 2105–2110.

Riechmann, J. L. and Ratcliffe, O. (2000) A genomic perspective on plant transcription factors. *Current Opinion in Plant Biology*, 3, 423–434.

Rieppel, O. (1993) The conceptual relationship of ontogeny, phylogeny, and classification. *Evolutionary Biology*, 27, 1–32.

Rothwell, G. W. (1995) The fossil history of branching: implications for the phylogeny of land plants, in *Experimental and Molecular Approaches to Plant Biosystematics* (eds P. Hoch and A. G. Stephenson), Monographs in Systematic Botany from the Missouri Botanical Garden 53, Missouri Botanical Garden, St Louis, MO, pp. 71–86.

Rothwell, G. W. (1999) Fossils and ferns in the resolution of land plant phylogeny. *Botanical Review (Lancaster)*, 65, 188–218.

Rowley, J. R. (1996) Exine origin, development and structure in pteridophytes, gymnosperms, and angiosperms, in *Palynology: Principles and Applications* (eds J. Jansonius and D. C. McGregor), American Association of Stratigraphic Palynologists Foundation, Vol. 1, Publishers Press, Salt Lake City, UT, pp. 443–462.

Rutishauser, R. (1999) Polymerous leaf whorls in vascular plants: developmental morphology and fuzziness of organ identities. *International Journal of Plant Sciences*, 160 (Suppl. 6), S81–S103.

Sakakibara, K., Nishiyama, T., Kato, M. and Hasebe, M. (2001) Isolation of homeodomain-leucine zipper genes from the moss *Physcomitrella patens* and the evolution of homeodomain-leucine zipper genes in land plants. *Molecular Biology and Evolution*, 18, 491–502.

Sanderson, M. J., Wojciechowski, M. F., Hu, J.-M., Sher Khan, T. and Brady, S. G. (2000) Error, bias, and long-branch attraction in data for two chloroplast photosystem genes in seed plants. *Molecular Biology and Evolution*, 17, 782–797.

Sattler, R. (1988) Homeosis in plants. *American Journal of Botany*, 75, 1607–1617.

Sattler, R. (1994) Homology, homeosis and process morphology in plants, in *Homology: The Hierarchical Basis of Comparative Biology* (ed. B. K. Hall), Academic Press, New York, pp. 424–475.

Sattler, R. and Rutishauser, R. (1997) The fundamental relevance of morphology and morphogenesis to plant research. *Annals of Botany (Oxford)*, 80, 571–582.

Scheres, B. (2000) Non-linear signalling for pattern formation? *Current Opinion in Plant Biology*, 3, 412–417.

Schneider, H. and Kenrick, P. (2001) An Early Cretaceous root-climbing epiphyte (Lindsaeaceae) and its significance for calibrating the diversification of polypodiaceous ferns. *Review of Palaeobotany and Palynology*, 115, 33–41.

Schneider-Poetsch, H.-J., Kolukisaoglu, U., Clapham, D. H., Hughes, J. and Lamparter, T. (1998) Non-angiosperm phytochromes and the evolution of vascular plants. *Physiologia Plantarum (Copenhagen)*, 102, 612–622.

Shindo, S., Ito, M., Ueda, K., Kato, M. and Hasebe, M. (1999) Characterization of MADS genes in the gymnosperm *Gnetum parvifolium* and its implication on the evolution of reproductive organs in seed plants. *Evolution and Development*, 1, 180–190.

Shutov, A. D., Braun, H., Chesnokov, Y. V. and Bäumlein, H. (1998) A gene encoding a vicilin-like protein is specifically expressed in fern spores: evolutionary pathway of seed storage globulins. *European Journal of Biochemistry*, 252, 79–89.

Smyth, D. (2000) A reverse trend – MADS functions revealed. *Trends in Plant Science*, 5, 315–317.

Soltis, P. S. and Soltis, D. E. (2000) Contributions of plant molecular systematics to studies of molecular evolution. *Plant Molecular Biology*, 24, 45–75.

Soltis, P. S., Soltis, D. S. and Chase, M. W. (1999) Angiosperm phylogeny inferred from multiple genes as a tool for comparative biology. *Nature*, 402, 402–404.

Speck, T. and Rowe, N. P. (1999) A quantitative approach for analytically defining size, growth form and habit in living and fossil plants, in *The Evolution of Plant Architecture* (eds M. H. Kurmann and A. R. Hemsley), Royal Botanic Gardens, Kew, pp. 447–479.

Stein, W. E. (1993) Modelling the evolution of the stelar architecture in vascular plants. *International Journal of Plant Sciences*, 154, 229–263.

Stewart, W. N. and Rothwell, G. W. (1993) *Paleobotany and the Evolution of Plants*, 2nd edn. Cambridge University Press, Cambridge.

Svensson, M. E., Johannesson, H. and Engström, P. (2000) The *LAMB1* gene from the clubmoss, *Lycopodium annotinum*, is a divergent MADS-box gene, expressed specifically in sporogenic structures. *Gene*, 23, 31–43.

Taylor, T. N. and Taylor, E. L. (1993) *The Biology and Evolution of Fossil Plants*. Prentice Hall, Englewood Cliffs, NJ.

Theißen, G. (2000) Evolutionary developmental genetics of floral symmetry: the revealing power of Linnaeus' monstrous flower. *BioEssasys*, 22, 209–213.

Theißen, G., Becker, A., Di Rosa, A., Kanno, A., Kim, J. T., Muenster, T., Winter, K.-U. and Saedler, H. (2000) A short history of MADS-box genes in plants. *Plant Molecular Biology*, 42, 115–149.

Theißen, G., Becker, A., Winter, K.-U., Münster, T., Kirchner, C. and Saedler, H. (2002) How the land plants learned their floral ABCs: the role of MADS-box genes in the evolutionary origin of flowers, in *Developmental Genetics and Plant Evolution* (eds Q. C. B. Cronk, R. M. Bateman and J. A. Hawkins), Taylor & Francis, London, pp. 173–205.

Thornton, J. W. and DeSalle, R. (2000) Gene family evolution and homology: genomics meets phylogenetics. *Annual Reviews of Genomics and Human Genetics*, 1, 41–73.

Tomlinson, P. B. (1984) Homology in modular organisms – concepts and consequences. Introduction. *Systematic Botany*, 9, 373.

Troll, W. (1937) *Vergleichende Morphologie der höheren Pflanzen*, Band 1, Teil 1. Gebrüder Bornträger, Berlin.

Troll, W. (1939) *Vergleichende Morphologie der höheren Pflanzen*, Band 1, Teil 2. Gebrüder Bornträger, Berlin.

Tsukaya, H. (2000) The role of meristematic activities in the formation of leaf blades. *Journal of Plant Research*, 113, 119–126.

Valentine, J. W. (2000) Two genomic paths to the evolution of complexity in body plans. *Paleobiology*, 26, 513–519.

Valentine, J. W., Jablonski, D. and Erwin, D. H. (1999) Fossils, molecules and embryos: new perspectives on the Cambrian explosion. *Development*, 126, 851–859.

Vergara-Silva, F., Martinez-Castilla, L. and Alvarez-Buylla, E. R. (2000) MADS-box genes: development and evolution of plant body plans. *Journal of Phycology*, 36, 803–812.

Wagner, G. P., Chiu, C.-H. and Laubichler, M. (2000) Developmental evolution as a mechanistic science: the inference from developmental mechanisms to evolutionary processes. *American Zoologist*, 40, 819–831.

Wagner, G. P. and Schwenk, K. (2000) Evolutionary stable configurations, functional integration and the evolution of phenotypic stability. *Evolutionary Biology*, 31, 155–217.

Wagner, P. J. (1995) Stratigraphic tests of cladistic hypotheses. *Paleobiology*, 21, 153–178.

Wagner, W. H. Jr. (1977) Systematic implications of the Psilotaceae. *Brittonia*, 29, 54–63.

Wagner W. H. Jr., Beitel, J. M. and Wagner, F. S. (1982) Complex venation patterns in the leaves of *Selaginella*: megaphyll-like leaves in lycophytes. *Science*, 218, 793–794.

Wardlaw, C. W. (1952) *Phylogeny and Morphogenesis*. Macmillan, London.

Wardlaw, C. W. (1965) *Organisation and Evolution of Plants*. Longmans, London.

Winter, K.-U., Becker, A., Kim, J. T., Saedler, H. and Theißen, G. (1999) MADS-box genes reveal that gnetophytes are more closely related to conifers than to flowering plants. *Proceedings of the National Academy of Sciences USA*, 96, 7342–7347.

Wray, G. A. (1999) Evolutionary dissociations between homologous genes and homologous structures, in *Homology* (eds G. R. Bock and G. Cardew), Novartis Foundation Symposium 222, Wiley, Chichester, pp. 189–203.

Wray, G. A. (2000) Peering ahead (cautiously). *Evolution and Development*, 2, 125–126.

Wray, G. A. and Lowe, C. J. (2000) Developmental regulatory genes and echinoderm evolution. *Systematic Biology*, 49, 28–51.

Zelditch, M. L. and Fink, W. L. (1996) Stability and innovation in the evolution of form. *Paleobiology*, 22, 247–250.

Zimmermann, W. (1959) *Die Phylogenie der Pflanzen*, 2nd edn. G. Fischer, Jena.

Zimmermann, W. (1965) *Die Telomtheorie*. G. Fischer, Stuttgart.

Chapter 18

The telome theory

Paul Kenrick

ABSTRACT

The telome theory states that the organs, appendages and vascular systems of land plants evolved from simple multicellular modules (telomes) that have been combined and modified in various ways. The theory combines a general hypothesis on the fundamental nature of the plant body with numerous, frequently controversial hypotheses on the origins of specific organ systems. An outline of the theory is presented and some of the main criticisms are discussed. It is argued that as an approach to investigating the evolution of form, the telome theory contributes to a general theory of standard parts or modularity in plants. By itself, it is neither complete nor entirely correct. The fundamental unit of the telome appears to be a real structure, at least in the early land plants of the Devonian Period. Certain hypotheses of organ evolution, in particular those related to leaves, integuments and sporophylls, remain valid. One of the main methodological shortcomings relates to how one might go about testing hypotheses of transformation. Several possible ways forward are suggested, including the use of developmental models, the study of transitional forms in the fossil record, and consistency with character distributions on cladograms. A molecular developmental approach could also provide critical tests of ideas on transformational homology by supplying new information on the relationships among organs.

18.1 Introduction

> To me telomes and phyllomes are so to speak the atoms of comparative morphology into which all organs are analysable.
>
> (Harris, 1940: 731)

The idea that organisms are composed of standard parts and that these can combine and metamorphose into different organs underpins the concept of homology and is fundamental to research in comparative biology (Riedl, 1978; Rieppel, 1988). In plants, as in other organisms, one can conceive of a hierarchy of such parts that spans several orders of magnitude from the molecular to the cellular to the organ level. Toward the apex of this 'parts pyramid' are structures such as leaves, petals, seeds and flowers which are not only composites of standard parts but which appear to be related among themselves. The telome theory attempts to explain the origin and development of complex plant organs from more basic multicellular units called

In *Developmental Genetics and Plant Evolution* (2002) (eds Q. C. B. Cronk, R. M. Bateman and J. A. Hawkins), Taylor & Francis, London, pp. 365–387.

telomes (Figure 18.1). These are standard parts that were first observed in fossils, and they can be seen in their unaltered form in the simple multicellular stem modules that make up the branching systems of the earliest land plants. Walter Zimmermann, the originator of the telome theory, argued that the remarkable diversity of organ systems in plants could be explained by the modification of telomes through the interplay of just a handful of elementary developmental processes (Zimmermann, 1930, 1938, 1965; for English summaries see Zimmermann, 1952; Wilson, 1953; Stewart, 1964). The appeal of the telome theory as an explanation of morphogenesis in higher plants lies in its economy of hypothesis and in its apparent applicability across a wide variety of organs and taxa.

The telome theory brings together a set of approaches, ideas and assumptions that represent the culmination of a movement that came to be known as the 'New

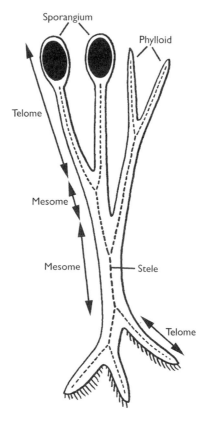

Figure 18.1 The vocabulary of the telome theory: telome, mesome, phylloid, sporangium. The telome is the fundamental supercellular building block in plants. It is a cylindrical axis that develops from an apical meristem. Anatomically, it is simple, comprising a single vascular cylinder (protostele). Telomes come in two types: vegetative telomes (phylloids) and fertile telomes (sporangia). Zimmermann recognised a fourth category, which he termed the mesome. A mesome is a single internode between two branching points. Telomes and mesomes are therefore developmental equivalents. All mesomes were telomes at some point in the ontogeny of the plant. Telomes become mesomes when the apical meristem divides. Redrawn and modified from Zimmermann (1959, Figure 39).

Morphology' (Thomas, 1932; Lam, 1948). This was a response by leading morphologists of the late nineteenth and early twentieth century to what were seen as the inadequacies of the old angiosperm-centred approach, in which interpretations of plant form were shoe-horned into one of three basic categories: stem, root and leaf. The telome theory challenged this morphological trinity by beginning an analysis of the evolution of form from the base rather than the apex of the phylogenetic tree. It presented a much more dynamic perspective, and one which was more compatible with the concept of an evolving world. This shift in approach was widely welcomed, and the telome theory became the paradigm within which much comparative morphology was interpreted and taught by generations of botanists. Zimmermann's ideas had their greatest impact on our understanding of the early evolution of fundamental organ systems and appendages in land plants. They found greatest favour among palaeobotanists, who could trace plausible transitions between telomic branching systems and other organs (e.g. leaves, seeds, pollen organs, lateral branches, roots) in the early fossil record. The telome theory has been widely used to explain the origins of the familiar organ systems of ferns, lycopods and gymnosperms.

Despite early success in pteridophytes and gymnosperms, the telome theory never really had a major impact on the field of angiosperm morphogenesis. It has also received sharp criticism due to widely recognised methodological and empirical shortcomings. One major concern is that it seems to be less a theory and more a lexicon of descriptive terms that can, with imagination, be applied to almost any structure and any proposed transformation series (Niklas, 1997: 268). In other words, it lacks predictive power. Others have argued that significant aspects of the theory are untestable or simply incorrect (Stein, 1998). A harsh critic might add that, as a theory, it describes everything but explains nothing. This chapter is an attempt to re-evaluate the conceptual content of the telome theory and to identify those elements of greatest relevance to an understanding of plant development.

18.2 An outline of the theory

The telome theory states that the land plant body evolved from simple multicellular units called telomes. These were combined in various ways and modified through the actions of a handful of developmental processes to produce the full diversity of plant organs, appendages and vascular architecture. The theory also postulates specific hypotheses of character transformation to explain the evolution of particular organs. Many of these were first proposed by Zimmermann, but they have been supplemented and modified by other researchers. The theory is explicitly phylogenetic in as much as it aimed to describe the historical development of morphology as accurately as possible. Zimmermann fully recognised the necessity of knowing at least the broad outline of relationships among plants. Information from the fossil record – in particular, data on the morphology of Palaeozoic fossils – was seen to be of crucial importance. Phylogenetic insights were also seen as a product of the application of the telome theory to particular problems. The telome theory therefore combines a general hypothesis on the fundamental nature of the plant body and the developmental processes that have sculpted it through time with numerous, frequently controversial hypotheses on the origins of specific organ systems.

18.2.1 Telome, mesome, phylloid and sporangium

Central to the theory is the notion of the 'telome' (Figure 18.1). In land plants, the telome is essentially a fundamental supercellular building block: an intermediate category between cell and 'fully-fledged' appendage or organ system (i.e. leaf, cone, complex stele). Conceptually, it is a cylindrical axis that develops from an apical meristem. Anatomically, it is simple, comprising a single vascular cylinder (protostele), a parenchymatous cortex, and an epidermis (Figure 18.2b). The concept of the telome was modelled on discoveries of early fossil plants from the Devonian Period, in particular the famous early land plants of the Rhynie Chert (Kidston and Lang, 1921). Telomes can be seen in their unmodified form in plant fossils such as *Cooksonia* and *Rhynia* (Figures 18.2, 18.3), which were built from a simple set of cylindrical elements that were repeated to give a dichotomously branching system. These examples of simple stick-like organisms seemed to provide a basic framework from which many of the familiar organs and appendages of land plants could be derived. Other more complex Devonian plants also appear to be built on a similar body plan, and the fossil record documents a series of morphological intermediates

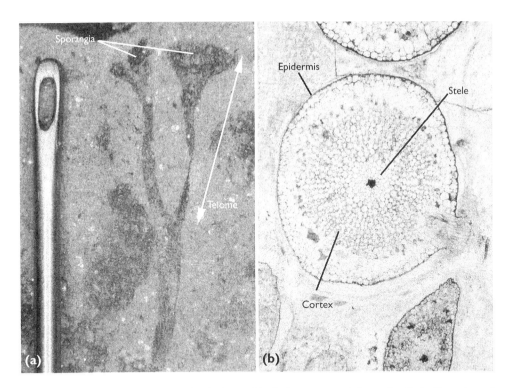

Figure 18.2 Telomes in early fossil land plants. (a) *Cooksonia pertoni* from the Early Devonian of England (The Natural History Museum, London: V58010). Compression fossil showing minute dichotomous axis with terminal sporangia. Sewing needle for scale on left. (b) Anatomy of a telome. *Rhynia gwynne-vaughanii* from the Early Devonian of Scotland (The Natural History Museum, London: Scott Collection 3133). Transverse section through minute permineralised branch illustrating complete soft tissue preservation at the cellular level.

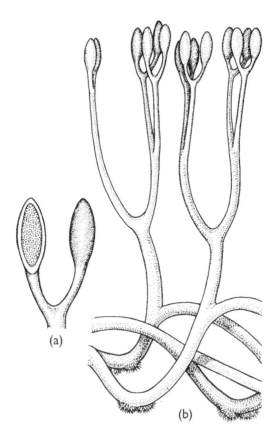

Figure 18.3 Reconstruction of *Aglaophyton* (*Rhynia*) *major* from Kenrick and Crane (1997a). (a) Details of sporangium including view of interior. (b) Reconstruction of habit.

between branching systems constructed from numerous telomes and certain types of plant organ (e.g. leaf, seed). Based on palaeobotanical insights such as these, Zimmermann proposed that land plant morphology was reducible to the 'atomic unit' of the telome. Plants are built from numerous simple standard parts that have been combined and modified in various ways to produce the vast panoply of organs and appendages seen today. This is the central tenet of the telome theory.

18.2.2 Elementary processes

If telomes are the fundamental building blocks of plants, then how did they evolve, and what processes have sculpted them into the leaves, stems, seeds and so forth of modern plants? Zimmermann's answer was to propose about a dozen specific and independent 'elementary processes' that gave rise to and operated on telomes (Table 18.1, Figure 18.4). He argued that all of the major organs and organ systems of plants are the outcome of elementary processes acting on their own or, more usually, in combination. By applying these processes to the simple branching systems of early land plants it is possible to transform them into more complex structures

Table 18.1 Elementary processes in the formation and modification of telomes (Zimmermann, 1952)

Process	Result
Telome origin	
1 Connection of cells through common cell walls	Filamentous growth
2 Rotation of plane of cell division	Branched filaments
3 Apical meristem	Cylindrical multicellular body
4 Changes to isomorphic life cycle (ancestral isomorphic life cycle an assumption of the theory)	Elaboration of either sporophytic (vascular plants) or gametophytic (bryophytes) phases of life cycle at expense of other
5 Tissue differentiation	Elaboration of ground tissue system and vascular system
Telome modification	
6 Overtopping	Stronger development of one branch in a dichotomy – distinction between main axis and lateral appendages
7 Planation	Switch from multidimensional to flattened branching
8 Syngenesis	Lateral fusion of telomes – implicated in webbing of leaves and development of complex vascular systems
9 Reduction	Loss of telomic units leading to loss of morphological complexity (e.g. bryophyte sporophyte)
10 Incurvation	Bending due to unequal growth of tissue on flanks of telome
11 Longitudinal differentiation	Development of serial homologues along an axis (e.g. leaves or pinnules)

(e.g. leaves, seeds) through a series of hypothetical intermediates. The telome theory therefore makes some general predictions about the sequence of character acquisition in plants and about certain processes operating at the apical meristem. It also emphasises the role of co-option and modification in the evolution of organs and organ systems. In the world of telomes, the evolution of a new structure never begins from scratch. It is always possible to conceive of ways of tinkering with or modifying existing structures.

18.2.3 Specific hypotheses of character transformation

This theme of co-option was further developed by Zimmermann and many others into various organ- and taxon-specific hypotheses of character transformation, couched in the language of telomes and the processes that transform them. The focus of this work has been to explain the specifics of how particular organs and organ systems have evolved. In developing the telome theory, Zimmermann put forward many hypotheses of this nature, including explanations for the evolution of the sporophyte in bryophytes, and for a broad range of basic structures in vascular plants such as leaves (megaphylls and microphylls), complex vascular systems, and spore-bearing reproductive structures. The original ideas of Zimmermann have been developed and added to over the years so that now there are telome-based hypothe-

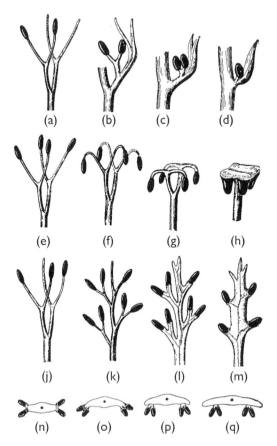

Figure 18.4 Telomic processes in action, from Zimmermann (1959: 47, Figure 60). (a–d) Origin of the lycopod (clubmoss) sporophyll by reduction from a telomic branching system. (e–h) Origin of the peltate sporophyll in sphenopsids (horsetails) via the processes of recurvation, reduction and fusion. (j–q) Origin of the leaf in ferns via the processes of overtopping and webbing to produce a blade.

ses for the evolution of many fundamental features of higher plants. The telome theory has been invoked in the origin of such diverse structures as the integument in seed plants (Andrews, 1961; Stewart and Rothwell, 1993; Kenrick and Crane, 1997a), the female reproductive organs of the conifers (Florin, 1951) and angiosperms (Thomas, 1931), the pollen organs of extinct pteridosperms (Halle, 1933), the synangium of *Psilotum* (Emberger, 1944), and the appendages of lycopods (Crane and Kenrick, 1997). The telome theory has been a very fruitful source of hypotheses on the origin and evolution of major organ systems in plants.

18.2.4 Life cycles

The telome theory also contained a specific and highly speculative hypothesis on the origins of land plant life cycles. Unlike other aspects of the theory, this hypothesis was developed in the complete absence of any empirical support, such as fossil and developmental evidence. Zimmermann supposed that the aquatic ancestors of land plants had an isomorphic alternation of generations in which both gametophyte and sporophyte had well-developed telomic branching systems. It was envisaged that the gamete- and spore-bearing phases of the life cycle underwent different fates following the transition to the land. In vascular plants, the gamete-producing phase was reduced whereas the spore-producing phase was elaborated. In bryophytes, life cycle evolution followed a different path. Both phases underwent some loss of morphology, but this was most marked in the sporophyte, which was reduced to a single telome and became parasitic on the gametophyte. The basic telome morphology was therefore forged in an aquatic environment, and its origin pre-dated the transition of plant life to land. According to Zimmermann, land plants inherited their basic body plan from their green algal ancestors.

18.2.5 Body plans and phylogeny

The telome theory was developed within the framework of a very specific phylogenetic hypothesis for land plants. Zimmermann was the author of an influential text on phylogenetic methods (Zimmermann, 1931; see retrospective in Donoghue and Kadereit, 1992), and he saw character transformations and hypotheses of taxon relationship as intimately interwoven and reciprocally illuminating. His own analysis indicated that land plants are a monophyletic group (Zimmermann, 1930). In outline, and using modern terminology, bryophytes are sister group to vascular plants. Within vascular plants Zimmermann recognised three major body plans that corresponded to three monophyletic groups: lycopods (clubmosses), sphenopsids (horsetails) and pteropsids (seed plants, together with ferns). The body plans of these three groups evolved from within a grade of extinct plants with telomic branching modelled on the early fossils *Rhynia* (now *Aglaophyton*) (Edwards, 1986; Remy and Hass, 1996; Figure 18.3) and *Psilophyton* (Banks *et al.*, 1975). The telome theory therefore makes important assumptions, or perhaps statements, about phylogenetic relationships. It also hypothesises the presence of four distinct body plans in land plants (three in vascular plants and one in bryophytes).

18.2.6 Hologeny

Zimmermann's principle of hologeny is relevant to a discussion of the developmental basis of the telome theory. Hologeny is the simple but important notion that character transformations are brought about by genetic changes, and that these are expressed during ontogeny (Zimmermann, 1965). In other words, we are dealing with heritable traits. Also, the evolution of organ systems is not simply a consequence of developmental changes appended to the end of ontogeny. Modifications are expected to occur and manifest themselves throughout ontogeny. These issues are not problematic today and will not be discussed further.

18.3 Does the telome theory have explanatory value and is it verifiable?

The telome theory falls into a category of ideas that attempts to explain order in living organisms in terms of the production and modification of standard parts (*sensu* Riedl, 1978). In this respect it suggests parallels with the kind of organisation which lies at the heart of metazoan body plans, in which cells and groups of cells are the standard parts of multicellular life (Riedl, 1978; Raff, 1996). In plants, standard parts are clearly recognisable at all hierarchical levels from the molecular through organelle, cell and organ. The novelty of the telome theory as applied to land plants is that it recognises a level that falls between cell and 'fully-fledged' organ – a level that is no longer recognisable in living species. Telomes are standard parts that have been combined, shaped and moulded through time into the full gamut of complex organs that characterise the modern plant. From developmental and evolutionary perspectives the telome theory is economical because the use of standard parts to make complex organs actually reduces the complexity of genome-level processes required to encode and control development. By using standard parts an organism can generate a more complex morphology than could be encoded in a one-to-one fashion by the genome, and it reduces the complexity of the command system during ontogeny (Raff, 1996). For example, the repeated deployment of a simple telome module with an apical meristem capable of dichotomising would lead to the self-assembly of complex branching systems, along the lines of those that can be generated artificially by simple computer algorithms (Niklas, 1982). This model for the origin of complexity from such underlying simplicity is one of the main strengths of a theory of standard parts, to which the telome theory contributes. The telome theory therefore has explanatory value, but is it verifiable?

18.3.1 Telomes

There is justification for regarding the fundamental unit of the telome as a real structure, at least in the early land plants of the Devonian Period (408 to 363 million years ago). Early fossils have simple dichotomous branching systems but no significant appendages (Gensel and Andrews, 1984; Rothwell, 1995; Kenrick and Crane, 1997a). Each node is composed of a single telomic unit (Figures 18.1, 18.2). The only significant variation is the presence of sporangia, and in many species these also bear a strong resemblance to stems. Telomes could therefore be seen as modified sporangia in which the spore precursors have been diverted to the production of vascular and ground tissue systems. Could these telomic systems provide the framework for the evolutionary origin of lateral appendages and stem anatomy? Certainly: fossil evidence shows that the degree of branching and its complexity increases with time, and the gradual specialisation of lateral branches to produce organs recognisable as leaves and cones appears to proceed through intermediate stages where telomic units are still clearly visible. Examples of this include the evolution of leaves (Stewart, 1964), pollen organs (Halle, 1933) and integuments (Stewart and Rothwell, 1993). In general terms, the telome theory as applied to lateral appendages appears to provide an accurate and consistent description of many features of the land plant during its early stages of evolution.

If telomic branching systems were the framework from which plant morphology has been built, it seems as though the processes that have shaped plant organs through evolutionary time have also erased the distinctive identity of telomes while at the same time creating a new set of standard parts such as leaves, seeds, cones and flowers. With the possible exception of bryophytes, the telome itself is not evident during ontogeny in living plants. In no way are telomic branching systems recapitulated during development. Does this mean that the telome theory is incorrect or that it is not open to verification through experimentation? The absence of evidence for telomes during plant ontogeny is not necessarily evidence of absence. Telomes are ancient structures, and they may have been modified to such an extent that their distinctive identity is no longer recognisable. In a general sense, experimental verification is therefore probably not possible, at least at the morphological level. There remains the possibility that the telomic structure of early land plants is somehow conserved in the underlying molecular mechanisms of plant development, but it is unclear what form this conservation might actually take.

18.3.2 Elementary processes

In a fundamental sense, morphology is process (Sattler, 1992; see also critique by Weston, 2000), and the processes that modify telomes are plausibly explained through actions of the apical meristem. Changes in the pattern and rate of cell division could in principle supply the repertoire of processes recognised by Zimmermann. One area of concern is the nature of the developmental basis of elementary processes. As noted by Stein (1998), from a developmental perspective these are actually poorly defined, and in at least one instance (formation of complex steles through syngenesis; see Section 18.4) are probably incorrect. It is easy to see how changes to meristematic activity might well result in processes such as overtopping, recurvation and planation (Table 18.1). However, in no instance do we actually *know* the developmental basis of any of Zimmermann's processes. Although specific processes are invoked in an explanation of how a particular organ evolved, their developmental basis remains unspecified. This is important because these process could have several plausible trajectories. For example, overtopping might result either from unequal dichotomy of the apex or from equal dichotomy followed by suppression of one branch. Clearly, these are not the same thing. Thus, at least one putative process (e.g. syngenesis) is doubtful, at least as it is applied to shoot evolution, others (e.g. overtopping) are plausible but might have more than one trajectory of development, and there may exist additional processes that one ought to consider (e.g. sterilisation: Crane and Kenrick, 1997). Furthermore, in applying elementary processes to hypothesise the evolution of particular organs it is usually possible to apply different combinations of processes to achieve the same result. In other words, there may be several plausible evolutionary scenarios for a particular organ. Zimmermann did not propose a method for testing competing hypotheses (Stein, 1998), hence the criticism of the theory being wholly descriptive and untestable. Several possible ways forward are suggested using specific hypotheses of character transformation.

18.3.3 Specific hypotheses of character transformation

Choosing among alternative hypotheses of organ transformation and moving beyond these to design rigorous tests is an aspect of the telome theory that is in need of substantial improvement (Stein, 1998). Indeed, this is its main methodological weakness. It is, however, also a problem of a more general nature in a theory of standard parts. Any hypothesis of transformation of one organ or part into another needs a criterion of plausibility with which to choose among competing hypotheses, as well as a rigorous method of testing.

Classical ideas of transformation have originated in various ways, being stimulated by hypotheses of phylogeny, theories of homology, the sequence of appearance in the fossil record, and arguments built around the plausibility of different scenarios of developmental change. Usually, the principle of parsimony (Farris, 1983) is invoked implicitly, both in the recognition of homology (pattern) and at the level of developmental plausibility (process). The most useful hypotheses of transformation are derived from patterns of taxic homology (*sensu* Patterson, 1982). From this perspective, they are *a posteriori* hypotheses of character change occurring between nested taxic homologies. Transformations are not supported if they are not congruent with the pattern of taxic homology. It is therefore possible to make use of phylogenetic context in devising a test (e.g. Bateman, 1994, 1999). A second approach to choosing among competing hypotheses of transformation is to apply the criterion of parsimony to the level of developmental process. The hypothesis that requires the least number of changes and transformations and which involves the fewest processes would be regarded as the most parsimonious. This method is very commonly applied at an intuitive level. But, in practice, quantifying the amount of change involved in competing hypotheses may not be straightforward. A third approach to testing hypotheses of transformation makes use of the fact that different hypotheses often imply different relationships among organs. For example, in the evolution of lycopod leaves (see Section 18.4.2) the sterilisation hypothesis predicts a close developmental relationship between leaves and sporangia, whereas the enation hypothesis suggests a closer relationship between leaves and trichomes, and the reduction hypothesis implies a relationship between leaves and lateral branches. Relationships among organs imply relationships among the underlying molecular mechanisms of development, which are in principle predictable and observable. Observations at a molecular level could provide an independent test of competing hypotheses on the origins of organ systems. The fourth possibility is to attempt to model the various proposed alternatives to search for inconsistencies with known developmental processes and the outcomes of these processes as measured against known living and fossil plants. This promising approach has not yet been widely applied, but an outline of how it might work in practice was developed by Stein (1998).

18.3.4 Phylogenetic framework

Zimmermann's general phylogenetic outline of plants holds up well. Modern studies support monophyly of land plants and vascular plants (Kenrick and Crane, 1997b; Kenrick, 2000; Pryer *et al.*, 2001; Schneider *et al.*, 2002). Three major clades of extant vascular plants are recognised and these broadly correspond to Zimmermann's three body plans (clubmoss, horsetail, fern + seed plant). There is general agreement that early land plants of the *Rhynia* and *Psilophyton* type are a basal grade of organisation within vascular plants. The relationships among major groups of bryophytes are still unclear, but it is likely that they are paraphyletic to vascular plants (Kranz *et al.*, 1995; Malek *et al.*, 1996; Lewis *et al.*, 1997; Hedderson *et al.*, 1998; Qiu *et al.*, 1998; Duff and Nickrent, 1999; Nishiyama and Kato, 1999). Many more details are known within these groups, but such changes that have occurred do not significantly alter the general pattern for land plants proposed by Zimmermann. Modern phylogenetic research therefore supports the general phylogenetic framework of the telome theory and vindicates the implicit assumption that such fundamental features of the land plant body as archegonium, antheridium, sporangium and spore are homologous among taxa.

18.3.5 Life cycle evolution in early land plants

Zimmermann's ideas about life cycles in early land plants and their aquatic ancestors were much more speculative and have fared less well. In general terms, he hypothesised that the ancestral life cycle of land plants was an isomorphic one, in which both gametophyte and sporophyte had well-developed telomic branching systems. Subsequent morphological evolution in bryophytes and vascular plants went in opposing directions. In bryophytes, the gamete-bearing part was elaborated and the spore-bearing part reduced to little more than a tiny parasitic capsule. In vascular plants, it was the gamete-bearing part that was reduced and the spore-bearing part that was hugely developed. New evidence of the phylogenetic relationships of green algae and land plants (Graham, 1993; McCourt *et al.*, 1996; Huss and Kranz, 1997; Chapman *et al.*, 1998) and direct evidence on life cycles in early fossils (Remy *et al.*, 1993; Kenrick, 1994) show that Zimmermann's ideas were correct only in part. These new data do not provide a critical test of the telome theory, because it is clearly consistent with more than one scenario of life cycle evolution. They do, however, have consequences for interpreting the direction or polarity of change in life cycle evolution.

There is no direct fossil evidence on the aquatic ancestors of land plants, but recent phylogenetic work has identified the closest living relatives as the charophycean algae (Graham, 1993; McCourt *et al.*, 1996; Chapman *et al.*, 1998). This finding is inconsistent with Zimmermann's ideas on an isomorphic life cycle in the algal ancestors, because all living multicellular members of the Charophyceae are strongly anisomorphic (Graham, 1993). There is no multicellular sporophyte in this group, let alone one with telomic branching, as Zimmermann predicted. In contrast to Zimmermann's assertion of an aquatic origin, the land plant sporophyte may well have evolved on land (Mishler and Churchill, 1985; Kenrick, 1994). Furthermore, phylogenetic studies of relationships among bryophytes and vascular plants, although still inconclusive in some respects, indicate that the simple spore-bearing phase in bryophytes is the plesiomorphic state for land plants (Kranz *et al.*, 1995; Malek *et al.*, 1996; Lewis *et al.*, 1997; Hedderson *et al.*, 1998; Qiu *et al.*, 1998; Duff and Nickrent, 1999; Nishiyama and Kato, 1999). In other words, the tiny parasitic sporophyte is not a reduced branching system as the telome theory presupposed. In fact, the converse appears to be the case. The bryophytic sporophyte may be the closest living structure that we have to Zimmermann's original concept of the unaltered telome.

With regard to vascular plants, at the time when Zimmermann wrote his synthesis there was no direct evidence on the life cycles of *Rhynia* and its relatives, but recent discoveries have shed much light on this question (Remy *et al.*, 1993; Kenrick, 1994). The telome theory hypothesised that plants like *Rhynia* would have a reduced gametophyte as in modern vascular plants. Remarkable recent discoveries by Remy and colleagues show that this idea is incorrect (Remy *et al.*, 1993). Unexpectedly, the life cycle in the earliest vascular plants was isomorphic. Unlike living vascular plants, both spore-bearing and gamete-bearing phases had well-developed telomic branching. This means that although Zimmermann was wrong about *Rhynia*, he was close to the mark about life cycle evolution in general within the vascular plant clade. His ancestral isomorphic life cycle was just shifted a little further up the phylogenetic tree. The telome theory seems to be correct in this respect: the vascular plant gametophyte has clearly undergone a spectacular loss of morphology and a drastic reduction in size, whereas the sporophyte has been the focus of most subsequent morphological elaboration.

18.4 Testing the telome theory

It is in the area of character transformations that the telome theory comes into sharpest focus and where the best opportunities for designing tests both of a specific and a general nature lie. The two examples given here illustrate an instance where the telome theory has been rejected and one in which a telomic interpretation is still valid. The first involves the evolution of complex stem vascular systems in higher plants. The second involves the evolution of the lycopod body plan.

18.4.1 Complex vascular systems – the telome theory rejected

One of the most problematic of Zimmermann's many hypotheses of organ transformation relates to the evolution of complex vascular systems in higher plants. Telomes have a simple vascular supply comprising a centrally located terete xylem strand surrounded by a zone of phloem (Figure 18.2b). The vasculature of most higher plants is, however, much more complex. The range of variation is broad, and seen in transverse section encompasses lobing of varying degrees, medullation of the pith, dissection and multiple steles (Ogura, 1972; Beck *et al.*, 1982). Zimmermann's explanation for this diversity was novel. He proposed that complex steles evolved through lateral fusion of multiple telomes (syngenesis; Table 18.1), each adding its own vascular system to produce a complex whole. Under this interpretation, the shoots of most higher plants are most definitely compound structures.

One well-worked example of the application of the process of syngenesis was developed by Delevoryas (1955) for the stem anatomy of extinct medullosan gymnosperms. Here it was argued that the polystelic condition that characterises medullosans evolved through lateral fusion of telome-like stems. A diagram was provided that detailed a series of forms leading from a *Psilophyton*-like ancestor with simple telomic branching through hypothetical intermediates involving the acquisition of secondary xylem and the fusion of stems, and ultimately the partial loss of individual stem identity. The end-product of syngenesis in this case was a polystelic stem anatomy in which the individual steles are the only remnants of the ancestral telomes. One prediction of such a scheme is that polystelic forms should be preceded by forms in which individual stem identity is more readily recognisable. These transitional forms should possess a kind of 'false trunk' composed of tightly bound or partially fused stems, which are still clearly recognisable as such (i.e. through the retention of tissues such as an epidermis).

As a general explanation of complex vascular systems the telome theory has several problems. First, there are no examples of complex vascular systems preceded by transitional forms with false trunks (i.e. trunks composed of partially fused stem elements) in the fossil record. Those plants that do produce false trunks represent comparatively derived groups. The trunks are generally formed from closely growing petioles (e.g. Osmundaceae: Ogura, 1972), roots (Cyatheaceae: Ogura, 1972), or stems bound together in a root mantle (e.g. *Tempskya*: Andrews and Kern, 1947). These taxa are neither closely related to early land plants with complex steles (e.g. Cladoxylales) nor do they precede them in the fossil record. Second, in plants with complex vascular systems there is no supporting developmental evidence for fusion of any kind (Stein, 1998). In fact, it is well established through developmental studies that stelar architecture is induced by hormonal gradients set up by shoot apices and lateral appendages (for discussion see Wight, 1987). Third, many of the complex stelar patterns observed in early land plants can be reproduced by models that simulate lateral appendages and their associated hormonal fields (Stein, 1993). There is therefore a broad coalition of data from systematics, the fossil record, experimental studies of living species, and modelling that is inconsistent with the telome theory. Furthermore, there is a more plausible alternative hypothesis for the origin of complex steles in higher plants (i.e. induction via meristem-mediated hormonal gradients). These would seem to be sufficient grounds for rejecting the telome theory as a plausible hypothesis in the evolution of complex steles.

18.4.2 Co-option and modification of sporangia to form leaves and ligules in lycopods

Lycopods are an ancient group of vascular plants with a very distinctive and highly conservative morphology (Jermy, 1990a, b; Øllgaard, 1990; Stewart and Rothwell, 1993). Small herbaceous species of the Devonian Period strongly resemble their modern relatives. So similar are they that *Asteroxylon* from the Early Devonian (Gensel and Andrews, 1984) would not look out of place in a modern temperate woodland. The group also encompasses many extinct forms, some of which were arborescent and important components of Late Palaeozoic ecosystems (Bateman *et al.*, 1992; DiMichele *et al.*, 1992; Bateman, 1994). Lycopods are the first recognisably modern plants in the fossil record. Their unique morphology makes them immediately recognisable and supports the widely held idea that their body plan is distinctively different from that of other plant groups. The origin of the lycopod body plan has been widely discussed, and in this context the lycopod leaf turns out to be a critical feature (see also Langdale *et al.*, 2002).

The lycopod leaf, termed a microphyll, is simple and unbranched (Figure 18.5). In herbaceous species it is usually small, often only a few millimetres long, but it can reach over a metre in length in some species of living *Isoetes* and in extinct arborescent forms. There is a simple vascular supply comprising a single vascular strand, except in

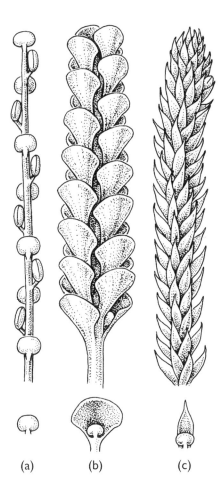

Figure 18.5 Strobili (above) and leaf-sporangium complexes (below) in lycopods (clubmosses). (a) Leafless extinct *Zosteropyllum myretonianum* (Edwards, 1975) from the Early Devonian of Europe. (b) Bract-bearing extinct *Adoketophyton subverticillatum* (Li and Edwards, 1992) from the Early Devonian of China. (c) Living *Huperzia selago* (Lycopodiaceae) (from Crane and Kenrick 1997, Figure 5).

a few extinct arborescent species that have a pair of parallel strands. Sporophylls bear a single sporangium in the leaf axil or on the adaxial surface. They otherwise strongly resemble vegetative leaves and are clearly serial homologues. In some derived groups leaves also bear a small flap of tissue on the adaxial surface near the leaf axil. This is called the ligule. The appendages of lycopods are thus very simple and they have been highly conserved throughout 400 million years of plant evolution. Is the lycopod leaf homologous with the leaves of other land plants and how did it evolve?

The homologies of leaves in vascular plants have been widely discussed and the current consensus is that the lycopod leaf is not homologous with the leaves of ferns and seed plants. This hypothesis has recently received strong corroboration from phylogenetic studies, which place lycopods sister group to euphyllophytes (i.e. the clade containing all other living vascular plants) (Kenrick and Crane, 1997b; Pryer *et al.*, 2001; Schneider *et al.*, 2002). Phylogenetic work that includes fossils shows that the stem groups of both the lycopod and euphyllophyte clades contain leafless vascular plants. In other words, the distinctive microphyll leaf is an autapomorphy of lycopods. How then did it originate?

There are three plausible hypotheses for the origin of the lycopod leaf, and these have been discussed in detail elsewhere (Crane and Kenrick, 1997) (Figure 18.6):

(1) *The reduction hypothesis*: Zimmermann (1959) suggested that the leaf was a reduced telomic branching system (Figure 18.6a). He based this hypothesis on the observation that many early land plants have lateral branching systems. Furthermore, one extinct group of early lycopods has forked leaves, which could be interpreted as vestigial evidence of branching.

(2) *Enation hypothesis*: Bower (1908) suggested that this leaf type evolved as a completely new outgrowth of the stem (Figure 18.6b). He pointed to evidence of simple, spine-like non-vascular appendages in many early land plants, which could be interpreted as intermediate stages in the evolution of a fully vascular leaf.

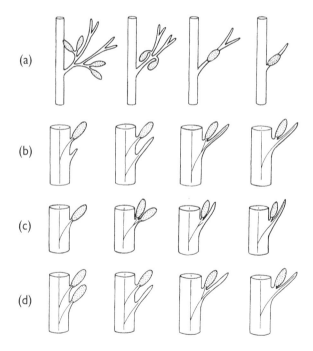

Figure 18.6 Four hypotheses of the origin of the lycopod sporophyll (sporangium-bearing leaf). (a) Reduction hypothesis (Zimmermann, 1959). Sporophyll evolves via reduction of a lateral branch system. (b) Enation hypothesis (Bower, 1908). Sporophyll originates as a completely new structure – an outgrowth of the stem. (c), (d). Two versions of the sterilisation hypothesis (Kenrick and Crane, 1997a). Sporophyll evolves via duplication and co-option of the sporangium. From Kenrick and Crane (1997a, Figure 7.21).

(3) *Sterilisation hypothesis*: Kenrick and Crane (1997a) suggested that the leaf evolved initially as a bract to the sporangium through duplication and sterilisation of one sporangium in a pair (Figure 18.6c, d). This bract was later co-opted to a photosynthetic function, producing a leaf. They emphasised ontogenetic, positional and anatomical similarities between sporangia and leaves and pointed to evidence of forms with sporangial-shaped bracts in the early fossil record (Figure 18.5). The ligule was also explained as a further duplication of the leaf pathway. Both the reduction hypothesis and the sterilisation hypothesis are telomic in nature because of their emphasis on co-option and modification of lateral branches or sporangia. How can one evaluate the plausibility of these competing hypotheses?

One approach is to ask whether the proposed transformations are consistent with the distribution of characters on the most parsimonious phylogenetic tree. One would expect the precursor state to the leaf in each case to circumscribe a more inclusive clade than the leaf itself. If Zimmermann's reduction hypothesis is correct, then lycopods should be a subgroup of a clade containing plants with branched laterals (leaf precursor). If Bower's enation hypothesis is correct, then lycopods should be a subgroup of a clade containing plants with simple, spine-like non-vascular appendages (leaf precursor). If the sterilisation hypothesis is correct, then lycopods should be a subgroup of a clade containing plants with sporangial bracts (leaf precursor). In reviewing the phylogenetic evidence, Crane and Kenrick (1997) concluded that both the enation and sterilisation hypotheses were plausible, but judged Zimmermann's reduction hypothesis unlikely.

Experimental verification of the sterilisation hypotheses is possible in principle if the underlying controls governing developmental pathways of the lycopod leaf, sporangium, and ligule systems could be identified at the genomic level. One possibility is that all appendages in lycopods are derived by duplication and co-option of the microsporangium developmental pathway (Figure 18.7). This could involve at least

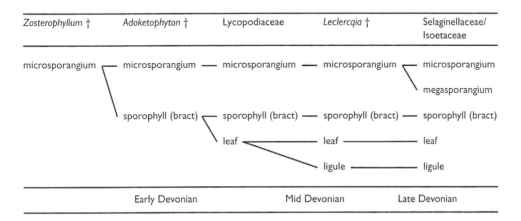

Figure 18.7 Hypothesis of duplication and co-option of the microsporangium developmental pathway in the origin of the lycopod body plan during the Devonian Period. Various other organs produced via further duplication and co-option of sporophyll and leaf pathways. This hypothesis has observational consequences in living Lycopodiaceae, Selaginellaceae and Isoetaceae. Modified from Crane and Kenrick (1997, Figure 6).

four duplications, two of which have observational consequences in living plants. The first duplication event is accompanied by sterilisation (suppression of the archaesporial developmental pathway) in one product to produce a sterile sporophyll or bract, an extinct character combination found in the fossil *Adoketophyton* (Figure 18.5). The second duplication involves the sporophyll (bract) and leads to the leaf. Thus, three duplications should be observable in living Lycopodiaceae (Figure 18.7). The third duplication involves the leaf and leads to the ligule. This extinct character combination is exemplified by the fossil *Leclercqia*. The final duplication involves the microsporangium again but this time without accompanying sterilisation and leads to the evolution of megasporangia. This model therefore predicts at least five duplications in living Selaginellaceae and Isoetaceae (Figure 18.7). If either the enation or reduction hypotheses were correct, there would be no reason to expect such a pattern of duplication events, at least as they apply to leaf and sporangium. In fact, if Bower's hypothesis of enation is correct, one might expect leaves to be more closely related at the molecular level to trichomes than to sporangia. If Zimmermann's hypothesis of reduction is correct, then a similarity between leaves and branching systems would be predicted.

Alternatively, there may not be a strict one-to-one correspondence, nor shared hierarchy, between morphologically defined 'parts' and 'subparts' and the underlying developmental processes. An increasingly sophisticated gene control mechanism may make use of the *same* developmental pathways expressed in *different contexts*. Under this hypothesis, duplication and modification of the structural genes does not occur. We would not expect to observe a family of related structural genes involved in the sporangium/sporophyll/leaf/ligule complex of lycopods. Rather, we would expect to see the same genes invoked in different contexts or at different times. Changes at the molecular level are to be expected only in the control genes. The observational consequence that this model of gene control has on testing the sterilisation, enation and reduction hypotheses are similar to those described for the gene duplication model described above. If the sterilisation hypothesis is correct, we would expect a similar set of structural genes to be active during development of individual elements in the sporangium/sporophyll/leaf/ligule complex of lycopods. Whereas the enation hypothesis indicates similarities between leaves and trichomes, and Zimmermann's hypothesis of reduction requires similarities between leaves and branching systems.

18.5 Conclusions

In his book *Order in Living Organisms* Riedl (1978) emphasised the key role played by modularity in the development of biological order. In complex metazoans – the focus of Riedl's work – information content measured in morphological terms is said to be orders of magnitude higher than that of the genome. In order to resolve this paradox, Riedl proposed that order arises via the deployment of a hierarchy of standard parts or modules. The use of standard parts facilitates the development of morphological complexity and simplifies the translation of genotype to phenotype during ontogeny. Modularity also enables the operation of three important internal evolutionary processes that affect development (Raff, 1996: 326): (1) Dissociation – the process of unlinking features within a system (e.g. heterochrony); (2) Duplication and divergence – the process by which morphological and molecular serial homologues are produced; and (3) Co-option – taking over an existing feature to serve a new function. These processes acting singly or in combinations are a potentially powerful source of evolutionary novelty. For example, duplication of a module followed by divergence may lead to it being co-opted to a different function (Baum and Donoghue, 2002). Thus, complex organs do not need to be built entirely from scratch. Modularity is an even stronger feature of development in plants, suggesting underlying similarities in the development of morphological complexity in several major branches of the tree of life.

The telome theory is not a complete approach to investigating the evolution of plant form. It is best understood in the context of a general theory of standard parts or modularity in plants, to which it contributes. As in metazoans, we can conceive of a hierarchy of parts in plants that spans several orders of magnitude from the molecular to the cellular to the organ level. Zimmermann recognised the explanatory power of a theory of standard parts, and he contributed a new element to the hierarchy of plant form, the telome module. This module is most clearly recognisable in the earliest land plants where it is a major component of the plant body. Zimmermann provided an elegantly simple set of hypothetical developmental mechanisms to explain how telomes could be combined and modified to generate many of the fundamental organs of land plants. This also went some way to explaining the underlying similarities among certain organs. The telome appears to be a real structure, but Zimmermann's development of telome-based hypotheses for every aspect of the plant body was overly ambitious. We have seen how it does not provide an accurate model for understanding the evolution of complex vascular systems in plants, and it is very probably incorrect in this respect. In other areas the application of telome-based explanations has been over-extended. A telomic approach to understanding morphology within ferns, the more derived gymnosperms, and angiosperms is simply inappropriate. The reason for this is that in these groups telomes have metamorphosed into a new and more diverse set of standard parts (e.g. leaves, seeds, sporophylls, petals, stamens). Here telomes are no longer part of the vocabulary of the morphologist because they no longer exist as discrete, identifiable entities. The concept of the telome does remain an important one for understanding the evolution of some fundamental organ systems in plants. Telomes are clearly recognisable as distinctive modules in the simple dichotomous branching systems of the earliest land plants. Zimmermann's ideas on leaf evolution through modification of telomic branching systems remain valid hypotheses.

The telome theory has several additional conceptual and methodological problems. Zimmermann's list of processes is incomplete and their causes in terms of meristematic activity are unspecified. A process could have more than one developmental cause, and different combinations of process could achieve the same result. From a methodological standpoint, Zimmermann did not provide a means of choosing among competing hypotheses of organ evolution. These major weaknesses often make the theory seem speculative or, at best, descriptive. Such criticisms are not however unique to the telome theory but could be levelled at any hypothesis of transformation of one organ system into another. How are we to know when and how co-option and modification of modules has occurred? Perhaps no single recipe will suffice, but one could seek evidence for transformation in development studies, transitional forms in the fossil record, and consistency with character distributions on cladogram topology. Because hypotheses of transformation imply relationships among organs and therefore also among their developmental pathways, molecular evidence may prove crucial in the future. In considering the evolution of leaves in lycopods, the three most plausible hypotheses imply quite different relationships among various organs (leaf, sporangium, ligule) that could in principle be tested.

Unfortunately, the molecular basis of development in these interesting basal plant groups is almost completely unknown. The upward outlook so beloved by the 'new morphologists' of the early twentieth century has yet to have a major impact in the field of plant developmental biology. Model organisms are still mostly angiosperms and mainly crop plants. Zimmermann might have appreciated the irony, and he would certainly have argued forcefully that we will never fully understand the evolution of plant form until we move further down the phylogenetic tree.

ACKNOWLEDGEMENTS

I thank Richard Bateman for encouraging me to write this chapter. Thanks also to William E. Stein and Robert Scotland for their thoughtful reviews.

REFERENCES

Andrews, H. N. (1961) *Studies in Paleobotany*. Wiley, New York.

Andrews, H. N. and Kern, E. M. (1947) The Idaho Tempskyas and associated fossil plants. *Annals of the Missouri Botanical Garden*, 34, 119–186.

Banks, H. P., Leclercq, S. and Hueber, F. M. (1975) Anatomy and morphology of *Psilophyton dawsonii*, sp. n. from the late Lower Devonian of Quebec (Gaspé), and Ontario, Canada. *Palaeontographica Americana*, 48, 77–127.

Bateman, R. M. (1994) Evolutionary–developmental change in the growth architecture of fossil rhizomorphic lycopsids: scenarios constructed on cladistic foundations. *Biological Reviews*, 69, 527–598.

Bateman, R. M. (1999) Architectural radiations cannot be optimally interpreted without morphological and molecular phylogenies, in *The Evolution of Plant Architecture* (eds M. H. Kurmann and A. R. Hemsley), Royal Botanic Gardens, Kew, London, pp. 221–250.

Bateman, R. M., DiMichele, W. A. and Willard, D. A. (1992) Experimental cladistic analysis of anatomically preserved lycopsids from the Carboniferous of Euramerica: an essay on paleobotanical phylogenetics. *Annals of the Missouri Botanical Garden*, 79, 500–559.

Baum, D. A. and Donoghue, M. J. (2002) Transference of function, heterotopy, and the evolution of plant development, in *Developmental Genetics and Plant Evolution* (eds Q. C. B. Cronk, R. M. Bateman and J. A. Hawkins), Taylor & Francis, London, pp. 52–69.

Beck, C. B., Schmid, R. and Rothwell, G. W. (1982) Stelar morphology of the primary vascular system of seed plants. *Botanical Review*, 48, 691–815.

Bower, F. O. (1908) *The Origin of a Land Flora*. Macmillan, London.

Chapman, R. L., Buchheim, M. A., Delwiche, C. F., Friedl, T., Huss, V. A. R., Karol, K. G., Lewis, L. A., Manhart, J., McCourt, R. M., Olsen, J. L. and Waters, D. A. (1998) Molecular systematics of the green algae, in *Molecular Systematics of Plants II: DNA Sequencing* (eds D. E. Soltis, S. S. Soltis and J. J. Doyle), Kluwer, Boston, pp. 508–540.

Crane, P. R. and Kenrick, P. (1997) Diverted development of reproductive organs: a source of morphological innovation in land plants. *Plant Systematics and Evolution*, 206, 161–174.

Delevoryas, T. (1955) The Medullosae – structure and relationships. *Palaeontographica*, B97, 114–167.

DiMichele, W. A., Hook, R. W., Beerbower, R., Boy, J. A., Gastaldo, R. A., Hotton III, N., Phillips, T. M., Scheckler, S. E., Shear, W. A. and Sues, H.-D. (1992) Paleozoic terrestrial ecosystems, in *Terrestrial Ecosystems Through Time: Evolutionary Paleoecology of Terrestrial Plants and Animals* (eds A. K. Behrensmeyer, J. D. Damuth, W. A. DiMichele, R. Potts, H.-D. Sues and S. L. Wing), University of Chicago Press, Chicago, pp. 205–325.

Donoghue, M. J. and Kadereit, J. W. (1992) Walter Zimmermann and the growth of phylogenetic theory. *Systematic Biology*, 41, 74–85.

Duff, J. R. and Nickrent, D. L. (1999) Phylogenetic relationships of land plants using mitochondrial small-subunit rDNA sequences. *American Journal of Botany*, 86, 372–386.

Edwards, D. (1975) Some observations on the fertile parts of *Zosterophyllum myretonianum* Penhallow from the Lower Old Red Sandstone of Scotland. *Transactions of the Royal Society of Edinburgh*, 69, 251–265.

Edwards, D. S. (1986) *Aglaophyton major*, a non-vascular land-plant from the Devonian Rhynie Chert. *Botanical Journal of the Linnean Society*, 93, 173–204.

Emberger, L. (1944) *Les Plantes Fossiles dans leurs Rapports avec les Végétaux Vivants*. Masson et Cie, Paris.

Farris, J. S. (1983) The logical basis of phylogenetic analysis, in *Advances in Cladistics 2* (eds N. I. Platnick and V. A. Funk), Columbia University Press, New York, pp. 7–36.

Florin, R. (1951) Evolution in *Cordaites* and conifers. *Acta Horti Bergiani*, 15, 285–388.

Gensel, P. G. and Andrews, H. N. (1984) *Plant Life in the Devonian*. Praeger, New York.

Graham, L. E. (1993) *Origin of Land Plants*. Wiley, New York.

Halle, T. G. (1933) The structure of certain fossil spore-bearing organs believed to belong to pteridosperms. *Kungliga Svenska Vetenskapsakademiens Handlingar*, 12, 1–103.

Harris, T. M. (1940) *Caytonia. Annals of Botany*, 4, 713–734.

Hedderson, T. A., Chapman, R. L. and Cox, C. J. (1998) Bryophytes and the origins and diversification of land plants: new evidence from molecules, in *Bryology for the Twenty-First Century* (eds J. W. Bates, N. W. Ashton and J. G. Duckett), Maney & Son, Leeds, pp. 65–77.

Huss, V. A. R. and Kranz, H. D. (1997) Charophyte evolution and the origin of land plants. *Plant Systematics and Evolution, Supplement*, 11, 103–114.

Jermy, A. C. (1990a) Isoetaceae, in *Pteridophytes and Gymnosperms. The Families and Genera of Vascular Plants* (eds K. U. Kramer and P. S. Green), Springer-Verlag, Berlin, pp. 26–31.

Jermy, A. C. (1990b) Selaginellaceae, in *Pteridophytes and Gymnosperms. The Families and Genera of Vascular Plants* (eds K. U. Kramer and P. S. Green), Springer-Verlag, Berlin, pp. 39–45.

Kenrick, P. (2000) The relationships of vascular plants. *Philosophical Transactions of the Royal Society of London*, B355, 847–855.

Kenrick, P. (1994) Alternation of generations in land plants: new phylogenetic and palaeobotanical evidence. *Biological Reviews*, 69, 293–330.

Kenrick, P. and Crane, P. R. (1997a) *The Origin and Early Diversification of Land Plants: A Cladistic Study*, Smithsonian Series in Comparative Evolutionary Biology, Smithsonian Institution Press, Washington.

Kenrick, P. and Crane, P. R. (1997b) The origin and early evolution of plants on land. *Nature*, 389, 33–39.

Kidston, R. and Lang, W. H. (1921) On Old Red Sandstone plants showing structure, from the Rhynie Chert Bed, Aberdeenshire. Part IV. Restorations of the vascular cryptogams, and discussion on their bearing on the general morphology of the pteridophyta and the origin of the organisation of land-plants. *Transactions of the Royal Society of Edinburgh*, 52, 831–854.

Kranz, H. D., Miks, D., Siegler, M.-L., Capesius, I., Sensen, W. and Huss, V. A. R. (1995) The origin of land plants: phylogenetic relationships among charophytes, bryophytes, and vascular plants inferred from complete small-subunit ribosomal RNA gene sequences. *Journal of Molecular Evolution*, 41, 74–84.

Lam, H. J. (1948) Classification and the new morphology. *Acta Biotheoretica*, 8, 107–154.

Langdale, J. A., Scotland, R. W. and Corley, S. B. (2002) A developmental perspective on the evolution of leaves, in *Developmental Genetics and Plant Evolution* (eds Q. C. B. Cronk, R. M. Bateman and J. A. Hawkins), Taylor & Francis, London, pp. 388–394.

Lewis, L. A., Mishler, B. D. and Vilgalys, R. (1997) Phylogenetic relationships of the liverworts (Hepaticae), a basal embryophyte lineage, inferred from nucleotide sequence data of the chloroplast gene *rbc*L. *Molecular Phylogenetics and Evolution*, 7, 377–393.

Li, C.-S. and Edwards, D. (1992) A new genus of early land plants with novel strobilar construction from the Lower Devonian Posongchong Formation, Yunnan Province, China. *Palaeontology*, 35, 257–272.

McCourt, R. M., Karol, K. G., Guerlesquin, M. and Feist, M. (1996) Phylogeny of extant genera in the family Characeae (Charales, Charophyceae) based on *rbc*L sequences and morphology. *American Journal of Botany*, 83, 125–131.

Malek, O., Lättig, K., Hiesel, R., Brennicke, A. and Knoop, V. (1996) RNA editing in bryophytes and a molecular phylogeny of land plants. *EMBO Journal*, 15, 1403–1411.

Mishler, B. D. and Churchill, S. P. (1985) Transition to a land flora: phylogenetic relationships of the green algae and bryophytes. *Cladistics*, 1, 305–328.

Niklas, K. J. (1982) Computer simulations of early land plant branching morphologies: canalization of patterns during evolution? *Paleobiology*, 8, 196–210.

Niklas, K. J. (1997) *The Evolutionary Biology of Plants*. University of Chicago Press, Chicago.

Nishiyama, T. and Kato, M. (1999) Molecular phylogenetic analysis among bryophytes and tracheophytes based on combined data of plastid coded genes and the 18S rRNA gene. *Molecular Biology and Evolution*, 16, 1027–1036.

Ogura, Y. (1972) *Comparative Anatomy of Vegetative Organs of the Pteridophytes*. Handbuch der Pflanzenanatomie, Band VII, Teil 3. Borntraeger, Berlin.

Øllgaard, B. (1990) Lycopodiaceae, in *The Families and Genera of Vascular Plants. Volume I, Pteridophytes and Gymnosperms* (eds K. U. Kramer and P. S. Green), Springer-Verlag, Berlin, pp. 31–39.

Patterson, C. (1982) Morphological characters and homology, in *Problems of Phylogenetic Reconstruction* (eds K. A. Joysey and A. E. Friday), Systematics Association Special Volume, Academic Press, London, pp. 21–74.

Pryer, K. M., Schneider, H., Smith, A. R., Cranfill, R., Wolf, P. G., Hunt, J. S. and Sipes, S. D. (2001) Horsetails and ferns are a monophyletic group and the closest living relatives to seed plants. *Nature*, 409, 618–622.

Qiu, Y.-L., Cho, Y., Cox, J. C. and Palmer, J. D. (1998) The gain of three mitochondrial introns identifies liverworts as the earliest land plants. *Nature*, 394, 671–674.

Raff, R. A. (1996) *The Shape of Life: Genes, Development, and the Evolution of Animal Form*. University of Chicago Press, Chicago.

Remy, W., Gensel, P. G. and Hass, H. (1993) The gametophyte generation of some early Devonian land plants. *International Journal of Plant Sciences*, 154, 35–58.

Remy, W. and Hass, H. (1996) New information on gametophytes and sporophytes of *Aglaophyton major* and inferences about possible environmental adaptations. *Review of Palaeobotany and Palynology*, 90, 175–194.

Riedl, R. (1978) *Order in Living Organisms*. John Wiley & Sons, Chichester.

Rieppel, O. (1988) *Fundamentals of Comparative Biology*. Birkhäuser, Basel.

Rothwell, G. W. (1995) The fossil history of branching: implications for the phylogeny of land plants, in *Experimental and Molecular Approaches to Plant Biosystematics* (eds P. C. Hoch and A. G. Stephenson), Missouri Botanical Garden, St Louis, pp. 71–86.

Sattler, R. (1992) Process morphology: structural dynamics in development and evolution. *Canadian Journal of Botany*, 70, 708–714.

Schneider, H., Pryer, K. M., Cranfill, R., Smith, A. R. and Wolf, P. G. (2002) Evolution of vascular plant body plans – a phylogenetic perspective, in *Developmental Genetics and Plant Evolution* (eds Q. C. B. Cronk, R. M. Bateman and J. A. Hawkins), Taylor & Francis, London, pp. 330–364.

Stein, W. E. (1993) Modelling the evolution of stelar architecture in vascular plants. *International Journal of Plant Sciences*, 154, 229–263.

Stein, W. E. (1998) Developmental logic: establishing a relationship between developmental process and phylogenetic pattern in primitive vascular plants. *Review of Palaeobotany and Palynology*, 102, 15–42.

Stewart, W. N. (1964) An upward outlook in plant morphology. *Phytomorphology*, 14, 120–134.

Stewart, W. N. and Rothwell, G. W. (1993) *Paleobotany and the Evolution of Plants*. Cambridge University Press, Cambridge.

Thomas, H. H. (1931) The early evolution of the angiosperms. *Annals of Botany*, 45, 647–672.

Thomas, H. H. (1932) The old morphology and the new. *Proceedings of the Linnean Society of London*, 176, 17–44.

Weston, P. H. (2000) Process morphology from a cladistic perspective, in *Homology and Systematics: Coding Characters for Phylogenetic Analysis* (eds R. Scotland and R. T. Pennington), Taylor & Francis, London, pp. 124–144.

Wight, D. C. (1987) Non-adaptive change in early land plant evolution. *Paleobiology*, 13, 208–214.

Wilson, C. L. (1953) The telome theory. *Botanical Review*, 19, 417–437.

Zimmermann, W. (1930) *Die Phylogenie der Pflanzen*. Fischer, Jena.

Zimmermann, W. (1931) Arbeitsweise der botanischen Phylogenetik und anderer Gruppierungswissenschaften, in *Handbuch der biologischen Arbeitsmethoden* (ed. E. Abderhalden), Urban & Schwarzenberg, Berlin, pp. 941–1053.

Zimmermann, W. (1938) Die Telometheorie. *Biologe*, 7, 385–391.

Zimmermann, W. (1952) Main results of the 'Telome Theory'. *The Palaeobotanist*, 1, 456–470.

Zimmermann, W. (1959) *Die Phylogenie der Pflanzen*. Fischer, Stuttgart.

Zimmermann, W. (1965) *Die Telomtheorie*. Fischer, Stuttgart.

Chapter 19

A developmental perspective on the evolution of leaves

Jane A. Langdale, Robert W. Scotland and Susie B. Corley

ABSTRACT

A major question in evolutionary biology is how leaves evolved. Morphologists have generally distinguished two types of leaf, referred to as microphyllous and mega-phyllous. Existing theories on the evolution of these two leaf types are based on anatomical, fossil and phylogenetic data. Here we discuss the evolution of micro-phylls and megaphylls in the context of information gained from recent developmental studies in model systems.

19.1 Introduction

Most extant land plant species (euphyllophytes) have megaphyllous leaves that are associated with the formation of parenchymatous areas, or leaf gaps, in the vascular cylinder of the stem (Gifford and Foster, 1988). The lamina of the megaphyll is characterised by complex venation patterns with extreme diversity in form and size. In contrast, lycophytes have microphyllous leaves that have a single, unbranched vascular strand and no leaf gap in the stem. The distinction between microphyllous and megaphyllous leaves has provided the context within which leaf evolution has been discussed; it has been suggested that the two leaf types evolved independently (Bower, 1935; Kenrick and Crane, 1997). However, certain observations question the traditional characterisation of both leaf types. For example, two species of the lycophyte *Selaginella* have megaphyllous-like leaves (Wagner *et al.*, 1982) and the euphyllophyte horsetails (*Equisetum*) have microphyllous leaves. As the lycophyte and euphyllophyte lineages were distinguished prior to the definition of microphylls and megaphylls, it is possible that phylogenetic considerations have influenced anatomical interpretation. Thus, by focusing on mechanisms that lead to the formation of leaves, rather than on mature leaf anatomy, we may be able to shed new light on our understanding of leaf evolution.

19.2 Mechanistic pathways

It is possible to contrast three pathways by which the developmental processes underpinning microphylls and megaphylls may have evolved, as summarised in Figure 19.1a–c. The first two possibilities consider that developmental mechanisms that produce megaphylls were recruited from mechanisms that produced micro-

In *Developmental Genetics and Plant Evolution* (2002) (eds Q. C. B. Cronk, R. M. Bateman and J. A. Hawkins), Taylor & Francis, London, pp. 388–394.

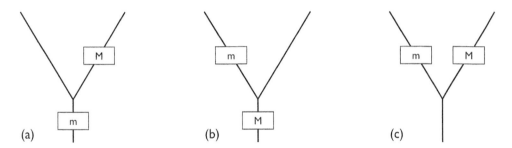

Figure 19.1 Three possible relationships of microphyllous (m) and megaphyllous (M) developmental mechanisms.

phylls or vice versa (Figure 19.1a–b). The third option is that the two leaf types evolved by adopting independent developmental mechanisms or by adopting the same developmental pathway independently (Figure 19.1c). The distribution of microphylls and megaphylls relative to the phylogeny of land plants containing fossil taxa suggests that microphylls and megaphylls evolved independently (Kenrick and Crane, 1997; Figure 19.2). The lycophyte lineage is associated with a putative stem lineage of leafless zosterophylls, whereas the euphyllophyte lineage is associated with the primarily leafless trimerophytes. Progymnosperm taxa that are associated with the seed plant lineage also contain leafless fossil plants. Thus, megaphylls may have

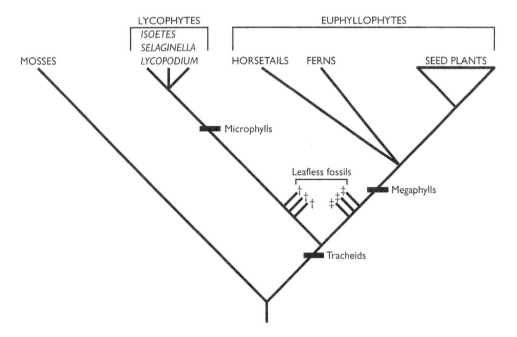

Figure 19.2 A simplified phylogeny of vascular plants to show the microphyll and megaphyll lineages. In the leafless fossils, zosterophylls are depicted by a single cross and trimerophytes are depicted by a double cross.

had at least two independent origins within euphyllophytes. Were different developmental pathways recruited each time?

19.3 Hypotheses of transformation

To explain the shoot-to-leaf transition in microphyll and megaphyll lineages, Bower proposed that microphylls evolved through 'enation' and megaphylls through 'planation'. The enation theory proposes that spine-like outgrowths from the shoot were vascularised to form the microphyllous leaf (Bower, 1935; Figure 19.3a). Support for this theory is provided by the fossil plant *Asteroxylon*, which may represent an intermediate stage between the non-vascularised enation and the vascularised microphyll (Stewart, 1983). However, the structure of *Asteroxylon* is also consistent with the theory put forward by Kenrick and Crane (1997), which suggested that microphylls evolved as a consequence of sporangial sterilisation. Megaphylls are thought to have evolved by unequal branching followed by specialisation (Figure 19.3b). It was proposed that one branch became laterally 'overtopped' by the other and that subsequent flattening or 'planation' of the overtopped branch, in combination with fusion or 'webbing' of the flattened region, resulted in formation of the megaphyllous leaf (Bower, 1935). If either the enation or sterilisation theories are correct,

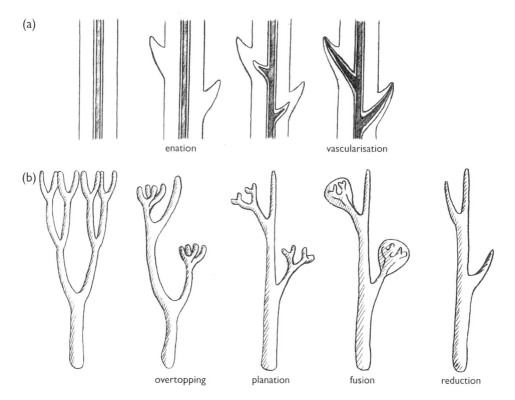

Figure 19.3 Schematic representation of the enation (a) and telome (b) theories. Adapted from Gifford and Foster (1988).

then presumably independent developmental mechanisms were recruited during microphyll and megaphyll evolution.

Despite convincing fossil evidence to the contrary, it has been hypothesised that microphylls evolved through an adaptation of the megaphyll developmental programme. In his telome theory, Zimmermann (1953; see Kenrick, 2002) proposed that microphylls were derived from megaphylls by a process of 'reduction' (Figure 19.3b). Although the reduction theory has not achieved widespread support, consideration of leaf form in horsetails gives the idea credence. The horsetails are nested within the euphyllophyte clade and fossils exhibit megaphyllous leaves. However, extant horsetails have microphyllous-like leaves. This observation tentatively suggests that the transition between megaphyllous and microphyllous leaves is possible. Two species of *Selaginella* that have leaves with branched veins (but no leaf gaps), despite *Selaginella* belonging to the microphyllous lycophyte clade (Wagner *et al.*, 1982), further suggest that a microphyllous to megaphyllous transition can also occur. Thus, the evolution of microphylls and megaphylls may have resulted from the modification of a single developmental pathway.

19.4 Leaf development in model systems

In the absence of an understanding of mechanism, it is hard to accurately predict how one developmental pathway can be converted or diverted into another. In recent years the developmental processes that operate during the transition from indeterminate shoot growth to the growth of a determinate lateral organ have started to be elucidated. Below, we outline these processes and apply our knowledge of them to assess the possible relationship between microphyllous and megaphyllous leaves.

In a number of angiosperm species it has been demonstrated that indeterminate shoot growth is dependent upon the action of a group of proteins encoded by *Knotted1*-like homeobox (*KNOX*) genes. Multiple copies of *KNOX* genes have been identified in many model plant species and in each case at least a subset of these genes is expressed in the shoot apical meristem (reviewed by Tsiantis and Langdale, 1998). Downregulation of *KNOX* gene expression on the flanks of the meristem is correlated with the initiation of leaf development (Jackson *et al.*, 1994). If this downregulation is not maintained in the leaf primordium normal development of the leaf is disrupted. The extent of this disruption depends on the species being examined, and on the *KNOX* gene family member that is ectopically expressed in the leaf (Smith *et al.*, 1992; Matsuoka *et al.*, 1993; Sinha *et al.*, 1993; Chuck *et al.*, 1996; Haraven *et al.*, 1996; Williams-Carrier *et al.*, 1997). As downregulation during leaf initiation occurs even in transgenic plants that constitutively express *KNOX* genes (Chuck *et al.*, 1996), it is likely that post-transcriptional processes operate at this stage. However, maintenance of repression within the leaf appears to be mediated by mechanisms involving a *MYB*-like gene. The orthologous *MYB*-like *ARP* genes – *ASYMMETRIC LEAVES1* (*Arabidopsis*) (Byrne *et al.*, 2000; Ori *et al.*, 2000; Semiarti *et al.*, 2001), *ROUGH SHEATH 2* (maize) (Timmermans *et al.*, 1999; Tsiantis *et al.*, 1999) and *PHANTASTICA* (*Antirrhinum*) (Waites *et al.*, 1998) – have been shown to repress *KNOX* gene expression in their respective species. In each case, loss of gene function results in normal leaf initiation but aberrant leaf development.

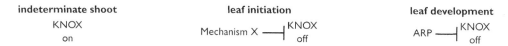

Figure 19.4 Molecular basis for the switch between indeterminate shoot and determinate leaf growth in maize. ARP denotes the *ASYMMETRIC LEAVES1, ROUGH SHEATH 2, PHAN-TASTICA* orthologue.

19.5 An evolutionary hypothesis

In its simplest form, the shoot-to-leaf transition can therefore be defined as *KNOX* on (indeterminate shoot), *KNOX* off by unknown mechanism 'X' (leaf initiation), and *KNOX* kept off by an ARP function (leaf development) (Figure 19.4). Subsequent variation in temporal and spatial expression patterns of individual *KNOX* genes could account for many of the leaf shapes observed in plants. In this regard, it is clearly important to determine whether the KNOX/X/ARP mechanism is used universally in land plants. If it is not, the independent evolution of microphylls and megaphylls may be confirmed. If the mechanism has been universally recruited, however, several hypotheses of leaf evolution could be put forward. We consider one of those possibilities here. Ancestral unbranched shoots would presumably have expressed *KNOX* genes throughout development and therefore initiation of a lateral outgrowth of any sort would have required repression of *KNOX* genes through a gain of mechanism X (Figure 19.5). If that outgrowth resulted in a branched shoot, then it is likely to have involved restoration of *KNOX* gene expression after initiation of lateral outgrowth. For the outgrowth to develop into a 'simple' microphyll it would have been necessary to maintain repression of all *KNOX* genes by recruiting an ARP function. For the outgrowth to develop into a 'complex' megaphyll it is likely that only some of the *KNOX* genes would have been repressed, while the expression of other *KNOX* genes would have been temporally and spatially regulated to produce a variety of leaf shapes. The development of either a microphyll or a megaphyll would therefore require a gain of

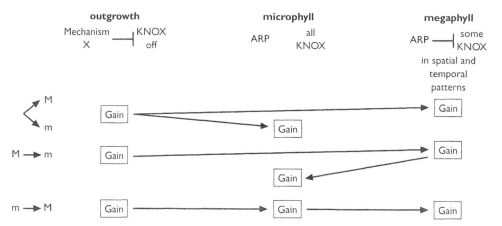

Figure 19.5 A developmental perspective on the origin of microphyllous (m) and megaphyllous (M) leaves. ARP denotes the *ASYMMETRIC LEAVES1, ROUGH SHEATH 2, PHANTASTICA* orthologue.

ARP function. Thus, in total, two gains of function would have been required if microphylls and megaphylls evolved independently (Figure 19.5). Similarly, two gains of function would have been necessary if microphylls were derived from megaphylls: the first to repress some *KNOX* genes and the second to repress all *KNOX* genes. However, if megaphylls were derived from microphylls, a gain of function (repression of all *KNOX* genes) would have been followed by a loss of function (loss of repression of some *KNOX* genes). Assuming that gene function is more easily lost than gained, an evolutionary pathway in which megaphyll developmental programs were derived from those used to produce microphylls would have been more economical than either of the alternative pathways.

19.6 Leaf development in lycophytes: the test

To test whether *KNOX/ARP* interactions play a role in the formation of microphylls, we are currently examining leaf development in *Selaginella* species. We have isolated *ARP* genes from two species of *Selaginella* and have amplified *KNOX* gene sequences. It remains to be seen whether *KNOX* and *ARP* genes interact in lycophytes and also whether any interactions are similar to those observed in angiosperms. However, data obtained from developmental studies will undoubtedly make an important contribution to our understanding of how leaves evolved.

REFERENCES

Bower, F. O. (1935) *Primitive Land Plants*. Macmillan Press, London.

Byrne, M. E., Barley, R., Curtis, M., Arroyo, J. M., Dunham, M., Hudson, A. and Martienssen, R. A. (2000) *ASYMMETRIC LEAVES1* mediates leaf patterning and stem cell function in *Arabidopsis*. *Nature*, 408, 967–971.

Chuck, G., Lincoln, C. and Hake, S. (1996) *KNAT1* induces lobed leaves with ectopic meristems when overexpressed in *Arabidopsis*. *Plant Cell*, 8, 1277–1289.

Gifford, E. M. and Foster, A. S. (1988) *Morphology and Evolution of Vascular Plants*, 3rd edn. W. H. Freeman, New York.

Haraven, D., Gutfinger, T., Parnis, A., Eshed, Y. and Lifschitz, E. (1996) The making of a compound leaf: genetic manipulation of leaf architecture in tomato. *Cell*, 84, 735–744.

Jackson, D., Veit, B. and Hake, S. (1994) Expression of maize *Knotted1*-related homeobox genes in the shoot apical meristem predicts patterns of morphogenesis in the vegetative shoot. *Development*, 120, 404–413.

Kenrick, P. (2002) The telome theory, in *Developmental Genetics and Plant Evolution* (eds Q. B. C. Cronk, R. M. Bateman and J. A. Hawkins), Taylor and Francis, London, pp. 365–387.

Kenrick, P. and Crane, P. R. (1997) *The Origin and Early Diversification of Land Plants*. Smithsonian Institution Press, Washington, DC.

Matsuoka, M., Ichikawa, H., Saito, A., Tada, Y., Fujimura, T. and Kano-Murakami, Y. (1993) Expression of a rice homeobox gene causes altered morphology of transgenic plants. *Plant Cell*, 5, 1039–1048.

Ori, N., Eshed, Y., Chuck, G., Bowman, J. L. and Hake, S. (2000) Mechanisms that control *knox* gene expression in the *Arabidopsis* shoot. *Development*, 127, 5523–5532.

Semiarti, E., Ueno, Y., Tsukaya, H., Machida, C. and Machida, Y. (2001) The *ASYMMETRIC LEAVES2* gene of *Arabidopsis thaliana* regulates formation of a symmetric lamina, establishment of venation and repression of meristem related homeobox genes in leaves. *Development*, 128, 1771–1783.

Sinha, N. R., Williams, R. E. and Hake, S. (1993) Overexpression of the maize homeobox gene, *Knotted-1*, causes a switch from determinate to indeterminate cell fates. *Genes and Development*, 7, 787–795.

Smith, L. G., Greene, B., Veit, B. and Hake, S. (1992) A dominant mutation in the maize homeobox gene, *Knotted-1*, causes its ectopic expression in leaf cells with altered fates. *Development*, 116, 21–30.

Stewart, W. N. (1983) *Paleobotany and the Evolution of Plants*. Cambridge University Press, Cambridge.

Timmermans, M. C. P., Hudson, A., Becraft, P. W. and Nelson, T. (1999) *ROUGH SHEATH2*: a *myb* protein that represses *KNOX* homeobox genes in maize lateral organ primordia. *Science*, 284, 151–153.

Tsiantis, M. and Langdale, J. A. (1998) The formation of leaves. *Current Opinions in Plant Biology*, 1, 43–48.

Tsiantis, M., Schneeberger, R., Golz, J. F., Freeling, M. and Langdale, J. A. (1999) The maize *rough sheath2* gene and leaf development programs in monocot and dicot plants. *Science*, 284, 154–156.

Wagner, W. H. Jr., Beitel, J. M. and Wagner, F. S. (1982) Complex venation patterns in the leaves of *Selaginella*: megaphyll-like leaves in lycophytes. *Science*, 218, 793–794.

Waites, R., Selvadurai, H. R. N., Oliver, I. R. and Hudson, A. (1998) The *PHANTASTICA* gene encodes a *myb* transcription factor involved in growth and dorsoventrality of lateral organs in *Antirrhinum*. *Cell*, 93, 779–789.

Williams-Carrier, R. E., Lie, Y. S., Hake, S. and Lemaux, P. G. (1997) Ectopic expression of the maize *kn1* gene phenocopies the *Hooded* mutant of barley. *Development*, 124, 3737–3745.

Zimmermann, W. (1953) Main results of the 'telome theory'. *The Paleobotanist*, 1, 456–470.

Chapter 20

An overview of seed plant leaf evolution

C. Jill Harrison, Quentin C. B. Cronk and Andrew Hudson

ABSTRACT

Although the leaves of seed plants share several characteristics that distinguish them from stems (they are lateral, determinate, dorsiventral and laminar structures that have axillary buds) they are generally believed to have evolved from a collection of stem-like structures by a series of distinct transitions. Each of these transitions is likely to have involved a novel mechanism of genetic control. Living and fossil intermediates support such a process for evolution of leaves in the sister group of seed plants, the ferns and their relatives (Moniliformopses), but comparable evidence for seed plants is lacking. Recent analysis of model angiosperms has begun to reveal the genetic control of leafy characteristics and, where tested, these appear largely conserved within angiosperms. While suggesting how some aspects of leafiness might have exploited developmental mechanisms operating in leafless ancestors, this provides little support for stepwise evolution of seed plant leaves from stems. Rather it suggests that the characteristics of leaves are interdependent and under common genetic control, more consistent with a radical origin of seed plant leaves.

20.1 Introduction

20.1.1 Characteristics of seed plant leaves

The overwhelming majority of seed plant leaves share several properties:

1   Leaves are produced on the flanks of a shoot apical meristem (SAM), so they become lateral to the shoot.
2   Leaves are usually flattened perpendicular to the shoot axis and may also have an asymmetrical distribution of specialised cell types along the same axis.
3   Unlike the shoot, which has the capacity for unlimited growth, leaves are determinate.
4   Leaves are associated with axillary buds that form either from the adaxial base of the leaf or from the stem adaxial to the leaf base.

20.1.2 Evolutionary steps needed to make a seed plant leaf

Seed plant leaves are believed to share an origin as a system of dichotomous branches, similar to the Devonian fossil plant *Rhynia* (Richardson, 1984; Stewart

In *Developmental Genetics and Plant Evolution* (2002) (eds Q. C. B. Cronk, R. M. Bateman and J. A. Hawkins), Taylor & Francis, London, pp. 395–403.

and Rothwell, 1993; Kenrick and Crane, 1997). In *Rhynia* the dichotomising stems terminated in sporangia. Dichotomy was in any plane, and both daughter branches followed the same fate. Megaphyll evolution from a *Rhynia*-like precursor is proposed to have involved several steps, formalised as part of Zimmermann's Telome Theory (Zimmermann, 1952; Kenrick, 2002).

1   Branches resulting from dichotomy of the shoot apical meristem (SAM) assumed different fates, so that one daughter branch could grow more than the other, overtopping it. We can imagine that this resulted in a major, possibly randomly determined axis of growth, here termed a pseudo-monopodium.
2   Differences in SAM fate were subsequently reinforced to give distinct apical and lateral meristems, resulting in a single continued axis of shoot growth, here termed a monopodium.
3   Lateral branches assumed a limited capacity for growth and became determinate.
4   Branching was limited to a plane perpendicular to the shoot axis.
5   Spaces between the flattened branches were infilled with lamina (webbing).

The sister group to seed plants, the Moniliformopses (including ferns and horsetails, which together with seed plants constitute the euphyllophytes: Kenrick and Crane, 1997; Pryer *et al.*, 2001) contain living and fossil intermediates that generally support these steps in evolution. Comparable evidence for seed plants is lacking, and all seed plants share the common characters associated with angiosperm leaves. Although likely to be derived from a lateral branching system, their evolution remains obscure.

Each step in evolution of a leaf is likely to have involved novel genetic control. Analysis has begun to reveal genetic mechanisms responsible for characteristics of leaves and other aspects of plant form in a limited range of model angiosperms. The most direct way to examine the evolution of these mechanisms would be by comparative analysis in a broader range of plants. In the meantime, however, we can ask whether knowledge gained from model angiosperms can reveal how leafy characters might have arisen in seed plants. Is stepwise acquisition of characters reflected in the existence of separate mechanisms controlling different aspects of leafiness? At a more general level, we can also examine whether a greater knowledge of angiosperm development can help suggest the mechanisms that might have been involved in early transitions from stems towards leaves.

20.1.3 What favoured the evolution of seed plant leaves?

Leaves have several advantages over a system consisting entirely of branches. Leaves have a high ratio of surface area to volume, which is efficient for photosynthesis and in shading competitors. They may also protect apices or developing reproductive structures. The pre-requisites to the evolution of the euphyllophyte leaf (pseudo-monopodial branching and lateral branches of limited growth) existed in the trimerophytes of the Lower Devonian (Kenrick, 2002). Other leafy characteristics may have been favoured by increasing need for photosynthetic efficiency and competition for light in the Upper Devonian. These advantages may have been countered by reduced capacity to lose heat in laminate leaves, which may explain their relatively late appearance towards the end of the Devonian, coincident with a dramatic increase in atmospheric CO_2 (Beerling *et al.*, 2001). Leaves may have also increased the necessity for stronger stems, and the appearance of more leafy characters also correlates with the phylogenetically iterative appearance of secondary thickening (Bateman *et al.*, 1998).

20.2 Early transitions in leaf evolution

The last common ancestor of both seed plants and the remaining euphyllophytes (Moniliformopses, including ferns and horsetails) was likely to have consisted of a monopodium with lateral branching systems of limited growth. It had therefore made the first transitions from a *Rhynia*-like predecessor, involving inequality of meristem branching. This resulted in, first, a pseudo-monopodial, and then a truly monopodial shoot with lateral meristems of limited growth.

20.2.1 Inequality of meristem fate

Differences between branch fates already existed in systems with equal dichotomous branching, as early branch pairs were sterile and indeterminate and later ones terminated in sporangia. Therefore, mechanisms promoting inequality of branches were likely to have been present in *Rhynia*-like plants and might have been deployed heterochronically in one of the two daughter meristems that resulted from a dichotomous split. In formation of a pseudo-monopodium, this might have been in response to a randomly determined inequality between daughter meristems (e.g. in their size) or as the result of competition or mutual inhibition. However, formation of a true monopodium requires a dedicated, indeterminate apical meristem.

20.2.2 Reinforcement of unequal fate

Model angiosperms provide examples of how meristem inequality can be reinforced. Several genetic systems are known to promote or repress meristem activity in model angiosperms, but function similarly in apical and axillary meristems. One exception is the transcription factor encoded by the *Teosinte branched1* (*Tb1*) gene of maize, which is expressed only in axillary SAMs and represses their development (Doebley *et al.*, 1997). Similarly, in the *Arabidopsis* inflorescence the distinction between the indeterminate inflorescence meristem and the determinate floral meristems that it produces laterally involves mutual repression between the *TERMINAL FLOWER 1* gene (expressed apically) and laterally expressed flower promoting genes including *LEAFY* (Ratcliffe *et al.*, 1999). The fate of each meristem depends on which gene it first expresses. However, neither case reveals the underlying difference between apical and axillary meristems responsible for differential gene expression.

20.2.3 Determinacy of lateral branches

Shoots have a capacity for unlimited, indeterminate growth whereas leaves are determinate. Therefore, evolution of the leaf is likely to have involved acquisition of mechanisms promoting determinacy in lateral branches. Model angiosperms provide few clues as to how this might have arisen, because determinacy of their shoots is normally associated with formation of a flower in which the meristem differentiates to form floral organs. Determinacy is therefore tightly linked to lateral organ fate. Determinacy is also apparent early in development of lateral organs; apical and lateral regions of leaf primordia commonly show higher rates of cell division than more central regions (Donnelly et al., 1999), and do not contain initial cells equivalent to those in the SAM that give rise to the rest of the shoot, as was elegantly demonstrated by clonal analysis of cell fate (e.g. Poethig and Sussex, 1985). Leaf determinacy is apparent from their inception, making it less simple to equate a leaf directly with an arrested branch system.

Genetic analysis provides a consistent view. Although altered activity of several genes can affect leafy characters, with one exception, they do not cause indeterminacy. The exception is indeterminacy (increased compounding) of the tomato leaf caused by expression of knotted1-like homeobox (knox) genes (Hareven et al., 1996; Tsiantis et al., 2002). In simple-leafed dicots and monocots, expression of knox genes is confined to cells of the SAM where they are required to promote indeterminacy (see below) and to internode initials. Expression is absent from leaf initials within the SAM and from developing leaf primordia (see references in Byrne et al., 2000). However, the involvement of knox genes in compound leaf development does not appear to be general; pea leaf primordia resemble those of simple-leafed plants in lacking knox expression (Hofer et al., 2001). Therefore, knox gene regulation of tomato leaf development might have involved secondary recruitment of mechanisms originally operating in the shoot.

Although determinacy in model dicots is usually associated with lateral organs, it may also result from differentiation of the SAM into other tissues, as in terminate seasonal growth in trees with sympodial shoots (e.g. Tilia cordata), from programmed death of SAM cells, or simply because the meristem ceases growth without differentiating. Loss of meristem activity can occur due to changes in activity of genes that are not involved in promoting organ fate. For example, reduced activity of the WUSCHEL transcription factor in Arabidopsis causes differentiation of the SAM into non-specific cell types (Laux et al., 1996). Even within angiosperms there are several mechanisms for determinacy. Shoots of Rhynia-like plants terminated in sporangia, indicating the existence of ancient mechanisms to promote determinacy which could have been recruited to control growth of lateral branches.

20.2.4 Specification of lateral organ fate in model angiosperms

Loss of knox gene expression marks cells within the meristem that will subsequently form lateral organs. Conversely, expression of phantastica-like myb genes is restricted to organ initial cells and primordia and interaction between these two classes of transcription factor genes appears important in specifying lateral organ fate (Langdale et al., 2002). Loss of phan activity in Antirrhinum allows ectopic knox gene expression and causes loss of two other leafy characters, lateral growth and dorsiventral asymmetry, although it is not clear whether these reflect independent roles of phan. Although this suggests that phan-like and knox genes may have a major role in specifying leaf fate, their interaction does not explain how the differences between lateral organs, the stem and SAM are first specified.

20.3 Further transitions in leaf evolution

20.3.1 Dorsiventrality and laminar growth

Most leaves show dorsiventral organisation. Most are flattened perpendicularly to the shoot axis (or, less commonly, parallel to it), and may consist of different cell types along this axis. Commonly cells specialised for light harvesting are adaxial (towards the apex), and those for gas exchange are abaxial. Asymmetry (*sensu* Waites and Hudson, 1995) is also common in the distribution of vascular and epidermal cell types.

These characteristics are proposed to reflect two transitions in evolution of megaphylls: restriction of branching to a single plane (planation) and webbing of the branch system to form a lamina. One explanation for the origin of planation is that it exploited polarity already present in the shoot, either along the apical–basal axis of growth or from the centre to the edge of the apical meristem. Evidence from model angiosperms suggests that asymmetry of leaves takes its cue from asymmetry in the meristem. Surgical experiments on potato which separated leaf initial cells from other apical cells resulted in formation of needle-like leaves with only abaxial cell types (Sussex, 1954). This suggested that a signal from the SAM was necessary to specify adaxial leaf identity. Analysis of the *Antirrhinum phantastica* mutant, which may also lack adaxial leaf cells, further suggested that interaction between adaxial and abaxial cells are necessary for lateral growth which flattens the leaf (Waites and Hudson, 1995). Leaves lacking adaxial cell types are also shorter than wild type, suggesting that growth in length might also be partially dependent on dorsiventral asymmetry. This means that at least two characteristics of the leaf (dorsiventrality and growth pattern) involve a common mechanism that may itself be related to polarity of the SAM or shoot.

Considerable support for this view is given by the behaviour of the *Arabidopsis* HD-ZIP III subfamily of genes, including *PHABULOSA* (*PHB*). These encode homeodomain transcription factors that may be activated by a putative lipid ligand (Ratcliffe *et al.*, 2000; McConnell *et al.*, 2001). Once activated, they specify adaxial organ fate and repress expression of genes needed for abaxial fate (Siegfried *et al.*, 1999; Eshed *et al.*, 2001; Kerstetter *et al.*, 2001). Plausibly, the activating ligand could come from the centre of the SAM, specifying adaxial organ fate in response to asymmetry in the SAM. This supports the idea that a lateral position was a prerequisite for planation.

HD-ZIP III genes have been identified in a moss, and therefore pre-date the origin of leaves (Sakibara *et al.*, 2001). In addition to organ initials, HD-ZIP III genes are also expressed in the developing phloem of *Arabidopsis*, and at least one is required for normal development of vascular tissues (Zhong and Ye, 1999). Therefore, the ancestral role for the family may have been to control vascular fate, perhaps in response to radial polarity in the shoot.

Because venation in most seed plant leaves is branched, it could correspond to a system of ancestral branches that have fused or been filled in by webbing. Analysis of vascular development has provided little support for this idea. Sachs (1981) suggested that vein branching results from induction by auxin. His canalisation hypothesis proposes that auxin is produced by developing leaves and transported to the stem. Higher auxin concentrations induce increased ability to transport auxin, there-fore causing files of cells to act as channels. Competition between neighbouring channels for auxin flow eliminates all but the strongest, which consequently show a fairly regular spacing, and these form veins (reviewed by Dengler and Kang, 2001). Patterns of venation reflect the pattern of auxin drainage. Their resemblance to shoot systems is therefore likely to be superficial.

20.3.2 Axillary meristem formation

Most euphyllophyte leaves, except ferns, are associated with secondary meristems in their axils. Genes required for axillary SAM formation have been identified, includ-ing *Lateral suppressor* (*Ls*), which promotes axillary SAM and petal formation in tomato, and is suggested to function in response to giberellin (Schumacher *et al.*, 1999). The *ls* mutant phenotype also reveals that axillary SAM formation can be uncoupled from leaf formation. Formation of axillary meristems may occur in response to adaxial leaf identity, or be regulated by the same process. For example, ectopic adaxial fate caused by *Arabidopsis* gain-of-function mutations in *PHB* leads to formation of ectopic SAMs (McConnell and Barton, 1998). Conversely, loss of adaxial identity in the leaf can also result in the loss of axillary meristems (Eshed *et al.*, 2001).

20.4 Are leafy characters separable genetically?

The view that is emerging from analysis of model angiosperms is that leafy char-acters are specified by interdependent genetic mechanisms and these, in turn, are linked to control of meristem activity. Loss of *phan* gene function in *Antirrhinum* reduces leaf growth, dorsiventral asymmetry and SAM activity. Similarly, HD-ZIP III genes might also have roles in promoting organ formation and SAM function that have yet to be revealed by loss-of-function mutations. There is currently little evidence of distinct genetic control of leafy characters that might reflect stepwise transitions from a branching system. However, this situation may change as the genetics of leaf development in angiosperms is further investigated and extended to a broader taxonomic range of plants.

20.5 Fuzzy morphology and natural variation

One consequence of the interdependence of leaf-like characters is that if one charac-ter is perturbed by mutation, others are also affected. If similar mutations in developmental regulatory genes are found in nature, perhaps we should expect a pleiotropic rather than a specific effect. This view is supported by two examples of organs of mixed nature.

The first example is the Cape Primrose, *Streptocarpus* (Gesneriaceae). In some species of *Streptocarpus* and other Old World Gesneriaceae caulescence is lost, so the plant body comprises a leaf-like organ which shows features usually restricted to stems. This organ arises by continued growth of a single cotyledon (Jong and Burtt, 1975; Möller and Cronk, 2001). Laminar growth occurs from meristematic regions at the base and sides of the organ (Jong and Burtt, 1975; Tsukaya, 1997). The plant is supported by a stem-like petiole, which also has a meristematic region at its base. Inflorescences are formed from a third meristematic region at the base of the midrib that is termed a groove meristem (Jong and Burtt, 1975). The combination of leaf-like features (dorsiventrality, laminar growth) merged with stem-like features (continued growth, formation of inflorescences) within an organ led to the naming of each aggregate unit as a 'phyllomorph' (Jong and Burtt, 1975).

A second example is the species of Podostemaceae that grow on streambeds. The roots of these plants form creeping flattened photosynthetic 'crusts', which give rise to lateral determinate shoots (Imaichi et al., 1999).

20.6 Conclusions

In model plants genes affecting leaf development often have pleiotropic effects, so morphology arising from their alteration can be unclear. In naturally occurring plants with atypical form morphology is also fuzzy, so distinction between organ systems may be blurred. The tight linkage between genetic events in the leaf and meristem seen so far in development, and the fuzziness of natural atypical morphologies, suggest that we may not be able to dissect the growth and evolution of the leaf into simple component parts, contrasting with the commonly held idea that regulatory tweaking of key developmental genes determines the evolution of morphology.

ACKNOWLEDGEMENTS

We thank Julie Hawkins and Richard Bateman for helpful comments, and Amanda Borking for proof-reading.

REFERENCES

Bateman, R. M., Crane, P. R., DiMichele, W. A., Kenrick, P. R., Rowe, N. P., Speck, T. and Stein, W. E. (1998) Early evolution of land plants: phylogeny, physiology, and ecology of the primary terrestrial radiation. *Annual Review of Ecology and Systematics*, 29, 263–292.

Beerling, D. J., Osborne, C. P. and Chaloner, W. G. (2001) Evolution of leaf-form in land plants linked to atmospheric CO_2 decline in the Late Paleozoic era. *Nature*, 410, 385–387.

Byrne, M. E., Barley, R., Curtis, M., Arroyo, J. M., Dunham, M., Hudson, A. and Martienssen, R. A. (2000) *Asymmetric leaves1* mediates leaf patterning and stem cell function in *Arabidopsis*. *Nature*, 408, 967–971.

Dengler, N. and Kang, J. (2001) Vascular patterning and leaf shape. *Current Opinion in Plant Biology*, 4, 50–56.

Doebley, J., Stec, A. and Hubbard, L. (1997) The evolution of apical dominance in maize. *Nature*, 386, 485–488.

Donnelly, P. M., Bonetta, D., Tsukaya, H., Dengler, R. E. and Dengler, N. G. (1999) Cell cycling and cell enlargement in developing leaves of *Arabidopsis*. *Developmental Biology*, 215, 407–419.

Eshed, Y., Baum, S. F., Perea, J. V. and Bowman, J. L. (2001) Establishment of polarity in lateral organs of plants. *Current Biology*, 11, 16: 1251–1260.

Hareven, D., Gutfinger, T., Parnis, A., Eshed, Y. and Lifschitz, E. (1996) The making of a compound leaf: genetic manipulation of leaf architecture in tomato. *Cell*, 84, 735–744.

Hofer, J., Gourlay, C., Michael, A. and Ellis, T. H. N. (2001) Expression of a class 1 *knotted1*-like homeobox gene is down-regulated in pea compound leaf primordia. *Plant Molecular Biology*, 45, 387–398.

Imaichi, R., Ichiba, T. and Kato, M. (1999) Developmental morphology and anatomy of the vegetative organs in *Malostriticha malayana* (Podostemataceae). *International Journal of Plant Sciences*, 160, 253–259.

Jong, K. and Burtt, B. L. (1975) The evolution of morphological novelty exemplified in the growth patterns of some Gesneriaceae. *New Phytologist*, 75, 297–311.

Kenrick, P. (2002) The Telome Theory, in *Developmental Genetics and Plant Evolution* (eds Q. C. B. Cronk, R. M. Bateman and J. A. Hawkins), Taylor & Francis, London, pp. 365–387.

Kenrick, P. and Crane, P. R. (1997) The origin and early evolution of plants on land. *Nature*, 389, 33–39.

Kerstetter, R. A., Bollman, K., Taylor, R. A., Bomblies, K. and Poethig, R. S. (2001) *KANADI* regulates organ polarity in *Arabidopsis*. *Nature*, 411, 706–709.

Langdale, J., Scotland, R. W. and Corley, S. B. (2002) A developmental perspective on the evolution of leaves, in *Developmental Genetics and Plant Evolution* (eds Q. C. B. Cronk, R. M. Bateman and J. A. Hawkins), Taylor & Francis, London, pp. 388–394.

Laux, T., Meyer, K. F. X., Berger, J. and Jürgens, G. (1996) The *WUSCHEL* gene is required for shoot and floral meristem integrity in *Arabidopsis*. *Development*, 122, 87–96.

McConnell, J. R. and Barton, M. K. (1998) Leaf polarity and meristem formation in *Arabidopsis*. *Development*, 125, 2935–2942.

McConnell, J., Emery, J., Eshed, Y., Bao, N., Bowman, J. and Barton, M. K. (2001) Role of *PHABULOSA* and *PHAVOLUTA* in determining radial patterning in shoots. *Nature*, 411, 709–713.

Möller, M. and Cronk, Q. C. B. (2001) Evolution of morphological novelty: a phylogenetic analysis of growth patterns in *Streptocarpus* (Gesneriaceae). *Evolution*, 55, 918–929.

Poethig, R. S. and Sussex, I. M. (1985) The cellular parameters of leaf development in tobacco – a clonal analysis. *Planta*, 165, 170–184.

Pryer, K. M., Schneider, H., Smith, A. R., Cranfill, R., Wolf, P. G., Hunt, J. S. and Sipes, S. D. (2001) Horsetails and ferns are a monophyletic group and the closest living relatives to seed plants. *Nature*, 409, 618–622.

Ratcliffe, O. J., Bradley, D. J. and Coen, E. S. (1999) Separation of shoot and floral identity in *Arabidopsis*. *Development*, 126, 1109–1120.

Ratcliffe, O. J., Riechman., J. L. and Zhang, J. Z. (2000) *INTERFASCICULAR FIBERLESS 1* is the same gene as *REVOLUTA*. *Plant Cell*, 12, 315–317.

Richardson, J. B. (1984) Paleobotany: the early evolution of leaves. *Nature*, 309, 749–750.

Sachs, T. (1981) The control of the patterned differentiation of vascular tissues. *Advances in Botanical Research*, 9, 151–262.

Sakibara, K., Nishiyama, T., Kato, M. and Hasebe, M. (2001) Isolation of homeodomain-leucine zipper genes from the moss *Physcomitrella patens* and the evolution of homeo-domain-leucine zipper genes in land plants. *Molecular Biology and Evolution*, 18, 491–502.

Schumacher, K., Schmitt, T., Rossberg, M., Schmitz, G. and Theres, K. (1999) The *Lateral suppressor* (*Ls*) gene of tomato encodes a new member of the VHIIID protein family. *Proceedings of the National Academy of Sciences, USA*, 96, 290–295.

Siegfried, K. R., Eshed, Y., Baum, S., Otsuga, D., Drews, G. N. and Bowman, J. L. (1999) Members of the *YABBY* gene family specify abaxial cell fate in *Arabidopsis*. *Development*, 126, 4117–4128.

Stewart, W. N. and Rothwell, G. W. (1993) *Paleobotany and the Evolution of Plants*. Cambridge University Press, Cambridge.

Sussex, I. M. (1954) Experiments on the cause of dorsiventrality in leaves. *Nature*, 174, 351–352.

Tsiantis, M., Hay, A., Ori, N., Kaur, H., Holtan, H., McCormick, S. and Hake, S. (2002) Developmental signals regulating leaf form, in *Developmental Genetics and Plant Evolution* (eds Q. C. B. Cronk, R. M. Bateman and J. A. Hawkins), Taylor & Francis, London, pp. 418–430.

Tsukaya, H. (1997) Determination of the unequal fate of cotyledons of a one-leaf plant, *Monophyllaea*. *Development*, 124, 1275–1280.

Waites, R. and Hudson, A. (1995) *Phantastica*: a gene required for dorsoventrality of leaves in *Antirrhinum majus*. *Development*, 121, 2143–2154.

Zhong, R. and Ye, Z. H. (1999) *IFL1*, a gene regulating interfascicular fibre differentiation in *Arabidopsis*, encodes a homeodomain-leucine zipper protein. *Plant Cell*, 11, 2139–2152.

Zimmermann, W. (1952) Main results of the 'Telome Theory'. *Paleobotanist*, 1, 456–470.

Chapter 21

Comparative developmental and molecular genetic aspects of leaf dissection

Stefan Gleissberg

ABSTRACT

The formation of marginal primordia occurs only during an early and brief phase in the development of dissected leaves. It reflects a transient competence of the leaf margin for organogenesis that precedes histogenesis. This chapter discusses organ formation competence of leaves in comparison to the shoot apical meristem and in relation to the action of *KNOX* genes. In addition, the developmental environment of the dissection process within the growing leaf is considered, referring to tissue maturation, differential growth and petiole–blade differentiation. A combination of comparative morphological and molecular genetic studies bears promise for further elucidation of morphogenesis of dissected angiosperm leaves.

21.1 Introduction

A major developmental means for leaf shape diversification during angiosperm evolution is the process of leaf dissection, by which new lateral growth axes are established along the margins of a leaf primordium. This process results in the leaf primordium becoming subdivided into a series of suborgans that may adopt a range of developmental fates. Primordia initiated from the lateral margins may develop as leaflets, lobes or as more inconspicuous marginal serrations on otherwise simple leaves. In contrast, truly simple leaves have only the single growth axis between tip and base established during leaf initiation at the shoot apical meristem (SAM). Both leaflets and lobes, and even serrations, can become dissected themselves, showing the potential for a (partial) iteration of the developmental program of the whole leaf.

A similarity between organ formation at the SAM and at leaf margins was first noted by Hofmeister (1868), who distinguished 'Sprossungen' (outgrowths) of different ranks, from shoots to leaves to trichomes. Jeune's concept of generation centres (e.g. Jeune, 1983; Jeune and Lacroix, 1993) can be applied to both organ formation at the SAM and at leaf margins (see also Cusset, 1986). Similarly, Hagemann (1970) and Hagemann and Gleissberg (1996) described organ formation as a process of meristem fractionation occurring both at the SAM and leaf margins. Because dissection of a leaf margin appears, to a certain degree, as a repetition of leaf initiation itself, primordial leaf margins may share some regulatory features with shoot apical meristems.

In *Developmental Genetics and Plant Evolution* (2002) (eds Q. C. B. Cronk, R. M. Bateman and J. A. Hawkins), Taylor & Francis, London, pp. 404–417.

Here I discuss the leaf margin's competence for lateral organ formation in several respects. First, I compare it to organ formation at the SAM in order to get a more general picture of organogenesis. Then current evidence for molecular control of leaf dissection in various angiosperms is reviewed, focusing on *Knotted1*-like homeobox (*KNOX*) genes. Finally, I summarise developmental circumstances under which leaf dissection occurs, helping to pose appropriate questions for future research.

21.2 Organ formation competence

21.2.1 Organ formation competence in higher plant shoots

Plants, in contrast to most animals, are permanently growing or 'open' organisms. Growth and the associated development of form (morphogenesis) is accomplished by cell proliferation and increase in cell volume. Cell proliferation is the defining characteristic for 'meristems' (Nägeli, 1858) and may or may not be associated with cell enlargement. As tissue starts to differentiate towards its final fate, cell proliferation slows down and ceases, and growth is increasingly accomplished by cell enlargement only. Recently, developmental geneticists have sometimes used the term 'meristem' only to refer to the SAM (e.g. Hake and Meyerowitz, 1998; Hudson, 1999). Traditionally, however, various types of meristems have been distinguished (e.g. Esau, 1977). One type are intercalary meristems that are intercalated between mature (fully differentiated and not growing) tissues. Intercalary meristems are responsible for organ elongation (e.g. of grass leaves or stem internodes in bamboo). Intercalary meristems, however, are not capable of initiating new organs; they only elongate existing organs. Organ formation occurs at the flanks of SAMs where leaf primordia (and axillary SAMs) arise.

In the context discussed here, organ formation refers solely to the establishment of a new growth direction relative to a pre-existing axis of growth. This process can be distinguished from the maintenance of an organ axis during subsequent growth and tissue differentiation. The process of leaf initiation from the SAM resembles the initiation of leaflets and serrations from the margin of leaf primordia. These regions of organ formation competence (OFC) have been termed 'blastozones' to distinguish them from other meristematic regions of the developing plant body (Hagemann and Gleissberg, 1996). Blastozones therefore represent only specific parts of the total population of proliferating cells. They show the cytohistological feature of eumeristems: small, isodiametric cytoplasma-rich cells with no apparent signs of tissue differentiation. The dense population of nuclei in blastozones could be important for integrative signalling between cells involved in the correct placement of organ initials. It is an essential feature of the open architecture of plants to restrict OFC to small, clearly defined regions separate from the fast-growing zones of enlargement and differentiation. Sachs (1991) suggested that slow growth could be a prerequisite for OFC at SAMs, preventing organs from being initiated in faster growing histogenetic meristems. Since newly established growth directions (a leaf, side shoot or leaflet) act as organisers for the arrangement of tissues, foremost of the developing vascular system, it is obviously important to prevent histogenetic regions such as intercalary meristems from initiating organs.

Despite the similarities between the processes of leaf initiation on the one hand

and leaflet and serration formation on the other hand, there are also important differences in apical and marginal blastozone function. First, apical blastozones are three-dimensionally and radially organised and frequently dome-shaped, while leaf marginal blastozones are two-dimensional and have a linear shape. The form of blastozones therefore anticipates the radial symmetry of the mature shoot axis and the planar symmetry of the leaves that are formed (Hagemann and Gleissberg, 1996). Second, SAMs have a fixed apical–basal zonation comprising a central zone, a peripheral or organogenetic zone, and a histogenetic zone. The central zone is not involved in organ formation but serves as a pool of stem cells. This assures an indeterminate function of the SAM. Cells of the central zone divide less frequently and are sometimes larger and more vacuolated than those of the adjacent peripheral zone involved in leaf initiation. Genes associated with stem cell function like *WUSCHEL* (*WUS*) and *CLAVATA1* (*CLV1*) are specifically expressed in the central zone. In addition, the adaxial side of leaf primordia contributes to maintenance of SAMs (Bowman and Eshed, 2000). Cells of the peripheral zone are smaller, less vacuolated and become involved in leaf initiation. Specifically expressed genes in the peripheral zone are *KNAT1* (Ori *et al.*, 2000) and in leaf anlagen *AINTEGUMENTA* (*ANT*) and *FILAMENTOUS FLOWER* (*FIL*). The differentiating stem proximal of the peripheral zone is not capable of organ formation. Leaf primordia lack this fixed apical–basal zonation. No central zone is present at the apex of leaf primordia, and this region frequently differentiates into a leaflet. OFC of leaf primordial margins is spatially less restricted, in many cases allowing for simultaneous leaflet and serration formation along the whole marginal periphery (Gleissberg, 1998a, b). Histogenesis and the associated loss of OFC proceeds, as discussed later, frequently not acropetally as in the SAM but exhibits diverse patterns.

21.2.2 Organ formation competence in other organisms with open organisation

The conceptual distinction of blastozones and meristems in general is also relevant because blastozones evolved multiple times with different internal organisations in various organism groups. A particularly interesting example is the plant body of Dasycladales, single-celled green algae with a main axis and whorls of lateral 'organs'. These algae have a distinct subapical blastozone from which lateral protrusions arise. These laterals (that may attain vegetative or reproductive functions) continue to initiate higher order protrusions but lose OFC during further rapid expansion. The similarity of their organisation to higher plant shoots is striking, given that these organisms are not internally chambered into cells (Sawitzky *et al.*, 1998). Other examples of convergent evolution of blastozones within a different anatomical organisation come from brown and red algae, other green algae (e.g. *Caulerpa*: Hagemann, 1992), and also from lichens (Cladoniaceae: Hammer, 2000). The blastozone concept permits comparison between these lineages of open-growth organisms.

21.3 Molecular components of organ formation competence

Since leaf dissection at marginal blastozones resembles organ formation at the SAM, there is a possibility that the underlying developmental programs may possess some similarity (see also Jackson, 1996; Hake and Meyerowitz, 1998; Goliber *et al.*, 1999).

21.3.1 KNOX *genes and leaf dissection*

This prediction is supported by the finding that class 1 *KNOX* genes have roles in OFC of both SAMs and leaf margins (Goliber *et al.*, 1999; Sinha, 1999). Initially, expression studies in *Zea mays* had shown that *KNOX* genes are specifically expressed in SAMs but are downregulated at sites where a next leaf primordium will be initiated (P0 sites) (Smith *et al.*, 1992). In maize and other grasses like *Oryza* leaves do not express *KNOX* at any time, correlated with the entire (undissected) nature of these monocot leaves (Jackson *et al.*, 1994; Sentoku *et al.*, 1999). This situation was found to be similar in the dicot *Arabidopsis* (Long *et al.*, 1996), although *Arabidopsis* leaf margins are toothed, not entire. The authors concluded that downregulation of *KNOX* genes would be required for the change from an indeterminate SAM developmental program to a determinate leaf program. A different pattern was later reported for tomato (*Lycopersicon esculentum* = *Solanum lycopersicon*), a species with compound-dissected leaves: *KNOX*-orthologues from tomato, *TKn1* and *LeT6*, are also expressed in the SAM, but there is apparently no downregulation upon leaf initiation as in *Arabidopsis* or *Zea*. Instead, expression continues to be present within developing leaf primordia specifically at marginal sites where leaflets are initiated.

In adjusting the interpretative model developed for the simple-leaved species, the tomato leaf was proposed to represent a mosaic organ with a 'partial indeterminate' nature (Chen *et al.*, 1997). However, although tomato leaves are different from simple leaves because they initiate leaflets and serrations, they clearly remain determinate organs. I would therefore prefer to view the role of *KNOX* genes not primarily in conjunction with organ indeterminacy, but rather with OFC; that is, growth direction establishment. From the expression data published for tomato so far there appears to be a clear correlation of *KNOX* expression with regions of OFC both in the SAM and leaf margins (Hareven *et al.*, 1996; Chen *et al.*, 1997; Parnis *et al.*, 1997). Upon leaf maturation, *KNOX* expression eventually ceases in these leaves.

In contrast to leaflet formation in tomato, tooth initiation in *Arabidopsis* appears not to be accompanied by *KNOX* expression (Lincoln *et al.*, 1994). This would suggest that formation of serrations relies on a developmental process distinct from leaflet formation. At this point, resolution of published gene expression data for *Lycopersicon*, where leaflets form tooth-like serrations, does not allow discrimination of *KNOX* expression correlated with leaflet formation on the one hand and tooth formation on the other. It would also be interesting to see whether relatives of *Arabidopsis* with more strongly dissected leaves show *KNOX* expression during leaflet and/or serration initiation.

Another source of evidence for the role of *KNOX* genes in leaf OFC are changes of expression levels. Overexpression in *Lycopersicon* has a striking effect on leaf morphology, leading to a dramatic enhancement of dissection (Hareven *et al.*, 1996; Chen *et al.*, 1997). This effect can be viewed as a prolongation of leaf margin OFC. A milder but similar effect is seen in transgenic *Arabidopsis* plants overexpressing the class1 *KNOX* gene *KNAT1*, in which incision depth and number of serrations are enhanced (Chuck *et al.*, 1996). This observation is interesting, given the fact that formation of serrations in wild-type plants appears not to require *KNOX* expression. Adventitious shoot meristems were also observed on the adaxial leaf surface in other plants overexpressing *KNOX*, for example in *Lycopersicon* and *Arabidopsis*. This indicates that, besides enhancement of dissection of the leaf margin, OFC may also be ectopically induced in these leaves. In contrast, in serration-less monocot leaves like in *Zea*, overexpression of the *KNOX* gene *Knotted1* in a dominant mutation elicits neither leaflet nor adventituous SAM formation, but only unorganised meristemoids on the upper leaf surface. Other factors are therefore likely to contribute to OFC.

Leaf margins and SAMs are differently affected by changes in *KNOX* expression levels, indicating important differences in the regulation of OFC between these two structures. For example, *KNOX* overexpression in *Arabidopsis* and *Lycopersicon* seems not to have a strong effect on organ formation at the SAM, despite enhancement of leaf dissection. In *KNOX* loss-of-function plants, OFC of the SAM is affected, but leaves, if formed, appear more or less normal. *Arabidopsis* plants mutant in *SHOOT MERISTEMLESS* (*STM*) are not able to initiate leaf primordia beyond the two cotyledons; apparently because the SAM is consumed by cotyledon formation (Barton and Poethig, 1993; Kaplan and Cooke, 1997; Bowman and Eshed, 2000). Effects are more subtle in a loss-of-function allele of the maize *Knotted1* gene (Kerstetter *et al.*, 1997), presumably because of redundant functions of paralogous *KNOX* genes. No information is available on *KNOX* loss-of-function effects in the compound-leaved model plant tomato, so it is not known whether leaf and/or leaflet initiation are perturbed.

PHANTASTICA-like *MYB*-genes (*PHAN* genes) have been shown to interact with *KNOX* genes. In *Antirrhinum*, *Zea* and *Arabidopsis* *KNOX* genes are found ectopically expressed in leaves of plants mutant in *PHAN* homologues. This implies that wild-type function of *PHAN*-like genes may be to prevent *KNOX* from being expressed in leaf primordia (Waites *et al.*, 1998; Timmermans *et al.*, 1999; Tsiantis *et al.*, 1999; Byrne *et al.*, 2000). Ectopic *KNOX* expression mediated by mutations in the *PHAN* homologue *ASYMMETRIC LEAVES1* (*AS1*) shows an enhancement of dissection reminiscent of *KNAT1* overexpressing transformants. This similarity is enhanced in plants double mutant for *AS1* or *AS2* and *SERRATE* (*SE*) or *PICKLE* (*PKL*) (Ori *et al.*, 2000). Organ formation and *KNOX* expression of the SAM are not affected, again pointing to differences in the regulation of organ formation between SAMs and leaves.

21.3.2 Conservation of leaf dissection programs in angiosperms

It is evident that our current information relating to *KNOX* genes in leaf dissection of angiosperms comprises no more than a few spot-tests. Similar expression patterns of *KNOX* genes in the monocot *Zea* and the rosid eudicot *Arabidopsis* first suggested that the role of *KNOX* genes in SAM development and leaf initiation might be conserved among angiosperms and that leaf initiation would generally require downregulation of *KNOX* in leaf anlagen (Smith and Hake, 1994). Data from the asterid eudicot *Lycopersicon* then suggested that leaf initiation can occur without *KNOX* downregulation at SAM flanks, and that the competence of primordial leaf margins for dissection might be achieved by *KNOX* expression at the leaf margins (Hareven *et al.*, 1996; Chen *et al.*, 1997; Parnis *et al.*, 1997). Preliminary *in situ* hybridisation data from a *KNOX* gene in *Eschscholzia californica* (basal eudicot clade = Ranunculidae; Papaveraceae) suggest an expression pattern similar to *Lycopersicon* (Gleissberg, unpubl. obs.). Figure 21.1 summarises different *KNOX* expression patterns in relation to OFC of the SAM and leaf margins. Figure 21.1a depicts the situation for entire-leaved taxa, where OFC is restricted to the SAM, and leaf initiation occurs in areas of *KNOX* downregulation. Figure 21.1b and c shows two ways by which OFC could be conferred to leaf margins: either *KNOX* expression is continuously expressed during leaf initiation at the SAM and later leaflet initiation at the leaf margin (Figure 21.1b), or, if organ initiation requires *KNOX* downregulation, it is transitorily downregulated during leaf initiation and, later, again upregulated at organ formation competent leaf margins. Sites of leaflet initiation at the margin would then again require downregulation within the marginal expression domain (Figure 21.1c). The pattern shown in Figure 21.1b has been reported for *Lycopersicon* only, and the pattern of Figure 21.1c has not yet been observed. However, it cannot be ruled out that at least one of the *Arabidopsis KNOX* gene family members shows such a pattern during late initiation of marginal serrations.

These results show the requirement to go beyond a few model species in the study of leaf dissection. Leaf dissection in different angiosperm clades might have evolved

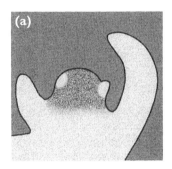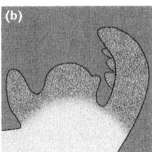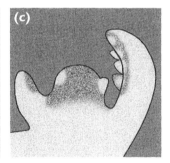

Figure 21.1 Expression of *KNOX* genes (denoted by pointillation) in species with entire (a) and dissected (b, c) leaves. (a) Expression in the SAM and downregulation upon leaf initiation (P0-stage) at the SAM flank (bright spot on the left). Three successively older primordia (P1-P3) are shown that do not display *KNOX* expression. This pattern is reported from *Zea mays* and *Arabidopsis thaliana*. (b) In *Lycopersicon esculentum* (*Solanum lycopersicon*), *KNOX* expression is not downregulated in leaf primordia, and accompanies leaflet formation during the organogenetic stage (primordium on the right). (c) Hypothetical pattern in which *KNOX* is initially downregulated during establishment of the leaf primordium at the SAM flank (P0–P1 stages), but is resumed conferring OFC to the leaf margin (P2 stage on the left). Leaflet formation is again associated with brief downregulation at sites of leaflet initiation (P3 stage on the right).

several times independently (Goliber *et al.*, 1999). Evolution of leaf dissection in some clades might have involved an extension of OFC into the primordial leaf margin via an accordingly altered *KNOX* expression pattern. The reverse process, an evolutionary loss of leaf OFC through a secondary restriction of *KNOX* function to the SAM, is also conceivable and might explain larger entire-leaved angiosperm clades (e.g. the monocots except Araceae, Caryophyllales within core eudicots). Secondarily entire leaves also evolved many times at the genus or species level, for example in *Dendromecon* (Papaveraceae) and several species of *Ranunculus* (Ranunculaceae). Goliber *et al.* (1999) estimated that dissected leaves evolved at least eleven times within angiosperms. However, it is not clear whether the earliest angiosperm leaves were entire or dissected, particularly if the occurrence of lobed or serrated leaf margins is considered as a delayed dissection program of the primordium. *Amborella*, currently widely viewed as the basalmost angiosperm (Qiu *et al.*, 1999), has simple leaves with serrated margins; *Ceratophyllum*, another near-basal angiosperm, has strongly dissected leaves. If dissected leaves represent the plesiomorphic state for angiosperms, multiple loss (rather than independent gains) of leaf OFC would be the primary process leading to the scattered distribution of dissected leaf taxa. In this case, one would also expect to find a higher degree of conservation of the developmental mechanism conferring leaf OFC.

That leaflet formation is not controlled by *KNOX* in all angiosperms was shown by Hofer and co-workers. In *Pisum sativum* (rosid eudicot clade; Fabaceae), leaflet formation is accompanied by an unrelated transcription factor, *UNI-FOLIATA* ('*UNI*': Hofer *et al.*, 1997). Loss-of-function of *UNI* converts the pinnate wild-type pea leaf into a simple or ternate leaf. Differences between various genotypes in the number of leaflets formed is correlated with duration and site of *UNI* expression during leaf development (Gourlay *et al.*, 2000). In contrast to *Lycopersicon*, a pea *KNOX* gene is not expressed in the leaflet forming primordia (Hofer *et al.*, 2001).

21.4 Developmental environment of leaf dissection

21.4.1 Three developmental phases of dissected leaf growth

Organ formation in leaves occurs only during a brief phase shortly after initiation of the leaf primordium (Jeune, 1983). Leaflet initiation may commence at a primordium length of 150–400 μm and segmentation (including formation of serrations) may be complete by a primordium length of less than 1 to 2 mm (Gleissberg, 1998a, b). This 'organogenetic phase' is preceded by the 'initial phase' (establishment of the leaf primordium) and followed by the more extensive 'histogenetic phase' of leaf growth during which rapid growth and tissue differentiation take place (Jeune, 1983). In leaves which lack leaflets but have marginal serrations, the organogenetic phase is delayed and commences at larger primordium sizes (Gleissberg, 1998b; Groot and Meicenheimer, 2000). Entire, undissected leaves lack an organogenetic phase.

Theoretically, if leaf initiation and leaf dissection are controlled by the same developmental pathway, OFC of the apical meristem could be directly conferred to the margin of the leaf primordium allowing leaflet formation. Alternatively, the developmental pathway imparting OFC could be 'turned off' upon leaf initiation and be later resumed during marginal segmentation. This could explain hetero-chronic shifts of leaf OFC from earlier to later phases as seen in simple-serrated (as compared to compound) leaves. The first case is supported by the continuous *KNOX* expression during leaf initiation and leaflet formation as seen in *Lycopersi-con* (discussed in Section 23.3.1). A third possibility would be that leaf initiation at the SAM and leaf dissection at the leaf margin are controlled by different develop-mental pathways and genes.

21.4.2 Marginal organ formation corresponds to primordium periphery dilatation

Leaf primordia generally show differential elongation along their axis. Acrotonic elongation is followed by acropetal leaflet initiation, while basitonic elongation is associated with a basipetal initiation sequence (Gleissberg, 1998a, b). It is possible that differential elongation serves as a trigger for segment initiation, such that a critical length of competent marginal tissue is necessary for an initiation event (Gleissberg, 1998a). This is also indicated by morphometric analyses that reveal a constant distance of the respective youngest lateral primordium from the leaf tip in acropetal leaves (*Ailanthus glandulosa*: Jeune, 1981) and from the leaf base in basipetal leaves (e.g. *Glechoma hederacea*: Jeune, 1981). Such a mechanism would be similar to the 'available space' mechanism discussed for leaf initiation at the SAM: 'A primordium becomes organised wherever competent cells are present in a sufficiently large mass' (Sachs, 1991: 156), although the growing competent area is radial symmetric at the shoot apex, whereas it is linear in leaves.

Leaf primordia grow preferentially in longitudinal direction in early phases (Poethig, 1997). If the primordial leaf margin is competent for organ formation, it may respond to differential elongation of the leaf axis by leaflet formation. With the onset of surface expansion of the blade an additional component contributes to margin dilatation; and if OFC is retained, this may cause the initiation of higher order segments that develop as serrations. In plants with simple but serrated leaves, OFC might begin only after blade dilatation has started.

21.4.3 Tissue differentiation antagonises marginal organ formation

Previous comparative developmental studies suggested an antagonistic interplay of organ-initiating and histologically-maturing regions in leaf primordia. Tissue mat-uration appears in distinct marginal (segment tips, petiole margin) and non-marginal domains (abaxial and adaxial faces), and the timing of these events appears to have an impact on the sites and patterns of segmentation in poppy leaves (Gleissberg, 1998a, b). Detailed developmental investigations would be desirable in *Lycopersi-con* to correlate *KNOX* expression with sites of leaflet and serration formation and to see if these areas of active organ formation are devoid of histogenesis (e.g. mar-ginal trichome formation). It could then be expected that the prolongation of the segmentation process seen in *KNOX*-overexpressing plants is correlated with a delay in marginal histogenesis.

21.4.4 Evolution of 'petiolated' versus 'non-petiolated' leaves

Two leaf types can be recognised on the basis of whether blade and petiole develop as clearly separated morphological units ('petiolated leaves') or whether breadth of lamina tissue decreases gradually toward the base of the leaf ('non-petiolated leaves': Figure 21.2). If petiolated leaves are dissected, the petiole anlage may show early trichome formation and associated loss of OFC, resulting in a restriction of the segmentation process to the distal blade part (Figure 21.2b). Leaves with a basipetal-pinnate or divergent pattern of segmentation are developmentally related to entire non-petiolated leaves and show a decrease of leaflet size toward the proximal end of the leaf. The development of these leaves may show, in contrast to petiolated leaves, no early trichome formation, and OFC may be retained in the proximal region of the leaf primordium (Figure 21.2a). As a consequence, elongation of this region is accompanied by segmentation, and the mature leaf does not show a clear border between blade and petiole (Gleissberg and Kadereit, 1999).

The degree to which leaves are longitudinally differentiated into blade and petiole is a major determinant of morphological diversity of angiosperm leaves.

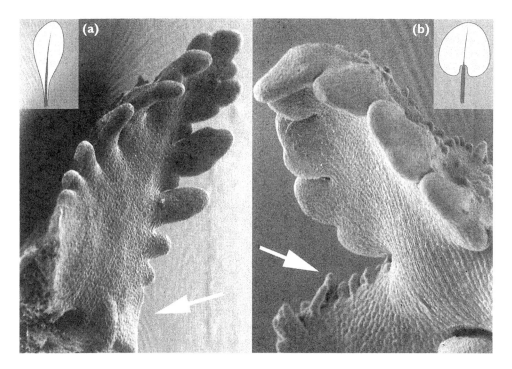

Figure 21.2 Leaf dissection in petiolated and non-petiolated leaves. (a) SEM of a *Chelidonium majus* leaf during leaflet formation. Note that basal portions, which in petiolated leaves give rise to the petiole, lack marginal trichomes (arrow), allowing leaflet formation to proceed far down the primordium. (b) The corresponding stage of a *Macleaya microcarpa* primordium with trichome formation at proximal margin portions; leaflet formation is restricted to a distinct distal portion of the leaf. The insets show outlines and midribs of corresponding undissected leaves.

Proximal–distal patterning also has an important impact on pattern of dissection. Namely, a basipetal dissection program will result in either a pinnate or a palmate/pedate architecture, depending on blade–petiole differentiation. Little is known about developmental regulators of blade–petiole differentiation. Earlier work has suggested that enhanced thickening growth of the leaf axis is an antagonist of lamina surface growth (Foster, 1936; Hagemann, 1970). According to this hypothesis, petiolated leaves would result from accentuated leaf-rib thickening in the proximal parts of leaf primordia (Figure 21.2b), while leaf-rib thickening would decrease gradually towards the tip in non-petiolated leaves (Figure 21.2a). The *rotundifolia* (*rot*) and *angustifolia* (*an*) mutants of *Arabidopsis* (Tsuge *et al.*, 1996; Kim *et al.*, 1998), other phenotypic classes of mutants (Serrano-Cartagena *et al.*, 1999; Berna *et al.*, 1999), and the recently described *leafy petiole* (*lep*) mutation (Graaff *et al.*, 2000) all differ from wild-type plants in the degree of blade–petiole differentiation, but an alteration of leaf-rib development associated with their phenotypes is not reported. In contrast, *asymmetric leaves1* (*as1*) mutants of *Arabidopsis* show an alteration of blade lamina formation associated with an altered midrib phenotype (Byrne *et al.*, 2000). AS1 is an orthologue of the *Antirrhinum PHANTASTICA* (*PHAN*) gene that is involved in both ad-/abaxial and proximo-distal patterning (Waites *et al.*, 1998; Hudson, 1999). Mutations in *PHAN* lead to loss of adaxial tissue, a condition found also in petioles of species with peltate blades (Gleissberg *et al.*, 1999). Therefore, genes controlling leaf dorsiventrality appear to play a significant role in proximo-distal patterning.

21.4.5 Leaf dissection in angiosperms: the case of Ranunculidae

An evolutionary study of petiolated-dissected versus non-petiolated-dissected leaves in Papaveraceae indicated a frequent change between these modes that does

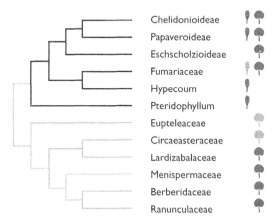

Figure 21.3 Phylogenetic relationships among Ranunculidae families (Qiu *et al.*, 1999) and distribution of petiolated versus non-petiolated leaves. Papaveraceae *s.l.* (thick lines, including Fumariaceae and *Pteridophyllum*) are sister to the remainder of Ranunculidae (thin lines). Outlines on the right indicate presence of non-petiolated and petiolated leaf types (left and right columns; grey colour refers to preliminary character coding). Non-petiolated leaves occur only within Papaveraceae *s.l.*

not allow us to determine the plesiomorphic condition for this group (Gleissberg and Kadereit, 1999). A preliminary morphological check of outgroup families, however, indicates that 'non-petiolated-dissected' leaves are characteristic for Papaveraceae *s.l.* and do not occur in the remainder of Ranunculales (Figure 21.3). Based on comparisons of mature leaf shapes, petiolate-dissected leaves of the (poly)ternate and palmate types are very common in Ranunculaceae, where they occur particularly in early-diverging taxa (*Aquilegia, Leptopyrum, Thalictrum*), but also occur in other subfamilies (Loconte *et al.*, 1995). There is a trend to acropetal leaves in several instances, but non-petiolate-dissected leaves with basipetal-pinnate or divergent modes of dissection apparently do not occur (see also Kürbs, 1973a, b). The same situation characterises Berberidaceae, as judged from mature morphology: polyternate taxa are found in presumably basal genera like *Nandina* and *Caulophyllum* but also some 'higher-branched' taxa like *Epimedium* and *Leontice* (Loconte *et al.*, 1995). Basipetal-pinnate or divergent leaves have apparently not evolved. Menispermaceae have mostly poorly dissected or entire leaves, yet some can also be classified as petiolated leaves. The same probably applies to the dissected leaves of Lardizabalaceae (Sugiyama and Hara, 1988) and the small families of Circaeasteraceae and Eupteleaceae. This distribution of leaf forms therefore indicates that non-petiolated leaves in Ranunculidae evolved only within Papaveraceae *s.l.*

21.5 Outlook

An integration of comparative developmental and molecular-genetic studies holds great promise for future studies of leaf development and evolution. Observations on developmental correlations from comparative studies in morphologically diverse yet related species complement hypotheses based on mutant analyses and may help to develop concepts of plant development. Conversely, morphological effects of altered gene expression in mutants and transgenic plants can be used to test and further develop morphological concepts, while providing insights into the molecular mechanisms acting in morphological evolution.

ACKNOWLEDGEMENTS

I thank Katharina Lehmphul for help with the figures, and an anonymous reviewer for helpful comments.

REFERENCES

Barton, M. K. and Poethig, R. S. (1993) Formation of the shoot apical meristem in *Arabidopsis thaliana*: an analysis of development in the wild type and in the *shoot meristemless* mutant. *Development*, 119, 823–831.

Berna, G., Robles, P. and Micol, J. L. (1999) A mutational analysis of leaf morphogenesis in *Arabidopsis thaliana. Genetics*, 152, 729–742.

Bowman, J. L. and Eshed, Y. (2000) Formation and maintenance of the shoot apical meristem. *Trends in Plant Science*, 5, 110–115.

Byrne, M. E., Barley, R., Curtis, M., Arroyo, J. M., Dunham, M., Hudson, A. and Martienssen, R. A. (2000) *ASYMMETRIC LEAVES1* mediates leaf patterning and stem stell function in *Arabidopsis*. *Nature*, 408, 967–971.

Chen, J.-J., Janssen, B.-J., Williams, A. and Sinha, N. (1997) A gene fusion at a homeobox locus: alterations in leaf shape and implications for morphological evolution. *Plant Cell*, 9, 1289–1304.

Chuck, G., Lincoln, C. and Hake, S. (1996) *KNAT1* induces lobed leaves with ectopic meristems when overexpressed in *Arabidopsis*. *Plant Cell*, 8, 1277–1289.

Cusset, G. (1986) La morphogenèse du limbe des Dicotylédones. *Canadian Journal of Botany*, 64, 2807–2839.

Esau, K. (1977) *Anatomy of Seed Plants* (2nd edn). Wiley, New York.

Foster, A. S. (1936) Leaf differentiation in angiosperms. *Botanical Reviews*, 2, 349–372.

Gleissberg, S. (1998a) Comparative analysis of leaf shape development in Papaveraceae-Papaveroideae. *Flora*, 193, 269–301.

Gleissberg, S. (1998b) Comparative analysis of leaf shape development in Papaveraceae-Chelidonioideae. *Flora*, 193, 387–409.

Gleissberg, S. and Kadereit, J. W. (1999) Evolution of leaf morphogenesis: evidence from developmental and phylogenetic data in Papaveraceae. *International Journal of Plant Sciences*, 160, 787–794.

Gleissberg, S., Kim, M., Jernstedt, J. and Sinha, N. (1999) The regulation of dorsiventral symmetry in plants, in *Biology of Biodiversity* (ed. M. Kato), Springer, Tokyo, pp. 223–241.

Goliber, T., Kessler, S., Chen, J.-J., Bharathan, G. and Sinha, N. (1999) Genetic, molecular, and morphological analysis of compound leaf development. *Current Topics in Developmental Biology*, 43, 259–290.

Gourlay, C. W., Hofer, J. M. I. and Ellis, T. H. N. (2000) Pea compound leaf architecture is regulated by interactions among the genes *UNIFOLIATA, COCHLEATA, AFILA* and *TENDRIL-LESS*. *Plant Cell*, 12, 1279–1294.

Graaff, E. van der, Den Dulk-Ras, A., Hooykaas, P. J. J. and Keller, B. (2000) Activation tagging of the *LEAFY PETIOLE* gene affects leaf petiole development in *Arabidopsis thaliana*. *Development*, 127, 4971–4980.

Groot, E. P. and Meicenheimer, R. D. (2000) Comparison of leaf plastochron index and allometric analyses of tooth development in *Arabidopsis thaliana*. *Journal of Plant Growth Regulation*, 19, 77–89.

Hagemann, W. (1970) Studien zur Entwicklungsgeschichte der Angiospermenblätter. Ein Beitrag zur Klärung ihres Gestaltungsprinzips. *Botanische Jahrbücher*, 90, 297–413.

Hagemann, W. (1992) The relationship of anatomy to morphology in plants: a new theoretical perspective. *International Journal of Plant Science*, 153, S38–S48.

Hagemann, W. and Gleissberg, S. (1996) Organogenetic capacity of leaves: the significance of marginal blastozones in angiosperms. *Plant Systematics and Evolution*, 199, 121–152.

Hake, S. and Meyerowitz, E. M. (1998) Growth and development – growing up green. *Current Opinion in Plant Biology*, 1, 9–11.

Hammer, S. (2000) Meristem growth dynamics and branching patterns in the Cladoniaceae. *American Journal of Botany*, 87, 33–47.

Hareven, D., Gutfinger, T., Parnis, A., Eshed, Y. and Lifschitz, E. (1996) The making of a compound leaf: genetic manipulation of leaf architecture in tomato. *Cell*, 84, 735–744.

Hofer, J., Gourlay, C., Michael, A. and Ellis, T. H. N. (2001) Expression of a class 1 *Knotted1*-like homeobox gene is down-regulated in pea compound leaf primordia. *Plant Molecular Biology*, 45, 387–398.

Hofer, J., Turner, L., Hellens, R., Ambrose, M., Matthews, P., Michael, A. and Ellis, N. (1997) *UNIFOLIATA* regulates leaf and flower morphogenesis in pea. *Current Biology*, 7, 581–587.

Hofmeister, W. (1868) Allgemeine Morphologie der Gewächse, in *Handbuch der physiologischen Botanik*, vol. 1. Engelmann, Leipzig, pp. 405–664.

Hudson, A. (1999) Axioms and axes in leaf formation? *Current Opinion in Plant Biology*, 2, 56–60.

Jackson, D. (1996) Plant morphogenesis: designing leaves. *Current Biology*, 6, 917–919.

Jackson, D., Veit, B. and Hake, S. (1994) Expression of maize *Knotted1*-related homeobox genes in the shoot apical meristem predicts patterns of morphogenesis in the vegetative shoot. *Development*, 120, 405–413.

Jeune, B. (1981) Modèle empirique du développement des feuilles de Dicotylédones. *Adansonia*, 4, 433–459.

Jeune, B. (1983) Croissance des feuilles de *Lycopus europaeus* L. *Beiträge zur Biologie der Pflanzen*, 58, 253–266.

Jeune, B. and Lacroix, C. R. (1993) A quantitative model of leaflet initiation illustrated by *Murraya paniculata* (Rutaceae). *Canadian Journal of Botany*, 71, 457–465.

Kaplan, D. R. and Cooke, T. J. (1997) Fundamental concepts in the embryogenesis of dicotyledons: a morphological interpretation of embryo mutants. *Plant Cell*, 9, 1903–1919.

Kerstetter, R. A., Laudencia-Chingcuanco, D., Smith, L. G. and Hake, S. (1997) Loss-of-function mutations in the maize homeobox gene, *Knotted1*, are defective in shoot meristem maintenance. *Development*, 124, 3045–3054.

Kim, G.-T., Tsukaya, H. and Uchimiya, H. (1998) The *ROTUNDIFOLIA3* gene of *Arabidopsis thaliana* encodes a new member of the cytochrome P-450 family that is required for the regulated polar elongation of leaf cells. *Genes and Development*, 12, 2181–2191.

Kürbs, S. (1973a) Vergleichend-entwicklungsgeschichtliche Studien an Ranunculaceen-Fiederblättern. Teil I. *Botanische Jahrbücher für Systematik*, 93, 130–167.

Kürbs, S. (1973b) Vergleichend-entwicklungsgeschichtliche Studien an Ranunculaceen-Fiederblättern. Teil II. *Botanische Jahrbücher für Systematik*, 93, 325–371.

Lincoln, C., Long, J., Yamaguchi, J. and Hake, S. (1994) A *Knotted1*-like homeobox gene in *Arabidopsis* is expressed in the vegetative meristem and dramatically alters leaf morphology when overexpressed in transgenic plants. *Plant Cell*, 6, 1859–1876.

Loconte, H., Campbell, L. M. and Stevenson, D. W. M. (1995) Ordinal and familial relationships of Ranunculid genera. *Plant Systematics and Evolution* (Suppl.), 9, 99–118.

Long, J. A., Moan, E. I., Medford, J. I. and Barton, M. K. (1996) A member of the KNOTTED class of homeodomain proteins encoded by the *SHOOTMERISTEMLESS* gene of *Arabidopsis*. *Nature*, 379, 66–69.

Nägeli, C. (1858) Das Wachsthum des Stammes und der Wurzel bei den Gefäßpflanzen und die Anordnung der Gefäßstränge im Stengel. *Beiträge zur wissenschaftlichen Botanik*, 1.

Ori, N., Eshed, Y., Chuck, G., Bowman, J. L. and Hake, S. (2000) Mechanisms that control *KNOX* gene expression in the Arabidopsis shoot. *Development*, 127, 5523–5532.

Parnis, A., Cohen, O., Gutfinger, T., Hareven, D., Zamir, D. and Lifschitz, E. (1997) The dominant developmental mutants of tomato, *Mouse-ear* and *Curl*, are associated with distinct modes of abnormal transcriptional regulation of a *Knotted* gene. *Plant Cell*, 9, 2143–2158.

Poethig, R. S. (1997) Leaf morphogenesis in flowering plants. *Plant Cell*, 9, 1077–1087.

Qiu, Y.-L., Lee, J., Bernasconi-Quadroni, F., Soltis, D. E., Soltis, P. S., Zanis, N., Zimmer, E. A., Chen, Z., Savolainen, V. and Chase, M. W. (1999) The earliest angiosperms: evidence from mitochondrial, plastid and nuclear genomes. *Nature*, 402, 404–407.

Sachs, T. (1991) *Pattern Formation in Plant Tissues*. Cambridge University Press, New York.

Sawitzky, H., Gleissberg, S. and Berger, S. (1998) Phylogenetic implications of patterns of cap development in selected species of *Acetabularia/Polyphysa* (Dasycladales, Chlorophyta). *Phycologia*, 37, 478–485.

Sentoku, N., Sato, Y., Kurata, N., Ito, Y., Kitano, H. and Matsuoka, M. (1999) Regional expression of the rice *KN1*-type homeobox gene family during embryo, shoot and flower development. *Plant Cell*, 11, 1651–1664.

Serrano-Cartagena, J., Robles, P., Ponce, M. R. and Micol, J. L. (1999) Genetic analysis of leaf form mutants from the *Arabidopsis* Information Service collection. *Molecular and General Genetics*, 261, 725–739.

Sinha, N. (1999) Leaf development in angiosperms. *Annual Review of Plant Physiology and Plant Molecular Biology*, 50, 419–446.

Smith, L. G., Greene, B., Veit, B. and Hake, S. (1992) A dominant mutation in the maize homeobox gene, *Knotted1*, causes its ectopic expression in leaf cells with altered fates. *Development*, 116, 21–30.

Smith, L. G. and Hake, S. (1994) Molecular genetic approaches to leaf development: *Knotted* and beyond. *Canadian Journal of Botany*, 72, 617–625.

Sugiyama, M. and Hara, N. (1988) Comparative study on early ontogeny of compound leaves in Lardizabalaceae. *American Journal of Botany*, 75, 1598–1605.

Timmermans, M. C. P., Hudson, A., Becraft, P. W. and Nelson, T. (1999) ROUGH SHEATH2: a *MYB* protein that represses *KNOX* homeobox genes in maize lateral organ primordia. *Science*, 284, 151–153.

Tsiantis, M., Schneeberger, R., Golz, J. F., Freeling, M. and Langdale, J. A. (1999) The maize *ROUGH SHEATH2* gene and leaf development programs in monocot and dicot plants. *Science*, 284, 154–156.

Tsuge, T., Tsukaya, H. and Uchimiya, H. (1996) Two independent and polarized processes of cell elongation regulate leaf blade expansion in *Arabidopsis thaliana* (L.) Heynh. *Development*, 122, 1589–1600.

Waites, R., Selvadurai, H. R., Oliver, I. R. and Hudson, A. (1998) The *PHANTASTICA* gene encodes a *MYB* transcription factor involved in growth and dorsoventrality of lateral organs in *Antirrhinum*. *Cell*, 93, 779–789.

Chapter 22

Developmental signals regulating leaf form

Miltos Tsiantis, Angela Hay, Naomi Ori, Hardip Kaur, Ian Henderson, Hans Holtan, Sheila McCormick and Sarah Hake

ABSTRACT

The formation of the shoot body of higher plants relies on the continuous organogenic and self-renewal activities of the shoot apical meristem. The exact nature of the developmental pathways governing shoot development is still poorly understood. Here, we discuss processes responsible for the transition from indeterminate meristematic identity to determinate leaf fate. Current evidence suggests that the precise mode of expression of *knotted1*-like homeobox (KNOX) transcription factors plays a central role both in meristem function and in acquisition of leaf identity. We discuss molecular genetic approaches to identify other factors that act in, or in conjunction with, the KNOX pathway to regulate shoot development. We also consider how modifications in developmental pathways defined by *KNOX* genes (i.e. *KNOX* genes, their upstream regulators and downstream effectors) could contribute to the generation of the distinct type of leaf form exemplified by dissected leaves.

22.1 Introduction

The formation of the shoot body of higher plants relies on the continuous organogenic and self-renewal activities of the shoot apical meristem (SAM). Despite recent progress, the exact nature of the developmental pathways governing shoot development is still poorly understood. Current evidence suggests that the precise mode of expression of *knotted1*-like homeobox (KNOX) transcription factors play a central role in both meristem function and acquisition of leaf identity. Thus, KNOX proteins are strongly expressed in the SAM where they have been shown to be required for meristem maintenance (Kerstetter *et al.*, 1997). The rapid down-regulation of *KNOX* proteins in a group of cells at the flank of the SAM is correlated with, and may be involved in, leaf cell fate specification. In contrast to species with simple leaves, compound leaves of tomato plants express *KNOX* genes (according to some comparative morphologists, subdivided leaves with distinct units (i.e. leaflets) should be termed 'dissected' rather than 'compound', since the latter is deemed to negate the equivalent leaf nature of both, but in this chapter the terms 'dissected' and 'compound' are used interchangeably; for a full discussion of this issue, see Kaplan, 2001). This expression pattern, in combination with the increased leaf dissection obtained by overexpressing *KNOX* genes in tomato, has led to the

In *Developmental Genetics and Plant Evolution* (2002) (eds Q. C. B. Cronk, R. M. Bateman and J. A. Hawkins), Taylor & Francis, London, pp. 418–430.

suggestion that differential regulation of *KNOX* gene expression may be responsible for the compound leaf morphology (Hareven *et al.*, 1996; Chen *et al.*, 1997; Parnis *et al.*, 1997; Janssen *et al.*, 1998). Very little is known about the upstream and downstream components of the *KNOX* developmental pathway and it is still unclear how differential regulation of this pathway is responsible for generating compound leaf morphology in tomato. It is also unclear whether differential regulation of *KNOX* genes is implicated in generating the compound leaf form in other species such as pea (Hofer *et al.*, 1997).

The effects of *KNOX* overexpression in *Arabidopsis* and tomato are comparable, revealing a strong correlation between increased lobing in the simple leaves of *Arabidopsis* (Chuck *et al.*, 1996) and increased dissection in the tomato leaf. This correlation is strengthened by analysis of mutants that show reduced leaflet number in tomato. Such mutants also decrease the degree of lobing of the leaflets still present, thus making the leaf margin more entire (e.g. the *procera* mutation: Van Tuinen *et al.*, 1999). In this chapter, we discuss efforts to uncover additional factors of the *KNOX* pathway. Further, we consider how taxon-specific differences in the function of this pathway may relate to distinct types of lateral organs exemplified by the compound leaf. Finally, we present several approaches to address the evolutionary role of *KNOX* genes in the control of leaf form.

22.2 Homeoboxes and beyond: in search of upstream factors

Since the initial characterization of the *knotted1* (*kn1*) mutant phenotype (Hake *et al.*, 1989) a wealth of evidence has accumulated suggesting that inappropriate expression of *KNOX* genes in leaves results in developmental aberrations. Study of recessive mutations that phenocopy effects of ectopic *KNOX* expression has been an important step in understanding how the KNOX pathway functions to control cell fates in the shoot body of plants (by KNOX pathway, we mean the array of genes that regulate and are regulated by KNOX proteins, i.e. upstream and downstream factors). The maize *rough sheath2* (*RS2*) gene, the *phantastica* (*PHAN*) gene of snapdragon and the *asymmetric1* (*AS1*) gene of *Arabidopsis* are each required for the correct compartmentalization of *KNOX* gene expression in their respective species (Waites *et al.*, 1998; Timmermans *et al.*, 1999; Tsiantis *et al.*, 1999; Byrne *et al.*, 2000). These genes encode related MYB transcription factors. The question as to whether RS2/PHAN/AS1-like proteins directly repress *KNOX* gene expression is still open.

Although *KNOX* genes are misexpressed in *rs2/phan/as1* mutant leaves, the initial down-regulation of *KNOX* that is associated with formation of the incipient leaf primordium is intact. This suggests that repression of *KNOX* might involve separate pathways, one which facilitates the initial down-regulation of *KNOX* during primordium formation at developmental stage P0, and a second which maintains *KNOX* genes in an off-state later in leaf development. It will therefore be of considerable interest to recover mutants that fail to down-regulate *KNOX* in the initial stages of development. These mutants may be severely compromised in initiating lateral organs.

Loci that control *KNOX* function in *Arabidopsis* may also play a role in determinacy at the leaf margin. For some time it has been known that ectopic *KNOX* expression results in deep leaf lobing that follows the serrations already present in the leaf margin of *Arabidopsis* (Lincoln *et al.*, 1994; Chuck *et al.*, 1996). Recently, the *serrate (se)* mutant has revealed a link between altered margin shape and responsiveness to KNOX. Ori *et al.* (2000) have shown that the *as1* and *as2* mutants of *Arabidopsis* display ectopic *KNOX* expression and limited leaf lobing. However, the *se as1* double mutant displays distinctly lobed leaf margins and frequently ectopic tissue outgrowth from the leaf sinuses, but the *se* mutation alone does not alter *KNOX* gene expression. This suggests that *se* changes the potential of the tissue to respond to the ectopic *KNOX* expression conditioned by the *as* loci, in a manner that results in deep lobing and production of ectopic structures at the leaf margins. Mutations in the *pickle (pkl)* locus also condition sensitization to ectopic *KNOX* expression (although in this case ectopic tissue production is not accompanied by enhanced lobing: Ori *et al.*, 2000). *SE* and *PKL* encode proteins which are thought to control chromatin structure, suggesting that the developmental potential of shoots of higher plants is regulated in part by chromatin remodelling. Given that the single *se* mutant displays more pronounced serrations at the leaf margin we can establish a conceptual link between the tendency to make the margins more indeterminate and responsiveness to *KNOX* expression.

These observations are likely to become increasingly important when considering the mechanisms responsible for leaflet elaboration in species with dissected–compound leaves. Data from comparative morphology suggest a very strong link between indeterminacy at the leaf margin and leaflet formation (Hagemann and Gleissberg, 1996; Gleissberg, 2002). Therefore, loci that can alter the developmental potential of the leaf margin are good candidates for being involved in elaborating compound leaf morphologies. To this end we are carrying out second site mutagenesis to identify loci that will enhance or suppress the margin phenotype of *se* mutants. While characterizing such mutants we have noted that several mutants with altered margins (*am*) display inappropriate *KNOX* expression only in the tips of leaf serrations (Figure 22.1; Plate 3e, f). This suggests that, even though leaf margins do not express *KNOX* genes, this marginal region has an increased potential to do so in mutations that directly or indirectly change leaf form.

22.3 Caveats in interpreting the effects of *KNOX* misexpression on leaf shape

AS1, PHAN and RS2 are negative regulators of *KNOX* gene expression in leaves of their respective species. However, despite the equivalence of action at the molecular level, the phenotypic outcomes of the corresponding mutations are, to some extent, divergent: *phan* mutants display radial leaves, whereas *rs2* and *as1* mutants do not. Several workers have speculated about this discrepancy, and have suggested two reasons that are not necessarily mutually exclusive.

According to one line of thought, the *PHAN* gene in *Antirrhinum* has an additional role (separate from repressing *KNOX* genes) in specifying the dorsiventral (D/V) asymmetry of leaves. In maize, such a role is either absent or redundant with a hypothetical RS2 duplicate factor (Timmermans *et al.*, 1999). Testing of this hypothesis will require identification of a loss-of-function mutation in such a factor and study of the double mutants with *rs2*. In any event, *as1* mutants of *Arabidopsis*

Figure 22.1 Arabidopsis mutants with altered margin *(am)* phenotypes display inappropriate *KNAT1* expression in the tips of leaf serrations. Expression of a *KNAT1*-promoter b-glu-curonidase (GUS) reporter gene fusion is assayed by incubation with a chromogenic GUS substrate that gives rise to a blue precipitate (arrowed). (a) *am1*; (b) *am2*; (c) wild type.

do not display radial leaves. This would suggest that *PHAN/AS1/RS2*-like genes do not have generalized roles in D/V asymmetry across the angiosperms. Of course, it is possible that this is the case in some species (e.g. *Antirrhinum*) but not in others, perhaps reflecting divergence in downstream targets of the genes. To answer this question fully, it will be necessary to obtain null alleles of *as1*, because it is possible that differences between *as1* and *phan* reflect differences in allelic severity. This is especially important, given the suggestion by Ori *et al.* (2000) that some aspects of the *as1* phenotype may be interpreted as minor perturbations along the D/V axis. Finally, because the radial leaves of *Antirrhinum* occur at higher nodes, it is possible that modifiers of *PHAN* action exist that enhance the *phan* phenotype in upper leaves. If this is the case, it will be interesting to determine whether such modifiers are associated with loci believed to control phase change transitions in plants (Telfer and Poethig, 1998).

An alternative view is that the apparent loss of D/V polarity in *phan* is an effect of ectopic *KNOX* expression. Differences between the *phan*, *as1* and *rs2* phenotypes could therefore reflect differences in the manner in which different species elaborate laminae. In *Antirrhinum*, KNOX-induced proximal to distal transformations could result in radialized leaves. This would occur if the distal part of the leaf lamina acquires ventral features of the more proximal petiole tissue, and would be equivalent to the blade (distal) to sheath (proximal) transformations that occur in *rs2* (Tsiantis *et al.*, 1999). Furthermore, the ectopic presence of *KNOX* genes within early leaf primordia might result in the aberrant reformation of a morphogenetic boundary defined by the area where cells that do not express *KNOX* are contiguous with *KNOX*-expressing cells in the meristem. This boundary may be involved in specifying the proximal–distal (P/D) axis of the primordium; its reformation may therefore result in ill-defined P/D axes in *phan* leaves. The possibility that *KNOX* expression affects both P/D and D/V axes also suggests that the formation of the two axes may be interdependent. Genetic crosses between loss-of-function *KNOX* mutants and *phan* mutants should reveal whether radiality in *phan* leaves is mediated by ectopic *KNOX* expression.

Given that *Arabidopsis* and *Antirrhinum* are both eudicots, it is surprising that their phenotypes are less reconcilable than those of *rs2* and *as1*. Possible explanations are the divergence of downstream targets such that *KNOX* expression in *Antirrhinum* leaves interferes with aspects of D/V axis formation (e.g. expression of the *YABBY* or *PHABULOSA* genes: McConnell and Barton, 1998; Siegfried *et al.*, 1999) and the presence of modifying loci. It is also possible that these differences highlight subtle contrasts between leaf development programs. For example, leaf primordia are dorsiventral from their inception (Sylvester *et al.*, 1996). However, there may be slight temporal differences in the elaboration of the D/V axis or cell division dynamics. Thus, ectopic *KNOX* expression would have different phenotypic outcomes depending on when *KNOX* expression occurred (a radial outcome may be less likely if the D/V axis is elaborated earlier). More research on comparative morphology and knowledge of cell division patterns during early leaf development (Donnelly *et al.*, 1999) should help resolve these issues.

Research on tomato has already highlighted the broad spectrum of phenotypes that can result from ectopic *KNOX* expression (Hareven *et al.*, 1996; Parnis *et al.*, 1997; Janssen *et al.*, 1998). These differences are thought to relate to the precise time, place and level of expression. Of particular interest are the *Mouse ear* and *Curl* mutations that condition aberrant transcription of the tomato *KNOX* gene *TKN2* and that result in distinct phenotypes (Chen *et al.*, 1997; Parnis *et al.*, 1997: Figure 22.2). A high proportion of upper leaves in *Mouse ear* plants are reduced to almost bladeless elongate lateral appendages. Similar phenotypes are obtained when *TKN2* is over-expressed under the control of the *35S* promoter (Parnis *et al.*, 1997). This suggests that ectopic *KNOX* expression alone can be sufficient to condition severe inhibition of lateral growth in dicot leaves.

22.4 Downstream effectors of the KNOX pathway

Little is known about the downstream effectors of the KNOX pathway. Connections have recently been made between KNOX action and growth regulators. Ectopic expression of *KNOX* genes in several species results in phenotypes similar to those resulting from increased cytokinin levels, including delayed senescence, reduced apical dominance, and ectopic meristematic activity (Estruch *et al.*, 1991; Li

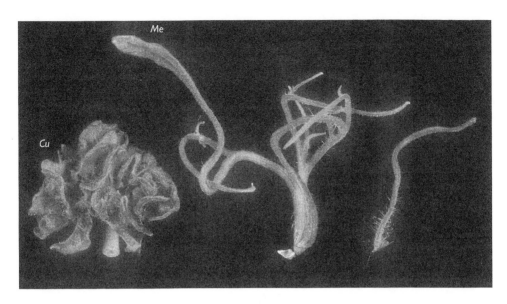

Figure 22.2 Dominant tomato mutants of the *Tkn2* locus, *Mouse ears (Me)* and *Curl (Cu)*, show very different leaf phenotypes.

et al., 1992). Recent studies have demonstrated that increased cytokinin levels do indeed occur in transgenic rice and tobacco plants over-expressing *KNOX* genes (Tamaoki *et al.*, 1997; Kusaba *et al.*, 1998). The same studies suggest that a large decrease in gibberellin (GA) levels occurs in *KNOX* over-expressing plants. However, decreased GA levels do not phenocopy the deeply lobed leaf phenotypes of *KNOX* overexpression in *Arabidopsis*, nor do increased cytokinin levels fully account for the whole spectrum of phenotypes encountered in *KNOX* over-expressing plants (Ori *et al.*, 1999). These results suggest that *KNOX* genes may act as global co-ordinators of multiple plant growth regulator (PGR) pathways in shoots.

To test the hypothesis that reduction of GA levels may be involved in mediating *KNOX* gene function we are using a transgenic *Arabidopsis* line harbouring a dexomethasone (DEX) inducible *kn1* construct. We have established that GA will strongly suppress the lobing leaf phenotypes resulting from inappropriate *kn1* expression (Figure 22.3). This suggests that at least a component of *KNOX* misexpression phenotypes in *Arabidopsis* may be mediated by inhibition of GA biosynthesis. Because *KNOX* expression in *Arabidopsis* leaves may result in decreased GA levels and GA can restore some of the effects of *KNOX* misexpression, part of *KNOX* function in the meristem may be to implement a low GA regime to allow normal meristem function. Conversely a high GA level would be necessary for leaf primordium differentiation.

As mentioned earlier, increased *KNOX* expression in the leaves of tomato leads to increased levels of dissection and formation of super-compound leaves (Hareven *et al.*, 1996; Chen *et al.*, 1997; Parnis *et al.*, 1997; Janssen *et al.*, 1998). If GA does indeed act antagonistically to *KNOX*, elevated GA signalling should reduce the level of tomato leaf dissection. Interestingly the *procera* mutation of tomato provides

1. (a) KNOX present (b) KNOX overexpression (c) GA hypersignalling

2. (a) KNOX absent (b) KNOX overexpression (c) GA application

Figure 22.3 KNOX function is partially mediated through a reduction in GA levels. 1. Tomato: (a) wild type; (b) *KNOX* overexpression; (c) *procera* mutant. 2. *Arabidopsis:* (a) wild type; (b) *35S::KN1:GR* with dexamethasone induction; (c) *35S::KN1:GR* with dexamethasone induction and GA application.

some insight into this question: *procera* plants show elevated levels of signalling through the GA pathway and display a reduction in both leaflet number and leaflet lobing (Figure 22.3). These phenotypes are also phenocopied by supplying GA to wild type tomato plants. This observation raises the possibility that the GA/*KNOX* interaction may be conserved across species and may be involved in regulating the level of leaf dissection.

22.5 The KNOX pathway and dissected–compound leaf form

The tomato leaf displays notable developmental flexibility in that several viable mutants exist with either reduced or increased compounding. The wild type tomato leaf, unlike simple leaves of other species, normally expresses *KNOX* genes and increased *KNOX* expression is known to increase the level of leaf dissection. This led to the suggestion that KNOX activity in tomato leaves is responsible for compound form in this species. It will be exciting to assess whether loss of function mutations in tomato *KNOX* genes have reduced levels of leaf dissection, as would be predicted.

Figure 22.4 Tomato mutants with altered leaf dissection: *multifolia* and *Mouse ears* show increased leaf dissection; *procera* and *entire* have reduced leaflet number and lobing. (a) *procera*; (b) wild type; (c) *multifolia*; (d) *entire*; (e) wild type; (f) *Mouse ears*.

The tomato *clausa (clau)* (Avivi *et al.*, 2000) and *multifolia (muf)* mutations are recessive and condition phenotypes similar to phenotypes arising from ectopic expression of *KNOX* genes in tomato leaves (Figure 22.4). Conversely, *procera* and *entire* condition reduced levels of leaf dissection. Characterization of genetic interactions, together with expression analysis of *KNOX* genes, should allow us to define the role of *KNOX* genes in specifying compound leaf architecture and to uncover novel factors involved.

To study the molecular mechanisms responsible for the different expression pattern of *KNOX* genes in tomato and *Arabidopsis*, we are studying the behaviour of an *Arabidopsis* KNOX-promoter reporter gene fusion in tomato and vice versa. The finding that KNOX exclusion from simple leaves has been consistently observed in diverse species suggests that the mechanism responsible for this expression pattern is ancient and robust. If differences between *KNOX* gene expression in tomato and *Arabidopsis* reflect the divergence of promoter sequences we might expect tomato *KNOX* promoters to direct reporter gene expression in *Arabidopsis* leaves. Conversely, activity of *Arabidopsis KNOX* promoters in tomato leaves may suggest that at least part of the genetic programme that directs *KNOX* expression in the *Arabidopsis* meristem is present in tomato leaves. This approach, which is made feasible by ability to genetically transform both *Arabidopsis* and tomato, will test the hypothesis that promoter structure contributes to species-specific expression of loci important for morphological evolution (Wang *et al.*, 1999). Transgenic tomatoes expressing the *GUS* reporter gene under the control of the promoter of the *Arabidopsis* KNOX gene KNAT1 show the same expression pattern as in *Arabidopsis* (i.e. meristem expression and full exclusion from leaves) (Figure 22.5). These experiments, in combination with traditional promoter analyses, will help determine the exact manner in which promoter cis-elements are modified through evolution to result in altered expression patterns of morphogenetically important genes.

Molecular genetic analyses in pea have suggested that *KNOX*-independent mechanisms can also contribute to compound leaf morphology. Hofer *et al.* (1997) demonstrated that specification of compound leaves in this species involves the pea orthologue of the *FLORICAULA/LFY* genes. The *FLORICAULA* (*FLO*) gene of *Antirrhinum* and its *Arabidopsis* orthologue *LEAFY* (*LFY*) encode transcription factors that promote the transition of indeterminate apical meristems to determinate floral meristems. Loss-of-function mutations in the pea orthologue *UNIFOLIATA* (*UNI*) convert compound leaves to simple ones. This finding led Hofer *et al.* (1997) to propose that UNI acts to maintain leaves in a transient state of indeterminacy that facilitates the prolonged organogenic activity. Similar explanations have been advanced for the roles of *KNOX* genes in compound leaf development. It therefore seems that an increased (prolonged) state of indeterminacy is required for specifying compound leaves; it also seems that the molecular determinants of this prolonged indeterminacy may differ in different species. It will therefore be of interest to determine which of the known components of the genetic pathways controlling determinacy in simple leaves are responsible for compound morphology in diverse species. In particular, it will be useful to understand whether specific forms of dissected morphology (e.g. pinnate, bipinate; for a consideration of the different types of dissected

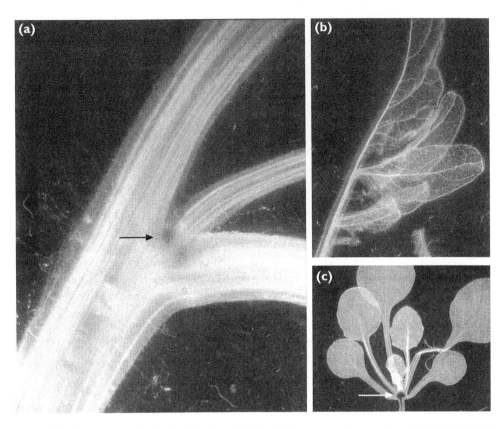

Figure 22.5 Expression of the *Arabidopsis* KNATI::GUS reporter in tomato. (a) KNATI::GUS expression (arrowed) in axillary meristems of tomato shoots; (b) no expression of KNATI::GUS in tomato leaves; (c) expression of KNATI::GUS in the shoot apical meristem (arrowed) but absence of expression in leaves of *Arabidopsis*.

leaf architecture see Sinha, 1999) are associated with specific molecular determinants. From a theoretical standpoint such studies will allow us to consider whether there are 'preferred' molecular routes via which specific types of dissected forms evolve. This in turn may suggest particular (as yet unknown) constraints that operate during the evolution of shoot form.

22.6 A way forward? General conclusions about comparative developmental genetics

As discussed above, even slight changes in the expression of *KNOX* genes can significantly alter shoot form. The fact that many of these deviant forms are often fertile means that there is a chance for novel morphologies to become fixed in a population. A good example of this process is the *hooded* mutant of barley which conditions misexpression of a barley *kn-1* orthologue and results in a novel floral structure (Müller *et al.*, 1995). These features could make the *KNOX* genes possible candidates as loci whose modified behaviour could be intimately related to the evolution of novel morphologies. There is evidence pointing towards a connection between persistence of *KNOX* expression in leaves and dissected–compound leaf morphology. In more general terms, comparative morphology suggests that leaf dissection is associated with an enhanced (prolonged) indeterminate potential of the leaf primordia.

On the basis of the above reasoning we are considering the following questions:

1   Does dissected morphology require the activity of *KNOX* genes in leaves? *KNOX* loss of function mutants in plants with compound leaves will be useful in addressing this question. Reducing activity using RNA interference (RNAi) technology may prove a valuable alternative in transformable but not genetically tractable systems.
2   How are upstream and downstream components of the KNOX pathway similar or divergent between simple and shoot dissected–compound leaves? Points of divergence may suggest areas of the development network as possible targets for modification by evolution.
3   What mechanisms specify persistence of *KNOX* expression in compound leaf primordia? This question can be approached using careful promoter swap experiments that would lead to the identification of specific cis-elements or trans-factor circuits that are remodelled during evolution (Yuh *et al.*, 2001)
4   Are other components of meristem function recruited in parallel to the *KNOX* pathway to specify compoundness?

Thus, through the comparative consideration of mutant phenotypes, gene expression patterns and promoter behaviour, viable hypotheses can be constructed regarding the mechanistic basis for the evolution of control of leaf shape. It is important that such hypotheses are tested *in planta*. One way to achieve this is via targeted misexpression studies using specific promoters. This should allow us to test to what extent a specific expression pattern of a gene can confer the morphology for which we suppose it to be responsible. For example, if it is known that a specific *KNOX* expression pattern is associated with dissected leaf morphology in one species, it would be informative to seek to reproduce the very same expression pattern in plants that normally have simple leaves. These experiments are more meaningful when very closely related species are used. For this reason it will be important to consider developing 'satellite'-model systems; that is, organisms that are closely related to a well-studied genetic model system but have significant differences in form. In our effort to study dissected leaf development we are focusing on small crucifers with compound leaves that are closely related to *Arabidopsis*.

Given the advent of genomics technologies it is exciting to consider the possibility of comparative expression profiling, or promoter trap expression studies in very closely related taxa with distinct morphological differences. It is therefore essential for such studies to be very clear regarding both the phylogenetic relationships of the species examined and the phylogenetic relationships between the genes compared. The wealth of data generated in this manner, together with the ability to manipulate expression of morphogenetically important genes according to predictions made by mutant phenotypes and expression patterns, should give some insight into how distinct plant morphologies arise in Nature. In fact, it could be argued that the comparative study of gene regulatory networks across taxa is imperative if we are to fully understand the complex way in which genetic pathways function during the development of organisms.

REFERENCES

Avivi, Y., Lev-Yadun, S., Morozova, N., Libs, L., Williams, L., Zhao, J., Varghese, G. and Grafi, G. (2000) Clausa, a tomato mutant with a wide range of phenotypic perturbations, displays a cell type-dependent expression of the homeobox gene LeT6/TKn2. *Plant Physiology*, 124, 541–552.

Byrne, M. E., Barley, R., Curtis, M., Arroyo, J. M., Dunham, M., Hudson, A. and Martienssen, R. A. (2000) ASYMMETRIC LEAVES1 mediates leaf patterning and stem cell function in *Arabidopsis*. *Nature*, 408, 967–971.

Chen, J.-J., Janssen, B.-J., Williams, A. and Sinha, N. (1997) A gene fusion at a homeobox locus: alterations in leaf shape and implications for morphological evolution. *Plant Cell*, 9, 1289–1304.

Chuck, G., Lincoln, C. and Hake, S. (1996) Knat1 induces lobed leaves with ectopic meristems when overexpressed in *Arabidopsis*. *Plant Cell*, 8, 1277–1289.

Donnelly, P. M., Bonetta, D., Tsukaya, H., Dengler, R. E. and Dengler, N. G. (1999) Cell cycling and cell enlargement in developing leaves of *Arabidopsis*. *Developmental Biology*, 215, 407–419.

Estruch, J. J., Prinsen, E., Van Onckelen, H., Schell, J. and Spena, A. (1991) Viviparous leaves produced by somatic activation of an inactive cytokinin-synthesizing gene. *Science*, 254, 1364–1367.

Gleissberg, S. (2002) Comparative developmental and molecular genetic aspects of leaf dissection, in *Developmental Genetics and Plant Evolution* (eds Q. C. B. Cronk, R. M. Bateman and J. A. Hawkins), Taylor & Francis, London, pp. 404–417.

Hagemann, W. and Gleissberg, S. (1996) Organogenetic capacity of leaves: the significance of marginal blastozones in angiosperms. *Plant Systematics and Evolution*, 199, 121–152.

Hake, S., Vollbrecht, E. and Freeling, M. (1989) Cloning *Knotted1*, the dominant morphological mutant in maize using *Ds2* as a transposon tag. *EMBO Journal*, 8, 15–22.

Hareven, D., Gutfinger, T., Parnis, A., Eshed, Y. and Lifschitz, E. (1996) The making of a compound leaf: genetic manipulation of leaf architecture in tomato. *Cell*, 84, 735–744.

Hofer, J., Turner, L., Hellens, R., Ambrose, M., Matthews, P., Michael, A. and Ellis, N. (1997) UNIFOLIATA regulates leaf and flower morphogenesis in pea. *Current Biology*, 7, 581–587.

Janssen, B. J., Lund, L. and Sinha, N. (1998) Overexpression of a homeobox gene, LeT6, reveals indeterminate features in the tomato compound leaf. *Plant Physiology*, 117, 771–786.

Kaplan, D. R. (2001) Fundamental concepts of leaf morphology and morphogenesis: a contribution to the interpretation of molecular genetic mutants. *International Journal of Plant Sciences*, 162, 3, 465–474.

Kerstetter, R. A., Laudencia-Chingcuanco, D., Smith, L. G. and Hake, S. (1997) Loss of function mutations in the maize homeobox gene, *knotted1*, are defective in shoot meristem maintenance. *Development*, 124, 3045–3054.

Kusaba, S., Kano-Murakami, Y., Matsuoka, M., Tamaoki, M., Sakamoto, T., Yamaguchi, I. and Fukumoto, M. (1998) Alteration of hormone levels in transgenic tobacco plants overexpressing the rice homeobox gene OSH1. *Plant Physiology*, 116, 471–476.

Li, Y., Hagen, G. and Guilfoyle, T. J. (1992) Altered morphology in transgenic tobacco plants that overproduce cytokinins in specific tissues and organs. *Developmental Biology*, 153, 386–395.

Lincoln, C., Long, J., Yamaguchi, J., Serikawa, K. and Hake, S. (1994) A *Knotted1*-like homeobox gene in *Arabidopsis* is expressed in the vegetative meristem and dramatically alters leaf morphology when overexpressed in transgenic plants. *Plant Cell*, 6, 1859–1876.

McConnell, J. R. and Barton, M. K. (1998) Leaf polarity and meristem formation in *Arabidopsis*. *Development*, 125, 2935–2942.

Müller, K., Romano, N., Gerstner, O., Garcia-Maroto, F., Pozzi, C., Salamini, F. and Rohde, W. (1995) The barley *Hooded* mutation caused by a duplication in a homeobox gene intron. *Nature*, 374, 727–730.

Ori, N., Juarez, M. T., Jackson, D., Yamaguchi, J., Banowetz, G. M. and Hake, S. (1999) Leaf senescence is delayed in tobacco plants expressing the maize homeobox gene *knotted1* under the control of a senescence-activated promoter. *Plant Cell*, 11, 1073–1080.

Ori, N., Eshed, Y., Chuck, G., Bowman, J. L. and Hake, S. (2000) Mechanisms that control *knox* gene expression in the *Arabidopsis* shoot. *Development*, 127, 5523–5532.

Parnis, A., Cohen, O., Gutfinger, T., Hareven, D., Zamir, D. and Lifschitz, E. (1997) The dominant developmental mutants of tomato, *Mouse-Ear* and *Curl*, are associated with distinct modes of abnormal transcriptional regulation of a *Knotted* gene. *Plant Cell*, 9, 2143–2158.

Siegfried, K. R., Eshed, Y., Baum, S. F., Otsuga, D., Drews, G. N. and Bowman, J. L. (1999) Members of the YABBY gene family specify abaxial cell fate in *Arabidopsis*. *Development*, 126, 4117–4128.

Sinha, N. (1999) Leaf development in angiosperms. *Annual Review of Plant Physiology and Plant Molecular Biology*, 50, 419–446.

Sylvester, A. W., Smith, L. and Freeling, M. (1996) Acquisition of identity in the developing leaf. *Annual Review of Cell and Developmental Biology*, 12, 257–304.

Tamaoki, M., Kusaba, S., Kano-Murakami, Y. and Matsuoka, M. (1997) Ectopic expression of a tobacco homeobox gene, NTH15, dramatically alters leaf morphology and hormone levels in transgenic tobacco. *Plant and Cell Physiology*, 38, 917–927.

Telfer, A. and Poethig, R. S. (1998) HASTY: a gene that regulates the timing of shoot maturation in *Arabidopsis thaliana*. *Development*, 125, 1889–1898.

Timmermans, M. C., Hudson, A., Becraft, P. W. and Nelson, T. (1999) ROUGH SHEATH2: a Myb protein that represses knox homeobox genes in maize lateral organ primordia. *Science*, 284, 151–153.

Tsiantis, M., Schneeberger, R., Golz, J. F., Freeling, M. and Langdale, J. A. (1999) The maize *rough sheath2* gene and leaf development programs in monocot and dicot plants. *Science*, 284, 154–156.

Van Tuinen, A., Peters, A. H. L. J., Kendrick, R. E., Zeevaart, J. A. D. and Koornneef, M. (1999) Characterisation of the procera mutant of tomato and the interaction of gibberellins with end-of-day far-red light treatments. *Physiologia Plantarum*, 106, 121–128.

Waites, R., Selvadurai, H. R., Oliver, I. R. and Hudson, A. (1998) The PHANTASTICA gene encodes a MYB transcription factor involved in growth and dorsoventrality of lateral organs in *Antirrhinum*. *Cell*, 93, 779–789.

Wang, R. L., Stec, A., Hey, J., Lukens, L. and Doebley, J. (1999) The limits of selection during maize domestication. *Nature*, 398, 236–239.

Yuh, C. H., Bolouri, H. and Davidson, E. H. (2001) Cis-regulatory logic in the endo16 gene: switching from a specification to a differentiation mode of control. *Development*, 128, 617–629.

Chapter 23

Evolutionary history of the monocot leaf

Paula J. Rudall and Matyas Buzgo

ABSTRACT

The supposedly fundamental differences between dicot and monocot leaves are based on a perception of developmental differences, both in early zonal differentiation and subsequent meristematic activity. The leaf base theory, which has been used as a model in developmental genetics, proposed that in monocot leaves the bulk of the lamina is derived from the proximal zone of the primordium and subsequent development is largely intercalary (hence the linear structure), in contrast to dicot leaves in which the lamina is initially derived from the distal zone and there is more pronounced marginal meristem activity. However, there is much overlap between monocot and dicot leaf types; laterally expanded petiolate laminas in magnoliids (e.g. Saururaceae) are morphologically and developmentally similar to those of basal monocots (Alismatales). The leaf base theory may be too simplistic as a model for leaf development and character coding. We take a less strictly typological view, arguing that meristematic activity in both dicots and monocots occurs in a highly plastic transition zone between precursor tip and sheath. Part of this transition zone forms an adaxial meristem which produces a unifacial petiole or blade in some monocots by suppression of the marginal meristem. Adaxial cross meristems in the transition zone give rise to, or are associated with, many other leaf structures, including peltate leaves and foliar appendages such as ligules and stipules. Ligules occur in certain monocot groups, most notably in the orders Poales, Zingiberales and Alismatales. They demarcate the junction between the sheath and lamina, and have therefore been used as markers in developmental studies. Stipules occur in many magnoliids, many eudicots and a few monocots (e.g. *Dioscorea* and *Smilax*). Some evidence contradicts the common belief that stipules and ligules have different developmental origins; both structures originate from positionally similar adaxial cross meristems. Furthermore, preliminary evidence suggests that the *liguleless* genes responsible for ligule formation in maize and rice may also be involved in dicot stipule development. The homologies and terminology of other types of foliar appendage (e.g. squamules, ocrea and auricles) also require further review in a developmental–genetic context.

23.1 Introduction

The leaf is a determinate organ, with an ultimate destiny from initiation, in contrast to the shoot apex, which has indeterminate growth and may continue to grow and

In *Developmental Genetics and Plant Evolution* (2002) (eds Q. C. B. Cronk, R. M. Bateman and J. A. Hawkins), Taylor & Francis, London, pp. 431–458.

produce other organs. Following inception, leaves of seed plants grow by means of both marginal and intercalary growth; indeed, the intercalary growth pattern largely differentiates angiosperm leaves from those of ferns, which grow mainly from marginal meristems (Doyle, 1978). Most authors (e.g. Hagemann, 1973) believe that the primitive angiosperm leaf was clearly differentiated into sheath, petiole and undivided blade. However, despite a long historical literature, the evolutionary origin and systematic homologies of the various different angiosperm leaf types remain contentious: morphological developmental studies in different taxa have led to contrasting hypotheses of developmental pathways.

Although there are many exceptions, adult monocot leaves are typically narrow and linear with parallel venation, basally ensheathing the stem, in contrast to the typical dicot leaf which has a well-defined petiole and elliptical blade with reticulate venation. Such monocot/dicot leaf divergence has been widely articulated, and may have systematic and predictive value, since morphological features that reliably define major clades are rare. For example, Prantl (1883) proposed a basiplastic (basally formative) system typical of monocot leaves, and a pleuroplastic (laterally formative) system typical of dicot leaves. Developmental geneticists (e.g. Tsiantis *et al.*, 1999) have related differences in gene expression between the asterid eudicots *Antirrhinum* and *Nicotiana* and the commelinoid monocot *Zea* to supposedly fundamental differences between dicot and monocot leaves.

These monocot/dicot differences represent a syndrome comprising two major factors: (1) early zonal differentiation and (2) subsequent meristematic activity following initial establishment growth. In this chapter we review the range of monocot leaf structures in relation to phylogeny and discuss the homologies of different leaf zones and appendages. Questions addressed include: is the ancestral monocot leaf unifacial linear, bifacial linear, or dicot-like, with a petiole and lamina? Are monocot leaf zones and appendages homologous with those of dicots? How strict is the definition of developmental leaf zones within monocots, and are the different types of ligule homologous? If fully integrated with ontogenetic and phylogenetic studies, developmental genetics will provide new perspectives on these key questions; however, both the models and the empirical data on which they are based require clarification.

23.2 Monocot phylogeny

Recent molecular analyses (e.g. Graham and Olmstead, 2000; Qiu *et al.*, 2000; Soltis *et al.*, 2000) have revised current thinking about relationships among early-branching angiosperms, and prompted a series of seminal papers evaluating character evolution in these taxa, especially with regard to reproductive characters (e.g. Endress and Igersheim, 1999; Floyd and Friedman, 2000; Sampson, 2000). Although there is now some consensus about the early-branching grade (*Amborella*, Nymphaeaceae and Illiciales), relationships between magnoliids, monocots and eudicots remain equivocal (Figure 23.1). However, it is now well established that dicotyledons in the old sense (i.e. basal angiosperms plus eudicots) are paraphyletic. Monocotyledons are a monophyletic group possibly embedded within the magnoliids (Soltis *et al.*, 1999); their closest relatives are therefore probably found among the magnoliids rather than the eudicots. Consequently, direct monocot/dicot comparisons

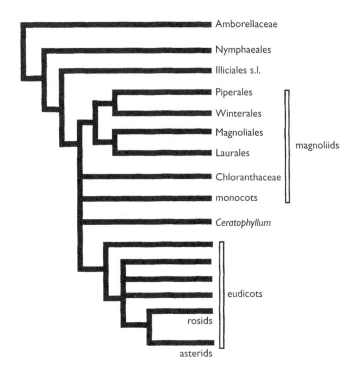

Figure 23.1 Diagram of relationships of major angiosperm clades, based on Qiu *et al.* (2000).

are too simplistic, especially between highly derived model organisms from both eudicots and monocots.

Cases in which monocot leaves show striking similarity to magnoliid leaves have previously been regarded as evidence of a phylogenetically basal position in monocots. For example, the yams (*Dioscorea* spp.), which have petiolate (occasionally stipulate) leaves with little or no sheath and an elliptical reticulately-veined lamina, have often been regarded as close to the ancestral monocots and their magnoliid precursors such as Piperales (e.g. Dahlgren *et al.*, 1985). Fossil Piperaceae are often difficult to distinguish from fossil *Smilax* or *Dioscorea* (e.g. Greenwood and Conran, 2000). Taylor and Hickey (1992) therefore used *Dioscorea* and *Smilax* as the monocot representatives in their analysis of relationships in basal angiosperms. They concluded that the ancestral angiosperms were rhizomatous or scrambling perennial herbs in which the leaf base was horizontally extended into a sheath which sometimes also developed into stipules.

Stewart and Rothwell (1993), Herendeen and Crane (1995) and Gandolfo *et al.* (2000) reviewed the early fossil history of monocot leaves. Gandolfo *et al.* (2000) rejected Doyle's (1973) list of characters that differentiate monocot leaves from those of gymnosperms and other angiosperms, because these characters, which were based on size of parallel veins, fine cross-venation and apical convergence of veins, are also found in the magnoliid families Saururaceae, Piperaceae and Barclayaceae and a fossil gnetalean, *Drewria*. Gandolfo *et al.* (2000) therefore considered the

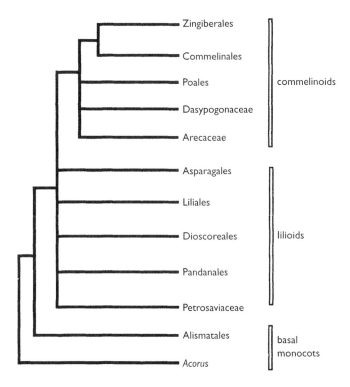

Figure 23.2 Diagram of relationships of major monocot clades, based on Angiosperm Phylogeny Group (1998), Chase *et al.* (2000) and Stevenson *et al.* (2000).

earliest unequivocal records of monocot leaves to be *Zingiberopsis* from the late Cretaceous (Maastrichtian), rather than the early Cretaceous fossil *Acaiaephyllum* (e.g. Doyle, 1973). There are also late Cretaceous and Palaeocene records of some basal monocots such as *Pistia* and *Cymodocea* (Herendeen and Crane, 1995).

Recent analyses of molecular sequence data (e.g. Chase *et al.*, 2000; Stevenson *et al.*, 2000) have identified several well-supported major monocot clades, grouped here for convenience into (1) basal monocots, (2) lilioid monocots and (3) commelinoid monocots (Figure 23.2). Although current evidence indicates that *Acorus* and Alismatales (including Araceae, Alismatidae and Tofieldiaceae) are basal, there remain two major unresolved polytomies in higher-level monocot systematics: (1) relationships between the commelinoid clade and the four lilioid clades (Asparagales, Dioscoreales, Liliales and Pandanales) plus Petrosaviaceae, and (2) relationships of the main groups within the commelinoid clade. The existing monocot tree does not therefore provide a useful optimisation framework for evolutionary patterns, with the exception of the basal monocots.

23.3 Leaf morphology

According to Zimmermann's (1930) telome theory, leaves of early land plants may have originated by unequal differentiation and subsequent homeotic modification of

Table 23.1 Glossary of leaf terminology used here. Italics denote German terms (e.g. Troll, 1939; Roth, 1949), many of them commonly used in English-language morphological literature

Term (and equivalent terms)	Definition/description
Mature leaf regions	
Lamina (leaf blade)	Main (distal) part of leaf; unifacial or bifacial. In a petiolate leaf, 'lamina' refers to the distal (normally flattened) blade, either simple or compound; in a non-petiolate leaf it refers to the entire leaf except sheath (although blade and sheath may be continuous and not clearly demarcated).
Petiole (*Stiel*)	Leaf stalk (between lamina and sheath); narrow and sometimes thickened; unifacial or bifacial.
Precursor tip (forerunner tip, *Vorläuferspitze*)	Thickened apex of leaf; sometimes glandular, sometimes unifacial.
Sheath	Lowermost (distal) part of leaf, ensheathing stem; always bifacial, derived from lower zone of primordium.
Appendages and specialised structures	
Auricles	Paired outgrowths at base of lamina.
Contraligule (tongue, false ligule, lingular, pseudoligule)	Ligule-like outgrowth at distal end of leaf sheath opposite leaf blade.
Hastula	'Crest' at junction of petiole and lamina.
Ligule (dorsal ligule)	Thin, often membraneous unvascularised outgrowth of the adaxial protoderm in region between sheath and petiole (or sheath and lamina if petiole absent).
Ochrea (ocrea)	Extension of sheath or petiole.
Pulvinus (callus)	Swelling or 'hinge' at either end of petiole.
Squamules (intravaginal squamules)	Small axillary appendages (often glandular, mucilage-secreting).
Stipules	Leaf sheath appendages, often paired and sometimes leafy and vascularised.
Primordial zones and meristems	
Adaxial meristem	Meristem formed in transition zone on the adaxial side of the leaf.
Cross meristem (*Transversalwulst, Ventralwulst*)	Transverse, prominent portion of the adaxial meristem, forming ligules, stipules and other structures.
Lower zone (*Unterblatt*, hypophyll)	Proximal part of leaf primordium.
Transition zone	Region of primordium between upper and lower zone (between sheath and precursor tip).
Upper zone (*Oberblatt*, hyperphyll)	Distal part of leaf primordium.

dichotomous branching systems (Stewart and Rothwell, 1993). Leaves of many seed plants have a bifacial base (sheath) which ensheaths part of the axis by lateral meristem activity (meristem incorporation) at an early stage of the leaf primordium. Rothwell (1982) postulated a heterochronic hypothesis for the origin of conifer leaves, in which a relatively minor genetic change resulted in suppression of frond development in favour of the development of simple leaves similar to the scale leaves at the bases of the branches in pteridosperms. Among extant seed plants the petiolate, reticulate-veined leaf of *Gnetum* is perhaps the closest analogue of the angiosperm leaf (Doyle, 1978; Stewart and Rothwell, 1993). Interestingly, the two closest relatives of *Gnetum* are *Ephedra*, with scale-like leaves, and *Welwitschia*, which has superficially monocot-like strap-shaped, parallel-veined leaves, although these develop rather differently from those of monocots, by means of a permanent basal intercalary meristem, with continuous distal erosion (Rodin, 1967a–c).

Analyses of the phylogeny of seed plants, including basal angiosperms (e.g. Taylor and Hickey, 1992; Nixon *et al.*, 1994), have scored the leaf base as either ensheathing or discrete. Ensheathing leaf bases occur in most monocots, and also in Gnetales, many magnoliids, such as Chloranthaceae, Piperaceae and Saururaceae, and some eudicots, such as Polygonaceae, Apiaceae and Plantaginaceae. However, this is possibly an arbitrary and artificial distinction; in some taxa (e.g. cycads: Stevenson, 1990), leaves are ensheathing during early ontogeny but not at maturity. The leaf sheath is reduced or absent in some basal angiosperms (e.g. *Amborella*, Myristicaceae and Winteraceae), many eudicots and a few monocots. In some monocots (especially palms), magnoliids (e.g. Saururaceae, Piperaceae and Magnoliaceae) and eudicots (e.g. Polygonaceae and Apiaceae), the sheath proliferates to form an ochrea. Some taxa have a sheath closed to a tube; for example, this occurs in *Allium* and some grasses and sedges.

The apical precursor tip (forerunner tip, or *Vorläuferspitze*) is the first part of the leaf to be established on the leaf primordium. In some taxa the precursor tip retains a thickened, sometimes unifacial structure until maturity, and may be glandular (e.g. in *Dioscorea sansibarensis*: Burkill, 1960), filiform or tendril-like (e.g. in *Littonia* and *Flagellaria*). However, the leaf apex in many monocots is bifacial (e.g. in *Sagittaria*: Bloedel and Hirsch, 1979) rather than unifacial and thickened, and in this case it is not always interpreted as a precursor tip, although this is again an arbitrary distinction.

There is a wide range of monocot leaf forms (Figures 23.3, 23.4, 23.5), summarised here as three types: (1) bifacial non-petiolate, either linear or elliptical (Figure 23.3c); (2) distinctly petiolate (Figures 23.3a, 23.4a, 23.5a, 23.5b); and (3) unifacial linear (Figure 23.4b). However, there are some intermediates; for example, in *Agave* (Kumar *et al.*, 1984), *Sansevieria*, *Ixiolirion* and *Homeria* there is a bifacial lamina and a thickened unifacial leaf tip (*Vorläuferspitze*). All three leaf types are widely distributed throughout the monocots (Table 23.2). Linear leaves (with more intercalary than marginal growth) are sometimes correlated with aquatic habitats, and may have evolved several times within monocots, many of which grow in wet habitats. Some magnoliids, such as Saururaceae, also show a strong affinity to water, although none of their extant species have linear leaves.

Bifacial leaves without a differentiated petiole may be either linear, as in many grasses (e.g. *Zea*) and Asparagales (e.g. *Anthericum*), or expanded and elliptical, as in many Liliales, such as *Alstroemeria*. In some taxa (e.g. grasses) the blade is well differentiated from the sheath, mainly because of the presence of a ligule; in others (e.g. *Anthericum, Hypoxis*) there is little distinction between sheath and blade in adult leaves.

Distinctly petiolate leaves occur in Araceae and many other Alismatales, some Pandanales (Cyclanthaceae and Stemonaceae), Dioscoreaceae, some Liliales (e.g. *Smilax, Trillium*), a few Asparagales (*Cyanastrum*) and several commelinoids, especially palms (Arecaceae) and Zingiberales. Petiolate leaves may have reticulate venation (e.g. in *Dioscorea*) or parallel cross-veins (e.g. *Stemona*). There is a continuous range between taxa with a distinct, thickened petiole and a narrow proximal region of the lamina. The petiole may be bifacial or unifacial. Some monocot leaves are

Figure 23.3 Monocot leaves: (a) *Ampelocalamus scandens,* leaf base, with ligular hairs at junction of sheath and petiole. (b) *Zingiber officinale,* ligule at junction of sheath and blade. (c) *Crinum pedunculatum,* plant with linear, expanded non-petiolate sheathing leaves. (d) *Dianella ensifolia,* semi-equitant leaves.

Figure 23.4 Monocot leaves: (a) *Serenoa repens,* leaf with hastula at junction of petiole and blade. (b) *Xeronema callistemon,* equitant leaves. (c) *Pothos macrocephalus,* leaves with expanded petioles.

Figure 23.5 Monocot leaves: (a) *Sagittaria sagittifolia*, sagittate (arrow-shaped) petiolate leaf. (b) *Smilax china*, leaf base with stipular tendrils at junction of petiole and sheath.

cordate/sagittate (e.g. *Sagittaria*, juvenile *Arisaema*, *Pothos*) or peltate (*Caladium*, *Alocasia*, *Anemopsis*); both types are also present in magnoliids (e.g. cordate leaves in *Houttuynia*, peltate leaves in *Peperomia*). In peltate leaves the petiole is attached to the lamina some distance from its margin (Franck, 1976). Peltate leaves appeared early in the fossil record, in the Cretaceous Albian, associated with aquatic habitats (Doyle, 1978).

The petiolate leaf, expanded lamina and reticulate venation are correlated features which have evolved several times in monocots. Cameron and Dickison (1998) showed that reticulate leaf venation has arisen independently in several different orchid lineages, largely (though not exclusively) in association with forest or forest-margin habitats. For example, within the vanilloid subtribe Pogoniinae, there is a probable trend from parallel venation and linear or narrowly elliptical non-petiolate leaves, to increased lamina width, decreased primary venation and increased secondary vascularisation, resulting in a net-like pattern, and finally petiolate leaves with reticulate venation, free vein endings and scattered stomata. Similarly, among Araceae, there is a trend from parallel to pinnate venation by lateral expansion of the lamina, and finally to truly reticulate venation in some Areae (Mayo *et al.*, 1997). In lilioid monocots (Table 23.2) there is significant correlation between an evergreen, climbing habit and petiolate leaves with reticulate venation (e.g. Conover, 1983).

Unifacial leaf blades have a bifacial sheath and unifacial lamina, either flattened (ensiform/equitant: e.g. Iridaceae, *Acorus*, *Xyris*, *Xeronema*) or rounded (terete: e.g.

Table 23.2 Leaf types in monocots. Information from Arber (1925), Burkill (1960: Dioscoreaceae), Dahlgren et al. (1985), Tomlinson (1990: palms), Kubitzki (1998a, b) and personal observations of the authors

Taxa	Leaf shape	Appendages
Basal monocots		
Acorus	Linear unifacial, equitant (ensiform).	
Alismatales	Juvenile and adult leaves often different in same plant (heterophylly or heteroblasty). Adult leaves linear bifacial in some alismatid families and exceptionally in Araceae; unifacial (equitant or terete) in Tofieldiaceae, some Juncaginaceae, Scheuchzeriaceae; otherwise distinctly petiolate with a variably broad lamina and linear or reticulate venation.	Ligules and auricles present in some alismatids; stipules reported in others and some Araceae. Squamules present in most alismatids except Scheuchzeriaceae, Tofieldiaceae and some Araceae.
Lilioid monocots		
Dioscoreales	In Nartheciaceae linear (bifacial or equitant); in Burmanniaceae linear, often scale-like in mychoheterotrophic taxa; in Dioscoreaceae petiolate with broad (sometimes compound) lamina with reticulate venation, sometimes with projecting, often glandular leaf tip (e.g. Dioscorea sansibarensis).	Some species of Dioscorea with stipules at petiole base, often modified into prickles; pulvinus often present at each end of petiole.
Pandanales	Linear bifacial in Pandanaceae and Velloziaceae, mostly petiolate in Cyclanthaceae and Stemonaceae, scale-like in Triuridaceae.	Hastulae sometimes present on petiolate in Cyclanthaceae.
Liliales	Mostly linear bifacial, often elliptical, occasionally petiolate (e.g. Smilax, Trillium).	Stipular tendrils present in Smilax.
Asparagales: higher asparagoids	Mostly linear bifacial, occasionally elliptical; sometimes petiolate (e.g. Hosta, Aspidistra, Eriospermum), sometimes reduced to scales (e.g. Aphyllanthes, Asparagus), sometimes unifacial terete (e.g. species of Allium), sometimes with unifacial precursor tip (e.g. Agave).	Ligules mostly absent, except ligular sheath in e.g. Allium.
Asparagales: lower asparagoids	Mostly linear bifacial, petiolate in Cyanastrum; sometimes unifacial equitant (e.g. some Iridaceae, Xeronema, some orchids), semi-equitant (e.g. Dianella), terete (e.g. Bobartia) or quadrangular (Xanthorrhoea), sometimes with unifacial precursor tip (e.g. Homeria).	Ligules mostly absent.

Commelinoid monocots

Arecales (Arecaceae: palms)	Juvenile leaves sometimes linear bifacial; adult leaves petiolate with broad lamina with wide range of shapes, sometimes dissected, plicate.	Hastulae sometimes present on petiole. Ligules sometimes present. Ochrea sometimes present.
Bromeliaceae	Linear, bifacial; rarely petiolate (e.g. some *Bromelia* species), sometimes with bifacial precursor tip (e.g. *Billbergia*).	Ligules absent or reduced.
Commelinales	In Commelinaceae usually elliptical, sometimes with short petiole; in Haemodoraceae and Philydraceae linear, unifacial and equitant; in Pontederiaceae juvenile leaves linear, adult leaves petiolate.	Ligules sometimes present in Pontederiaceae.
Dasypogonaceae	Linear, bifacial.	Ligules absent.
Hanguanaceae	Petiolate with broad lamina.	Ligules absent.
Rapateaceae	Linear, bifacial or unifacial. Distinct sheath, petiole and blade in some genera (e.g. *Saxofridericia*).	Ligules rarely present.
Poales	Linear or elliptical (some bamboos petiolate), mostly bifacial, except unifacial ensiform (in *Anarthria*, some Juncaceae, some Xyridaceae) or terete (some Juncaceae); scale-like in *Ecdeiocolea* and most Restionaceae, elliptical with a leaf-tip tendril in *Flagellaria*.	Ligules or ligule homologues present in Joinvilleaceae, Juncaceae, some Restionaceae, some Xyridaceae, most Poaceae and some Cyperaceae, absent from Eriocaulaceae, *Flagellaria*, Typhaceae, Hydatellaceae, Mayacaceae and some Cyperaceae.
Zingiberales	Petiolate with broad lamina and short sheath; petiole indistinct or absent in some Heliconiaceae and Zingiberaceae, sometimes with precursor tip (e.g. *Heliconia*).	Ligules sometimes present (e.g. Costaceae, Zingiberaceae). Pulvinus (callus) present at top of petiole in Marantaceae.

some *Allium* species, some *Triglochin* species). Ensiform or terete leaves occur in a wide range of monocots, including *Acorus*, some Alismatales (e.g. Tofieldiaceae, Scheuchzeriaceae), some Dioscoreales (Nartheciaceae, except *Aletris*), some Asparagales (most Iridaceae, *Xeronema*, *Lanaria*, some Agavaceae, e.g. *Hosta*, some Convallariaceae, e.g. *Sansevieria* and *Dracaena*) and some commelinoids (all Haemodoraceae, Philydraceae, Rapateaceae; also some Xyridaceae, e.g. *Xyris* and *Achlyphila*), Juncaceae, *Anarthria* (Anarthriaceae) and some grasses. Within Iridaceae, the ensiform leaf takes various forms, such as plicate (zig-zag), as in *Tigridia* and *Babiana*, or grooved, as in the unique and distinctive *Crocus* leaf, which is secondarily bifacial, with a central keel and two lateral arms (Rudall, 1990). Using a recent molecular phylogeny of Iridaceae (Reeves *et al.*, 2001), it is clear that thick ensiform leaves (e.g. *Isophysis*, *Diplarrhena*, *Patersonia* and *Pillansia*) are the plesiomorphic condition in Iridaceae, and plicate leaves have evolved at least twice, within subfamilies Iridoideae (Tigridieae) and Ixioideae (e.g. *Babiana*, *Tritonia*), also evidenced from leaf histology (Rudall, 1990). Some Hemerocallidaceae (*Agrostocrinum*, *Dianella*, *Excremis*, *Phormium* and *Thelionema*) have 'semi-equitant' (*schwertformig*) leaves, in which the basal and upper regions are bifacial and sheathing, but between these parts there is a short, flattened, isobilateral region which presumably corresponds to a petiole (Figure 23.3d).

Alternative ensiform leaf types exist in *Acorus*, both in hibernating scale leaves, which have a short lamina and conical sheath, and leaves on the spathe base (Kaplan, 1973b; M. Buzgo, unpubl. obs.). In a 'normal' ensiform leaf the adaxial surface is lost by apparent 'folding'. Alternatively, the ensiform shape of the spathe lamina forms by leaf 'rotation' (Figure 23.6). The base of the spathe lamina is trigonal. Distally, one edge of the trigonal leaf expands laterally and the angle on the opposite margin becomes less acute, resulting in ensiformity. These two routes to ensiformity in *Acorus* indicate that ensiform leaf shape may be derived from more conventional linear bifacial leaves.

23.4 Heteroblasty and leaf plasticity

Williams-Carrier *et al.* (1997) demonstrated that expression of *knotted1 (kn1)*-like homeobox genes in *Nicotiana* and *Arabidopsis* resulted in greater and different phenotype alteration than similar experiments on maize, and concluded that a fundamental difference of plasticity exists between monocots and dicots, dicots retaining a more flexible, less determined state. However, leaves of many monocots, especially in the basal clades, retain a high degree of developmental plasticity, depending on factors such as environment (water level, light) and life cycle (Bloedel and Hirsch, 1979; Cook, 1996). For example, in some Alismataceae, the range of variation includes (1) laminate submerged leaves with an exclusive basal intercalary growth, (2) floating leaves with ovally-expanded laminas possessing parallel venation and limited apical proliferation, and (3) emergent leaves with an expanded, cordate or sagittate lamina and reticulate venation.

Similarly, there is a range of variation within some taxa with unifacial leaves. For example, some clones of *Acorus* form seasonally diverse types of leaves: scale leaves in autumn and foliar leaves in spring. Furthermore, within a single spathe of *Acorus* it is possible to observe a range from a trigonal leaf blade with a bifacial triangular

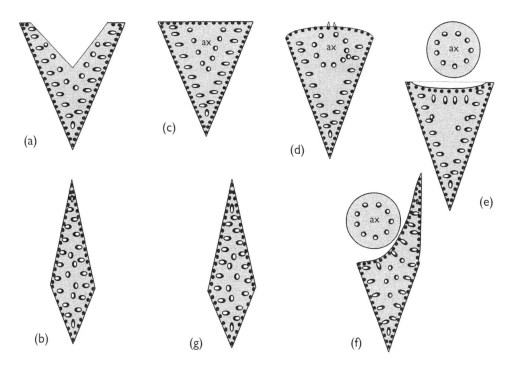

Figure 23.6 Acorus sp. Diagrammatic cross sections of leaf and inflorescence. (a–b). Foliage leaf: (a) Sheath. (b) Lamina. (c–g). Inflorescence and spathe: (c) Mid region of peduncle. (d) Distal region of peduncle, just beneath spadix; axial vasculature arranged as a ring. (e) Region above separation of spadix and triangular spathe, vascular bundles along adaxial side derived from axial bundles at separation. (f) One margin of spathe expands along spadix, the opposite rim attains an obtuse angle and becomes a lateral midrib. (g) Expanded margin median adaxial, on outer side, opposite obtuse rim; new lateral midrib formed; spathe now similar to ensiform, symmetrical part of foliage leaf (as in b). ax = axial vasculature, white circles = xylem (flanked by black phloem), small black circles = sclerenchyma.

transverse section to a typical ensiform leaf (Figure 23.6). Within a single family, Juncaginaceae, there is a range of lamina types from bifacial linear to bifacial trigonal and unifacial terete. In some taxa ensiform leaf development ranges from terete to ensiform from seedling to adult leaves, for example in *Acorus calamus* (Kaplan, 1970, 1973b; Tillich, 1995), *Tigridia, Iris* (Iridaceae: Rudall, 1990) and *Xyris* (Tillich, 1995; Sajo and Rudall, 1999).

Such plasticity is not entirely consistent with the controversial maturation schedule hypothesis (Freeling, 1992; Muehlbauer *et al.*, 1997), which states that cells pass through a series of competency stages from sheath to blade, responding to a range of differentiation signals before eventual maturation. Thus, in maize, all cells have the potential to express the sheath fate, but only distal cells progress to the blade competency stage, with ligule and auricle development as an intermediate stage. Muehlbauer *et al.* (1997) used the *liguleless3* homeobox (*knox*) gene to demonstrate a continuum of phenotype transformations from normal to bladeless leaves. On the other hand, in many monocots some distal cells mature precociously: the distal precursor tip is the first-formed part of the lamina.

23.5 Foliar appendages

Foliar appendages are characteristic of some angiosperms (Tables 23.1, 23.2). Structures known as ligules occur in a wide range of plants, including pteridophytes such as *Selaginella* and *Isoetes* in which they secrete mucilage. However, such terminology should be treated with caution; these structures are not comparable with the monocot ligule. Cavot Abrigeon and Lemoigne (1978) analogised pteridophyte ligules with monocot squamules (Table 23.1), rather than with the monocot ligule. Ligules occur in certain monocot groups, most notably in the orders Poales and Zingiberales (Figure 23.3a, b; Plate 3d), and also in the early-branching monocots (Alismatales). They are rare in the lilioid orders (Table 23.2), although rudimentary ligules are present in many lilioid taxa (e.g. *Allium*) well illustrated by Roth's drawings of leaf primordia (Figure 23.7). In some Poales (e.g. maize), both ligules and auricles are present. There are several records of ligule-less (eligulate) grass mutants (summarised by Napp Zinn, 1973). For example, there are ligulate and eligulate forms of various *Triticum* species; also, some recessive mutants of *Hordeum vulgare* and *Zea mays* are eligulate, and there have been comparative studies of leaf ontogeny of mutant forms of *Oryza sativa*, in which the ligules and auricles are suppressed.

Ligules are important in the context of understanding leaf morphology in monocots because they demarcate the junction between the two axial domains, the sheath and lamina, and can be used as markers in developmental studies (Tsiantis and Langdale, 1998). However, diverse structures have been named 'ligule', not necessarily reflecting homology. Philipson (1935) examined ligule development in grasses and reviewed the early literature. He concluded that the ligule consists of an outgrowth of the adaxial leaf epidermis, in some cases continuous with an extended free upper border of the sheath. Taxa which lack the laminar surface outgrowth but have an extended sheath border are sometimes regarded as stipulate rather than ligulate. Tran van Nam (1968, 1971) described 'dorsal ligules' extending around the abaxial side of the lamina in some grasses, and a 'latent ligule' in *Echinochloa* (Poaceae), in which there is no epidermal outgrowth, but differences in epidermal cell morphology. Other ligule (or stipule) homologues include the contraligule, which is a ligular sheath extension common in Cyperaceae (Dahlgren *et al.*, 1985; Camelbeke and Goetghebeur, 1999). In Cyperaceae, Camelbeke and Goetghebeur (1999) identified three generic groups based on the ligule character: (1) ligule always absent, (2) ligule always present, and (3) ligule variable: present or absent in different species.

Tomlinson (1990) discussed the homology of the hastula, a scale-like appendage at the junction of petiole and lamina in palms (Figure 23.4a) (and also Cyclanthaceae), usually adaxial but sometimes also abaxial. Some authors have homologised the hastula with a ligule; and indeed in non-petiolate palms it is ligule-like. However, Tomlinson (1990) regarded this homology as inappropriate for several reasons, especially because a ligule normally occurs at the base of the petiole rather than the top, and also because some palms have both ligule and hastula. Tran van Nam (1974) homologised the callus of Marantaceae with the grass ligule. Other foliar appendages/elaborations (e.g. squamules, ocrea, auricles) are listed in Table 23.1; their terminology requires review after testing of homologies using molecular developmental studies. Further comparative studies on ligules and ligule homologues are needed, especially in commelinoid monocots closely related to the grasses.

Stipules are lateral appendages of the leaf base. Stipules occur in many magnoliids (e.g. *Lactoris, Houttuynia, Chloranthus*, some Piperaceae and Magnoliaceae) and many eudicot families (e.g. Begoniaceae, Betulaceae, Cistaceae, Euphorbiaceae, Fabaceae, Malpighiaceae, Malvaceae, Passifloraceae, Platanaceae, Hamamelidaceae, Rosaceae, Sterculiaceae, Tiliaceae, Ulmaceae, Urticaceae and Violaceae). Some authors believe that true stipules are absent from monocots (e.g. Tillich, 1998). However, others interpret lateral outgrowths of the leaf sheath as stipules in some monocot taxa, such as *Dioscorea* (Burkill, 1960), *Hydrocharis* and *Potamogeton* (Dahlgren *et al.*, 1985) or as stipule-like lobes in others, such as some Cyperaceae and Poaceae (Dahlgren *et al.*, 1985). Troll (1939) and many other authors have interpreted the tendrils of *Smilax* as stipules (Figure 23.5b).

It is commonly believed that dicot stipules and monocot ligules have different developmental origins (e.g. Baum, 1998). However, the few comparative ontogenetic studies available (e.g. Roth, 1949) indicate that this is often not the case; both originate from positionally similar adaxial cross meristems (see below). Interestingly, Mooney and Freeling (1997) suggested that the *liguleless* genes responsible for ligule formation in maize and rice may also be involved in dicot stipule development. Although observation of gene expression may not conclusively demonstrate homology in all cases (e.g. McLellan *et al.*, 2002), if *liguleless* genes were implicated in the development of both stipules and ligules, these structures could be considered homologous.

23.6 Leaf ontogeny

The angiosperm leaf arises near the shoot apex from a group of founder cells (Poethig, 1984) following *knotted1*-like gene suppression (reviewed by Tsiantis and Langdale, 1998). The leaf primordium (Figures 23.7, 23.8) is rather similar to the primordium of an axillary shoot; indeed, the initial meristem bulge often comprises both leaf and axillary shoot, for example in *Hesperis* (Hagemann, 1963). The shape of the shoot apex sometimes reflects the ultimate nature of the leaf, since ensiform-leafed monocot taxa with a long unifacial region and short sheathing base (e.g. *Acorus* and most Iridaceae) have a relatively shallow apical meristem compared with taxa with a short unifacial tip and long sheathing base (e.g. *Dracaena, Sansevieria, Homeria* and *Hosta*), where the apex is more convex (Stevenson, 1973; Kaplan, 1975; Rudall, 1990). According to Kaplan (1970, 1973b) differences in leaf form may be related to shoot apex dimensions: leaves formed on larger (older) axes exhibit a more protracted period of embryonic growth and hence reach a larger ultimate size because of the production of a greater cellular capital for final expansion. Most monocots have a broad stem apex with a primary thickening meristem (Rudall, 1991), a feature that occurs rarely in other angiosperms (with some exceptions, e.g. Cactaceae), although this requires further review.

Hudson and Waites (1998) suggested a possible monocot/dicot divergence in which monocot leaves are dorsiventrally flattened at emergence, whereas dicot leaves initiate as non-flattened primordia. However, this hypothesis requires further testing on more taxa; angiosperm leaf primordia are normally regarded as dorsiventral from inception (Franck, 1976; Steeves and Sussex, 1989). Most models of monocot/dicot divergence relate to subsequent development. Development of the leaf primordium into a bifacial hood-like structure (hood leaf or *Kapuzenblatt*: Roth, 1949) is extremely rapid in monocots, as the base partially or wholly encircles the stem, forming the leaf sheath, sometimes even in a closed tube. There is therefore early division into two clear zones: sheath and leaf tip. At this stage the leaf primordium forms a hood over the shoot apex (Roth, 1949), following abaxial cell elongation (e.g. Periasamy and Muruganathan, 1985).

Arber (1950) pointed out that leaf ontogeny in flowering plants is consistent with the partial shoot theory that the shoot (leaf and stem together) represents a unified structure, rather than leaf and stem each representing discrete units. The partial shoot theory builds on Zimmermann's (1930) telome theory, of which an essential

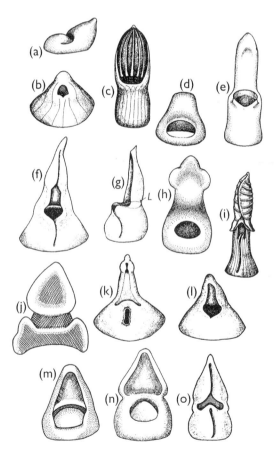

Figure 23.7 Drawings of leaf primordia in monocots (a–i, k, l) and magnoliids (j, m–o), after Roth (1949). (a) *Acorus* (Acoraceae). (b), (c) *Veratrum* (Melanthiaceae). (d), (e) *Allium* (Alliaceae). (f) *Pothos* (Araceae). (g) *Bambusa* (Poaceae). (h) *Pontederia* (Pontederiaceae). (i) *Caladenia* (Araceae). (j) *Victoria amazonica* (Nymphaeaceae). (k), (l) *Arisaema* (Araceae). (m–o). *Houttuynia* (Saururaceae). L = ligule.

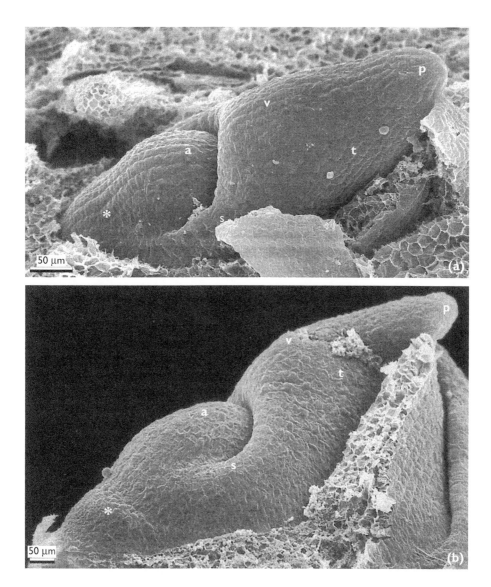

Figure 23.8 *Acorus calamus* leaf primordia. (a) is younger than (b). Key: p = precursor tip, t = transition zone, v = adaxial meristem ('ventral meristem'), s = sheath margins, a = axial apical meristem, * = prophyll of lateral shoot. Scale bar = 50 μm.

element is that leaves are primarily radial and stemlike, and dorsiventrality is a secondary development.

23.7 Phyllode theory

Kaplan (1973a, 1975) compared development of monocot and dicot leaves to test the then prevalent theory advocated by Arber (1925) and others that monocot leaves represent a modified petiole (the phyllode theory), in which the leaf blade has been suppressed. Kaplan compared monocot leaf development with that of some eudicot taxa with phyllodic leaves, such as *Acacia* and some Apiaceae (Umbelliferae). He demonstrated that ensiform leaves are indeed developmentally similar, not as a result of blade suppression but reflecting an alternative course of blade morphogenesis, where the adaxial surface remains meristematic much longer than the abaxial (i.e. by an adaxial meristem: see below). In this interpretation he differed from Roth (1949), Thielke (1949) and other workers, who regarded the unifacial leaf as an entirely new, sympodial structure, with a secondary leaf apex which suppresses the primary apex, followed by adaxial meristem (*ventralmeristem*) activity in the region below it. Kaplan considered the sympodium concept unnecessarily complicated, as it obscured unequivocal morphological relationships between bifacial and unifacial appendages. He concluded that the unifacial part of the monocot leaf is the homologue of the dicot lamina, and not a modified petiole. However, Kaplan's highly rational and factually-based rebuttal of both the phyllode theory and sympodial theory of leaf development has unintentionally tended to obscure some important observations and leaf developmental concepts proposed by Troll (1939) and Roth (1949).

23.8 Leaf base theory

The 'two zone' (*Unterblatt/Oberblatt*) model of leaf development was first proposed by Eichler (1861) and later widely adopted for all flowering plants (Troll, 1939). In this model, the proximal (basal) zone of the primordium (*Unterblatt*, or hypophyll) gives rise to the leaf sheath and stipules, and the distal (upper) zone (*Oberblatt*, or hyperphyll) to the lamina (Table 23.1). The dicot petiole may arise from either the *Unterblatt* or the *Oberblatt* (Hagemann, 1970).

The leaf base theory (Knoll, 1948) proposed that leaves of monocots are fundamentally different from those of dicots in that the bulk of the monocot lamina is derived from the proximal (lower) zone of the primordium (*Unterblatt*, or hypophyll), and the dicot lamina from the distal (upper) zone (*Oberblatt*, or hyperphyll) (Figure 23.9). Following initial establishment and early zonal demarcation, subsequent monocot leaf development is often largely restricted to basal regions by intercalary growth, hence the linear structure, whereas many dicotyledons have more pronounced marginal meristems. Kaplan (1973a) tentatively supported the leaf base theory using comparative observations on early leaf development in some eudicots (e.g. *Ribes*) and some monocots with either unifacial leaves or a *Vorläuferspitze* (e.g. *Sansevieria*: Convallariaceae *s.l.*; *Billbergia pyramidalis*: Bromeliaceae; *Heliconia*: Heliconiaceae), but cautioned that further developmental studies were needed on other groups, especially taxa lacking a *Vorläuferspitze*. The leaf base theory has subsequently become widely cited and used as a model in developmental genetics (e.g. Tsiantis *et al.*, 1999), because clear domain boundaries are indicative of genetically defined regional cell identities which direct localised growth.

However, as Kaplan (1973a) cautioned, further studies have indeed shown that some monocot laminas differ from the 'typical' monocot leaf developmental pattern in that the lamina develops from the upper (rather than lower) leaf zone. This is particularly so in basal monocots such as *Arisaema* (Periasamy and Muruganathan, 1986) and *Sagittaria* (Bloedel and Hirsch, 1979), but also in some more derived taxa

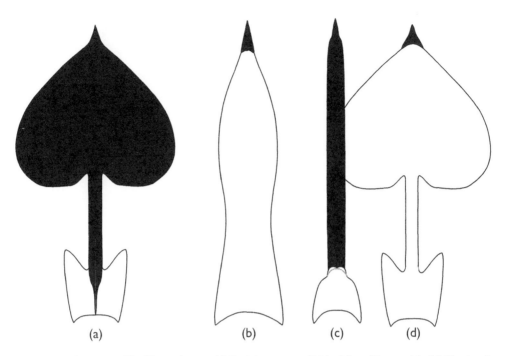

Figure 23.9 Diagram of leaf base theory. (a) Petiolate magnoliid leaf (e.g. *Houttuynia*). (b) Textbook monocot leaf (e.g. *Agave*). (c) Unifacial monocot leaf (e.g. *Acorus*). (d) Petiolate monocot leaf (e.g. *Dioscorea*). Black = hyperphyll, white = hypophyll.

such as *Smilax* (Bharathan, 1996). In these taxa the lamina is formed by the zone that forms the precursor tip in other monocots, and developmental patterns are therefore more similar to those found in some magnoliids, such as Piperales. More ontogenetic studies are required in a wider range of monocot taxa, using anatomical methods as well as SEM. Bharathan (1996) attempted to code these characters and explain these discrepancies in a phylogenetic context, suggesting that development of the lamina from the lower leaf zone may be a feature that evolved within the monocots.

23.9 Adaxial meristem and transition zone concept

While acknowledging the existence of early-established upper and lower zones, we regard their use as a model for subsequent development and character coding as unnecessarily simplistic and prefer to take a less strictly typological view, specifically that meristematic activity occurs in a highly plastic transition zone between precursor tip and sheath (Figure 23.10). The hood-like base of the leaf primordium is generally identified with the sheath, derived from the lower zone, and its apex may develop into a precursor tip, derived from the upper zone. In many monocots the precursor tip differentiates into mature cells precociously, while the rest of the leaf primordium remains meristematic (e.g. in *Dioscorea*: Periasamy and Muruganathan,

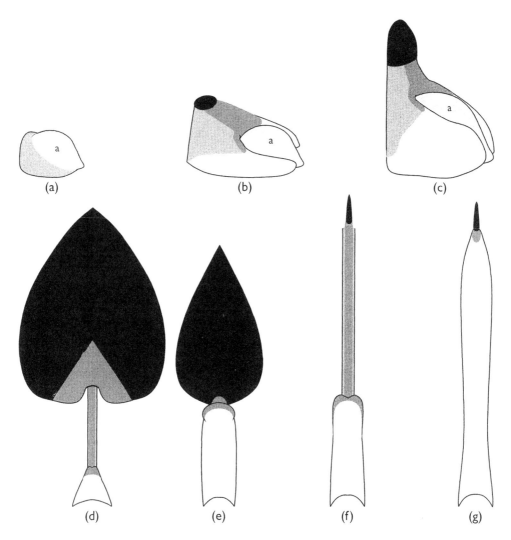

Figure 23.10 Diagram of transition zone theory. (a–c). Early stages of leaf primordia, from initiation (a) to establishment of lamina (c). (d–e) Diagrams of mature leaves. (d) Petiolate-laminate peltate (e.g. *Caladium*). (e) Non-petiolate laminate (e.g. *Pothos, Bambusa*). (f) Unifacial (e.g. *Acorus, Triglochin, Xeronema, Allium*). (g) Bifacial linear (e.g. *Agave*). a = apical meristem; black = precursor tip (and regions derived from this); light grey = transition zone (and derived regions); dark grey = adaxial meristem ('ventral meristem') and derived regions; white = sheath.

1985). Such meristematic activity often occurs in a transition zone between the precursor tip and the sheath, in which both adaxial meristem formation and intercalary growth take place; this constitutes the upper zone (*Oberblatt*) in the sense of Kaplan (1973a, 1975).

The transition zone concept is not new. According to Hagemann (1973), most morphologists (including Eichler, 1861) believe that the dicot petiole results from intercalary growth in a transition zone. Many authors (e.g. Troll, 1939; Roth, 1949) have demonstrated the existence of an adaxial meristem (*Ventralmeristem*) in this transition zone, which ultimately gives rise to the petiole and blade by intercalary and marginal growth. The adaxial meristem is part of the transition zone, bound basally to the apex of the sheath and apically to the base of the precursor tip.

A strong adaxial meristem in the bifacial leaf primordium results in formation of a unifacial petiole in some monocots (e.g. *Sagittaria, Arisaema, Calla, Caladium*: Roth, 1949; Bloedel and Hirsch, 1979; Periasamy and Muruganathan, 1986), magnoliids (e.g. *Houttuynia*: Roth, 1949) and eudicots (Franck, 1976), by suppression of the marginal meristem. In *Dioscorea pentaphylla*, which has palmately compound petiolate leaves, the leaflets initiate as separate lateral primordia on the marginal meristem, and subsequently the rachis is formed by an adaxial meristem towards the base of the primordium (Periasamy and Muruganathan, 1985).

Several authors (e.g. Roth, 1949; Troll and Meyer, 1955; Kaplan, 1973a, 1975; Stevenson, 1973) have demonstrated that an adaxial (ventral) meristem in the transition zone forms the unifacial blade in *Acorus* and other unifacial-leafed taxa where a bifacial phase between tip and sheath is suppressed. In *Acorus*, ensiformity appears only after the precursor tip and the sheath are formed, in a region between the top of the sheath and the precursor tip (i.e. the adaxial side of the transition zone). The ensiform portion is initially short, but elongates by intercalary growth, and expands dorsiventrally to the lamina, with emphasis on the adaxial side.

A central argument for the adaxial meristem and transition zone concept is the formation of one or more cross (transverse) meristems, clearly depicted as transverse bulges in some of Roth's (1949) drawings of leaf primordia (Figure 23.7). Adaxial cross meristems in the transition zone give rise to or are associated with many other leaf structures, including peltate leaves and foliar appendages such as ligules and stipules. According to Hagemann (1973), such cross meristems are derived from adaxial ingrowth of the marginal meristem. The cross meristem often appears after lateral sheath expansion by meristem incorporation, and is spatially continuous with the lateral meristem of the sheath.

In taxa such as *Hedychium* (Zingiberaceae) and *Pothos* (Araceae), the leaf primordium is constricted at the distal end of the sheath (Figure 23.4c). In others (e.g. *Potamogeton*) the lateral sheath margins expand beyond the constriction as stipule-like lobes. There is sometimes prominent meristem proliferation on the adaxial side (instead of lateral) of this constriction which can give rise to an adaxial cross zone at the distal end of the sheath, resulting in continuous sheath margins at the sheath apex or even a ligule (Roth, 1949). Similar structures are found in some dicots (Roth, 1949), where an adaxial cross zone forms between two separate lateral stipule primordia (e.g. in *Drosera* and *Victoria*), or sometimes overtops these as a single median stipule (e.g. in *Fragaria, Potentilla*, Polygonaceae, *Houttuynia, Anemopsis*). Dicot median stipules and monocot ligules may therefore be homologous, at least in some cases. However, the homologies and timing of ligule development may not be uniform among all monocot taxa. Muehlbauer *et al.* (1997) recorded ligule initiation at plastochron 7 seedling leaves of wild-type maize as a row of periclinal adaxial surface divisions at the blade-sheath boundary, which up to that stage was demarcated only by the auricles.

The peltate leaf blade is formed in the same way in monocots and dicots, and is correlated with sagittate–cordate leaves by the presence of a persistent adaxial cross meristem between the posterior leaf lobes (Roth, 1949; Bloedel and Hirsch, 1979; Periasamy and Muruganathan, 1986). This laminar cross zone is distinct from the cross zone at the apex of the sheath, but derives also from an adaxial meristem in the transition zone (Figure 23.10). According to Troll (1939), the major ontogenetic difference between unifacial and peltate leaves is that in unifacial leaves the adaxial meristem extends along the length of the upper part of the primordium, whereas in peltate leaves it remains localised (see also Franck, 1976). The palm hastula at the junction of the petiole (Figure 23.4a) may be formed in the same way, although this inference requires testing.

23.10 Conclusions

23.10.1 What is the ancestral monocot leaf type?

The ancestral monocot leaf could be unifacial linear, bifacial linear, or dicot-like, with a petiole and lamina. All three types occur in basal monocots. The putatively basal taxa of many monocot clades have unifacial (terete or ensiform) laminas. For example, unifacial laminas occur in the putatively first-branching monocot, *Acorus* (e.g. Chase *et al.*, 2000), some Alismatales (e.g. *Scheuchzeria*, *Triglochin*, *Tofieldia*), some Nartheciaceae (the basal family of Dioscoreales: Caddick *et al.*, 2002), and several lower Asparagales (e.g. Rudall *et al.*, 1997) (Table 23.2). The lamina of *Butomus* is linear bifacial (trigonal), but similar in transverse section to the lamina of some *Triglochin* species and the spathe base of *Acorus* (Figure 23.6). Hence, using existing phylogenies, the parallel-veined lamina developing from a region immediately beneath the precursor tip may indeed be the ancestral monocot type. On the other hand, laterally expanded petiolate laminas in magnoliids (e.g. Saururaceae) and basal monocots (Alismatales) are morphologically and developmentally similar. Because of the close relationship of these groups, it seems possible that these leaf structures may be homologous.

23.10.2 Are monocot leaf zones and structures homologous with those of dicots?

To answer this question, the contrasting hypotheses outlined above require further testing using developmental studies, especially on basal monocots and magnoliids. Basal monocots such as *Sagittaria* (Alismataceae) and *Agalonema* and *Pothos* (Araceae) are easily cultivated and develop rapidly, and therefore offer ideal model organisms to study varying leaf differentiation programs. In contrast to the leaves of more highly derived monocots (e.g. Poaceae), where features such as ligules, sheath closure and plications vary between taxa (see above), in basal monocots the wide range of leaf types in closely related taxa, together with leaf heteroblasty and plasticity, indicate that only a few developmental parameters have to alter to change one of these leaf shapes into another. This is illustrated by the *Acorus* series from ensiform to trigonal outlined above and *Sagittaria* leaf stages from scale leaves to linear and sagittate leaves (Bloedel and Hirsch, 1979).

23.10.3 How strict is the definition of developmental leaf zones within monocots?

The leaf base theory of monocot leaf development is a typological concept, and may be too simplistic as a model for leaf development and character coding, since the adaxial meristem region is highly plastic. Similarly, other morphological concepts such as leaf unifaciality may be misleading in a developmental context. Leaf parts become unifacial in some taxa by means of strong adaxial growth, which requires dorsiventrality. In *Acorus* the median, adaxial expansion of the upper leaf part corresponds with lateral expansion of the sheath, not with the adaxial side of the sheath, as might be expected for adaxial expansion of a bifacial leaf primordium.

Character coding of meristematic regions is notoriously difficult. Developmental genetics shows us that the same genes may be active in different parts of different organisms. Furthermore, ectopic expression of the same gene in different organisms may lead to different structural modifications. In monocots, many different leaf structures arise from closely adjacent regions of meristematic tissue within the transition zone, indicating that they may have evolved by spatial reorganisation of already existing structures, or homeosis, a process that has been explored widely in relation to floral evolution, but less frequently in vegetative structures (Bateman and DiMichele, 2002; Baum and Donoghue, 2002). Sattler's concepts of continuum morphology, which acknowledges both gradations between typical structures and partial homology, are highly relevant in the context of developmental genetics (e.g. Weston, 2000).

23.10.4 What are the homologous leaf structures?

Several authors have attempted to elucidate organ relationships by means of comparative ontogenetic studies, on the basis that similarities in early developmental stages may reveal structural homologies (e.g. Roth, 1949; Kaplan, 1975). Some of the more recent ontogenetic studies have been performed in a phylogenetic context, although mostly on floral organs (e.g. Buzgo and Endress, 2000). Integration of these approaches with developmental genetics may help to improve our understanding of organ and tissue homologies and resolve several outstanding questions, especially since organ identity is controlled by a limited number of genes (Gustafsson and Albert, 1999). Such an integrated approach may also provide reciprocal illumination for character coding and eventual resolution of higher level relationships, such as the major polytomies outlined above (Figures 23.1, 23.2). Morphological terminology also requires review after testing of homologies by means of molecular developmental studies, using expression patterns of genes such as *rough sheath2* (*PHANTASTICA*) on taxa with a peltate lamina, unifacial petiole or unifacial (ensiform) leaves.

Potentially testable hypotheses include (1) the classical theory that the precursor tip (*Vorläuferspitze*) in monocots corresponds to the lamina in other angiosperms; (2) that stipules (or stipule-like structures) in monocots are homologous to stipules in other angiosperms: they are induced by orthologues of genes responsible for stipule formation in other angiosperms; and (3) that adaxial meristem activity in the transition zone of developing leaf primordia corresponds to lateral expansion of the sheath, as a continuation of meristem incorporation. Mutations of genes known to cause defective sheaths in model organisms may cause reduction of ligule formation in ligulate leaves, or adaxial expansion in unifacial portions of ensiform leaves, both of which occur at about the same stage of leaf development (i.e. before elongation of the upper part, while the precursor tip/leaf apex becomes distinct). This is why the basal portion of the adaxial meristem occurs at the apical end of the sheath, often resulting in a structure continuous with or similar to the sheath margins (as a ligule). The expansion of the adaxial meristem into petiole and lamina is by intercalary elongation of the transition zone, after the genes for lateral meristem incorporation (sheath expansion) are active. *PHAN* and *KNOX* genes suppress margins in maize (Waites and Hudson, 1995; Waites *et al.*, 1998; Tsiantis *et al.*, 1999). A good test of these putative homologies would be whether they suppress unifaciality or margin formation in *Acorus*.

ACKNOWLEDGEMENTS

We are grateful to Favio González, Richard Bateman and Quentin Cronk for helpful discussion.

REFERENCES

Angiosperm Phylogeny Group (1998) An ordinal classification for the families of flowering plants. *Annals of the Missouri Botanical Garden*, 85, 531–553.

Arber, A. (1925) *Monocotyledons*. Cambridge University Press, Cambridge.

Arber, A. (1950) *The Natural Philosophy of Plant Form*. Cambridge University Press, Cambridge.

Bateman, R. M. and DiMichele, W. A. (2002) Generating and filtering major phenotypic novelties: neoGoldschmidtian saltation revisited, in *Developmental Genetics and Plant Evolution* (eds Q. C. B. Cronk, R. M. Bateman and J. A. Hawkins), Taylor & Francis, London, pp. 109–159.

Baum, D. A. (1998) The evolution of plant development. *Current Opinion in Plant Biology*, 1, 79–86.

Baum, D. A. and Donoghue, M. J. (2002) Transference of function, heterotopy, and the evolution of plant development, in *Developmental Genetics and Plant Evolution* (eds Q. C. B. Cronk, R. M. Bateman and J. A. Hawkins), Taylor & Francis, London, pp. 52–69.

Bharathan, G. (1996) Does the monocot mode of leaf development characterize all monocotyledons? *Aliso*, 14, 271–279.

Bloedel, C. A. and Hirsch, A. M. (1979) Developmental studies of the leaves of *Sagittaria latifolia* and their relationship to the leaf base theory of monocotyledonous leaf morphology. *Canadian Journal of Botany*, 57, 420–434.

Burkill, I. H. (1960) The organography and the evolution of Dioscoreaceae, the family of the yams. *Botanical Journal of the Linnean Society*, 56, 319–412.

Buzgo, M. and Endress, P. K. (2000) Floral structure and development of Acoraceae and its systematic relationships with basal angiosperms. *International Journal of Plant Science*, 161, 23–41.

Caddick, L. R., Rudall, P. J., Wilkin, P., Hedderson, T. A. J. and Chase, M. W. (2002) Phylogeny and circumscription of Dioscoreales inferred from morphological and molecular data. *Botanical Journal of the Linnean Society* (in press).

Camelbeke, K. and Goetghebeur, P. (1999) The ligule, a new diagnostic character in *Scleria* (Cyperaceae). *Systematics and Geography of Plants*, 68, 73–84.

Cameron, K. M. and Dickison, W. C. (1998) Foliar architecture of vanilloid orchids: insights into the evolution of reticulate venation in monocotyledons. *Botanical Journal of the Linnean Society*, 128, 45–70.

Cavot Abrigeon, E. and Lemoigne, Y. (1978) La ligule chez les pteridophytes fossiles et actuelles. *Bulletin de la Société Linnée, Lyon*, 47, 506–516, 581–594.

Chase, M. W., Soltis, D. E., Olmstead, R. G. *et al.* (1993) Phylogenetics of seed plants: an analysis of nucleotide sequences from the plastid gene *rbcL*. *Annals of the Missouri Botanical Garden*, 80, 528–580.

Chase, M. W., Soltis, D. S., Soltis, P. S., Rudall, P. J., Fay, M. F., Hahn, W. H., Sullivan, S., Joseph, J., Molvray, M., Kores, P. J., Givnish, T. J., Sytsma, K. J. and Price, J. C. (2000) Higher-level systematics of the monocotyledons: an assessment of current knowledge and a new classification, in *Monocots – Systematics and Evolution, Proceedings of the Second International Conference on the Comparative Biology of the Monocotyledons, Monocots II* (eds K. L. Wilson and D. A. Morrison), CSIRO, Collingwood, pp. 3–16.

Conover, M. V. (1983) The vegetative morphology of the reticulate-veined Liliiflorae. *Telopea*, 2, 401–412.

Cook, C. D. K. (1996) *Aquatic Plant Book*, 2nd edn. SPB Academic Publishing, The Hague.

Dahlgren, R. M. T., Clifford, H. T. and Yeo, P. F. (1985) *The Families of the Monocotyledons*. Springer-Verlag, Berlin.

Doyle, J. A. (1973) The monocotyledons: their evolution and comparative biology. V. Fossil evidence on early evolution of the monocotyledons. *Quarterly Review of Biology*, 48, 399–413.

Doyle, J. A. (1978) Origin of angiosperms. *Annual Review of Ecology and Systematics*, 9, 365–392.

Eichler, A. W. (1861) *Zur entwickelungsgeschichte des Blattes mit besonderer Berücksichigung der Nebeblatt-Bildungen*. N. G. Elwert'sche, Universitäts-Buchhandlung, Marburg.

Endress, P. K. and Igersheim, A. (1999) Gynoecium diversity and systematics of the basal eudicots. *Botanical Journal of the Linnean Society*, 130, 305–393.

Floyd, S. K. and Friedman, W. E. (2000) Evolution of endosperm developmental patterns among basal flowering plants. *International Journal of Plant Sciences*, 161, S57–S81.

Franck, D. H. (1976) The morphological interpretation of epiascidiate leaves – an historical perspective. *Botanical Review*, 42, 345–388.

Freeling, M. (1992) A conceptual framework for maize leaf development. *Developmental Biology*, 153, 44–58.

Gandolfo, M. A., Nixon, K. C. and Crepet, M. A. (2000) Monocotyledons: a review of their early Cretaceous record, in *Monocots – Systematics and Evolution, Proceedings of the Second International Conference on the Comparative Biology of the Monocotyledons, Monocots II* (eds K. L. Wilson and D. A. Morrison), CSIRO, Collingwood, pp. 44–51.

Graham, S. W. and Olmstead, R. G. (2000) Utility of 17 chloroplasat genes for inferring the phylogeny of the basal angiosperms. *American Journal of Botany*, 87, 1712–1730.

Greenwood, D. R. and Conran, J. G. (2000) The Australian Cretaceous and Tertiary monocot fossil record, in *Monocots – Systematics and Evolution, Proceedings of the Second International Conference on the Comparative Biology of the Monocotyledons, Monocots II* (eds K. L. Wilson and D. A. Morrison), CSIRO, Collingwood, pp. 52–59.

Gustafsson, M. H. G. and Albert, V. A. (1999) Inferior ovaries and angiosperm diversification, in *Molecular Systematics and Plant Evolution* (eds P. M. Hollingsworth, R. M. Bateman and R. J. Gornall), Taylor and Francis, London, pp. 403–431.

Hagemann, W. (1963) Weitere Untersuchungen zur Organisation des Sprossscheitelmeristems; der Vegetationspunkt traubiger Floreszenzen. *Botanische Jahrbücher*, 82, 273–315.

Hagemann, W. (1970) Studien zur Entwicklungsgeschichte der Angiospermenblatter. Ein Beitrag zur Klarung ihres Gestaltungsprinzips. *Botanische Jahrbücher*, 90, 297–413.

Hagemann, W. (1973) The organization of shoot development. *Revtista de Biologia (Lisbon)*, 9, 43–67.

Herendeen, P. S. and Crane, P. R. (1995) The fossil history of the monocotyledons, in *Monocotyledons: Systematics and Evolution* (eds P. J. Rudall, P. J. Cribb, D. F. Cutler and C. J. Humphries), Royal Botanic Gardens, Kew, London, pp. 1–22.

Hudson, A. and Waites, R. (1998) Early events in leaf development. *Seminars in Cell and Developmental Biology*, 9, 207–211.

Kaplan, D. R. (1970) Comparative foliar histogenesis in *Acorus calamus* and its bearing on the phyllode theory of monocotyledons leaves. *American Journal of Botany*, 57, 331–361.

Kaplan, D. R. (1973a) The monocotyledons: their evolution and comparative biology, VII. The problem of leaf morphology and evolution in the monocotyledons. *Quarterly Review of Biology*, 48, 437–457.

Kaplan, D. R. (1973b) Comparative developmental analysis of heteroblastic leaf series of axillary shoots of *Acorus calamus* L. (Araceae). *La Cellule*, 69, 253–290.

Kaplan, D. R. (1975) Comparative developmental evaluation of the morphology of unifacial leaves in the monocotyledons. *Botanische Jahrbücher*, 95, 1–105.

Knoll, F. (1948) Bau, Entwicklung und morphologische Bedeutung unifazialer Vorlauferspitzen an Monokotylenblattern. *Österreichische Botanische Zeitschrift*, 95, 163–193.

Kubitzki, K. (1998a) The families and genera of vascular plants. III. Flowering plants – Monocotyledons – Lilianae. Springer-Verlag, Berlin.

Kubitzki, K. (1998b) The families and genera of vascular plants. IV. Flowering plants – Monocotyledons – Alismatanae and Commelinanae. Springer-Verlag, Berlin.

Kumar, V., Awasthi, D. K. and Murty, Y. S. (1984) Shoot apex, leaf development and unifacial tip in *Agave wightii* Drumm. et Prain (Agavaceae). *Botanical Magazine, Tokyo*, 97, 437–446.

McLellan, T., Shephard, H. L. and Ainsworth, C. (2002) Identification of genes involved in evolutionary diversification of leaf morphology, in *Developmental Genetics and Plant Evolution* (eds Q. C. B. Cronk, R. M. Bateman and J. Hawkins), Taylor and Francis, London, pp. 315–329.

Mayo, S. J., Bogner, J. and Boyce, P. C. (1997) *The Genera of Araceae*. Royal Botanic Gardens, Kew, London.

Mooney, M. and Freeling, M. (1997) Using regulatory genes to investigate the evolution of leaf form. *Maydica*, 42, 173–184.

Muehlbauer, G. J., Fowler, J. E. and Freeling, M. (1997) Sectors expressing the homeobox gene *liguleless3* implicate a time-dependent mechanism for cell fate along the proximal-distal axis of the maize leaf. *Development*, 124, 5097–5106.

Napp Zinn, K. (1973) *Anatomie des Blattes. II. Blattanatomie der Angiospermen. A. Entwicklunggsgeschichtliche und topographische Anatomie des Angiospermenblattes, vols 1 and 2.* Gebrüder Borntraeger, Berlin and Stuttgart.

Nixon, K. C., Crepet, W. L., Stevenson, D. and Friis, E. M. (1994) A re-evaluation of seed plant phylogeny. *Annals of the Misssouri Botanical Garden*, 81, 484–533.

Periasamy, K. and Muruganathan, E. A. (1985) Ontogeny of palmately compound leaves in angiosperms. 2. *Dioscorea pentaphylla*. *Indian Botanical Contractor*, 2, 75–84.

Periasamy, K. and Muruganathan, E. A. (1986) Ontogeny of palmately compound leaves in angiosperms. 3. *Arisaema* spp. *Proceedings of the Indian Academy of Sciences*, 96, 475–486.

Philipson, W. R. (1935) The development and morphology of the ligule in grasses. *New Phytologist*, 34, 310–325.

Poethig, R. S. (1984) Cellular parameters of leaf morphogenesis in maize and tobacco, in *Contemporary Problems in Plant Anatomy* (eds R. A. White and W. C. Dickison), Academic Press, New York, pp. 235–238.

Prantl, K. (1883) Studien über Wachstum, Verzwigung und Nervatur der Laubblätter. *Berichte der Deutschen Gesellschaft*, 1, 280.

Qiu, Y.-L., Lee, J., Bernasconi-Quadroni, F., Soltis, D. E., Soltis, P. S., Zanis, M., Zimmer, E. A., Chen, Z., Savolainen, V. and Chase, W. (2000) Phylogeny of basal angiosperms: analyses of five genes from three genomes. *International Journal of Plant Sciences*, 161, S3–S27.

Reeves, G., Chase, M. W., Goldblatt, P., Rudall, P. J., Fay, M. F., Cox, A. V., LeJeune, B. and Souza-Chies, T. (2001) Molecular systematics of Iridaceae: evidence from four plastid DNA regions. *American Journal of Botany*, 88, 2074–2087.

Rodin, R. J. (1967a) Leaf anatomy of *Welwitschia*. *American Journal of Botany*, 45, 90–103.

Rodin, R. J. (1967b) Leaf structure and evolution in American species of *Gnetum*. *Phytomorphology*, 16, 56–68.

Rodin, R. J. (1967c) Ontogeny of foliage leaves in *Gnetum*. *Phytomorphology*, 17, 118–128.

Roth, I. (1949) Zur Entwicklungsgeschichte des Blattes, mit besonderer Berucksichtigung von Stipular- und Ligularbildungen. *Planta*, 37, 299–336.

Rothwell, G. W. (1982) New interpretation of the earliest conifers. *Review of Palaeobotany and Palynology*, 37, 7–28.

Rudall, P. J. (1990) Comparative leaf morphogenesis in Iridaceae. *Botanische Jahrbücher*, 112, 241–260.

Rudall, P. J. (1991) Lateral meristems and stem thickening growth in monocotyledons. *Botanical Review*, 57, 150–163.

Rudall, P. J., Furness, C. A., Chase, M. W. and Fay, M. F. (1997) Microsporogenesis and pollen sulcus type in Asparagales (Lilianae). *Canadian Journal of Botany*, 75, 408–430.

Sajo, M. G. and Rudall, P. J. (1999) Development of ensiform leaves and other vegetative structures in *Xyris*. *Botanical Journal of the Linnean Society*, 130, 171–182.

Sampson, F. B. (2000) Pollen diversity in some modern magnoliids. *International Journal of Plant Sciences*, 161, S193–S210.

Soltis, P. S., Soltis, D. E. and Chase, M. W. (1999) Angiosperm phylogeny inferred from multiple genes as a tool for comparative biology. *Nature*, 402, 402–403.

Soltis, D. E., Soltis, P. S., Chase, M. W., Mort, M. E., Albach, D. C., Zanis, M., Savolainen, V., Hahn, W. H., Hoot, S. B., Fay, M. F., Axtell, M., Swensen, S. M., Prince, L. M., Kress, W. J., Nixon, K. C. and Farris, J. S. (2000) Angiosperm phylogeny inferred from 18S rDNA, *rbcL*, and *atpB* sequences. *Botanical Journal of the Linnean Society*, 133, 381–461.

Steeves, T. A. and Sussex, I. M. (1989) *Patterns in Plant Development*. Cambridge University Press, Cambridge.

Stevenson, D. W. (1973) Phyllode theory in relation to leaf ontogeny in *Sansevieria trifasciata*. *American Journal of Botany*, 60, 387–395.

Stevenson, D. W. (1990) Morphology and systematics of the Cycadales. *Memoirs of the New York Botanical Garden*, 57, 8–55.

Stevenson, D. W., Davis, J. I., Freudenstein, J. V., Hardy, C. R., Simmons, M. P. and Specht, C. D. (2000) A phylogenetic analysis of the monocotyledons based on morphological and molecular character sets, with comments on the placement of *Acorus* and *Hydatella*, in *Monocots – Systematics and Evolution, Proceedings of the Second International Conference on the Comparative Biology of the Monocotyledons, Monocots II* (eds K. L. Wilson and D. A. Morrison), CSIRO, Collingwood, pp. 17–24.

Stewart, W. N. and Rothwell, G. W. (1993) *Palaeobotany and the Evolution of Plants*, 2nd edn. Cambridge University Press, Cambridge.

Taylor, D. W. and Hickey, L. J. (1992) Phylogenetic evidence for the herbaceous origin of angiosperms. *Plant Systematics and Evolution*, 180, 137–156.

Thielke, C. (1949) Beiträge zur Entwicklungsgeschichte unifazialer Blätter. *Planta*, 36, 154–177.

Tillich, H. J. (1995) Seedlings and systematics in monocotyledons, in *Monocotyledons:*

Systematics and Evolution (eds P. J. Rudall, P. J. Cribb, D. F. Cutler and C. J. Humphries), Royal Botanic Gardens, Kew, London, pp. 303–352.

Tillich, H. J. (1998) Development and organisation, in *The Families and Genera of Vascular Plants. III. Flowering Plants – Monocotyledons – Lilianae* (ed. K. Kubitzki), Springer-Verlag, Berlin, pp. 1–19.

Tomlinson, P. B. (1990) *The Structural Biology of Palms*. Clarendon Press, Oxford.

Tran, T. T. H. (1968) La notion de ligule latente. *Bulletin Société Botanique, France*, 115, 63–76.

Tran van Nam, T. T. H. (1971) La ligule dorsiventrale des Graminées. *Bulletin Société Botanique, France*, 118, 639–657.

Tran van Nam, T. T. H. (1974) Sur le callus des marantacees. *Bulletin Société Botanique, France*, 121, 97–108.

Troll, W. (1939) *Vergleichende Morphologie der höheren Pflanzen. I. Vegetationsorgane*. Gebrüder Borntrager, Berlin.

Troll, W. and Meyer, H. J. (1955) Entwicklungsgeschichtliche Untersuchungen über das Zustandekommen unifazialer Blattstrukturen. *Planta*, 46, 286–360.

Tsiantis, M. and Langdale, J. A. (1998) The formation of leaves. *Current Opinion in Plant Biology*, 1, 43–48.

Tsiantis, M., Schneeberger, R., Golz, J. F., Freeling, M. and Langdale, J. A. (1999) The maize *rough sheath2* gene and leaf development programs in monocot and dicot plants. *Science*, 284, 154–156.

Waites, R. and Hudson, A. (1995) *Phantastica*: a gene required for dorsoventrality of leaves in *Antirrhinum majus*. *Development*, 121, 2143–2154.

Waites, R., Selvadurai, H. R. N., Oliver, I. R. and Hudson, A. (1998) The *PHANTASTICA* gene encodes a MYB transcription factor involved in growth and dorsoventrality of lateral organs in *Antirrhinum*. *Cell*, 93, 779–789.

Weston, P. (2000) Process morphology from a cladistic perspective, in *Homology and Systematics* (eds R. Scotland and R. T. Pennington), Taylor and Francis, London, pp. 124–144.

Williams-Carrier, R. E., Yung, S. L., Hake, S. and Lemaux, P. E. (1997) Ectopic expression of the maize *kn1* gene phenocopies the Hooded mutant of barley. *Development*, 124, 3737–3745.

Zimmermann, W. (1930) *Die Phylogenie der Pflanzen: Ein Überblick über Tatsachen und Proleme*. Fischer, Jena.

Chapter 24

Diatoms: the evolution of morphogenetic complexity in single-celled plants

Eileen J. Cox

ABSTRACT

Aspects of cell wall development within the diatoms are reviewed, with particular reference to the taxonomic interpretation of shared or contrasting morphologies and symmetry. Evidence for shared developmental pathways within the Naviculaceae is presented. However, the same taxa lie in different positions on the suggested developmental trajectories, depending on the structures being considered. In another case, similar terminal morphologies result from contrasting ontogenies. The development and control of cell symmetry is discussed in relation to current knowledge of cytoplasmic organisation and constraints on wall development. Members of the Cymbellales are presented as a suitable group in which to investigate the evolution of contrasting valve symmetries, as well as modifications in striation and raphe structure. The potential of diatoms as model organisms to further our understanding of the role of the cytoskeleton in morphogenesis is explored.

24.1 Introduction

Diatoms are not only key primary producers in many aquatic ecosystems (Werner, 1977) but are also the most species-rich algal group (Mann and Droop, 1996). They are characterised by the possession of highly structured, bipartite, siliceous cell walls and occur throughout the world in marine and fresh waters, planktonic and benthic habitats. The beauty and regularity of their cell wall morphology fascinated early light microscopists, so diatoms are, in this respect, perhaps the best-known group of unicellular algae. They are also increasingly important indicators of water quality, the reliability of which depends on accurate identification, preferably to species level (Cox, 1991; Prygiel, 1991). Diatom recognition and classification still rest largely on features of wall morphology (Round *et al.*, 1990), and tend to assume that wall morphology is constant and species-specific. However, the ways in which wall structure and shape are determined and their development regulated remain largely unknown. This chapter reviews some recent data on biraphid diatoms and presents ontogenetic and evolutionary hypotheses that might be amenable to testing. In particular, clarifying the role of the cytoskeleton and its relationship to the developing cell walls may be crucial to understanding how wall morphogenesis is controlled. Some of the peculiarities of diatom structure and reproduction may also offer novel means to investigate the developmental genetics

In *Developmental Genetics and Plant Evolution* (2002) (eds Q. C. B. Cronk, R. M. Bateman and J. A. Hawkins), Taylor & Francis, London, pp. 459–492.

of the actin and microtubular cytoskeleton in relation to cell polarity and wall deposition.

24.2 Valve structure and formation

24.2.1 Structure

One of the major distinctions within diatoms rests on the symmetry of the valves. (The cell wall, or frustule, comprises two valves and a series of linking elements, the girdle bands.) Two basic symmetry patterns can be recognised: radial, from a central 'point' or annulus of symmetry, and bilateral, with a more or less axial pattern centre. These correspond to the traditional major systematic separation of centric (Centrales) and pennate (Pennales) diatoms. The second group can be further subdivided into those with a raphe system (a pair of slits along the axis) and those without, separating motile from non-motile groups (for a good general account of diatom structure and symmetry see Round *et al.*, 1990). Despite the differences in basic pattern, all diatoms show semi-conservative wall formation (daughter cells inherit one parental valve and form one new valve after mitosis) within silica deposition vesicles (SDVs), and several ontogenetic patterns can be recognised. Pickett-Heaps *et al.* (1990) provided a detailed account of the cell biology of valve morphogenesis, but did not address the evolution of different structures or symmetries in any systematic context.

24.2.2 Vegetative wall formation

Except under infrequent circumstances (e.g. initial cell, drought or nutrient deficiency), cell wall formation occurs after mitosis and cytokinesis (Geitler, 1963; von Stosch and Kowallik, 1969) within membrane-bound vesicles (SDVs) inside the parent cell (Drum and Pankratz, 1964). (On the rare occasions when cytokinesis does not occur, one of the nuclei degenerates: see Section 24.2.4.) Two sibling valves are formed simultaneously, one in each new cell, so that all diatom frustules have one older (epi-) and one younger (hypo-) valve. The hypovalves may also be very slightly smaller than the epivalves, because they are formed within the confines of the parent cell. In the long term, this usually leads to reduction in the average size of cells in a population and may also modify valve shape (Geitler, 1932). Sexual reproduction leads to the restoration of maximum cell size (Figure 24.1).

In pennate diatoms, the SDV begins as a long, very narrow tube, in a specific position related to the ultimate symmetry of the valve (Pickett-Heaps *et al.*, 1990). A rib of silica is rapidly formed in this tube, followed, in raphid diatoms, by a smaller rib beside it. The space between these double ribs will become the raphe slit as the first rib recurves to meet the shorter second rib (Figure 24.2). Thus, there is an inherent asymmetry in all raphid diatoms; the Voigt discontinuity (an irregularity in the striae on the secondary side of the valve) marks the position where the two ribs meet (Mann, 1981b).

As the raphe slits form, lateral outgrowths (virgae) begin to develop and subsequently cross-connections (vimines) are formed, defining pores within the striae (Figure 24.3). Various types of pore occlusion (Ross *et al.*, 1979; Mann, 1981a) and

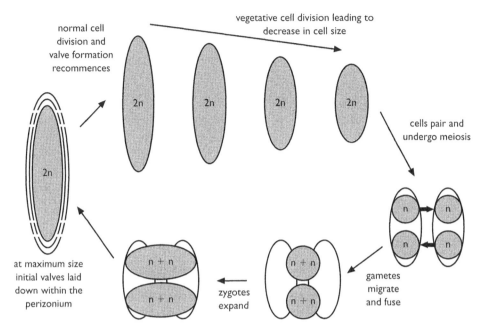

Figure 24.1 Diagram showing the life history of a pennate diatom. Vegetative division leads to a decrease in cell size. Maximum cell size is restored after sexual reproduction. With the exception of the initial cell which lays new valves down within the perizonium (far left of diagram), valves are formed within the confines of a parent cell. Note that fusion of meiotic nuclei only occurs during or after auxospore expansion.

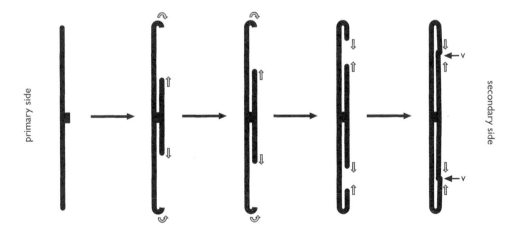

Figure 24.2 Diagram showing the formation of the raphe slits in naviculoid diatoms. The small black arrows (v) indicate the position of the Voigt discontinuity, where the ribs meet.

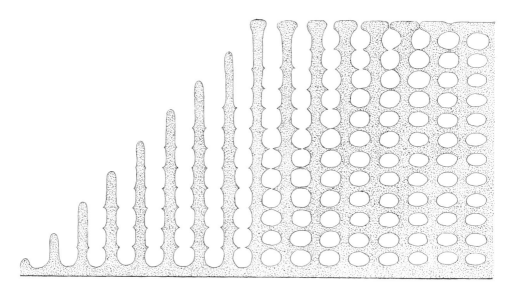

Figure 24.3 Diagram showing the sequence of virga and vimines formation (reading from left to right).

pore arrangements within striae (Cox and Ross, 1981) have been recognised and are used as taxonomic characters (see Section 24.3). Thus, Mann (1981a) discriminated between finely perforate hymenate occlusions (=ricae *sensu* Ross *et al.*, 1979) which occupy an entire pore, and volate occlusions, where a flap of silica extends from the pore wall. Mann first stated that these two occlusion types do not occur together in a single frustule (Mann, 1981a, 1982), when the terms refer to the most delicate structures in a pore, but then gave examples of species possessing both types of occlusion! Coarser occlusions, termed cribra, have also been recognised (Ross *et al.*, 1979), with which both volate and hymenate occlusions may be found. Given the possible combinations of occlusion type, and the fact that each type of occlusion can occur across either the outer or inner pore opening, there is scope for considerable variation in pore substructure in diatoms. Furthermore, within the raphid diatoms, variation in the precise sequence of valve formation, and in the predominant direction of silica deposition, has recently been documented (Cox, 1999b). In the majority of raphid species studied, silica deposition was mainly distal to proximal (i.e. towards the cytoplasm), but in *Haslea* sp. it was mainly proximal to distal (cf. predominantly proximal to distal deposition in centric diatoms: Schmid and Volcani, 1983; Schmid, 1986; Pickett-Heaps *et al.*, 1990). However, in each case, the finest pore occlusions, the hymenes, were the last part of the valve to be formed.

24.2.3 Cytological controls of pattern production in diatoms

All eukaryotic cells possess three-dimensional networks of cytoskeletal components involving microtubules and microfilaments. Microtubules are known to affect the control of cell shape and morphogenesis in animal and plant cells but, compared with naked protists, diatoms have a rather poor cytoskeleton (Schmid, 1994), presumably because silica is providing mechanical strength and support. This can also be interpreted as an energetic economy; energy is only required during wall formation, not for its maintenance, and silica is energetically less costly but metabolically less versatile as a wall material than organic alternatives (Raven, 1983). This implies that, after completion of the cell wall (microfilament bundles adjacent to the SDV during silicification, persisting until completion of valve formation: Pickett-Heaps *et al.*, 1979), the diatom cytoskeleton is primarily involved with cytoplasmic processes. Within the frustule, the protoplast attaches to the valves at specific points by locally thickened organic layers and electron dense material (Schmid, 1994), remaining attached at these points if the cell is plasmolysed. In raphid diatoms, the raphe is the region where the protoplast is most firmly anchored. Schmid (1994) also suggests that morphological features in the rigid wall serve as spatial determinants that allow the protoplast to order its three-dimensional activity.

In diatoms the microtubular cytoskeleton components are depolymerised during prometaphase and must be re-established after cleavage and prior to, or during, valve formation (Schmid, 1994). The last formed girdle band of the hypovalve seems to be the reference region for the cytoskeleton, establishing the cleavage furrow, and thus analogous to the microtubule (MT) preprophase band in higher plant cells. After postcleavage in pennate diatoms, the SDV appears in the region equivalent to the former spindle equator, usually at the site of the prospective valve centre (Edgar and Pickett-Heaps, 1984). Valve morphogenesis can then proceed, the outline, topography and pattern of the valves being controlled by different factors.

While the valve outline of vegetative cells is determined by the mother cell wall (cf. Figure 24.4), valve topography is moulded by the cleavage furrow and shaped by a combination of turgor, local contractions and tension between the new membrane and cytoplasmic cortex. Sibling valve development may be interactive or non-interactive (Mann, 1984a, 1994), according to the degree, or lack, of mutual influence they exert on each other. Valve pattern depends upon the cell's ability to localise silica to form different domains.

Various hypotheses have been presented to explain pattern development in the diatoms. Pickett-Heaps *et al.* (1979) suggested that silica might be precipitated onto fine polysaccharide filaments, whereas Schmid (1980) described expansion of the SDV within a fibrillar scaffold. Meanwhile, Robinson and Sullivan (1987) postulated a model that did not require a polysaccharide matrix, but invoked control by the cell cytoskeleton. The latter would move proteins containing silica nucleation sites (spanning the silicalemma) around in a genetically determined manner, similar to the spectrin/ankyrin/band 3 system in red blood cells (Branton *et al.*, 1981). There is evidence that fibres or other organic material closely associated with, or occasionally within, the SDV prevent silica precipitation in localised areas that then become slits or holes in the valve wall (Pickett-Heaps *et al.*, 1979; Edgar and Pickett-Heaps, 1984; Pickett-Heaps *et al.*, 1990; Pickett-Heaps, 1998). The raphe fibre, the template for the raphe slit, is associated with MTs along its entire length, such that the application of MT-inhibitors leads to a disrupted raphe slit (Schmid, 1980). Chambers may also be moulded around mitochondria or spacer vesicles (Pickett-Heaps *et al.*, 1979; Edgar, 1980; Schmid *et al.*, 1996). The close association between the SDV of the chamber roof and mitochondria in developing *Pinnularia* valves is presumed to prevent silicification in this area (Schmid *et al.*, 1996). On the other hand,

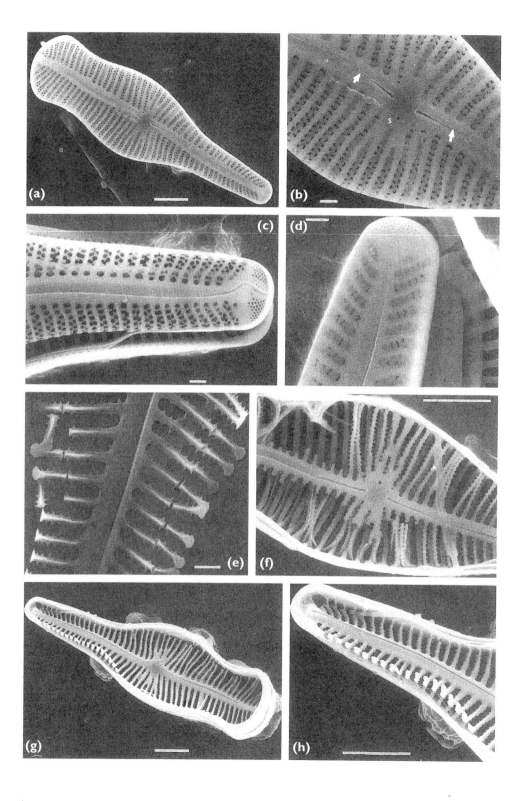

Gordon and Drum (1994) suggest that chemical and physical considerations can account for pattern in diatoms, although their models for silica deposition match the morphology of centric diatoms better than pennate ones.

More recently, there have been investigations of the organic components in, and associated with, diatom walls (Kröger and Sumper, 1998; van de Poll *et al.*, 1999; Kröger *et al.*, 2000). Kröger *et al.* (2000) showed that long-chain polyamines and silaffins induce rapid *in vitro* precipitation of silica from silicic acid solution. Different silica nanosphere morphologies were generated by different chain length polyamines and by the synergistic interaction of these polyamines and silaffins; however, none of these arrays resembled the patterns of pores found in diatoms. Kröger and Sumper (1998) suggested that the formation of silica biominerals requires unique protein structures and properties, and that diatoms have evolved novel amino acids and protein domains to deal with this. Schmid *et al.* (1996) argued against a purely chemical–physical model, suggesting that it is inconsistent with the tight association of the SDV membrane around the condensing silica throughout wall development. They also suggested that MTs may play a stabilising role, while (actin) filaments and associated proteins, in combination with the cleavage plasmalemma and its skeleton, are crucial to the patterning process. The fact that silicification can be uncoupled from pattern formation in wall elements supports the hypothesis that pattern-creation is mediated by the plasma membrane and the cytoskeleton (Schmid *et al.*, 1996).

24.2.4 Peculiarities of initial cell formation

Unlike vegetative valves that are formed within the parent cell after cytokinesis, initial valves are laid down sequentially (epivalve then hypovalve) within the perizonium, following acytokinetic mitoses in the fully expanded auxospore (Thaler, 1972; Mann, 1989a). The organic wall of the zygote (which may bear silica scales: Kaczmarska *et al.*, 2000) splits as the auxospore begins to expand, leaving caps at the poles of the cell, while lightly silicified, transverse perizonial rings are formed around the elongating auxospore. Whereas the central perizonial ring (primary band) is usually a complete hoop, the others are split rings, forming a suture along one side of the auxospore. The initial epivalve is formed under this suture, opposite the initial hypovalve (Mann, 1982, 1984d, 1989a). Mitosis precedes the formation of each valve, but this is not accompanied by cytokinesis and one nucleus is lost

---

Figure 24.4 SEMs of valves of *Gomphonema truncatum* at different stages of development. (a) External view of mature valve, showing broader head and narrower foot poles. (b) Detail of central part of valve, showing raphe slits (arrows), stigma (s) and striae. Note that the curved pores are arranged in single rows (uniseriate) near the raphe slit, but in offset double rows (biseriate) nearer the valve margin. (c) External view of foot pole as the vimines are fusing. (d) Mature foot pole. (e) Internal view showing early stage of development; virgae have extended the width of the valve face and vimines are starting to form. (f) Central area showing external view of developing hypovalve (some of the virgae have broken), with internal view of mature epivalve below. (g) Internal view of developing valve showing the constraints on shape of the epivalve. (h) Detail showing how virgae are deflected when they abut the mature epivalve. Scale bars represent 5 μm for (a) and (f–h) and 1 μm for (b–e).

after each mitosis (Geitler, 1963; von Stosch and Kowallik, 1969). The functional nucleus is closely associated with the forming valve, as in vegetative cells. The nucleus must therefore traverse the cell between the formation of the initial epi- and hypovalves. It has been suggested (Schmid, 1994) that acytokinetic mitoses precede initial valve formation because cytokinesis is precluded by the lack of cytoskeletal anchors in the auxospore (plant cells require a functional surrounding wall in order to divide: Meyer and Abel, 1975; Schindler et al., 1989; Suzuki et al., 1998). A less well-organised cytoskeleton may exert fewer constraints on valve morphogenesis, while differences in the cytoplasmic environment under the auxospore wall compared to within a vegetative cell may also contribute to the disparities between the initial and normal vegetative valves (see Section 24.5.2).

24.3 Systematic significance accorded to morphological variation

With the development of SEM it was recognised that the ultrastructure of diatom frustules, especially the raphe slits and their associated thickenings, pore types and arrangement, can vary within taxa sharing the same symmetry, or be common to different symmetry groups. However, in spite of some attempts to evaluate structure and symmetry as taxonomic criteria (Cox, 1979; Kociolek and Stoermer, 1988), only gradually are shape and symmetry being seen as poorer guides to relationships than structure (Cox, 1982; Mann and Stickle, 1997; Prasad et al., 2000). Protoplast characters, such as chloroplasts and pyrenoids, provide additional support for recent systematic revisions (Cox, 1987, 1993; Mann, 1989b; Cox and Williams, 2000).

The form of the raphe sternum (axial rib of silica containing the raphe fissures) and the paths of terminal fissures are usually regarded as generic characters, though authors differ in the range of structure they consider permissible within a genus (Lange-Bertalot et al., 1996; Witkowski et al., 1998; Cox, 1999a). For example, within the Naviculales, Navicula has internal raphe slits that open laterally in a narrow rib for most of their length, usually with an accessory rib present beside the raphe on the primary side of the valve (Figure 24.5a–g) (Cox, 1999a). Haslea and Gyrosigma have the same type of raphe, but in Haslea the accessory rib usually develops a flange that partially overlaps the raphe rib (Figure 24.5h) (Cox, 1977; Massé et al., 2001). There may also be a shorter central rib on the secondary side. In Gyrosigma, on the other hand, the accessory rib does not overlap the raphe rib and there is usually a shorter rib on the secondary side (Figure 24.5i) (Cox, 1977, 1979).

Similarly, the type of pore occlusion (e.g. hymenate, cribrate or volate: Mann, 1981a) is usually characteristic of particular genera, but the significance of uni- or biseriate striae (one or two rows of pores, respectively, between pairs of virgae: cf. Cox and Ross, 1981) is more contentious. Dawson (1974) argued that some taxa with biseriate striae should be removed from Gomphonema and placed in Gomphoneis. However, most authors (Krammer and Lange-Bertalot, 1986; Kociolek and Stoermer, 1989) do not recognise this generic transfer because Gomphoneis is usually defined by having marginal laminae as well as biseriate striae. Gomphonema thus includes species with uniseriate or biseriate striae, while some species have striae that are uniseriate near the raphe slits but biseriate towards the valve margins (Figure 24.4a–d: Cox, 1999b). Lange-Bertalot et al. (1996) removed several taxa

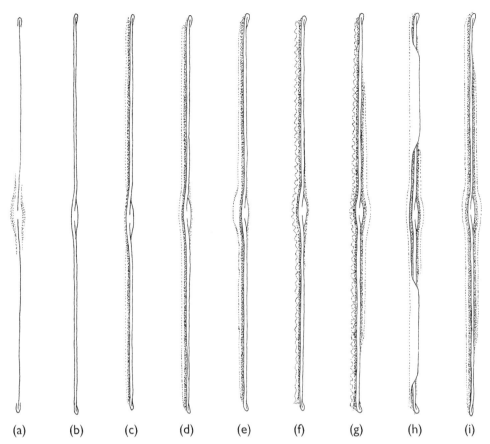

(a)　(b)　(c)　(d)　(e)　(f)　(g)　(h)　(i)

Figure 24.5 Drawings (not to scale) to show variation in internal raphe structure and accessory ribs in the Naviculales. (a–g). Variation in distinctness of the raphe sternum and accompanying accessory ribs in *Navicula*. (a) Only slight thickening around the central raphe endings and at the polar helictoglossae, no accessory rib. (b–f). The raphe sternum is well defined, but the raphe slit opens towards the secondary side, except at the centre and poles; accessory rib development is beginning as two partial ribs (c), then a continuous rib (d), which may become thicker at the centre (e), or impinge on the virgae and have an undulate distal margin (f), or with a additional slight rib on the secondary side (g). (h) Raphe sternum and accessory ribs in *Haslea*, showing how the primary accessory rib may form a flange that overlaps part of the raphe. (i) Raphe sternum and accessory ribs in *Gyrosigma* (note that the sigmoid curvature of the raphe is not shown).

('*Hippodonta*' spp.) from *Navicula* (not recognised by Cox, 1999a), some of which have biseriate striae, unlike *Navicula*, but others have uniseriate striae, like *Navicula*. Similar double pores also occasionally occur in *Navicula* species (Cox, 1999a). Within the araphid pennates there is further evidence that the switch between uni- and biseriate striae occurs fairly easily. Several *Synedra* species have a mixture of uniseriate and biseriate striae (Jüttner *et al.*, 2000), while clones of '*Centronella reicheltii*' with uniseriate striae have been shown to produce valves with biseriate striae, or to mix the two arrangements in a single valve (Schmid, 1997).

Meanwhile, one of the features separating *Haslea* from *Navicula* is that the external pore openings form continuous slits in the former but not in the latter (Cox, 1979; Round *et al.*, 1990).

Other morphological features that have been used systematically include the presence or absence of hyaline (i.e. solid) or thickened hyaline areas (fascia and stauros), lateral and axial laminae, alveolate striae, longitudinal canals and apical pore fields. There has been an implicit assumption that if two structures appear the same in mature valves they are homologous. However, a cladistic study of taxa possessing a stauros (a transverse thickened hyaline area in LM) (Cox and Williams, 2000) revealed that such taxa are not closely related to each other, indicating that the stauros is not a homologous character. In addition, morphogenetic studies (Cox, 2001) have shown that transverse hyaline areas can be formed in different ways (Figure 24.6), supporting Cleve's (1894) distinction between a fascia and a stauros.

24.4 Morphological variation may suggest developmental pathways

The morphological variation in some of the raphe and pore characters mentioned above could be interpreted as stages along developmental pathways. Within *Navicula*, the thickening of the raphe sternum ranges from slight to strong, from narrow throughout to broader at the centre. The accessory rib varies from absent to discontinuous to continuous to thicker at the centre (Figure 24.5). Morphogenetic evidence from studies of *Haslea* shows that the accessory rib begins to form alongside each raphe slit once the virgae and vimines are complete. Initially, the rib is interrupted at the centre of the valve, but later becomes continuous (Cox, 1999b). Unpublished

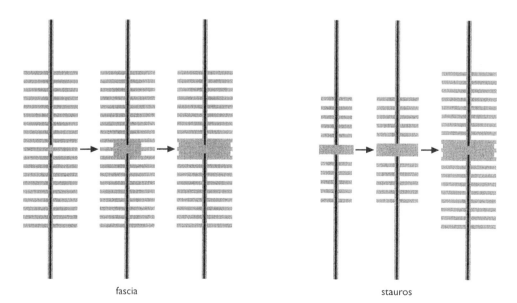

fascia   stauros

Figure 24.6 Diagram showing the developmental routes by which transverse hyaline areas (fascia and stauros) may be formed.

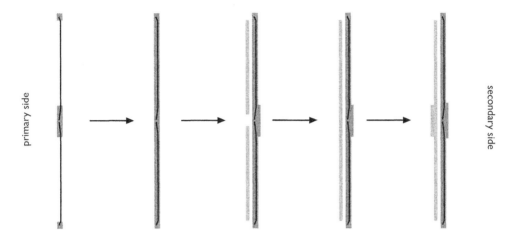

Figure 24.7 Diagram showing the hypothetical sequence of raphe sternum and accessory rib development in the Naviculales.

data on *Navicula gregaria* also indicate that the accessory rib begins alongside each raphe slit and then thickens towards the centre of the valve. This would suggest that the morphological range absent > discontinuous > continuous > thicker at the centre reflects the developmental pathway of the accessory rib (Figure 24.7). Assuming that the rib on the secondary side of the raphe is formed later than that on the primary side, Figure 24.8 shows the position of some naviculoid genera along the possible developmental pathway of the raphe system.

With respect to the external pore openings, each pore has its own external linear opening in most *Navicula* species, whereas in some *Gyrosigma* species the external openings can extend over several pores, and in *Haslea* external longitudinal slits extend throughout the length of the valves over many pores (Figure 24.8). A few *Navicula* and most '*Hippodonta*' species have double pores. Studies on *Haslea* (Cox, 1999b) show that silica is added along the outer surface of the fused vimines and

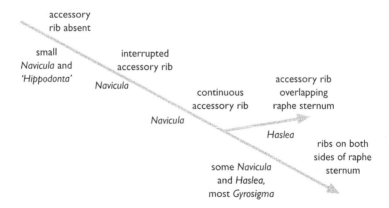

Figure 24.8 Postulated developmental pathway for raphe and rib development in the Naviculales.

then accretes to form strips with continuous longitudinal slits between them. If some external cross-connections are formed, this could allow two or three areolae (pores) to share a single external slit. Cross-connections over every virga would produce a slit for each areola, and connections across the middle of each areola would produce double pores. Some *Navicula* species (Cox, 1999a) have longitudinal grooves between adjacent pore openings that suggest they could be formed in this way, while the occurrence of occasional double pores suggests that additional cross-connections can readily occur. It is therefore possible to postulate a developmental pathway from *Haslea* to '*Hippodonta*' with respect to pore structure (Figure 24.9). However, it should be noted that this trajectory is in the opposite direction to that for accessory rib development with respect to the genera.

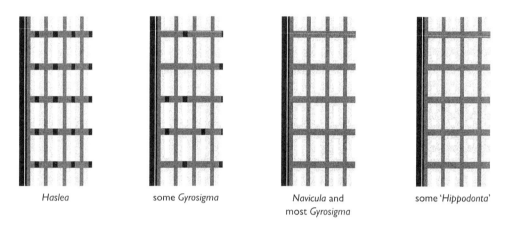

| *Haslea* | some *Gyrosigma* | *Navicula* and
most *Gyrosigma* | some '*Hippodonta*' |

Figure 24.9 Diagram showing the different degrees of external pore occlusion within the Naviculales. The network of virgae and vimines is shown in black, with the accretion of silica to form the linear external openings in grey.

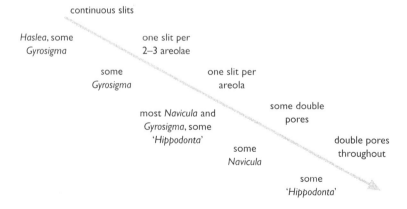

continuous slits

Haslea, some *Gyrosigma*

one slit per 2–3 areolae

some *Gyrosigma*

one slit per areola

most *Navicula* and *Gyrosigma*, some '*Hippodonta*'

some double pores

some *Navicula*

double pores throughout

some '*Hippodonta*'

Figure 24.10 Postulated developmental pathway for pore development in the Naviculales.

24.5 Symmetry and structure

24.5.1 Shifts in symmetry

As mentioned above, the discovery that valve construction can be similar across different symmetry groups, and the recognition that raphid diatoms are intrinsically asymmetrical about their apical axis, are leading to the re-evaluation of symmetry as a taxonomic criterion. Thus, bilaterally symmetrical and dorsiventral taxa are now included in the same genus (Cox, 1982; Round *et al.*, 1990; Prasad *et al.*, 2000), and taxa with shared symmetry are being separated into different genera, based on valve structure (Round *et al.*, 1990). Cells can also be asymmetrical about their transverse plane, producing heteropolar cells in which one end is broader than the other, often accompanied by the presence of specialised pore fields at the narrower end. It is assumed that the primitive condition is isopolar and bilaterally symmetrical.

Mann (1983) recognised that, with respect to the inherent asymmetry of the raphid diatom valve, two types of frustule symmetry occur in raphid diatoms. These are cis symmetry, when the primary sides of both valves are on the same side of the frustule (mirror symmetry), and trans symmetry, when the primary sides of the valves are on opposite sides of the frustule (diagonal symmetry) (Figure 24.11). Mann investigated cis and trans symmetry in a range of raphid diatoms and found that particular species produce either all cis cells or a mixture of cis and trans cells. A species cannot form only trans cells because sibling valves form in the same orientation; their primary sides are on the same side of the frustule, because the daughter nuclei lie on the same side of the parent cell (Figure 24.11). The production of cis only or cis plus trans cells is linked to whether the nucleus always divides on the same side of the cell or oscillates between one side and the other (Mann, 1984b). If the nucleus does not oscillate, only cis cells will be formed and the potential to develop stronger valve asymmetry is introduced. Of the taxa investigated by Mann (1983), cis cells occur predominantly in the Cymbellales, which includes several genera with markedly or slightly dorsiventral valves (Table 24.1). Cis cells were also observed in the similarly laterally asymmetrical genus *Hantzschia* (Bacillariales), as well as the laterally symmetrical *Lyrella* and *Petroneis* (Lyrellales). In discussing the development of dorsiventrality in otherwise symmetrical genera, Mann and Stickle (1997) suggest that the evolution of dorsiventrality is unlikely to occur without substantial reorganisation of cell cycle controls or internal polarity. They predict that

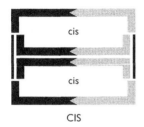 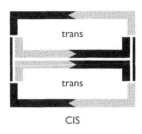 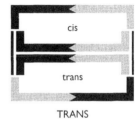

Figure 24.11 Diagram showing the formation of different types of cell symmetry. Cell symmetry of the parent cells is indicated in capital letters, that of the daughter cells in lower case.

such constraints should be absent in genera such as *Lyrella*, which includes at least one laterally asymmetrical species, because it exhibits constant cis symmetry and thus seems preadapted to evolving dorsiventrality. In other words, the absence of nuclear oscillation is the essential precursor to developing overt dorsiventrality. Several genera within the Cymbellales have constant cis symmetry but laterally symmetrical valve outlines (e.g. *Anomoeoneis*, *Placoneis* and *Gomphonema*). It would also be interesting to investigate whether all members of genera such as *Climaconeis*, which has some dorsiventral representatives (ventral side is the primary side), have constant cis symmetry or only those which have developed lateral asymmetry. The girdle differs in width from side to side in genera such as *Cymbella* and *Encyonema*, whereas this does not occur in *Climaconeis* (D. Mann, pers. comm. 2001). The latter could therefore demonstrate cis/trans dimorphism.

The inherent dorsiventrality of the raphe system results in the external terminal fissures and the Voigt discontinuities occurring on the secondary side of the valve (Figure 24.2). The accompanying polarity of the protoplast in non-oscillating cells results in the nucleus lying on the primary side of the valves, and in taxa with a single chloroplast, the centre of the chloroplast (and the pyrenoid) on the secondary side. However, with respect to valve outline, the more convex margin in a dorsiventral species can be either the primary or the secondary side of the valve. Thus, in *Encyonema* the primary side is wider than the secondary, whereas in *Cymbella* the converse is the case (Figure 24.13) (Mann, 1981b). Because the positions of the nucleus and chloroplast are fixed in relation to the primary side of the valve, this means that the centre of the chloroplast lies against the less convex margin in *Encyonema*, the more convex in *Cymbella*. Hence, the primary side of a valve is not inevitably broader or narrower than the secondary, but is determined by some other developmental control. MTs and the MT organising centre are closely associated with both the nucleus and the primary side of the developing valve. The MTOC is located close to the SDV as the raphe fissure is formed, but also tightly associated with the nucleus via MTs (Pickett-Heaps *et al.*, 1990). Mann (1984a) suggested that the cymbelloid shape might have arisen only once, the contrasting ventral and dorsal curvatures between *Encyonema* and *Cymbella* respectively, arising by polarity reversals in which the protoplast, and orientation of the valve secreting machinery 'swapped sides'. However, differences in valve structure between *Encyonema* and *Cymbella* would suggest that, once a particular lateral asymmetry has been established, it is retained.

Another change in symmetry within the raphid diatoms is from isopolar to heteropolar, where one apex is broader than the other; such cells often have modified pores at their narrower end where they attach to the substratum. Again, most heteropolar raphid diatoms belong to the Cymbellales (Table 24.2), but they also include representatives of some other orders, such as *Peronia* and *Actinella* in the Eunotiales, *Mastogloia* in the Mastogloiales, *Pseudogomphonema* in the Naviculales, *Gomphonitzschia* and *Gomphotheca* in the Bacillariales, *Rhopalodia* in the Rhopalodiales and *Surirella* in the Surirellales. Whereas the development of lateral asymmetry seems to be linked to the loss of nuclear oscillation, in theory heteropolarity should not be dependent on cis symmetry, although both *Gomphonema* and *Didymosphenia* show cis symmetry. However, cell polarity must be established in order to produce one narrower end with specialised pores (Figure 24.4) through which polysaccharide is secreted for attachment. Once a heteropolar cell is formed with specialised secretory pores at

Table 24.1 Cis and trans frustule symmetry in raphid genera (cf. Figure 24.11), showing systematic position (based on Round *et al.*, 1990) and valve shape. Data on cell symmetry taken from Mann (1983) and Mann and Stickle (1988)

Order	Family	Genus	Valve shape	Frustule symmetry
Lyrellales	Lyrellaceae	*Lyrella*	symmetrical	cis
		Petroneis	symmetrical	cis
Mastogloiales	Mastogloiaceae	*Mastogloia*	symmetrical	cis + trans
		Aneumastus	symmetrical	cis + trans
Cymbellales	Rhoicospheniaceae	*Rhoicosphenia*	heteropolar	cis
	Anomoeoneidaceae	*Anomoeoneis*	symmetrical	cis
	Cymbellaceae	*Placoneis*	symmetrical	cis
		Cymbella	dorsiventral	cis
		Encyonema	dorsiventral	cis
	Gomphonemataceae	*Gomphonema*	heteropolar	cis
		Didymosphenia	heteropolar[1]	cis
Naviculales	Cavinulaceae	*Cavinula*	symmetrical	cis + trans
	Amphipleuraceae	*Amphipleura*	symmetrical	cis + trans
		Frustulia	symmetrical	cis + trans
	Brachysiraceae	*Brachysira*	symmetrical	cis + trans
	Neidaceae	*Neidium*	symmetrical	cis + trans
	Sellaphoraceae	*Sellaphora*	symmetrical	cis + trans
		Fallacia	symmetrical	cis + trans
	Pinnulariaceae	*Pinnularia*	symmetrical	cis + trans
		Caloneis	symmetrical	cis + trans
		Pinnunavis	symmetrical	cis + trans
	Diploneidaceae	*Diploneis*	symmetrical	cis + trans
	Naviculaceae	*Navicula*	symmetrical	cis + trans[2]
	Pleurosigmataceae	*Pleurosigma*	symmetrical	cis + trans
		Gyrosigma	symmetrical	cis + trans
	Stauroneidaceae	*Stauroneis*	symmetrical	cis + trans
		Craticula	symmetrical	cis + trans
Bacillariales	Bacillariaceae	*Hantzschia*	dorsiventral	cis
		Nitzschia	symmetrical	cis + trans

Notes:
1 Valve may also be slightly dorsiventral.
2 No trans cells were recorded for *Navicula capitata* but only three cis cells were found among 200 cells examined.

one end, its offspring are almost certainly constrained to maintain that polarity, because of the probable localisation of cytoplasmic organelles associated with polysaccharide secretion. Polarity in zygotes of brown algae has been linked to pH gradients and polar distribution of mitochondria (Gibbon and Kropf, 1991), which may in turn be linked to the point of sperm entry, although subject to modification by external factors (Kropf *et al.*, 1999; Cove, 2000). Immediately after auxospore formation in diatoms, there is no obvious extrinsic determinant of cell polarity, yet initial cells, while often somewhat irregular, are usually already heteropolar in shape (Mann, 1984d; Passy-Tolar and Lowe, 1995). However, initial cells of *Didymosphenia* have been observed in which the apices are virtually the same size, although differentiation of the apical pore field is discernible at one end (Cox, unpubl. obs.).

Table 24.2 Systematic position (based on Round *et al.*, 1990) of raphid genera mentioned in this paper. Valve symmetry within each genus is indicated: symmetrical = symmetrical about apical and transapical axes; dorsiventral = symmetrical about transapical axis; heteropolar = symmetrical about apical axis

Order	Family	Genus	Valve shape
Eunotiales	Eunotiaceae	*Eunotia*	dorsiventral
		Semiorbis	dorsiventral
		Actinella	heteropolar
	Peroniaceae	*Peronia*	heteropolar
Lyrellales	Lyrellaceae	*Lyrella*	symmetrical–dorsiventral[2]
Cymbellales	Rhoicospheniaceae	*Rhoicosphenia*	heteropolar[1]
		Campylopyxis	heteropolar[1]
		Cuneolus	heteropolar[1]
		Gomphoseptatum	heteropolar
		Gomphonemopsis	heteropolar
	Cymbellaceae	*Placoneis*	symmetrical
		Cymbella	dorsiventral
		Brebissonia	symmetrical
		Encyonema	dorsiventral
		Gomphocymbella	heteropolar + dorsiventral
	Gomphonemataceae	*Gomphonema*	heteropolar
		Didymosphenia	heteropolar (+dorsiventral)
		Gomphoneis	heteropolar
		Reimeria	dorsiventral
Naviculales	Berkeleyaceae	*Climaconeis*	symmetrical–dorsiventral[2]
	Scoliotropidaceae	*Biremis*	symmetrical–dorsiventral[2]
	Naviculaceae	*Navicula*	symmetrical
		(*Hippodonta*)	symmetrical
		Pseudogomphonema	heteropolar
		Seminavis	dorsiventral
		Haslea	symmetrical
	Pleurosigmataceae	*Toxonidea*	dorsiventral
		Gyrosigma	sigmoid
Thalassiophysales	Catenulaceae	*Catenula*	dorsiventral
		Amphora	dorsiventral
		Undatella	dorsiventral
	Thalassiophysaceae	*Thalassiophysa*	dorsiventral
Bacillariales	Bacillariaceae	*Hantzschia*	dorsiventral
		Cymbellonitzschia	dorsiventral
		Gomphonitzschia	heteropolar
		Gomphotheca	heteropolar

Notes:
1 Valves are also genuflexed.
2 Although some representatives are dorsiventral, the majority of taxa are symmetrical.

24.5.2 Escaping from the confines of a rigid cell wall

Because new valves are normally laid down within the confines of the parent cell, valve shape can undergo only limited change during vegetative cell division. The outline of a developing hypovalve is largely determined by the outline of the parent wall. Thus, as the SDV reaches the girdle region, virgal growth must either cease or curve to form the valve mantle (Figure 24.4). However, in theory, major shape changes could be introduced at the initial cell stage when there are fewer constraints on shape and size, and a shape change introduced in the initial frustule would then be propagated through many generations. The generation of different shaped cells, (e.g. triradiate or bipolar, bi-angular, triangular or quadrangular) occurs in several centric genera, such as *Biddulphia*, *Bellerochea* and *Lithodesmium* (von Stosch, 1977, 1982; Schmid, 1997), but is much rarer among pennate diatoms. The best known examples are the triradiate '*Centronella reicheltii*' (Schmid, 1997), *Phaeodactylum* and a few other fragilarioid diatoms (Geitler, 1939). Von Stosch (1965) and Roshchin (1994) have also reported the occurrence of triradiate forms in *Achnanthes*, a monoraphid diatom.

The initial valves of the majority of pennate diatoms are not strikingly different shapes, but are often atypical of the taxon, and 'normal' valve formation can require one or more cell divisions to establish (Thaler, 1972; Mann, 1984d, 1989a; Schmid, 1984a; Cohn *et al.*, 1989). Initial valves are often more rounded, with more or less irregular pore arrangements and somewhat dysfunctional raphe systems (Cohn *et al.*, 1989; Mann, 1989a; Jewson and Lowry, 1993; Passy-Tolar and Lowe, 1995). In *Fragilariforma virescens* (Ralfs) Williams initial valves are more heavily silicified, with more diffuse pore fields and apical rather than trans-apical rimoportulae. Thus, considerable reorientation of features occurs between the initial and 'normal' vegetative valves (Williams, 2001), including structural differences reminiscent of those induced by low doses of microtubule inhibitors. Interaction between adjacent sibling cells may also be crucial, particularly if cell turgor plays a role (Geitler, 1932; Schmid, 1984b). Thus, although the shape and size of initial valves are less constrained than for normal valves, if vegetative valves require the parent cell environment to adopt their 'correct', stable morphology, any permanent shift in shape or symmetry is presumably a function of the parent cell rather than a random event.

But this still begs two key questions: (1) How does the cell 'know' what shape to form and control the relative widths of its valves? (2) How is the polarity of a heteropolar species set? As mentioned in Section 24.5.1, lateral asymmetry and heteropolarity are observed in initial cells, if less strongly than in the subsequent vegetative valves, but we know nothing about internal gradients or cytological asymmetry in initial cells. Unlike zygote development in fucoid algae, polarity cannot be induced by sperm entry because raphid diatoms have morphologically isogamous reproduction. Furthermore, when heteropolar cells pair, some line up 'head to toe', rather than head pole beside head pole (Geitler, 1932). Geitler (1958) found that some races of *Gomphonema* species paired in the same orientation and presumably (although he did not report this) the auxospores retained that orientation. But other races invariably paired 'head to toe' (Geitler, 1958), requiring that polarity is reset in the auxospore. There is similar variation in pairing behaviour between dorsiventral taxa (Geitler, 1975) with some taxa consistently pairing with their primary sides adjacent, whereas others do not. Thus, cell symmetry must be an intrinsic feature established in the initial cell rather than determined by mode of gamete pairing, and consolidated after the first cytokinetic mitosis. Detailed studies of cytoplasmic organisation in the expanding zygotes and the initial cells would help to determine how and when polarity is established (see Section 24.7.4).

24.6 A suitable case for treatment?

In spite of their diversity of cell shape and symmetry, members of the Cymbellales (*sensu* Round *et al.*, 1990) show considerable similarity in cytoplasmic features and cell construction, and few would disagree that they are closely related. All have a single chloroplast with a central pyrenoid and cis symmetry in those taxa for which it has been investigated. Most genera (excluding *Rhoicosphenia*, *Anomoeoneis*) have volate pores, stigmata are often present and many have apical pore fields. However, a phylogenetic investigation by Kociolek and Stoermer (1988) (Figure 24.12) did not show generic groupings in clades congruent with Round *et al.*'s (1990) generic arrangement. Symmetry was an important character in Kociolek and Stoermer's (1988) analysis (although not defined in relation to the primary side of the valve: cf. Figure 24.13), whereas Round *et al.* (1990) placed more emphasis on valve structure. The failure to recognise that dorsal and ventral were not ontogenetically homologous is one reason why *Cymbella*, *Reimeria* and *Encyonema* fall near each other in Kociolek and Stoermer's (1988) analysis (Figure 24.12). Nevertheless, subsequently (but again with the same assumption that dorsiventrality can be treated as a homologous character), Kociolek and Stoermer (1993) showed that *Gomphocymbella ancyli* is closely related to *Cymbella* whereas East African *Gomphocymbella* species are closely related to *Gomphonema*. Thus, treating *Gomphocymbella* as two taxa, and including *Cymbella diluviana*, which resembles *Placoneis* but has apical pore fields, the distribution of combinations of characters in the order (Table 24.3)

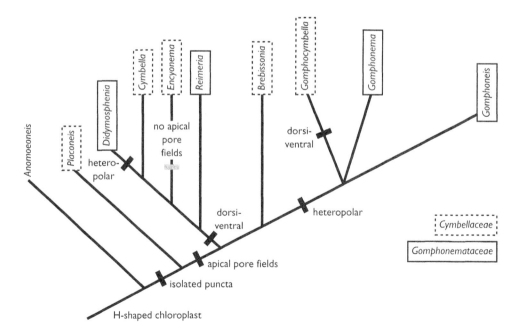

Figure 24.12 The relationships of the Cymbellales based on a cladistic analysis by Kociolek and Stoermer (1988), showing some of the characters used. The systematic position of genera based on Round *et al.* (1990) is shown.

Table 24.3 Table showing the distribution of some valve features among genera of the Cymbellales. Dorsiventral I = primary side dorsal; dorsiventral 2 = primary side ventral; intermissio = visible interruption between the internal central raphe endings (absence of an intermissio means that there has been an overgrowth of silica, obscuring the central deflections: cf. Cox, 1999b)

Taxon	Symmetry		Apical pore fields	Striae	Intermissio
	apical	lateral			
Brebissonia	isopolar	symmetrical	two	biseriate	present
Cymbella	isopolar	dorsiventral 2	two	uniseriate	absent
Cymbella diluviana	isopolar	dorsiventral I	two	uniseriate	present
Didymosphenia	heteropolar	(dorsiventral 2)	one	uniseriate	absent
Encyonema	isopolar	dorsiventral I	none	uniseriate	present
Gomphocymbella (most)	heteropolar	dorsiventral I	one	uniseriate	present
Gomphocymbella ancyli	heteropolar	dorsiventral 2	one	uniseriate	absent
Gomphoneis	heteropolar	symmetrical	one	biseriate	present
Gomphonema	heteropolar	symmetrical	one	uni-/biseriate	present
Placoneis	isopolar	symmetrical	none	uni-/biseriate	present
Reimeria	isopolar	dorsiventral I	two	uniseriate	present

allows several evolutionary questions to be asked (*C. diluviana* has been considered an intermediate between *Reimeria* and *Encyonema*: Kociolek and Stoermer, 1990). Have particular valve symmetries been acquired (and/or lost) on more than one occasion? Are the dorsiventral taxa more closely related to each other, or to the isopolar or heteropolar taxa? Is structure, for example those of the internal raphe system and internal stigma openings, more phylogenetically informative than shape and symmetry?

Based on the various combinations of symmetry, presence or absence of apical pore fields, stria type and internal central raphe endings, Figure 24.13 postulates a scenario that could explain the evolution of the genera. This scenario assumes an isopolar ancestor and the retention of lateral asymmetry once it has developed; dorsiventral 1 (primary side wider) cannot switch directly to dorsiventral 2 (secondary side wider). The development of marginal laminae is presumed to be similar to the development of alveolae in *Pinnularia* (Cox, 1999b). Interestingly, although stigma internal openings were not a major consideration in developing this scenario, taxa with more complex stigmata such as *Cymbella*, *Didymosphenia* and *G. ancycli* have fallen together on the right-hand side of Figure 24.13. These taxa also have hidden

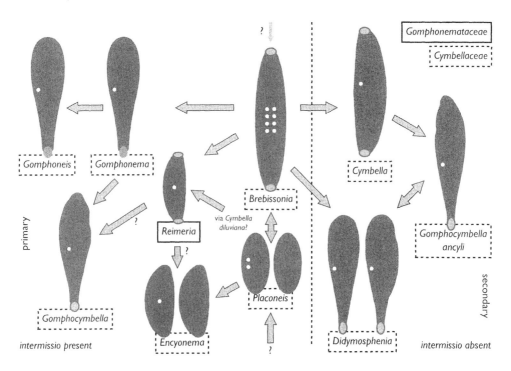

Figure 24.13 Diagram showing possible routes by which the different genera within the Cymbellales could have been derived. This assumes an isopolar ancestor, with or without apical pore fields at each pole. It has also been assumed that once one type of lateral asymmetry evolves it is retained. Dorsiventrality and position of stigmata are shown in relation to primary and secondary sides of the valve. Genera to the left of the vertical division have an intermissio (visible interruption in the raphe slits), those to the right do not. Lines around generic names indicate the families to which they were allocated by Round *et al.* (1990).

internal central raphe endings (= intermissio absent). Genome sequencing would be a logical approach to investigate this evolutionary scenario and would also reveal the degree of genetic difference between the various shapes and symmetries within the group.

24.7 Potential of diatoms as model organisms for developmental studies

24.7.1 Model organisms

'What people want is a system that allows them to study an interesting phenomenon with ease' (D. Mann, pers. comm. 2001). Ease is often synonymous with easy to obtain and grow; the organisms which best exemplify the phenomenon of interest may not be the easiest to handle. Thus, physiological studies have generally been performed on small, rapidly growing algal taxa, and *in vitro* grazing experiments and ecotoxicological tests have been performed on laboratory 'weeds' rather than more typical members of aquatic communities. The basic pattern of raphid diatom morphogenesis was derived from '*Navicula pelliculosa*' (Chiappino and Volcani, 1977), obtained from the Indiana Culture Collection, but originally isolated by R. A. Lewin in 1949 (Reimann *et al.*, 1966). (This taxon is neither a member of *Navicula s.s.* nor was it correctly identified. It should more correctly be known as *Fistulifera saprophila*.) The use of this diatom rather than any other reflected its availability, ease of culture and utilisation in a variety of physiological and biochemical studies (e.g. Lewin, 1953; Reimann *et al.*, 1966; Sullivan, 1976), rather than any intrinsic merit for morphogenetic studies. *Arabidopsis* has been adopted as a model higher plant for genetic studies because its life cycle can be completed within weeks rather than months or years, and genetic mutants are readily generated (Smyth, 1990). Yet some workers (Coen and Meyerowitz, 1991; Bradley *et al.*, 1993) prefer *Antirrhinum* as a subject for the investigation of the genetic control of floral development because its genetic and physiological attributes make it amenable to study. These attributes include large flowers that are easily emasculated, crossed and studied (Bradley *et al.*, 1993). Thus, the benefits afforded by larger flowers may outweigh the disadvantages of a longer life cycle when studying floral development. Organisms which are less easily cultivated may be appropriate model systems if they allow particularly interesting questions to be answered. Few diatoms are laboratory weeds, but the diversity and regularity of their highly structured walls invites the investigation of intrinsic pattern control by the cell (Table 24.4).

24.7.2 Diatoms in relation to other plants

Although diatoms are unusual, even among the Chromophyta, in using silica rather than cellulose or other organic compounds to build their cell walls, incorporation of silica into structural elements is not unique to them. Other chromophyte algae, for example members of the Chrysophyceae, produce siliceous scales that ultimately overlay the plasmalemma (McGrory and Leadbeater, 1981). Horsetails (*Equisetum*) accumulate relatively large amounts of silica (Kaufman *et al.*, 1981) and even higher plants such as grasses may incorporate silica within their cell walls to give structural

Table 24.4 Summary of the potential advantages and disadvantages of using diatoms to investigate the control of wall shape, symmetry and pattern

Advantages	Disadvantages
Unicellular	
can be cultured under defined conditions	not all amenable to culture
	long-term maintenance in culture often difficult because of cell size reduction
cells do not need to be dissected out for observation	cells may tend to lie in one orientation only
	relevant orientation may require careful selection of taxa
Cell cycle	
short generation times (vegetative division)	
cultures can be synchronised to produce many cells at same time	
Large cells	
large cells allow LM observation of cytoplasmic reorganisation	some large species difficult to grow in culture
longer division time may facilitate observation of wall morphogenesis	larger species divide less frequently
Wall patterning	
species-specific, under genetic control, precise and regular	cells diploid, production of wall mutants probably difficult
readily seen with LM and SEM	may be obscured by chloroplasts in live cells
shape and symmetry can vary between closely related taxa	
Symmetry and polarity	
polarity in initial cell not established via external stimulus	initial cells difficult to obtain
dorsiventrality linked to position of nucleus and lack of nuclear oscillation	lateral asymmetry in some valves linked to shifts in SDV position during valve formation
Phenotypic plasticity	
occurs in some species	poorly documented
Establishment of cytoskeleton	
auxospore > initial cell > vegetative cell transition permits study in relation to regularity of patterning	difficult stages to obtain, probably require development of new techniques
different cytoskeleton components may be involved in morphogenesis and maintenance of cellular activity	unproven
Cytoskeleton visualisation	
LM techniques (fluorescence and confocal laser scanning) have been developed to visualise MTs and actin	not yet shown to work for diatoms
fixation techniques may reduce chloroplast fluorescence	chloroplasts in cells may autofluoresce, interfering with visualisation

rigidity (Jones *et al.*, 1963; Sangster and Parry, 1981). (Silica is also used as skeletal material by various other unrelated protist groups, including radiolarians and testate amoebae.) If the possession of a rigid cell wall has reduced the energetic costs of cytoskeleton maintenance, the detectable interphase diatom cytoskeleton may be restricted to anchoring the protoplast within the wall and to controlling organelle distribution and motility (in raphid diatoms). As in other plant cells, the cytoskeleton will also be involved in nuclear movements during mitosis, positioning of the mitotic spindle and cytokinesis. In both cases, the position of the nucleus is critical in determining where cytokinesis will occur (Pickett-Heaps *et al.*, 1990, 2000; Lloyd, 1991), although there are some differences in spindle microtubule arrangement and behaviour. Mitosis in diatoms resembles more closely that in cellular slime moulds than other plants (Fuge, 1982). Thus, unlike diatoms, there is no apical anchoring of the spindle microtubules in higher plants and chromosomes slide along the microtubules during anaphase rather than being moved primarily by the sliding of the tubules of the different half-spindles as the poles are pushed apart. In diatoms there is some sliding of the chromosomes along the spindle (Pickett-Heaps and Bajer, 1977) by means of kinetochore tubules (Cande *et al.*, 1985; Worderman *et al.*, 1986). Diatoms are also unusual among plants and algae in having a cleavage furrow lined by microfilaments (MFs) probably containing actin, indicated by the sensitivity of cleavage to cytochalasin (Pickett-Heaps and Spurck, 1982). At other times, MFs are crucial for cytoplasmic flow and the transport of chemicals and organelles around a cell.

If that part of the cytoskeleton concerned with vegetative wall morphogenesis is formed only when required (Pickett-Heaps *et al.*, 1979), *in vivo* studies of the control of pattern formation in diatoms can be focused on the few hours during the cell cycle that valves are being formed. Correspondingly, it should be possible to identify permanent aspects of the cytoskeleton controlling other cellular activities. As generation times in diatoms are usually short (1–3 days) and synchronisation of cultures can be achieved relatively easily (Darley and Volcani, 1971), many cells can be produced at the desired developmental stage for investigation. Furthermore, if cellular control of valve morphogenesis is reduced by a poorly developed cytoskeleton in the auxospore, the investigation of auxospore, initial and first vegetative cell wall formation may be crucial to understanding the relationship between wall deposition and cytoskeletal integrity.

24.7.3 Advantages of using the exception rather than the rule

Because diatom walls are so precisely and elaborately structured, even small modifications in pattern can be informative. The consistent morphological variation between similar but non-interbreeding populations (demes) of *Sellaphora pupula* (Kützing) Mereschkowsky is a manifestation of the genetic differences between them (Mann, 1984c, 1989c). Alternatively the phenotypic variation seen in a few taxa such as *Navicula gregaria* Donkin (Cox, 1995) indicates that the environment can induce morphological modifications within certain constraints, and might be used to investigate the control of pattern under different conditions. The discrepancies between initial and normal vegetative valves offer evidence that the integrity of the cytoskeleton is critical to wall formation (Schmid, 1994). Each example could be investigated to determine the extent of the MT and MF components of the cytoskeleton and their relationship to the SDV, alongside investigations of the molecular controls of actin and tubulin assembly and disassembly and gene expression.

It has been suggested (Baluška *et al.*, 2000) that the actin cytoskeleton evolved in close association with the plasma membrane, while the MT cytoskeleton remained closely associated with the eukaryotic nucleus. Cortical MTs polarise the growth of higher plant cells by controlling the orientation of cellulosic microfibrils, but the actin-enriched cell cortex domains represent a more primitive cell periphery organisation. Changes in the organisation of actin and MT cytoskeletons during cell division have been investigated in fission yeasts (Brunner and Nurse, 2000) and have shown that actin is located in growing cell ends, whereas MTs generally extend along the cell axis. This suggests that MTs play a role in global cellular positioning mechanisms, whilst actin is more localised. However, actin is considered to be the main cytoskeletal element in budding yeasts, providing the structural basis for cell morphogenesis and polarity development (Bretscher, 1991; Li *et al.*, 1995). The effects of profilin on actin have been used to show that actin plays a key role in the morphogenetic process in *Micrasterias* (Holzinger *et al.*, 1997), although, unlike in higher plants, microtubules do not participate in the morphogenetic process and do not orient the cell wall microfibrils.

Actin commonly binds at the cytoplasmic face of adhesion sites. Attachment between the actin cytoskeleton and cell wall probably involves spectrins and integrins, immunorelated homologues of which have been demonstrated in plant cells (de Ruijter and Emons, 1999), where they may be involved in organelle motility, tip morphogenesis and apex-directed contraction (Wyatt and Carpita, 1993). Using antigens, Reuzeau *et al.* (1997) identified proteins that seem to correspond to the cytoskeleton anchoring transmembrane protein β_1 integrin, and the cytoskeletal protein spectrin, intermediate filaments and F-actin. β_1 integrin was seen as dots on the plasma membrane (adhesion sites), and with spectrin on the endomembrane system. Given the close association of MTs and MFs (actin) with the SDV in diatoms during silicification, and the emerging association of actin and spectrin at adhesion sites, the silicification model suggested by Robinson and Sullivan (1987) (see Section 24.2.3) warrants further investigation.

While multiple copies of actin genes have been found in members of the Streptophyta *sensu* Kenrick and Crane (1997) (angiosperms, gymnosperms, charophytes, bryophytes and ferns) (An *et al.*, 1999; Bhattacharya *et al.*, 2000), only single copies have been found in many protists (red algae, most green algae, chromophytes and fungi) (Bhattacharya *et al.*, 1991; An *et al.*, 1999). Unusually, six distinct actin genes have been reported for the chromophyte *Emiliania huxleyi* (Bhattacharya *et al.*, 1993), but it is probable that diatoms will be like other chromophytes and contain a single actin gene. Multiple copies of genes are thought to allow a more diverse pattern of gene regulation rather than simply producing large amounts of actin (Moniz de Sá and Drouin, 1996), therefore investigating gene regulation in taxa with a single copy of the gene should be more straightforward. Choosing a unicellular organism with clearly defined variation in wall pattern offers the possibility of studying differential control within a single cell.

The use of *in vivo* visualisation techniques and confocal scanning electron microscopy (Lloyd *et al.*, 1992) allows cytoskeleton dynamics to be explored noninvasively and non-destructively. Tubulin antibodies can be used to visualise MTs (Goddard *et al.*, 1994), while actin filaments can be visualised *in vivo* with rhodamine phalloidin (Cleary, 1995; Hall, 1998). Integrin and spectrin can also be fluorescently labelled (Reuzeau *et al.*, 1997). It is known that members of the Rho family of GTPases are involved in signal transduction pathways, linking receptors on the plasma membrane and organisation of the actin cytoskeleton (Hall and Nobes, 2000), as well as regulating the assembly of F-actin in cell movement (Chung *et al.*, 2000). Inhibitors of the different GTPases can be used to see the effect each has on assembly and organisation of the actin cytoskeleton, and to wall morphogenesis, while mutant Rho family members have been shown to modify the cytoskeleton (Chung *et al.*, 2000). Diatoms could be useful model organisms in this context because many can be readily cultured, and the precision and diversity of wall pattern would facilitate comparative studies. Because diatoms are unicellular organisms there is no need to section or dissect out material for investigation, and species with larger cells can be selected to facilitate microscopical examination. Thus, over a century ago, Lauterborn (1896) was able to describe mitosis in diatoms in considerable detail, observations that have been confirmed by later workers (Pickett-Heaps *et al.*, 1984).

24.7.4 Advantages of unicells in studying the development of polarity

Using unicells to study the development of polarity excludes the complication of cell-to-cell interaction and communication, i.e. the multicellular environment within which polarity and dorsiventrality must often be investigated. In unicells the constraints on shape and symmetry can be investigated more directly, and the effects of the extracellular environment excluded. The use of fucoid zygotes (Quatrano, 1978) has revealed that polarity is determined by intracellular gradients that may subsequently be fixed in relation to an external stimulus. However, in *Cystoseira*, in the absence of any other gradients, the point at which the sperm enters the egg cell determines the site of rhizoid formation (Knapp, 1931). (Similarly, the position of sperm entry defines the posterior of the embryo in the nematode *Caenorhabditis elegans*, directing the process of cytoplasmic flow: Golden, 2000.) Once the zygote has become polarised, subsequent segregation of nuclei into the polar regions results in them developing in different microenvironments, within which selective gene expression can occur (Jacobs and Shapiro, 1998). In contrast, there is no evidence that external gradients (or, in many cases, the mode of mother cell pairing or gamete fusion in heteropolar diatoms), determines the polarity of the resulting zygote (see Section 24.5.2). (However, orientation of the apical axis of the expanding zygote may be determined by the mode of mother cell pairing: Round *et al.*, 1990.)

In other unicells in which external stimuli are not known to be important asymmetry is achieved by the asymmetrical distribution of cell fate determinants and consequent changes in gene expression patterns (Jacobs and Shapiro, 1998). Thus, preferential accumulation of the transcriptional repressor ASH1p in *Saccharomyces cerevisiae*, which mediates interconversion of mating types, is determined by directed transport of ASH1 mRNA along actin filaments. In *C. elegans* PAR proteins are asymmetrically distributed in the single-celled embryo (Bowerman and Shelton,

1999), probably polarising the actin cytoskeleton (Golden, 2000). Although the pathways by which *par* genes act are poorly defined, molecular identities show that the mechanisms regulating anterior–posterior (AP) polarity are conserved in different organisms (Bowerman and Shelton, 1999). Similar mechanisms could therefore be expected to operate in diatoms.

A role for cytoskeletal elements in determining polarity in *Fucus* is suggested by the disruption of axis fixation in the presence of cytochalasin, provided the axis has not already become fixed (Quatrano, 1978; Quatrano *et al.*, 1985, 1991). Brawley and Robinson (1985) showed that microfilaments are required for the development and maintenance of endogenous cytoplasmic currents during the polarisation of fucoid zygotes. Significant changes occur in the cortical distribution of F-actin during early embryogenesis in *Pelvetia*, with preferential localisation at the rhizoid tip. Cytochalasin B inhibits the localised secretion of the polysaccharide fucoidin (Quatrano, 1978) by reversibly inhibiting photopolarisation and by blocking Golgi vesicle transport (Brawley and Robinson, 1985). MTs are not involved in the establishment of the polar axis in brown algal zygotes (Kropf, 1994) but are critical in positioning the nucleus, and in morphogenesis of the zygote tip. This parallels the importance of the MC (MTOC) in positioning the nucleus in diatoms (Pickett-Heaps *et al.*, 1990).

In both lower and higher plants the cytoskeleton has been implicated in gravitropism (Kiss, 2000), and actin MFs have also been shown to be involved in perception/transduction, although MTs are involved in moss protonema gravitropism (Schwuchow *et al.*, 1990). Within plant cells, strands of cytoplasm have tensegral properties (rigidity plus tension and force generation) due to the close interaction of MTs and actin generate polarised activity (Pickett-Heaps *et al.*, 2000). The possession of an internal cytoskeleton which powers the movement of substrates, enzymes, solutes, macromolecules, vesicles and organelles around the cytoplasm is particularly important for plant cells, where the cytoplasm is often peripheral around a large central vacuole. Again, large-celled diatoms could be useful model systems for the study of polarised cells, and the role of the cytoskeleton in redistributing a variety of substances and organelles around the cell. The attached habit of many heteropolar diatoms depends upon the localised secretion of polysaccharide through apical pore fields, whereas motile diatoms primarily secrete polysaccharide through the raphe slits during movement, and tube-dwelling species may additionally release polysaccharide between their girdle bands (Cox, 1981).

24.7.5 *Where is genetic control being exerted?*

Genetic mechanisms (regulated by PAR proteins) that control asymmetric cell division have been conserved among prokaryotes, microbial eukaryotes and higher organisms (Jacobs and Shapiro, 1998). Similarly, anterior–posterior patterning in all animals is governed by Hox cluster genes, while axis formation in frog eggs, limb development, and pattern formation in insect wings are established by common signalling pathways (Kim and Fraser, 1998). Throughout, the MF and MT cytoskeleton plays a crucial role in morphogenesis, mitosis and motility, and interacts with membrane-bound proteins, which have been well established for animal cells and undoubtedly exist in plant cells (Pickett-Heaps *et al.*, 2000). The genetic controls and pathways involved in regulating the cytoskeleton are being elucidated in model organisms such as *Dictyostelium, Saccharomyces, Caulobacter, Caenorhabditis* and mouse fibroblasts (Hall, 1998; Kim and Fraser, 1998; Bowerman and Shelton, 1999; Chung *et al.*, 2000). Yet, in spite of intensive study, there is, for example, still no interpretable correlation between genotype and phenotype in yeast using morphological mutants (Pickett-Heaps *et al.*, 2000). One highly unusual diatom, *Phaeodactylum tricornutum*, exists in different forms according to habitat, but only produces a single siliceous valve in its oval cell form (Round *et al.*, 1990). It has been shown that different morphs of a single clone of this species produce different antigenicity patterns (Gutenbrunner *et al.*, 1994). The authors suggest that these immunological differences could be produced by post-translational changes in wall polypeptides, presumably a response to the contrasting growth conditions. Given the molecular and mechanismal similarities across kingdoms, techniques developed for many of the model organisms (Johnson and Pringle, 1990; Lloyd *et al.*, 1992; Goddard *et al.*, 1994; Cleary, 1995; Reuzeau *et al.*, 1997; Hall, 1998; Kost *et al.*, 1998; Mathur *et al.*, 1998) could be adapted for the study of polarity development and morphogenetic control in diatoms, with the highly structured cell walls providing a clear expression of cellular activity.

24.8 Concluding remarks

The diversity and complexity of cell wall structure in diatoms has fascinated microscopists for more than 150 years, yet even today little is known of the mechanisms and controls on their production. As unicells that can be grown under controlled conditions, diatoms have great potential as experimental organisms, and the development of new microscopical and immunological techniques facilitate their study *in vivo*. They offer the possibility of studying how genetic information is translated into the precise control of a rigid siliceous structure via the cytoskeleton, and to the factors that can modify or disrupt wall production. Their well-known structural diversity allows a plethora of comparisons in vegetative cells, while the peculiarities of their life cycle permit investigation of the gradual re-establishment of the cytoskeleton after sexual reproduction.

ACKNOWLEDGEMENTS

Thanks are due to Dawn Rose, Pat Sims and David Williams for discussions on various aspects of diatom systematics and development, and to Richard Bateman and David Mann for comments on drafts of this chapter.

REFERENCES

An, S. S., Möpps, B., Weber, K. and Bhattacharya, D. (1999) The origin and evolution of green algal and plant actins. *Molecular Biology and Evolution*, 16, 275–285.

Baluška, F., Volkmann, D. and Barlow, P. W. (2000) Actin-based domains of the 'cell periphery complex' and their associations with polarized 'cell bodies' in higher plants. *Plant Biology*, 2, 253–267.

Bhattacharya, D., Aubry, J., Twait, E. C. and Jurk, S. (2000) Actin gene duplication and the evolution of morphological complexity in land plants. *Journal of Phycology*, 36, 813–820.

Bhattacharya, D., Stickel, S. K. and Sogin, M. L. (1991) Molecular phylogenetic analysis of actin genic regions from *Achlya bisexualis* (Oomycota) and *Costaria costata* (Chromophyta). *Journal of Molecular Evolution*, 33, 525–536.

Bhattacharya, D., Stickel, S. K. and Sogin, M. L. (1993) Isolation and molecular phylogenetic analysis of actin coding regions from the prymnesiophyte alga, *Emiliania huxleyi*, using reverse transcriptase and PCR methods. *Molecular Biology and Evolution*, 10, 689–703.

Bowerman, B. and Shelton, C. A. (1999) Cell polarity in the early *Caenorhabditis elegans* embryo. *Current Opinion in Genetics and Development*, 9, 390–395.

Bradley, D., Carpenter, R., Elliott, R., Simon, R., Romero, J., Hantke, S., Doyle, S., Mooney, M., Luo, D., McSteen, P., Copsey, L., Robinson, C. and Coen, E. (1993) Gene regulation of flowering. *Philosophical Transactions of the Royal Society of London, Series B*, 339, 193–197.

Branton, D., Cohen, C. M. and Tyler, J. (1981) Interaction of cytoskeletal proteins on the human erythrocyte membrane. *Cell*, 24, 324–332.

Brawley, S. H. and Robinson, K. R. (1985) Cytochalasin treatment disrupts the endogenous currents associated with cell polarization in fucoid zygotes: studies of the role of F-actin in embryogenesis. *Journal of Cell Biology*, 100, 1173–1184.

Bretscher, A. (1991). Microfilament structure and function in the cortical cytoskeleton. *Annual Review of Cell Biology*, 7, 337–374.

Brunner, D. and Nurse, P. (2000) New concepts in fission yeast morphogenesis. *Philosophical Transactions of the Royal Society of London, Series B*, 355, 873–877.

Cande, W. Z., McDonald, K., Wordeman, L. and Coltrin, D. (1985) *In vitro* anaphase spindle elongation using isolated diatom spindles. *Cell Motility and the Cytoskeleton*, 5, 169–170.

Chiappino, M. L. and Volcani, B. E. (1977) Studies on the biochemistry and fine structure of silica shell formation in diatoms. VII. Sequential cell wall development in the pennate *Navicula pelliculosa*. *Protoplasma*, 93, 205–221.

Chung, C. Y., Lee, S., Briscoe, C., Ellsworth, C. and Firtel, R. A. (2000) Role of Rac in controlling the actin cytoskeleton and chemotaxis in motile cells. *Proceedings of the National Academy of Sciences of the USA*, 97, 5225–5230.

Cleary, A. L. (1995) F-actin redistributions at the division site in living *Tradescantia* stomatal complexes as revealed by microinjection of rhodamine-phalloidin. *Protoplasma*, 185, 152–165.

Cleve, P. T. (1894) Synopsis of the naviculoid diatoms. Part 1. *Kongliga Svenska Vetensk-Akademiens Handlingar*, 26, 1–194.

Coen, E. S. and Meyerowitz, E. M. (1991) The war of the whorls: genetic interactions controlling flower development. *Nature*, 353, 31–37.

Cohn, S. A., Spurck, T. P., Pickett-Heaps, J. D. and Edgar, L. A. (1989) Perizonium and initial valve formation in the diatom *Navicula cuspidata* (Bacillariophyceae). *Journal of Phycology*, 25, 15–26.

Cove, D. J. (2000) The generation and modification of cell polarity. *Journal of Experimental Botany*, 51, 831–838.

Cox, E. J. (1977) Raphe structure of naviculoid diatoms. *Nova Hedwigia Beiheft*, 54, 261–274.

Cox, E. J. (1979) Symmetry and valve structure in naviculoid diatoms. *Nova Hedwigia Beiheft*, 64, 193–206.

Cox, E. J. (1981) Mucilage tube morphology of three tube-dwelling diatoms and its diagnostic value. *Journal of Phycology*, 17, 72–80.

Cox, E. J. (1982) Taxonomic studies on the diatom genus *Navicula* Bory. IV. *Climaconeis* Grun., a genus including *Okedenia inflexa* (Bréb.) Eulenst. *ex* De Toni and members of *Navicula* sect. *Johnsoniae sensu* Hustedt. *British Phycological Journal*, 17, 147–168.

Cox, E. J. (1987) *Placoneis* Mereschkowsky: the re-evaluation of a diatom genus originally characterized by its chloroplast type. *Diatom Research*, 2, 145–157.

Cox, E. J. (1991) What is the basis for using diatoms as monitors of river quality?, in *Use of Algae for Monitoring Rivers* (eds B. A. Whitton, E. Rott and G. Friedrich), Universität Innsbruck, Innsbruck, pp. 33–40.

Cox, E. J. (1993) Diatom systematics: a review of past and present practice and a personal view for future development. *Nova Hedwigia Beiheft*, 106, 1–20.

Cox, E. J. (1995) Morphological variation in widely distributed diatom taxa: taxonomic and ecological implications, in *Proceedings of the 13th International Diatom Symposium* (eds D. Marino and M. Montresor), Biopress, Bristol, pp. 335–345.

Cox, E. J. (1999a) Studies on the diatom genus *Navicula* Bory. VIII. Variation in valve morphology in relation to the generic diagnosis based on *Navicula tripunctata* (O. F. Müller) Bory. *Diatom Research*, 14, 207–237.

Cox, E. J. (1999b) Variation in patterns of valve morphogenesis between representatives of six biraphid diatom genera (Bacillariophyceae). *Journal of Phycology*, 35, 1297–1312.

Cox, E. J. (2001) What constitutes a stauros? A morphogenetic perspective, in *Festschrift for Horst Lange-Bertalot* (eds R. Jahn, J. P. Kociolek, A. Witkowski and P. Compère), 1, 303–316.

Cox, E. J. and Ross, R. (1981) The striae of pennate diatoms, in *Proceedings of the 6th Symposium on Recent and Fossil Diatoms* (ed. R. Ross), Koetz, Koenigstein, pp. 267–278.

Cox, E. J. and Williams, D. M. (2000) Systematics of naviculoid diatoms: the interrelationships of some taxa with a stauros. *European Journal of Phycology*, 35, 273–282.

Darley, W. M. and Volcani, B. E. (1971) Synchronized cultures: Diatoms, in *Methods in Enzymology*, vol. 23A (ed. A. San Pietro), Academic Press, New York, pp. 85–96.

Dawson, P. A. (1974) Observations on species transferred from *Gomphonema* C. A. Agardh to *Gomphoneis* Cleve. *British Phycological Journal*, 9, 75–82.

de Ruijter, N. C. A. and Emons, A. M. C. (1999) Actin-binding proteins in plant cells. *Plant Biology*, 1, 26–35.

Drum, R. W. and Pankratz, H. S. (1964) Post mitotic fine structure of *Gomphonema parvulum*. *Journal of Ultrastructure Research*, 10, 217–223.

Edgar, L. A. (1980) Fine structure of *Caloneis amphisbaena* (Bacillariophyceae). *Journal of Phycology*, 16, 62–72.

Edgar, L. A. and Pickett-Heaps, J. D. (1984) Valve morphogenesis in the pennate diatom *Navicula cuspidata*. *Journal of Phycology*, 20, 47–61.

Fuge, H. (1982) Mitosespindeln – Evolution verschiedener Systeme. *Biologie in unserer Zeit*, 12, 161–167.

Geitler, L. (1932) Der Formwechsel der pennaten Diatomeen. *Archiv für Protistenkunde*, 78, 1–226.

Geitler, L. (1939) Gameten- und Auxosporenbildung von *Synedra ulna* im Vergleich mit anderen pennaten Diatomeen. *Planta*, 30, 551–566.

Geitler, L. (1958) Notizen über Rassenbildung, Fortpflanzung, Formwechsel und morphologische Eigentümlichkeiten bei pennaten Diatomeen. *Österreichische Botanische Zeitschrift*, 105, 408–442.

Geitler, L. (1963) Alle Schalenbildungen der Diatomeen treten als Folge von Zell- oder Kernteilungen auf. *Berichte der Deutschen Botanischer Gesellschaft*, 75, 393–396.

Geitler, L. (1975) Formwechsel, sippenspezifischer Paarungsmodus und Systematik bei einigen pennaten Diatomeen. *Plant Systematics and Evolution*, 124, 7–30.

Gibbon, B. C. and Kropf, D. L. (1991) pH gradients and cell polarity in *Pelvetia* embryos. *Protoplasma*, 163, 43–50.

Goddard, R. H., Wick, S. M., Silflow, C. D. and Snustad, D. P. (1994) Microtubule components of the plant-cell cytoskeleton. *Plant Physiology*, 104, 1–6.

Golden, A. (2000) Cytoplasmic flow and the establishment of polarity in *C. elegans* 1-cell embryos. *Current Opinion in Genetics and Development*, 10, 414–420.

Gordon, R. and Drum, R. W. (1994) The chemical basis of diatom morphogenesis. *International Review of Cytology*, 150, 243–372.

Gutenbrunner, S. A., Thalhamer, J. and Schmid, A. M. M. (1994) Proteinaceous and immunochemical distinctions between the oval and fusiform morphotypes of *Phaeodactylum tricornutum* (Bacillariophyceae). *Journal of Phycology*, 30, 129–136.

Hall, A. (1998) Rho GTPases and the actin cytoskeleton. *Science*, 279, 509–524.

Hall, A. and Nobes, C. D. (2000) Rho GTPases: molecular switches that control the organization and dynamics of the actin cytoskeleton. *Philosophical Transactions of the Royal Society of London, Series B*, 355, 965–970.

Holzinger, A., Mittermann, I., Laffer, S., Valenta, R. and Meindl, U. (1997) Microinjection of profilins from different sources into the green alga *Micrasterias* causes transient inhibition of cell growth. *Protoplasma*, 199, 124–134.

Jacobs, C. and Shapiro, L. (1998) Microbial asymmetric cell division: localization of cell fate determinants. *Current Opinion in Genetics and Development*, 8, 386–391.

Jewson, J. H. and Lowry, S. (1993) *Cymbellonitzschia diluviana* Hustedt (Bacillariophyceae): habitat and auxosporulation. *Hydrobiologia*, 269/270, 87–96.

Johnson, D. I. and Pringle, J. R. (1990) Molecular characterization of CDC42, a *Saccharomyces cerevisiae* gene involved in the development of cell polarity. *Journal of Cell Biology*, 111, 143–152.

Jones, L. H. P., Milne, A. A. and Wadham, S. M. (1963) Studies of silica in the oat plant. II. Distribution of silica in the plant. *Plant and Soil*, 30, 71–80.

Jüttner, I., Cox, E. J. and Ormerod, S. (2000) New or poorly known diatoms from Himalayan streams. *Diatom Research*, 15, 237–262.

Kaczmarska, I., Bates, S. S., Ehrman, J. M. and Léger, C. (2000) Fine structure of the gamete, auxospore and initial cell in the pennate diatom *Pseudo-nitzschia multiseries* (Bacillariophyta). *Nova Hedwigia*, 71, 337–357.

Kaufman, P. B., Dayanandan, P., Takeoka, Y., Bigelow, W. C., Jones, J. D. and Iler, R. (1981) Silica in shoots of higher plants in *Silicon and Siliceous Structures in Biological Systems* (eds T. L. Simpson and B. E. F. Volcani), Springer-Verlag, New York, pp. 409–449.

Kenrick, P. and Crane, P. R. (1997) The origin and early evolution of plants on land. *Nature*, 389, 33–39.

Kim, S. K. and Fraser, S. E. (1998) Pattern formation and developmental mechanisms. Converging views of diverging pathways. *Current Opinion in Genetics and Development*, 8, 383–385.

Kiss, J. Z. (2000) Mechanisms of the early phases of plant gravitropism. *Critical Reviews in Plant Sciences*, 19, 551–573.

Knapp, E. (1931) Entwicklungsphysiologische Untersuchungen an Fucaceen-Eiern. *Planta*, 14, 731–751.

Kociolek, J. P. and Stoermer, E. F. (1988) A preliminary investigation of the phylogenetic relationships among the freshwater, apical pore field-bearing cymbelloid and gomphonemoid diatoms (Bacillariophyceae). *Journal of Phycology*, 24, 377–385.

Kociolek, J. P. and Stoermer, E. F. (1989) Phylogenetic relationships and evolutionary history of the diatom genus *Gomphoneis*. *Phycologia*, 28, 438–454.

Kociolek, J. P. and Stoermer, E. F. (1990) *Navicula diluviana* or *Cymbella diluviana*? Ultrastructure and systematic position of an enigmatic diatom, in *Proceedings of the 10th International Diatom Symposium* (ed. H. Simola), Koeltz, Koenigstein, pp. 173–182.

Kociolek, J. P. and Stoermer, E. F. (1993) The diatom genus *Gomphocymbella* O. Müller: taxonomy, ultrastructure and phylogenetic relationships. *Nova Hedwigia Beiheft*, 106, 71–91.

Kost, B., Spielhofer, P. and Chua, N. H. (1998) A GFP-mouse talin fusion protein labels plant actin filaments *in vivo* and visualizes the actin cytoskeleton in growing pollen tubes. *Plant Journal*, 16, 363–401.

Krammer, K. and Lange-Bertalot, H. (1986) *Bacillariophyceae. Süsswasserflora von Mitteleuropa 2 (4)*. G. Fischer, Stuttgart.

Kröger, N., Deutzmann, R., Bergsdorf, C. and Sumper, M. (2000) Species-specific polyamines from diatoms control silica morphology. *Proceedings of the National Academy of Sciences of the United States of America*, 97, 14133–14138.

Kröger, N. and Sumper, M. (1998) Diatom cell wall proteins and the cell biology of silica biomineralization. *Protist*, 149, 213–219.

Kropf, D. L. (1994) Cytoskeletal control of cell polarity in a plant zygote. *Developmental Biology*, 165, 361–371.

Kropf, D. L., Bisgrove, S. R. and Hable, W. E. (1999) Establishing a growth axis in fucoid algae. *Trends in Plant Science*, 4, 490–494.

Lange-Bertalot, H., Metzeltin, D. and Witkowski, A. (1996) *Hippodonta* gen. nov. Umschreibung und Begründung einer neuen Gattung der Naviculaceae. *Iconographia Diatomologica*, 4, 247–275.

Lauterborn, R. (1896) *Untersuchungen über Bau, Kernteilung und Bewegung der Diatomeen*. Engelmann, Leipzig.

Lewin, J. C. (1953) *Physiology of* Navicula pelliculosa *(Bréb.) Hilse*. PhD thesis, Yale University.

Li, R., Zheng, Y. and Drubin, D. G. (1995) Regulation of cortical actin cytoskeleton assembly during polarized cell growth in budding yeast. *Journal of Cell Biology*, 128, 599–615.

Lloyd, C. W. (1991) Cytoskeletal elements of the phragmosome establish the division plane in vacuolated higher plants, in *The Cytoskeletal Basis of Plant Growth and Form* (ed. C. W. Lloyd), Academic Press, London, pp. 245–257.

Lloyd, C. W., Venverloo, C. J., Goodbody, K. C. and Shaw, P. J. (1992) Confocal laser microscopy and three-dimensional reconstruction of nucleus-associated microtubules in the division plane of vacuolated plant cells. *Journal of Microscopy*, 166, 99–109.

McGrory, C. B. and Leadbeater, B. S. S. (1981) Ultrastructure and deposition of silica in the Chrysophyceae, in *Silicon and Siliceous Structures in Biological Systems* (eds T. L. Simpson and B. E. F. Volcani), Springer-Verlag, New York, pp. 201–230.

Mann, D. G. (1981a) Sieves and flaps: siliceous minutiae in the pores of raphid diatoms, in *Proceedings of the 6th Symposium on Recent and Fossil Diatoms* (ed. R. Ross), Koltz, Koenigstein, pp. 279–300.

Mann, D. G. (1981b) A note on valve formation and homology in the diatom genus *Cymbella. Annals of Botany*, 47, 267–269.

Mann, D. G. (1982) Structure, life history and systematics of *Rhoicosphenia* (Bacillariophyta). II. Auxospore formation and perizonium structure of *Rh. curvata. Journal of Phycology*, 18, 264–274.

Mann, D. G. (1983) Symmetry and cell division in raphid diatoms. *Annals of Botany*, 52, 573–581.

Mann, D. G. (1984a) An ontogenetic approach to diatom systematics, in *Proceedings of the 7th International Diatom Symposium* (ed. D. G. Mann), Koeltz, Koenigstein, pp. 113–144.

Mann, D. G. (1984b) Protoplast rotation, cell division and frustule symmetry in the diatom *Navicula bacillum. Annals of Botany*, 53, 295–302.

Mann, D. G. (1984c) Observations on copulation in *Navicula pupula* and *Amphora ovalis* in relation to the nature of diatom species. *Annals of Botany*, 54, 429–438.

Mann, D. G. (1984d) Structure, life history and systematics of *Rhoicosphenia* (Bacillariophyta). V. Initial cell and size reduction in *Rh. curvata* and a description of the Rhoicospheniaceae fam. nov. *Journal of Phycology*, 20, 544–555.

Mann, D. G. (1989a) On auxospore formation in *Caloneis* and the nature of *Amphiraphia* (Bacillariophyta). *Plant Systematics and Evolution*, 163, 43–52.

Mann, D. G. (1989b) The diatom genus *Sellaphora*: separation from *Navicula*. *British Phycological Journal*, 24, 1–20.

Mann, D. G. (1989c) The species concept in diatoms: evidence for morphologically distinct, sympatric gamodemes in four epipelic species. *Plant Systematics and Evolution*, 164, 215–237.

Mann, D. G. (1994) The origins of shape and form in diatoms: the interplay between morphogenetic studies and systematics, in *Shape and Form in Plants and Fungi* (eds D. S. Ingram and A. Hudson), Academic Press, London, pp. 17–38.

Mann, D. G. and Droop, S. J. M. (1996) Biodiversity, biogeography and conservation of diatoms, in *Biogeography of Freshwater Algae* (ed. J. Kristiansen), *Hydrobiologia*, 336, 19–32.

Mann, D. G. and Stickle, A. J. (1988) Nuclear movements and frustule symmetry in raphid pennate diatoms. In *Proceedings of the 9th International Diatom Symposium*, F. Round (ed.), 281–291. Biopress, Bristol and Koeltz Scientific Books, Koenigstein.

Mann, D. G. and Stickle, A. J. (1997) Sporadic evolution of dorsiventrality in raphid diatoms, with special reference to *Lyrella amphoroides* sp. nov. *Nova Hedwigia*, 65, 59–77.

Massé, G., Rincé, Y., Cox, E. J., Allard, G., Belt, S. and Rowland, S. (2001) *Haslea salstonica* sp. nov. and *Haslea pseudostrearia* sp. nov., two new marine diatoms from the Salcombe estuary, Devon, U.K. *Comptes Rendus de l'Academie des Sciences, Life Sciences*, 324, 617–626.

Mathur, J., Szabados, L., Schaefer, S., Grunenberg, B., Lossow, A., Jonas-Straube, E., Schell, J., Koncz, C. and Koncz-Kalman, Z. (1998) Gene identification with sequenced T-DNA tags generated by transformation of *Arabidopsis* cell suspension. *Plant Journal*, 13, 707–716.

Meyer, Y. and Abel, W. D. (1975) Importance of the cell wall for cell division and in the activity of the cytoplasm in cultured tobacco protoplasts. *Planta*, 123, 33–40.

Moniz de Sá, M. and Drouin, G. (1996) Phylogeny and substitution rates of angiosperm actin genes. *Molecular Biology and Evolution*, 13, 1198–1212.

Passy-Tolar, S. I. and Lowe, R. L. (1995) *Gomphoneis mesta* (Bacillariophyta). II. Morphology of the initial frustules and perizonium ultrastructure with some inferences about diatom evolution. *Journal of Phycology*, 31, 447–456.

Pickett-Heaps, J. D. (1998) Cell division and morphogenesis of the centric diatom *Chaetoceros decipiens* (Bacillariophyceae) II. Electron microscopy and a new paradigm for tip growth. *Journal of Phycology*, 34, 995–1004.

Pickett-Heaps, J. D. and Bajer, A. S. (1977) Mitosis: an argument for multiple mechanisms achieving chromosomal movement. *Cytobios*, 19, 171–180.

Pickett-Heaps, J. D., Gunning, B. E. S., Brown, R. C., Lemmon, B. E. and Cleary, A. L. (2000) The cytoplast concept in dividing plant cells: cytoplasmic domains and the evolution of spatially organized cell division. *American Journal of Botany*, 86, 153–172.

Pickett-Heaps, J., Schmid, A. M. M. and Edgar, L. A. (1990) The cell biology of diatom valve formation. *Progress in Phycological Research*, 7, 1–168.

Pickett-Heaps, J., Schmid, A. M. M. and Tippit, D. H. (1984) Cell division in diatoms: a translation of part of Robert Lauterborn's treatise of 1896 with some modern confirmatory observations. *Protoplasma*, 120, 132–154.

Pickett-Heaps, J. D. and Spurck, T. P. (1982) Studies on kinetochore function in mitosis. I. The effects of colchicine and cytochalasin on mitosis in the diatom *Hantzschia amphioxys*. *European Journal of Cell Biology*, 28, 77–82.

Pickett-Heaps, J. D., Tippit, D. H. and Andreozzi, J. A. (1979) Cell division in the pennate diatom *Pinnularia*. IV. Valve morphogenesis. *Biologie Cellulaire*, 35, 199–206.

Poll, W. H. van de, Vrieling, E. G. and Gieskes, W. W. C. (1999) Location and expression of frustulins in the pennate diatoms *Cylindrotheca fusiformis*, *Navicula pelliculosa* and *Navicula salinarum* (Bacillariophyceae). *Journal of Phycology*, 35, 1044–1053.

Prasad, A. K. S. K., Riddle, K. A. and Nienow, J. A. (2000) Marine diatom genus *Climaconeis* (Berkeleyaceae, Bacillariophyta): two new species *Climaconeis koenigii* and *C. colemaniae* from Florida Bay, USA. *Phycologia*, 39, 199–211.

Prygiel, J. (1991) Use of benthic diatoms in surveillance of the Artois–Picardie basin hydrobiological quality, in *Use of Algae for Monitoring Rivers* (eds B. A. Whitton, E. Rott and G. Friedrich), Universität Innsbruck, Innsbruck, pp. 89–96.

Quatrano, R. S. (1978) Development of cell polarity. *Annual Review of Plant Physiology*, 29, 487–510.

Quatrano, R. S., Brian, L., Aldridge, J. and Schultz, T. (1991) Polar axis fixation in *Fucus* zygotes: components of the cytoskeleton and extra-cellular matrix. *Development*, S1, 11–16.

Quatrano, R. S., Griffing, L. R., Huber-Walchli, V. and Doubet, S. (1985) Cytological and biochemical requirements for the establishment of a polar cell. *Journal of Cell Science Supplement*, 2, 129–141.

Raven, J. A. (1983) The transport and function of silicon in plants. *Biological Reviews*, 58, 179–207.

Reimann, B. E. F., Lewin, J. C. and Volcani, B. E. (1966) Studies on the biochemistry and fine structure of silica shell formation in diatoms. II. The structure of the cell wall of *Navicula pelliculosa* (Bréb.) Hilse. *Journal of Phycology*, 2, 74–84.

Reuzeau, C., Doolittle, K. W., McNally, J. G. and Pickard, B. G. (1997) Covisualization in living onion cells of putative integrin, putative spectrin, actin, putative intermediate filaments, and other proteins at the cell membrane and in an endomembrane sheath. *Protoplasma*, 199, 173–197.

Robinson, D. H. and Sullivan, C. W. (1987) How do diatoms make silicon biominerals? *Trends in Biochemical Sciences*, 12, 151–154.

Roshchin, A. M. (1994) Dvudomnoe vorsproizvedenie *Achnanthes longipes* Ag. (Bacillariophyta). *Algologia*, 4, 22–29.

Ross, R., Cox, E. J., Karayeva, N. I., Mann, D. G., Paddock, T. B. B., Simonsen, R. and Sims, P. A. (1979) An amended terminology for the siliceous components of the diatom cell. *Nova Hedwigia Beiheft*, 64, 511–533.

Round, F. E., Crawford, R. M. and Mann, D. G. (1990) *The Diatoms: Biology and Morphology of the Genera*. Cambridge University Press, Cambridge.

Sangster, A. G. and Parry, D. W. (1981) Ultrastructure of silica deposits in higher plants, in *Silicon and Siliceous Structures in Biological Systems* (eds T. L. Simpson and B. E. F. Volcani), Springer-Verlag, New York, pp. 383–407.

Schindler, M., Meiners, S. and Cheresh, D. A. (1989) RGD-dependent linkage between plant cell wall and plasma membrane: consequences for growth. *Journal of Cell Biology*, 108, 1955–1965.

Schmid, A. M. (1980) Valve morphogenesis in diatoms. A pattern-related filamentous system in pennates and the effect of APM, colchicine and osmotic pressure. *Nova Hedwigia*, 33, 811–847.

Schmid, A. M. (1984a) Wall morphogenesis in *Thalassiosira eccentrica*: comparison of auxospore formation and the effect of MT-inhibitors, in *Proceedings of the 7th International Diatom Symposium* (ed. D. G. Mann), Koeltz, Koenigstein, pp. 47–70.

Schmid, A. M. (1984b) Tricornate spines in *Thalassiosira eccentrica* as a result of valve-modelling, in *Proceedings of the 7th International Diatom Symposium* (ed. D. G. Mann), Koeltz, Koenigstein, pp. 71–95.

Schmid, A. M. (1986) Wall morphogenesis in *Coscinodiscus wailesii* Gran and Angst. II.

Cytoplasmic events of valve morphogenesis, in *Proceedings of the 8th International Symposium on Recent and Fossil Diatoms* (ed. M. Ricard), Koeltz, Koenigstein, pp. 293–314.

Schmid, A. M. M. (1994) Aspects of morphogenesis and function of diatom cell walls with implications for taxonomy. *Protoplasma*, 181, 43–60.

Schmid, A. M. M. (1997) Intraclonal variation of the tripolar pennate diatom 'Centronella reicheltii' in culture: strategies of revision to the bipolar *Fragilaria*-form. *Nova Hedwigia*, 65, 27–45.

Schmid, A. M. M., Eberwein, R. K. and Hesse, M. (1996) Pattern morphogenesis in cell walls of diatoms and pollen grains: a comparison. *Protoplasma*, 193, 144–173.

Schmid, A. M. M. and Volcani, B. E. (1983) Wall morphogenesis in *Coscinodiscus wailesii* Gran and Angst. 1. Valve morphology and development of its architecture. *Journal of Phycology*, 19, 387–402.

Schwuchow, J., Sack, F. D. and Hartmann, E. (1990) Microtubule distribution in gravitropic protonemata of the moss *Ceratodon*. *Protoplasma*, 159, 60–69.

Smyth, D. R. (1990) *Arabidopsis thaliana*: a model plant for studying the molecular basis of morphogenesis. *Australian Journal of Plant Physiology*, 17, 323–331.

Stosch, H. A. von (1965) Manipulierung der Zellgrösse von Diatomeen im Experiment. *Phycologia*, 5, 21–44.

Stosch, H. A. von (1977) Observations on *Bellerochea* and *Streptotheca* including descriptions of three new planktonic diatom species. *Nova Hedwigia Beiheft*, 54, 113–166.

Stosch, H. A. von (1982) On auxospore envelopes in diatoms. *Bacillaria*, 5, 127–156.

Stosch, H. A. von and Kowallik, K. (1969) Der von L. Geitler aufgestellte Satz über die Notwendigkeit einer Mitose für jede Schalenbildung von Diatomeen. Beobachtungen über die Reichweite und Überlegungen zu seiner zellmechanischen Bedeutung. *Österreichische Botanische Zeitung*, 116, 454–474.

Sullivan, C. W. (1976) Diatom mineralization of silicic acid. I. $Si(OH)_4$ transport characteristics in *Navicula pelliculosa*. *Journal of Phycology*, 12, 390–396.

Suzuki, K., Itoh, T. and Sasamoto, H. (1998) Cell wall architecture prerequisite for the cell division in the protoplasts of white poplar, *Populus alba* L. *Plant and Cell Physiology*, 39, 632–638.

Thaler, F. (1972) Beitrag zur Entwicklungsgeschichte und zum Zellbau einiger Diatomeen. *Österreichische botanische Zeitschrift*, 120, 313–347.

Werner, D. (1977) Introduction with a note on taxonomy, in *The Biology of Diatoms* (ed. D. Werner), *Botanical Monographs*, 17, 1–17.

Williams, D. M. (2001) Comments on the structure of 'post-auxospore' valves of the species *Fragilariaforma virescens*, in *Festschrift for Horst Lange-Bertalot* (eds R. Jahn, J. P. Kociolek, A. Witkowski and P. Compère), 1, 103–117.

Witkowski, A., Lange-Bertalot, H. and Stachura, K. (1998) New and confused species in the genus *Navicula* (Bacillariophyceae) and the consequences of restricted generic circumscription. *Cryptogamie, Algologie*, 19, 83–108.

Worderman, L., McDonald, K. L. and Cande, Z. W. (1986) The distribution of cytoplasmic microtubules throughout the cell cycle of the centric diatom *Stephanopyxis turris*: their role in nuclear migration and positioning the mitotic spindle during cytokinesis. *Journal of Cell Biology*, 102, 1688–1698.

Wyatt, S. E. and Carpita, N. C. (1993) The plant cytoskeleton-cell-wall continuum. *Trends in Cell Biology*, 3, 413–417.

Chapter 25

Identifying the genetic causes of phenotypic evolution: a review of experimental strategies

David A. Baum

ABSTRACT

The central challenge in evolutionary developmental genetics is the identification of the genes responsible for heritable phenotypic differences between species. Expression-based studies, whether candidate gene driven or genomic in structure, are useful for determining whether a gene is involved in the production of a derived character. However, such approaches are prone to false positive results and do not easily distinguish genes that caused the phenotype to evolve ('primary' genes) from those that are downstream of the causal genes ('secondary' genes). Genetic approaches, whether involving crosses or transformation, are well suited to identifying primary genes but also have limitations. In the case of classical genetic approaches, one can only work with interfertile study organisms and it is usually difficult to clone the genetic locus of interest. Transformation-based approaches, whether candidate gene driven or using a shotgun strategy, are powerful but may pose significant technical challenges and are prone to false negative results. Nonetheless, through the use of diverse research strategies, it should be possible to pinpoint the molecular basis of phenotypic evolution in a number of cases and then use that information to clarify the types of changes to development that are most important in causing species differences.

25.1 Introduction

Achieving the long-sought synthesis of evolutionary and developmental biology requires a clear understanding of how genetic changes within populations accumulate to give rise to the phenotypic differences between species. Despite the fact that all derived alleles must at some time have existed as polymorphisms within populations (Purugganan, 2000), we may not be able to extrapolate directly from the population level to an understanding of species-level evolution (Arthur, 2000). The allelic variation that underlies phenotypic polymorphism within populations may not be representative of the kinds of mutations that accumulate to differentiate species. This follows because macroevolution could be dominated by classes of variation that are exceedingly rare within populations. Therefore, the challenge of macroevolutionary genetics is to identify the genetic basis of phenotypic evolution above the species level to determine whether it differs from the causes of phenotypic differentiation within populations. In particular, it is desirable to establish whether

In *Developmental Genetics and Plant Evolution* (2002) (eds Q. C. B. Cronk, R. M. Bateman and J. A. Hawkins), Taylor & Francis, London, pp. 493–507.

the genetic determinants at these two levels differ in the magnitude of their pheno-typic effects, in their genomic locations (e.g. regulatory versus coding regions), or in the classes of mutations involved (e.g. substitutions versus insertion/deletion events).

There are as yet very few empirical studies that have successfully cloned genes that are causally responsible for phenotypic differences between species (Wheeler *et al.*, 1991; Sucena and Stern, 2000). In plants there are no recorded cases involving natural species, although genes have been cloned that appear to be responsible for traits that have evolved during crop domestication (e.g. Doebley *et al.*, 1997; Purug-ganan *et al.*, 2000).

There have been several recent reviews of plant developmental evolution, ranging from general overviews (Kellogg, 1996; Baum, 1998; Purugganan, 1998) to discus-sions of particular organ systems (e.g. Goliber *et al.*, 1999) or gene families (Theißen *et al.*, 2000). However, only one of these (Kellogg, 1996) provided much discussion of research strategy. My main aims in this chapter, therefore, are to summarize and evaluate research strategies for plant macroevolutionary genetics and to point the more evolutionary reader towards case studies that exemplify the use of key experi-mental techniques.

I will begin by discussing the kinds of genetic variation we are interested in and review the phylogenetic structure of the problem. Then I will consider the available experimental approaches, which range from comparative expression analysis to clas-sical and reverse genetic approaches.

25.2 Primary and secondary genes

A gene that is *involved* in the evolution of a particular aspect of the phenotype is a gene whose expression or activity changed coincident with the evolution of a derived character state and which serves to explain some aspects of the derived state. It is likely that for any given phenotype, many genes will have been involved in pheno-typic evolution. This inference follows because, even if there was one key event, such as a modification in the expression of one developmental regulatory gene, all genes that are genetically downstream will also come to be differently expressed. The ulti-mate challenge in the study of phenotypic evolution is, thus, not just to identify genes involved in the evolution of derived characters but also to identify those specific genetic changes that *caused* individual organisms to manifest the derived rather than the ancestral phenotype. I will follow the terminology of McLellan *et al.* (2002) and refer to the genes causing phenotypic evolution as primary genetic factors, and genes that translate those genetic differences into differences in pheno-types as secondary genetic factors. While the identification of secondary genetic factors is useful because they clarify how phenotypic (as opposed to genotypic) evolution happens, the real challenge is to identify primary genes.

25.3 Phylogeny and the choice of study system

Our aim is to identify the primary and secondary genetic changes involved in the transition from an ancestral to a derived character-state (Figure 25.1). Ideally this would be achieved by comparing organisms sampled just before and just after the evolutionary event of interest. However, except in very rare cases (microbial evolution in the laboratory, recent changes in domesticates) such organisms will be unavailable for study. Instead, one is usually working with extant species that manifest the inferred ancestral and derived states (Figure 25.1). For shorthand, I will refer to study species that manifest the ancestral condition as ancestral models, and study species bearing the derived condition as derived models. The use of the term 'model' does not imply that these species will have been studied extensively by geneticists, but refers to the fact that the species in question are chosen to represent the ancestral or derived character state, respectively.

For a given characteristic there may be many living species that manifest the ancestral and derived character states and one needs criteria for selecting particular species to use as models. The ideal ancestral model is one whose divergence from the lineage leading to the derived model occurred just before the evolutionary transition under study. This follows because one thereby minimizes the total amount of evolutionary time that separates the two models and, all things being equal, time is expected to correlate with the number of confounding background changes. However, this requirement could be offset if the species that would make the best ancestral model based on phylogenetic position turns out to be an unsuitable experimental organism (e.g. it is difficult to grow, has a large genome, or is not suitable for

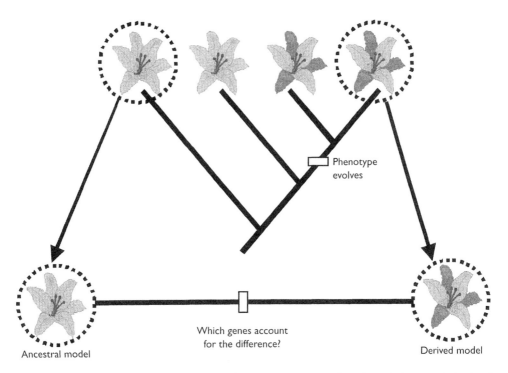

Figure 25.1 An example to illustrate the phylogenetic structure of studies of the genetic basis of phenotypic evolution. Suppose we have identified a phenotypic transition between a unipartite and bipartite perianth. Our aim is to identify the genes that caused the evolution of the derived phenotype (bipartite perianth). In practice, we address this question by identifying the genetic causes of the phenotypic difference between selected living species typifying the ancestral and derived conditions.

transformation) or if it has diverged markedly in the traits of interest. In the case of species showing the derived phenotype they will all be equidistant, in terms of time, from the ancestral model. Therefore, the choice of derived model will depend only on practical issues and the extent to which the species has undergone subsequent phenotypic evolution.

In principle, the methods that will be discussed in this chapter are appropriate for any discrete character that has a genetic basis and varies among living species. However, there is generally a trade-off between the perceived importance of a character and the likelihood that one will be able to successfully identify its primary genetic cause. A trait that characterizes a major clade and that is attributed some special significance, for example as a putative key innovation, is likely to be difficult to study. Major morphological innovations are usually perceived as such because they represent complex transitions with a large 'gap' between ancestral and derived models. This implies that there will be multiple interacting genetic factors, making it very difficult to identify primary genes. Furthermore, even if one or a few key genes had been responsible for the evolution of the trait, the fact that a feature characterizes a major radiation tends to push its origin far back in time, increasing the problems of confusing background evolution. Thus, it is often the case that those features whose origin we would most like to understand (e.g. the flower) are those that are most difficult to study.

Such a perspective should not be used to argue against studying the developmental genetics of major characters such as flowers. Such work is needed to help us envisage how such complex features could have originated by simple genetic changes. However, we also should not be under any delusion that we will soon know the specific genetic changes that resulted in the evolution of the flower. Thus, as a complement to studies of major evolutionary innovations, we need studies of less significant, but more easily studied, characters. Our best chances for identifying primary genes are probably in studies of the minor characteristics that differentiate closely related species (Stern, 2000).

25.4 Comparative gene expression

It is likely that any perceptible phenotypic differences between the ancestral and derived models will entail differences in the expression of genes. Although this will not always be true, it is reasonable to assume that information on expression has the potential to shed light on a gene's role in character evolution. This can be addressed by focusing on specific genes identified beforehand as candidates for being primary or secondary genetic factors, or by screening the genome for genes showing a suggestive expression profile.

25.4.1 Candidate genes approaches

Candidate genes are generally identified by reference to developmental genetic research conducted in a well-studied 'model' species like *Arabidopsis thaliana*, *Antirrhinum majus*, *Nicotiana tabacum* or *Zea mays*. One is most secure in developing candidate gene hypotheses when the prior genetic work is in a species that is closely related to the species being studied or, better still, when a genetic model system is the ancestral or derived model.

Caution should be exercised to ensure that a gene selected for further study could indeed be responsible for the phenotypic transition being studied and that this hypothetical role is amenable to experimental refutation. It goes without saying that just because gene X is needed for the proper development of structure Y does not mean that X played a role in the evolutionary origin or modification of Y. For example, consider the transition from bilobed to entire petals in *Clarkia lingulata* (Lewis and Roberts, 1956). It is quite likely (but not shown) that the *Clarkia* orthologue of *APETALA3* (*AP3*) is critical for the establishment of petal identity in *Clarkia*, as it is in many other eudicots (reviewed in Bowman, 1997). Nonetheless, this gene is not a good candidate for the evolution of petal shape because there is no evidence that manipulation of the activity or expression of *AP3* can influence petal shape. Therefore, a well-conceived candidate gene hypothesis needs more information than nominal importance in the development of the feature under study. One will be able to make the strongest *a priori* case when modification of gene activity/expression causes an ancestral model to manifest the derived phenotype, or a derived model to manifest the ancestral phenotype.

To illustrate these principles at work I will briefly summarize three projects from my own laboratory that utilize candidate gene hypotheses and explain why these hypotheses are plausible.

(1) *Jonopsidium acaule* differs from most other Brassicaceae, including *Arabidopsis*, in producing flowers from the axils of rosette leaves rather than on elongated inflorescences. A candidate gene for this effect is the floral meristem identity gene *LEAFY* (*LFY*), because rosette flowers are induced in *Arabidopsis* plants overexpressing *LFY* (Weigel and Nilsson, 1995). The hypothesis we have explored is that expression of *LFY* is greater in level, greater in spatial extent, and/or is initiated earlier in *J. acaule* than in the ancestral model, *Arabidopsis* (Shu *et al.*, 2000).

(2) *Mohavea confertiflora* has only two stamens at maturity due to the abortion of three adaxial stamens (Endress, 1998). In contrast, species assigned to the traditional genus *Antirrhinum* (within which *Mohavea* is nested; Oyama, unpubl. data) abort only one stamen and, hence, have four stamens at maturity. A candidate gene for this effect is *CYCLOIDEA* (*CYC*), because *cyc* mutants in the ancestral model *Antirrhinum majus* fail to abort adaxial stamens and the CYC gene product is locally expressed in the region of the adaxial stamen primordium (Luo *et al.*, 1996). The hypothesis being tested (Hileman, unpubl. data) is that the additional stamen abortion in *Mohavea* is caused by expanded expression of *CYC* to include the three adaxial stamen primordia.

(3) *Pavonia strictiflora* is one of a few species in Malvaceae that has a much reduced corolla, attracting pollinators (probably hummingbirds) with expanded red sepals (calyx) and bracteoles (epicalyx). Candidate genes for this derived phenotype are the B-group MADS-box genes (orthologues of *APETALA3/DEFICIENS* and *PISTILLATA/GLOBOSA*), because ectopic expression of *AP3* and *PI* in the calyx of *Arabidopsis* causes them to become petaloid (Krizek and Meyerowitz, 1996). The hypothesis being tested is that in *P. strictiflora* one or both B-group genes are expressed in the calyx and epicalyx (Yen, unpubl. data).

In each of these examples, expression data would tend to reject the candidate gene hypothesis if the pattern and level of expression in the ancestral models (*Arabidopsis thaliana*, *Antirrhinum majus* and *Hibiscus rosa-sinensis*) is identical to that observed in the derived models (*Ionopsidium acaule*, *Mohavea confertiflora* and *Pavonia strictiflora*). However, it would remain a formal possibility that the derived phenotype was caused by changes in the activity of the candidate gene rather than its expression.

Gene expression comparison can be conducted at either the RNA or protein levels. The most commonly-used approaches are northern hybridization (e.g. Kramer and Irish, 1999), RT-PCR (e.g. Svensson *et al.*, 2000), *in situ* hybridization (e.g. Shu *et al.*, 2000) and immunolocalization (e.g. Sinha and Kellogg, 1996; Kramer and Irish, 1999). If one finds that the candidate gene is expressed differently in the ancestral and derived models, as predicted under the candidate gene hypothesis, then one can take this as positive evidence for a role in the character's evolution. This claim will be strongest if there is a 1:1 correlation between gene expression and production of the property, both among the tissues within each species and between species. However, it should be borne in mind that this interpretation depends upon the assumption that the gene in question does indeed act in the same way as in the model systems. Ideally, this inference should be evaluated rather than simply assumed.

It may be challenging to formally study gene action in wild species but in many cultivated species mutant populations are available. Furthermore, as more and more species become amenable to plant transformation (Hall, 1999; Bent, 2000), it is becoming increasingly feasible to introduce constructs that will either reduce or increase expression of the gene of interest. Enhanced expression can be achieved by inserting additional copies of the gene (e.g. Blázquez *et al.*, 1997) or by using constitutive viral promoters (e.g. Weigel and Nilson, 1995). Conversely, loss-of-function phenotypes can be generated using antisense constructs (e.g. Bell *et al.*, 1995; Haldrup *et al.*, 1999; Yoshizumi *et al.*, 1999; Kater *et al.*, 2001), overexpression constructs that induce transcriptional silencing (e.g. Que *et al.*, 1997; Haldrup *et al.*, 1999), or constructs that contain inverted repeats that produce double-stranded RNA hairpin structures, which tend to induce transcriptional silencing (Chuang and Meyerowitz, 2000; Smith *et al.*, 2000). All these constructs can be designed to include inducible promoters or protein domains, allowing the manipulation of expression/activity during development (e.g. Simon *et al.*, 1996; Yoshizumi *et al.*, 1999).

If the data support a candidate gene hypothesis, the next step would be to evaluate whether *cis*-regulatory changes at that locus are the primary genetic cause of the phenotype or whether the primary genetic factor is upstream (acting on the candidate gene via direct or indirect *trans*-regulation). If, in contrast, the expression data reject the candidate gene hypothesis, two possibilities should be considered: first, the gene is the primary genetic factor but the genetic changes are in gene activity rather than expression or, second, the gene is not involved in the evolution of the derived character state. In the former case one predicts that, if the gene is a transcription factor or part of a signal transduction cascade, genes downstream should show modified expression, a prediction that could be tested by further studies of gene expression. In the latter case one would wish to look at the expression of genes in parallel pathways or downstream of the original candidate. Alternatively, one could abandon the candidate gene approach and, instead, screen the genome for genes that show the predicted expression profile.

25.4.2 Genomic approaches

Genomic approaches to the study of character evolution involve searching through the genome for genes showing expression patterns consistent with a role in character evolution. Specifically, one attempts to identify genes that are expressed in the modified structure in the derived model but not in the unmodified structure in the ancestral model, or vice versa. Several techniques permit such a genome-wide perspective, including differential display (e.g. Heck *et al.*, 1995), subtractive hybridization (e.g. Wen *et al.*, 1999), and microarrays (e.g. Cavalieri *et al.*, 2000; Schenk *et al.*, 2000). While used extensively to look for genes differentially expressed in different tissues or in response to different environmental cues, these approaches have not been used successfully to study plant species diversity. However, as microarray approaches become progressively easier to implement, cheaper, and more reliable (Lockhart and Winzeler, 2000), there are certain to be attempts to use this technique for evolutionary studies. Therefore, it is sensible to consider the strengths and weaknesses of genome-wide expression profiling in macroevolutionary genetics.

While a genomic approach has the advantage of not depending on prior information from genetic analysis in other species, it is likely to generate numerous false positives. This is because there could be many genes whose expression profiles satisfy the predicted pattern but which do not have any bearing on the evolution of the derived character. These genes might be random differences of no consequence or they could play a role in other phenotypic differences between the species under study.

Given the likelihood of many false positives it would be sensible to develop criteria for evaluating the likelihood that a gene having the predicted expression profile is a primary genetic factor for the trait under study. As a first cut, one might focus on genes that, based on work in genetic model systems, could plausibly be active in the generation of the derived phenotype. Additionally, if multiple genes from the same pathway show the same pattern the best candidates will be those near the top of the regulatory hierarchy. In theory, it may be possible to infer the regulatory relationships between genes in one of the model species by looking at their tendency to be co-expressed in different tissues, at different ontogenetic stages, and in different growth conditions (Kim *et al.*, 2000). A final way to rank candidate genes would be based on a comparison of gene sequences between the two species and nearby relatives. Specifically, genes that show accelerated evolution, an excess of replacement substitution, or reduced polymorphism in the derived model relative to the ancestral model may be good candidates for further study because the anomalous patterns of molecular evolution could reflect a historical shift in gene function associated with the origin of the derived character (e.g. Hanson *et al.*, 1996; Wang *et al.*, 1999; White and Doebley, 1999). However, given that these three methods for ranking candidate genes are all prone to errors, it may prove beneficial in the long run to develop efficient methods with which to assess the causal role of all genes that have suggestive expression profiles.

25.5 Identifying primary genetic factors

Ultimately, the best way to show that a candidate gene is a primary genetic cause is through a controlled genetic experiment. If gene A is the cause of the phenotypic difference between the ancestral and derived models, replacement of gene A from the ancestral model (A_a) with the gene copy from the derived model (A_d) should result in the production of the derived phenotype (Figure 25.2). In principle, such an experiment can be achieved using either classical genetics (i.e. introgression) or transformation. Each of these approaches has limitations and potential difficulties of interpretation that merit brief exploration.

25.5.1 Classical genetic approaches

In its purest manifestation, the introgression approach involves making hybrids between the two model species followed by repeated backcrossing with the ancestral model species. Through the use of genetic markers it is possible to generate progeny that have the ancestral model genotype except that they are homozygous for the candidate locus from the derived model (A_d/A_d). If these plants show the derived phenotype then the hypothesis that the candidate gene was responsible for the derived character is supported. This approach has been used, for example, to identify regions that explain derived morphological characteristics of maize relative to teosinte (Doebley and Stec, 1993; Dorweiler *et al.*, 1993).

In cases where one does not have a candidate locus, the same basic approach can

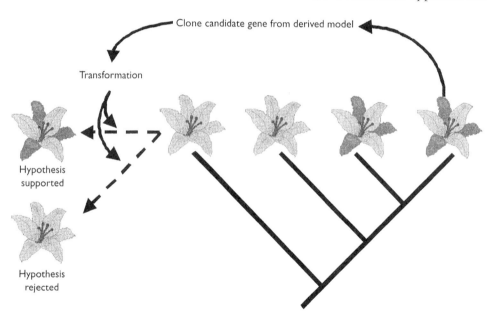

Figure 25.2 The transformational strategy for testing a candidate gene hypothesis. The candidate gene (with regulatory sequences) is cloned from the derived model and introduced into the ancestral model. If the transformants manifest the derived phenotype the candidate gene hypothesis would be supported. If the transformants manifest the ancestral phenotype (or some other phenotype) the hypothesis would tend to be rejected (see text). However, if the hypothesis predicts that the evolutionary transition involved loss-of-function mutations, the individual of the ancestral model undergoing transformation would need to be a mutant at the endogenous locus. Other reasons why a negative (or positive) result can be misleading are discussed in the text.

be used simultaneously for all parts of the genome using statistical methods to keep track of the phenotypic effects of different regions. This approach, termed quantitative trait locus (QTL) mapping, is very powerful and is becoming used widely in evolutionary biology (McLellan *et al.*, 2002). In plants, QTL mapping has identified major loci involved in the domestication of maize (Doebley and Stec, 1993) and for floral differences between wild species of *Mimulus* (Bradshaw *et al.*, 1996, 1998) and *Microseris* (Vlot *et al.*, 1992; Gailing *et al.*, 1999).

The main disadvantage of classical genetic approaches is that, without massive sample populations and a dense genetic map, one is unlikely to be able to localize the genetic effect to a particular gene, let alone to a specific modification of that gene or its flanking sequences. Furthermore, these approaches are limited to cases in which ancestral and derived models are interfertile, which greatly restricts the species pairs and thus the characters that can be studied.

25.5.2 Transformational approaches

The simplest transformation strategy would be to clone the candidate gene (A_d) from the derived model, including all 5' and 3' *cis*-regulatory regions, and introduce it into the ancestral model species to see whether the resulting transgenic lines exhibit the derived phenotype (Figure 25.2). If A_d was dominant to A_a (e.g. because A_d has a broader expression domain) one might hope to obtain the derived phenotype even in a wildtype background. Ideally, however, transformation should use lines that have a non-functional copy of the endogenous gene, so that one can observe the effect of A_d in the ancestral genetic background without having to worry about the presence of A_a.

If one obtained the derived phenotype by placing the derived gene in an ancestral genotypic background (whether mutant or wildtype), one could make a strong case for the gene being a primary genetic determinant of the evolutionary transition under study. It is, however, conceivable that a secondary genetic difference could mimic a primary difference as a result of genetic assimilation. Imagine that a change in the promoter of gene X, causing the gene to be expressed in the calyx as well as the corolla, was responsible for the expansion of red pigment from the corolla to the calyx of *Pavonia strictiflora*. Suppose that its target gene, A, was ancestrally only expressed in the presence of gene product X. If, after the transfer of pigment to the calyx, the promoter of A picked up an enhancer element that caused it to be expressed in the calyx regardless of X, that change would be more or less neutral and could easily become fixed. In that case, introduction of A_d into the ancestral model could result in a pigmented calyx, despite the fact that gene A was not the evolutionary cause of the red calyx. Nonetheless, while a theoretical possibility, such a confounding result is probably relatively rare because it depends upon a series of individually improbable events. Furthermore, in principle, such cases of genetic assimilation could be identified by conducting transformation experiments with genes located upstream of the initial candidate.

If one obtained a positive result from the transformation experiments it would be possible to conduct a series of further experiments to precisely localize the region of the insert that is responsible for the effect. In particular, one could evaluate whether the effect is mediated by *cis*-regulatory elements or by the coding sequence itself by making chimeric constructs containing, for example, the 5′ non-transcribed region of A_a and the coding region of A_d. Additionally, one could examine the properties of the A_d promoter by fusing it to a suitable reporter. Through such manipulations one could potentially home in on the actual nucleotide basis of phenotypic evolution.

The basic transformation approach just described is candidate gene driven. However, a transformational strategy could, in principle, circumvent this restriction. For example, it ought to be possible to use shotgun transformation to generate multiple transgenic lines of the ancestral model that contain random genomic fragments from the derived model. These lines could then be screened for the derived character-state. While this strategy would seem capable of pinpointing the primary genetic factors underlying phenotypic evolution, it has yet to be used, primarily because of technical hurdles such as obtaining sufficiently high transformation efficiencies and developing efficient ways to screen large numbers of transgenic lines.

The transformation approach, with or without candidate genes, implicitly assumes that the derived character is encoded by few genes of large phenotypic effect. This follows because, if the phenotype were encoded by many genes of small effect, the impact of those individual genes would be too small to detect in transgenic plants. There remains controversy as to how much of plant evolution occurs via genes of large effect, but it is incontrovertible that at least some character evolution involves genes of large effect (e.g. Gottlieb, 1984; Bradshaw *et al.*, 1998; Doebley and Lukens, 1998). Therefore, while the transformation approach cannot be expected to work in all cases, it represents a suitable strategy for studying many phenotypes.

The transformation approach is prone to yielding false negative results in which a primary gene fails to induce the derived phenotype. This could happen for practical reasons, such as using a construct that is too short (missing important regulatory elements) or too long (including fragments of flanking sequences that affect expression). Additionally, it is easy to imagine a negative or ambiguous result due to changes in the genetic background. For example, genomic co-evolution could lead to concerted evolution of enhancer elements and transcription factors (Dover, 2000) or there could be changes in the mechanisms of gene regulation.

The probability that background genetic changes will confound transformation experiments is dependent on the time elapsed since the ancestral and derived model species diverged. While one might expect few difficulties working with closely related models, such as *Brassica* and *Arabidopsis* or potato and tomato, it will presumably be a problem if one was comparing an angiosperm and a moss. What we do not know is how distantly related our models can be without greatly compromising our ability to learn something meaningful about the role of a candidate gene in phenotypic evolution. Whatever that limit is, it should be borne in mind that the threshold will probably be much lower for studies of evolutionary change than for studies of evolutionary conservation. Showing, for example, that an *Antirrhinum* gene can rescue an *Arabidopsis* mutation when driven by the *Arabidopsis* promoter demonstrates conservation of gene activity (Irish and Yamamoto, 1995). It is likely to be much harder to use such an approach to determine if a particular gene is the cause of differences between asterids and rosids.

Despite the problems associated with the transformational approach, it has great potential for evolutionary developmental genetic studies. The best example is a study by Wheeler *et al.* (1991) which suggested that species differences in *Drosophila* mating songs are due to changes at the *PER* locus. In plants, Samach *et al.* (1997) used such a strategy to try to understand differences in the regulation of B-group MADS-box genes between *Arabidopsis* and *Petunia*. We can hope that, given recent advances in transformation technology (Hall, 1999; Bent, 2000) and improved knowledge of plant genomes, increasing numbers of researchers will use a transformation strategy to study plant phenotypic evolution.

25.6 Conclusions and future prospects

Despite the availability of a variety of approaches for identifying the genetic basis of character evolution, there are several good reasons why researchers might shy away from this area. The experimental methods needed are often risky, technically demanding and/or expensive. Candidate gene approaches can potentially yield results in short order, but they are prone to negative results, which tend to be difficult to publish. Conversely, approaches that circumvent candidate genes such as comparative microarray analysis, QTL mapping or shotgun transformation are likely to yield publishable results, but their time horizon is generally too long for conventional funding sources. So why, one might wonder, are so many researchers currently venturing into this field?

What has become widely appreciated is that, yes, identifying the genetic causes of character evolution is difficult but, at the same time, it is tremendously important to try. We will never fully understand phenotypic evolution until we have determined how closely the genetic changes that cause character evolution resemble the genetic variation that is found in extant populations. Thus, given the methodological and theoretical advances summarized in this chapter, and the associated improved prospects of success, more and more scientists consider the career risks worthwhile. Additionally, even when one fails to pinpoint the genes one is looking for, valuable insights can be gained. For example, although QTL studies have not yet resulted in the identification of cloned genes causing morphological evolution (with the noticeable exception of genes involved in the evolution of the maize architecture: Doebley *et al.*, 1997; White and Doebley, 1999), they have greatly clarified the genetic architecture of morphological diversity. Likewise, while we are still a long way from understanding what genetic changes caused the evolution of the flower, studies of floral organ and meristem identity genes in divergent angiosperm lineages and in gymnosperms (e.g. Kramer and Irish, 1999; Sundström *et al.*, 1999; Becker *et al.*, 2000; Frohlich and Parker, 2000) have helped us to develop theories that are subject to further experimental verification (e.g. Frohlich and Parker, 2000; Frohlich, 2002).

The central mission of this chapter is to review the pros and cons of different methods for studying the genes that underlie phenotypic evolution, and to provide an entrée into the relevant literature. However, a subtler but no less heartfelt aim is to promulgate the message that it is worth trying. Through the collaboration of evolutionary biologists and developmental geneticists, or through the efforts of individuals who manage to bridge these two fields, there is every reason to believe that rapid progress will be made. Indeed, I am optimistic that, in one or two decades, enough case studies will have been completed for us to have sound data for developing realistic theories to explain the molecular macroevolution of plant phenotypes.

ACKNOWLEDGEMENTS

I gratefully acknowledge the colleagues who provided useful feedback on this manu-script: Richard Bateman, David Des Marais, Veronica Di Stilio, Daniel Fulop, Lena Hileman, Dianella Howarth, Elena Kramer, Andrew Murray, Ryan Oyama and one anonymous reviewer. I also thank the editors for arranging such an engaging confer-ence and for being willing to consider this unsolicited submission.

REFERENCES

Arthur, W. (2000) The concept of developmental reprogramming and the quest for an inclu-sive theory of evolutionary mechanisms. *Evolution and Development*, 2, 49–57.

Baum, D. A. (1998) The evolution of plant development. *Current Opinion in Plant Biology*, 1, 79–86.

Becker, A., Winter, K.-U., Meyer, B., Saedler, H. and Theißen, G. (2000) MADS-box gene diversity in seed plants 300 million years ago. *Molecular Biology and Evolution*, 17, 1425–1434.

Bell, E., Creelman, R. A. and Mullet, J. E. (1995) A chloroplast lipoxygenase is required for wound-induced jasmonic acid accumulation in *Arabidopsis*. *Proceedings of the National Academy of Sciences of the United States of America*, 92, 8675–8679.

Bent, A. F. (2000) Arabidopsis *in planta* transformation. Uses, mechanisms, and prospects for transformation of other species. *Plant Physiology*, 124, 1540–1547.

Blázquez, M. A., Soowal, L. N., Lee, I. and Weigel, D. (1997) *LEAFY* expression and flower initiation in *Arabidopsis*. *Development*, 124, 3835–3844.

Bowman, J. L. (1997) Evolutionary conservation of angiosperm flower development at the molecular and genetic levels. *Journal of Bioscience*, 22, 515–527.

Bradshaw, H. D., Otto, K. G., Frewen, B. E., McKay, J. K. and Schemske, D. W. (1998) Quantitative trait loci affecting differences in floral morphology between two species of monkeyflower (*Mimulus*). *Genetics*, 149, 367–382.

Bradshaw, H. D., Wilbert, S. M., Otto, K. G. and Schemske, D. W. (1996) Genetic mapping of floral traits associated with reproductive isolation in monkeyflowers (*Mimulus*). *Nature*, 376, 762–765.

Cavalieri, D., Townsend, J. P. and Hartl, D. L. (2000) Manifold anomalies in gene expression in a vineyard isolate of *Saccharomyces cerevisiae* revealed by DNA microarray analysis. *Proceedings of the National Academy of Sciences of the United States of America*, 97, 12369–12374.

Chuang, C.-F. and Meyerowitz, E. M. (2000) Specific and heritable genetic interference by double-stranded RNA in *Arabidopsis thaliana*. *Proceedings of the National Academy of Sciences of the United States of America*, 97, 4985–4990.

Doebley, J. and Lukens, L. (1998) Transcriptional regulators and the evolution of plant form. *Plant Cell*, 10, 1075–1082.

Doebley, J. and Stec, A. (1993) Inheritance of the morphological differences between maize and teosinte: comparison of results for two F2 populations. *Genetics*, 134, 559–570.

Doebley, J., Stec, A. and Hubbard, L. (1997) The evolution of apical dominance in maize. *Nature*, 386, 485–488.

Dorweiler, J., Stec, A., Kermicle, J. and Doebley, J. (1993) *Teosinte glume architecture 1*: A genetic locus controlling a key step in maize evolution. *Science*, 262, 233–235.

Dover, G. (2000) How genomic and developmental dynamics affect evolutionary processes. *BioEssays*, 22, 1153–1159.

Endress, P. K. (1998) *Antirrhinum* and Asteridae – evolutionary changes of floral symmetry. *Society for Experimental Biology Symposium Series*, 51, 133–140.

Frohlich, M. W. (2002) The Mostly Male theory of flower origins: summary and update regarding the Jurassic pteridosperm Pteroma, in *Developmental Genetics and Plant Evolution* (eds Q. C. B. Cronk, R. M. Bateman and J. A. Hawkins), Taylor & Francis, London, pp. 85–108.

Frohlich, M. W. and Parker, D. S. (2000) The mostly male theory of flower evolutionary origins: from genes to fossils. *Systematic Botany*, 25, 155–170.

Gailing, O., Hombergen, E.-J. and Bachmann, K. (1999) QTL mapping reveals specific genes for the evolutionary reduction of microsporangia in *Microseris* (Asteraceae). *Plant Biology* 1, 219–225.

Goliber, T., Kessler, S., Chen, J. J., Bharathan, G. and Sinha, N. (1999) Genetic, molecular, and morphological analysis of compound leaf development. *Current Topics in Developmental Biology*, 43, 259–290.

Gottlieb, L. D. (1984) Genetics and morphological evolution in plants. *American Naturalist*, 123, 681–709.

Haldrup, A., Naver, H. and Scheller, H. V. (1999) The interaction between plastocyanin and photosystem I is inefficient in transgenic *Arabidopsis* plants lacking the PSI-N subunit of photosystem I. *Plant Journal*, 17, 689–698.

Hall, R. D. (ed.) (1999) *Plant Cell Culture Protocols*. Humana Press Inc., Totowa, NJ.

Hanson. M. A., Gaut, B. S., Stec, A. O., Fuerstenberg, S. I., Goodman, M. M., Coe, E. H. and Doebley, J. F. (1996) Evolution of anthocyanin biosynthesis in maize kernels: the role of regulatory and enzymatic loci. *Genetics*, 143, 1395–1407.

Heck, G. R., Perry, S. E., Nichols, K. W. and Fernandez, D. E. (1995) *AGL15*, a MADS domain protein expressed in developing embryos. *Plant Cell*, 7, 1271–1282.

Irish, V. F. and Yamamoto, Y. T. (1995) Conservation of floral homeotic gene function between *Arabidopsis* and *Antirrhinum*. *Plant Cell*, 7, 1635–1644.

Kater, M. M., Franken, J., Carney, K. J., Colombo, L. and Angenent, G. C. (2001) Sex determination in the monoecious species cucumber is confined to specific floral whorls. *Plant Cell*, 13, 481–493.

Kellogg, E. A. (1996) Integrating genetics, phylogenetics and developmental biology, in *Impact of Plant Molecular Genetics* (ed. B. W. S. Sobral), Birkhäuser, Boston, pp. 159–172.

Kim, S. C., Dougherty, E. R., Chen, Y. D., Sivakumar, K., Meltzer, P., Trent, J. M. and Bittner, M. (2000) Multivariate measurement of gene expression relationships. *Genomics*, 67, 201–209.

Kramer, E. M. and Irish, V. F. (1999) Evolution of genetic mechanisms controlling petal development. *Nature*, 399, 144–148.

Krizek, B. A. and Meyerowitz, E. M. (1996) The *Arabidopsis* homeotic genes *APETALA3* and *PISTILLATA* are sufficient to provide the B class organ identity function. *Development*, 122, 11–22.

Lewis, H. and Roberts, M. R. (1956) The origin of *Clarkia lingulata*. *Evolution*, 10, 126–138.

Lockhart, D. J. and Winzeler, E. A. (2000) Genomics, gene expression and DNA arrays. *Nature*, 405, 827–836.

Luo, D., Carpenter, R., Vincent, C., Copsey, L. and Coen, E. (1996) Origin of floral asymmetry in *Antirrhinum*. *Nature,* 383, 794–799.

McLellan, T., Shephard, H. L. and Ainsworth, C. C. (2002) Identification of genes involved in evolutionary diversification of leaf morphology, in *Developmental Genetics and Plant Evolution* (eds Q. C. B. Cronk, R. M. Bateman and J. A. Hawkins), Taylor & Francis, London, pp. 315–329.

Purugganan, M. D. (1998) The molecular evolution of development. *Bioessays,* 20, 700–711.

Purugganan, M. D. (2000) The molecular population genetics of regulatory genes. *Molecular Ecology*, 9, 1451–1461.

Purugganan, M. D., Boyles, A. L. and Suddith, J. I. (2000) Variation and selection at the *CAULIFLOWER* floral homeotic gene accompanying the evolution of domesticated *Brassica oleracea*. *Genetics*, 155, 855–862.

Que, Q. D., Wang, H.-Y., English, J. J. and Jorgensen, R. A. (1997) The frequency and degree of cosuppression by sense chalcone synthase transgenes are dependent on transgene promoter strength and are reduced by premature nonsense codons in the transgene coding sequence. *Plant Cell*, 9, 1357–1368.

Samach, A., Kohalmi, S. E., Motte, P., Datla, R. and Haughn, G. W. (1997) Divergence of function and regulation of class B floral organ identity genes. *Plant Cell*, 9, 559–570.

Schenk, P. M., Kazan, K., Wilson, I., Anderson, J. P., Richmond, T., Somerville, S. C. and Manners, J. M. (2000) Coordinated plant defense responses in *Arabidopsis* revealed by microarray analysis. *Proceedings of the National Academy of Sciences of the United States of America*, 97, 11655–11660.

Shu, G., Amaral, W., Hileman, L. C. and Baum, D. A. (2000) *LEAFY* and the evolution of rosette flowering in violet cress (*Jonopsidium acaule*, Brassicaceae). *American Journal of Botany*, 87, 634–641.

Simon, R., Igeno, M. I. and Coupland, G. (1996) Activation of floral meristem identity genes in *Arabidopsis*. *Nature*, 384, 59–62.

Sinha, N. R. and Kellogg, E. A. (1996) Parallelism and diversity in multiple origins of C-4 photosynthesis in the grass family. *American Journal of Botany*, 83, 1458–1470.

Smith, N. A., Singh, S. P., Wang, M.-B., Stoutjesdijk, P. A., Green, A. G. and Waterhouse, P. M. (2000) Total silencing by intron-spliced hairpin RNAs. *Nature*, 407, 319–320.

Stern, D. L. (2000) Perspective: Evolutionary developmental biology and the problem of variation. *Evolution*, 54, 1079–1091.

Sucena, E. and Stern, D. L. (2000) Divergence of larval morphology between *Drosophila sechellia* and its sibling species caused by cis-regulatory evolution of *ovo/shaven-baby*. *Proceedings of the National Academy of Sciences of the United States of America*, 97, 4530–4534.

Sundström, J., Carlsbecker, A., Svensson, M. E., Svenson, M., Johanson, U., Theißen, G. and Engström, P. (1999) MADS-box genes active in developing pollen cones of Norway spruce (*Picea abies*) are homologous to the B-class floral homeotic genes in angiosperms. *Developmental Genetics*, 25, 253–266.

Svensson, M. E., Johannesson, H. and Engstrom, P. (2000) The *LAMB1* gene from the clubmoss, *Lycopodium annotinum*, is a divergent MADS-box gene, expressed specifically in sporogenic structures. *Gene*, 253, 31–43.

Theißen, G., Becker, A., Di Rosa, A., Kanno, A., Kim, J. T., Munster, T., Winter, K.-U. and Saedler, H. (2000) A short history of MADS-box genes in plants. *Plant Molecular Biology*, 42, 115–149.

Vlot, E. C., Van Houten, W. H. J., Mauthe, S. and Bachmann, K. (1992) Genetic and nongenetic factors influencing deviations from five pappus parts in a hybrid between *Microseris douglasii* and *M. bigelovii* (Asteraceae, Lactuceae). *International Journal of Plant Science*, 153, 89–97.

Wang, R.-L., Stec, A., Hey, J., Lukens, L. and Doebley, J. (1999) The limits of selection during maize domestication. *Nature*, 398, 236–239.

Weigel, D. and Nilsson, O. (1995) Developmental switch sufficient for flower initiation in diverse plants. *Nature*, 377, 495–500.

Wen, C. K., Smith, R. and Banks, J. A. (1999) *ANI1*: a sex pheromone-induced gene in *Ceratopteris* gametophytes and its possible role in sex determination. *Plant Cell*, 11, 1307–1317.

Wheeler, D. A., Kyriacou, C. P., Greenacre, M. L., Yu, Q., Rutila, J. E., Rosbash, M. and Hall, J. C. (1991) Molecular transfer of a species-specific behavior from *Drosophila simulans* to *Drosophila melanogaster*. *Science*, 251, 1082–1085.

White, S. E. and Doebley, J. F. (1999) The molecular evolution of *terminal ear1*, a regulatory gene in the genus *Zea*. *Genetics*, 153, 1455–1462.

Yoshizumi, T., Nagata, N., Shimada, H. and Matsui, M. (1999) An arabidopsis cell cycle-dependent kinase-related gene, *CDC2b*, plays a role in regulating seedling growth in darkness. *Plant Cell*, 11, 1883–1895.

Appendix I
Glossary

abaxial: pertaining to the side oriented away from the axis (stem) in lateral organs; in a plant with standard upright stem orientation this normally corresponds to the lower side of **dorsiventral** lateral organs such as leaves and **petals** (cf. **adaxial**).

ABC model: an influential model of gene action in flower development whereby organs are determined as a result of three overlapping expression domains: C-class genes alone specify carpels, stamens are specified by B- and C-class genes acting together, petals are determined by A- and B-class genes acting together, and sepals are determined by A-class genes alone. ABC genes are **MADS-box** genes.

Ac/Ds: the autonomous transposon Ac encodes a transposase protein; an active form of this element is required to mobilise the derivative element Ds (see also Spm/dSpm).

acropetal: toward the apex (of stem or other organ) (cf. **basipetal**).

actinomorphy: radial symmetry, most typically of a flower (cf. **zygomorphy**).

adaxial: pertaining to the side oriented toward the axis (stem) in lateral organs; in a plant with standard upright stem orientation this normally corresponds to the upper side of **dorsiventral** lateral organs such as leaves and petals.

adventitious: organs (typically stems or roots) produced in non-standard positions; for example, buds generated from stems or leaf margins rather than leaf axils, or roots generated from stems rather than the radicle.

AFLP (Amplified Fragment Length Polymorphism): a PCR-based technique incorporating ligation and restriction steps, capable of generating from nanogram quantities of DNA a great diversity of DNA fragments, many of which will be polymorphic in populations and are therefore useful for genetic mapping.

aleurone: epidermal layer of the triploid endosperm in grasses.

anlage: a prospective structure or initial; for example, the plastochron zero (P0) primordium in the shoot apical meristem (SAM).

anthophyte (euanthial) theory: hypothesis of the evolution of the angiosperm bisexual flower from a bisexual strobilus of an ancestral gymnosperm, first formulated by Arber and Parkin in 1908 (cf. **pseudanthial theory**).

apomorphy: derived ('descendant') state of a character that shows one or more relatively primitive ('ancestral') states in other comparable taxa (cf. **plesiomorphy**).

asymmetric: flower that wholly lacks planes of symmetry (e.g. *Canna*); confused by some authors with **zygomorphic** (monosymmetric) flowers that show a **dorsiventral** axis of asymmetry, but have a plane of left–right mirror symmetry.

auxospore: a special diatom cell that expands in a controlled way before producing new silica cell walls; usually develops from a zygote after sexual reproduction.

bacterial artificial chromosome (BAC): a long piece of foreign DNA inserted into a bacterium in order to maintain large clones.

basipetal: away from the apex (of stem or other organ) (cf. **acropetal**).

bifacial: possessing an **abaxial** and an **adaxial** surface with contrasting properties.

blastozone: region of cell division in the leaf.

candidate gene: hypothesised to be the **primary gene** responsible for a trait in a particular plant, based on knowledge of its homologue in a better-known 'model' organism.

capitulum: a tight head of sessile flowers, most commonly found in Asteraceae.

carboxy terminus: the end of a protein with the free carboxyl group.

cis-regulation: regulation of a gene by control regions at the same locus (such as the gene's own promoter) (cf. **trans-regulation**).

colinearity: having the same linear order of genes along the chromosome.

competence: acquired ability to enter a particular developmental pathway.

corymb: a flat-topped racemose flower, usually resulting from greater growth of the pedicels of the lower flowers, bringing them to the same horizontal plane as the upper flowers.

C-value paradox: peculiar feature of plant genomes that their sizes range over many orders of magnitude, from *c.* 100 Mb to over 10,000 Mb.

cyme: a determinate compound inflorescence, each axis of which terminates in a flower (cf. **raceme**).

determinate: with a definite, genetically programmed, ultimate size and shape (cf. **indeterminate**).

determinate growth: growth which ends because all stem-cells have differentiated as a result of the completion of growth (the determinate cessation of an organism's developmental history when it reaches its adult form is characteristic of higher animals as opposed to **indeterminate** higher plants, although individual plant organs often show determinate growth and some plants, such as the fossil rhizomorphic lycopsids, exhibit unusually high degrees of determinacy).

dichasial: of a **cyme** producing two new axes at each branching (cf. **monochasial**).

dichotomous saltation: saltation driven by mutation of at least one gene within a single ancestral lineage to generate a novel daughter lineage.

didynamous: paired; in asterid flowers this refers to the possession of four stamens in two pairs, two long and two short.

discoid: pertaining to, or possessing, a disk: an aggregation of tubular florets in Asteraceae.

distal: away from the axis.

distichous: positioned in two opposing vertical rows; for example, the florets of some grasses on the inflorescence axis (cf. **monostichous**).

DNA microarray: large number of spots of minute quantities of DNA (usually, cDNAs of known identity from a pre-existing library) which may be hybridised with labelled cDNAs from a particular tissue; the spots that hybridise thus giving a 'snapshot' of patterns of gene expression in that tissue.

dorsal: see **dorsiventral**.

dorsiventral (dorsal, ventral): commonly (but grammatically incorrectly) spelled 'dorsoventral', this initially zoological term refers to the plane of asymmetry bisecting the 'back' (dorsum) and the 'stomach' (venter), but was co-opted for the **adaxial/abaxial** plane in plant organs; the venter ('stomach') was originally viewed as the side closest to the axis (adaxial), but the term has become increasingly commonly used in the opposite sense; thus, the specifically botanical terms adaxial and abaxial are clearer and preferable.

down-regulation: decrease in the number and/or rate of production of a gene transcript (or other specific feature of an organism) (cf. **up-regulation**).

ectopic (= exotopic): outside the usual position; of a developmental process such as organogenesis, histogenesis or gene expression occurring on an unusual organ or in an unusual position (caution should be exercised in using this term when the apparent ectopic phenomenon is a result of partial homeotic transformation of the location concerned, in which case the ectopic expression is in fact an endotopic indication of **homeosis**).

EMS (ethylmethane sulfonate): a mutagen frequently used for disrupting DNA sequences to produce **loss-of-function** mutations.

enation: non-vascularised outgrowth of superficial tissues, usually of a stem.

endodermal sheath: a layer of cells encircling and forming a relatively impermeable barrier around a **stele**.

ensiform: sword-shaped and vertically not horizontally flat, as in *Iris*: a characteristic form of **unifacial** leaf that is **equitant** (by over-development of the abaxial side).

epi-allele: allele resulting from an **epimutation** or **paramutation**.

epigenetic silencing: down-regulation of gene activity, typically involving methylation (**transposons** are frequently silenced by methylation of the transposon ends; the presence of a transposon in the promoter may result in novel 'epi-alleles' that differ in expression levels).

epigenetics: alteration of gene expression by means other than DNA mutations in that gene (epigenetic processes can greatly increase the range of allelic variants derived from a single DNA sequence).

epimutation: heritable change in gene expression but not gene sequence; this usually takes place by abnormally high methylation of a gene, producing loss of function, which may then be heritable for many generations (though its function can be reset by meiosis).

epiphyll: see **hyperphyll**.

equitant: of a once-folded structure with the median axis uppermost (with respect to the axis, i.e. abaxial), often overlapping other similar structures (derived from the Latin for saddle); see **ensiform**.

eudicots: the largest extant clade of angiosperms, characterised by three symmetrically placed pollen apertures (colpi) or aperture arrangements derived from this condition.

euphyll (= megaphyll): leaf homologous to a stem system, as in all angiosperm leaves.

eustele: stele consisting of a ring of discrete vascular bundles surrounding a central pith; a non-homologous feature of horsetails, some gymnosperms and most dicot angiosperms.

exaptation: a feature evolved to fulfil a particular function but then subsequently co-opted to fulfil another distinct function (see **preadaptation**).

gain-of-function: a mutation leading to the acquisition of function, often extended spatial or temporal gene expression, and therefore dominant (cf. **loss-of-function**).

genetic recall: reversal to a plesiomorphic character state due to loss of supression, where apomorphic characteristics are due to inactivation or possibly reactivation of a gene or genes.

gynostemium: structure formed by partial or complete fusion of the androecium and gynoecium; for example, the column of the orchid flower.

heterochrony: phyletic change in the timing or duration of a developmental process (cf. **heterotopy**).

heterodimer: protein in active form consisting of two joined protein products of different loci (cf. **homodimer**).

heterotopy: phyletic change in the topological position of a developmental process, such as the origin of an organ at a new location on the organism and/or in a different cell layer (cf. **heterochrony**).

homeobox: distinctive nucleotide sequence first identified in fruit-flies but subsequently shown to be present in many key developmental genes across the biotic kingdoms, encoding the **homeodomain**.

homeodomain: a conserved DNA-binding region of a protein that is encoded by a homeobox region of a gene (this is characteristic of an important set of transcription factors).

homeoheterotopy: at least partial phyletic replacement of a pre-existing structure by a strongly contrasting structure (complete replacement of the structure constitutes **homeosis** *s.s.*) (cf. **neoheterotopy**).

homeosis: change of developmental fate altering one organ wholly into another organ.

homodimer: active form of protein consisting of two joined protein products generated from the same locus (cf. **heterodimer**).

hopeful monster: an individual showing a profound phenotypic change from its parent(s) that demonstrably reflects a genetic modification.

hyperphyll (= epiphyll): upper part of leaf from which the petiole and blade are developed (cf. **hypophyll**).

hypophyll: part of leaf adjacent to the axis that eventually forms the leaf scar and from which the stipule develops (cf. **hyperphyll**).

indeterminate: without a definite final size or shape; growth is terminated by environmental factors rather than by an internal genetic programme (cf. **determinate**).

indeterminate growth: continuation of an organism's developmental history when it reaches its adult form (at no point do all stem-cells differentiate); such growth is characteristic of higher plants and some plant organs, as opposed to **determinate** higher animals.

initial cell: the first walled cell produced after expansion of the **auxospore** in a diatom.

key innovation: often viewed as any putatively functional character that appears to stimulate diversification, but better defined as a **synapomorphy** acquired by a lineage immediately prior to an **evolutionary radiation** that is demonstrably a much greater stimulus to that radiation than any other synapomorphies acquired on the same phylogenetic branch.

lamina (= leaf blade): main, distal part of leaf, either **unifacial** or bifacial (in a **petiolate** leaf, 'lamina' refers to the distal blade, either simple or compound; in a non-petiolate leaf it refers to the entire leaf except the **sheath**).

latent homology: see **underlying synapomorphy**.

ligule: a flap of tissue; in grasses a projection from the top of the leaf sheath, in derived lycopods such as *Isoetes* internal to the leaf base and non-homologous with that of grasses.

linkage analysis: procedure for genetic mapping, important for gene-finding in organisms that share **colinearity** with previously mapped model organism genomes.

loss-of-function: a mutation leading to a reduction in gene function, often reduced or absent spatial or temporal gene expression, therefore recessive (cf. **gain-of-function**).

MADS-box: a conserved DNA region (encoding a MADS domain) characteristic of a class of plant transcription factor encoding genes with (in plants) other recognisable (I, K and C) motifs.

megaphyll: a leaf phyletically derived from a reduced stem system, characteristic of the **euphyll** clade (cf. **microphyll**).

microphyll: a leaf not phyletically derived from a reduced stem system but more likely from an **enation** (cf. **megaphyll**).

Miniature Inverted Repeat Transposons (MITEs): a family of DNA-based transposons that are often found in positional association with genes.

modifier: a gene altering the expression of another gene at another locus.

modulation: alteration of gene expression by changes in other genes forming an interacting network (the means by which many developmental states and organismal morphologies can result from similar gene complements).

monochasial: of a **cyme** producing a single new axis at each branching (cf. **dichasial**).

monostichous: positioned in a single vertical row, as in the florets of some grasses (cf. **distichous**).

monosymmetry: term for standard **zygomorphy** formulated by Endress in 1999 for flowers with a single bilateral plane of symmetry.

neoheterotopy: a phyletic shift of a structure to a new location on an organism not previously occupied by a distinct organ (cf. **homeoheterotopy**).

ontogeny: development of an organism.

orthologous: two or more genes owing their sequence similarity to gradual divergence in different taxa (cf. **paralogous**).

pappus: a group of modified sepals often taking the form of a ring of silky hairs, e.g. in Asteraceae.

paralogous: two or more genes owing their sequence similarity to a gene duplication event (cf. **orthologous**).

paramutation: allele whose expression is altered (sometimes by methylation) as a result of the influence of its homologous allele.

parasaltation: a genetic modification that is expressed as a profound phenotypic change across two to several generations and results in a potentially independent evolutionary lineage.

pericarp: fruit wall, such as the outer covering of the grain in grasses.

perizonium: wall of the diatom **auxospore** consisting of lightly silicified hoops or scales that can variously constrain auxospore expansion.

petals: stamen-derived or bract-derived organs between the **sepals** and stamens, often serving to attract pollinating animals.

petiole (= stiel): leaf stalk linking **lamina** and **sheath**; narrow and sometimes secondarily thickened, **unifacial** or **bifacial**.

phenocopy (= ecophenotypy): a phenotypic modification of an organism that mimics genotypically based differences between organisms (for example, when induced to become prostrate by growing in windy conditions, upright broom (*Cytisus scoparius*) phenocopies broom plants of var. *prostratus* that are genotypically prostrate).

phylogenetic prepattern: an ancestral pattern of gene expression or gene function that predisposes the evolution of particular traits in some but not all of the descendants; such a character may constitute an **underlying synapomorphy** (= apomorphic tendency).

phylogeny: pattern of evolutionary history of a set of biological entities, most commonly species.

planation: compression into a single plane.

plesiomorphy: primitive ('ancestral') state of a character that shows one or more relatively derived ('descendant') states in other comparable taxa (cf. **apomorphy**).

pollination syndrome: a collection of plant features that together are characteristic of a particular mode of pollination.

pollinium: a cohesive mass containing all the pollen grains found in a single anther locule.

polyternate: of compound leaves which develop by repeated division of lobes into three.

preadaptation: a pre-existing feature of an organism that increases the probability of further (typically adaptive) changes to that feature (see **exaptation**).

prepattern: pattern of gene expression during ontogeny that predisposes tissues to particular developmental fates (see **phylogenetic prepattern**).

primary gene: a gene whose inherited modifications (perhaps changes in **cis-regulation**) result in alterations to developmental pathways and is held primarily responsible for a specific case of morphological diversification (cf. **secondary gene**).

procambial strand: an axial row of undifferentiated but pre-determined cells from which the vascular tissue eventually forms.

process morphology: morphological thinking that places emphasis on developmental processes and holds that morphological character states are not discrete entities but points on a developmental continuum.

prospecies: a putatively recently evolved lineage possessing the essential intrinsic properties of a taxonomic species but yet to achieve acceptable levels of abundance and especially of longevity.

protoxylem poles: locations of the first-formed protoxylem elements of a vascular bundle.

proximal: towards the axis.

pseudanthial theory: hypothesis of the evolution of the angiosperm flower from a combination of unisexual strobili of a gymnospermous ancestor, first formulated by Wettstein in 1907 (cf. **anthophyte theory**).

QTL (= quantitative trait locus): a locus to which variation in a quantitative character maps, based on co-segregation of the quantitative character with markers, indicating the presence at, or near, that locus of a gene that makes a significant contribution to that (typically multigenic) character.

raceme: an **indeterminate** inflorescence consisting of a stem with lateral flowers, the oldest flowers being at the base and younger flowers being near the growing tip, as in *Antirrhinum* (cf. **cyme**).

radiate: bearing ray florets in Asteraceae.

radiation (evolutionary): often used for a generally perceived increase in biodiversity, but better defined more precisely as a large surplus in the rate of natality over the rate of mortality for species and/or character states within a specified clade over a specified time interval.

redundancy: strictly, the state describing genes whose loss has no fitness consequence, but usually applied to genes that are very similar in function to other genes, resulting in a single gene-knockout phenotype that is at best difficult to observe.

reticulate saltation: saltation driven by allopolyploidy and thus blending the entire genomes of two ancestral lineages.

retrotransposon: a class of **transposon** that replicates and transposes by means of reverse transcription of RNA and the insertion of the resultant DNA into the genome; a large part of the genome of some plants is composed of retrotransposon-derived DNA.

rhizophore: a leafless branch in the clubmoss *Selaginella* performing functions more characteristically associated with roots.

rimoportula: a tube through the silica wall of a diatom frustule, usually opening internally by a slit and externally by a simple aperture.

saltation: a genetic modification that is expressed as a profound phenotypic change across a single generation and results in a potentially independent evolutionary lineage.

secondary gene: a gene that shows expression patterns correlated with morphological differences between organisms, and may show differences in sequence between those organisms, but this has not given rise directly to the morphological variation observed, as the significant differences in expression are **trans-regulated** (secondary genes are often the downstream target of a **primary gene**).

sepals: modified bracts protecting the flower early in development, the outermost whorl of the **eudicot** flower, specified by A function but not B or C function in the **ABC model**.

serrature: serrate (saw-like) toothing of a leaf margin.

sheath: lowermost (distal) part of leaf, ensheathing stem; always bifacial, derived from the proximal zone of primordium.

sister group: cladistic term for the clade or species most closely related to another specified clade or species.

solenostele: a vascular cylinder with widely separated leaf gaps and phloem to both the interior and exterior.

spikelet: the basic unit of the grass inflorescence comprising two bracts (glumes) subtending one or more florets.

Spm/dSpm: the autonomous transposon Spm encodes a transposase protein; an active form of this element is required to mobilise the derivative element dSpm (see also **Ac/Ds**).

sporangium: spore-producing organ.

stele: axial cylinder of tissue within which the vascular tissue develops.

stigma(ta): in diatoms, one (or more) pores that differ structurally from the pores of the striae, usually situated near the central area, between the proximal ends of the central striae and the central raphe endings.

stipule: leaf sheath appendage, often paired and sometimes leafy and vascularised.

taxic homology: synapomorphic feature characterising a monophyletic group.

taxonomic species: one or more (typically many more) populations separated from all other comparable populations by phenotypic, and putative genotypic, discontinuities that may reflect one or more isolating mechanisms operating over a considerable period of time.

TCP-box: a conserved DNA sequence motif characteristic of a gene family (TCP-genes) that control a number of growth processes (often by inhibiting cell division in axillary meristematic regions); it is named from the three well-characterised genes which possess the motif (*TEOSINTE BRANCHED1* (maize), *CYCLOIDEA* (snapdragon) and *PROLIFERATING CELL FACTOR1* (rice)

telome: a fundamental super-cellular building block; an intermediate category between cell and fully-fledged appendage or organ system, comprising a vascular cylinder, a parenchymous cortex and an epidermis.

tepals: modified leaves forming a perianth around the stamens and carpels in the angiosperm flower, but not differentiated into a **sepal** whorl and a **petal** whorl as in **eudicots** (characteristic of many 'petaloid' monocots such as irises).

teratos (plural, terata): an individual showing a profound phenotypic change from its parent(s) irrespective of whether the underlying cause is genetic or ecophenotypic.

transcription factor: a protein that is involved in initiating transcription of a gene.

transformational homology: fundamental evolutionary similarity of otherwise contrasting features.

transposase: a protein encoded by a **transposon** and causing transposition of a DNA sequence to a new location in the genome.

transposon (= transposable element): a DNA sequence capable of insertion at non-homologous regions of the chromosome (transposition).

trans-regulation: regulation of a gene by genes or control regions at a different locus (such as upstream regulatory genes); cf. **cis-regulation**.

trichome: plant hair; outgrowth consisting of one or more cells of epidermal origin, sometimes branched and/or secretory.

underlying synapomorphy: a cryptic property of a whole clade which promotes the evolution of a particular apomorphy in some but not all of the members of the clade (apomorphic tendency).

unifacial: with a single morphological surface, usually homologous with the **abaxial** surface.

up-regulation: increase in the number and/or rate of production of a gene transcript (or other specific feature of an organism) (cf. **down-regulation**).

ventral: see dorsiventral.

zygomorphy: bilateral symmetry (= **monosymmetry** of Endress), most typically applied to flowers.

Appendix 2
Genetic nomenclature

Although different genetic nomenclatures have been developed for different model organisms there are some shared features. The names of genes, loci and mutants are generally italicised, but the names of proteins are not. Genetic nomenclature used in this volume generally follows the *Arabidopsis* usage, but where necessary may conform to the standards for individual organisms, and so is not uniform throughout. Nomenclature for *Arabidopsis* is followed for organisms for which there is no standard nomenclature. Standard nomenclatures for the model organisms *Arabidopsis* and Maize are tabulated below (further standardisation across taxa is desirable).

Chimeric promoter-gene constructs, in which the promoter of one gene is exchanged for another, are indicated as follows: *KNAT1::GUS* signifies a *KNAT1* promoter driving transcription of a *β-glucuronidase* (*GUS*) reporter gene. *35S::PHYB-GFP* represents the cauliflower mosaic virus (CaMV) 35S RNA promoter (35S) driving transcription of a phytochrome B (*PHYB*) and green fluorescent protein (*GFP*) fusion construct (allowing the PHYB protein to be localised by fluorescence imaging of the fusion protein).

	Arabidopsis	Maize
mutants (and abbreviations) genes or loci	italics lower case, e.g. *terminal flower* (*tfl*) italicised and capitalised, e.g. *TERMINAL FLOWER1*	italics lower case, e.g. *defective kernel* (*dek*) italicised but not capitalised, e.g. *defective kernel12*. When used as a noun the first letter capitalised, e.g. "the mutation in *Defective kernel12*"
abbreviations of genes or loci	most recently-described genes given three-letter abbreviations, e.g. *TFL1*	most recently-described genes given three-letter abbreviations, e.g. *dek12*
additional loci	if more than one locus bears the same gene name the loci are numbered sequentially, e.g. *TFL1*, *TFL2*	historically the first locus to receive that name would not have a numerical suffix but additional loci were numbered sequentially. Thus, we have *shrunken* (*sh*), *shrunken2* (*sh2*), and *shrunken4* (*sh4*). Maize geneticists are now encouraged to designate a new locus as 1 and to refer to, e.g. *shrunken1* or *sh1*
mutant alleles	different mutant alleles numbered sequentially after a hyphen, e.g. *tfl1* mutant alleles would be *tfl1-1* and *tfl1-2*; if there is only one allele at a locus that allele is not given a number and is referred to only *tfl1*	different mutant alleles numbered sequentially after a hyphen; the mutation by which the gene was identified is the reference allele, e.g. *bz1-Ref* or *bz1-R*, but also sometimes referred to as *bz1-1*
dominant alleles/wild type alleles	dominance not indicated in the name wild type + or wt	dominance indicated by capitalising the first letter of the name, e.g. *Knotted1*
proteins	capitalised but not italicised, e.g. TFL1	capitalised but not italicised, e.g. KN1

Copyright acknowledgements

Figure 5.2 from Kellogg, E. A. (2000) The grasses: a case study in macroevolution, *Annual Review of Ecology and Systematics*, 31, 217–238 with permission. © Annual Reviews www.AnnualReviews.org

Figure 5.3 from Soreng, R. J. (1990) Chloroplast–DNA phylogenetics and biogeography in a reticulating group: study in *Poa* (Poaceae), *American Journal of Botany*, 77, 1383–1400 with permission.

Figures 6.1 and 6.2 from Frohlich, M. W. and Parker, D. S. (2000) The mostly male theory of flower evolutionary origins: from genes to fossils, *Systematic Biology*, 25, 155–170 © Taylor & Francis.

Figure 7.1 from Bateman, R. M. and DiMichele, W. A. (1994) Heterospory: the most iterative key innovation in the evolutionary history of the plant kingdom, *Biological Reviews* 69, 345–417 with permission of Cambridge University Press.

Figure 12.1a from Coen, E. S. and Nugent, J. M. (1994) Evolution of flowers and inflorescences, *Development Supplement*, S107–S116 with permission of The Company of Biologists Ltd.

Figure 12.1b from Trow, A. H. (1912) On the inheritance of certain characters in the common groundsel, *Senecio* vulgaris, and its segregates. *Journal of Genetics*, 2, 239–276 with permission of Cambridge University Press.

Figure 13.1 from Weberling, F. (1989) *Morphology of Flowers and Inflorescenses*, p. 10, with permission of Eugen Ulmer Gmbh & Co, Stuttgart.

Figure 13.2a from Soltis, P., Soltis, D. and Chase, M. (1999) Angiosperm phylogeny inferred from multiple genes as a tool for comparative biology, *Nature*, 402, 402–404 with permission of *Nature*. © Macmillan Magazines Limited.

Figure 13.2b from Endress, P. (1997) *Antirrhinum* and Asteridae – evolutionary changes of floral symmetry, *Symposia for the Society for Experimental Biology*, 51, 133–140 with permission of the Society for Experimental Biology.

Figure 13.3 (part of) from Luo, D., Carpenter, R., Copsey, L., Vincent, C., Clark, J. and Coen, E. (1999) Control of organ asymmetry in flowers of *Antirrhinum*, *Cell*, 99, 367–376 with permission of Elsevier Science.

Figure 13.3 (part of) and Figure 13.4 from Coen, E. and Nugent, J. (1994) Evolution of flowers and inflorescences, *Development Suppl.*, 107–116, 113 with permission of The Company of Biologists Ltd.

Figure 13.5 from Doebley, J. (1992) *Trends in Genetics*, vol. 8, pp. 302–307 with permission of Elsevier Science.

Figures 18.1 and 18.4 reproduced from Zimmerman, W. (1959) *Die Phylogenie der Pflanzen*, Gustav Fischer, Stuttgart, 777pp. with permission of Urban & Fischer Verlag GmbH & Co. KG, München.

Figure 18.3 from Kenrick, P. and Crane, P. R. (1997) *The Origin and Early Diversification of Land Plants: A Cladistic Study. Smithsonian Series in Comparative Evolutionary Biology.* Smithsonian Institution Press, Washington, 441pp., with permission.

Figures 18.5, 18.6 and 18.7 from Crane, P. R. and Kenrick, P. (1997) Diverted development of reproductive organs: a source of morphological innovation in land plants, *Plant Systematics and Evolution*, 206, 161–174. © Springer-Verlag GmbH & Co.

Figure 22.3b from Hareven, D., Gutfinger, T., Parnis, A., Eshed, Y. and Lifschitz, E. (1996) The making of the compound leaf: genetic manipulation of leaf architecture in tomato, *Cell*, 84, 735–744 with permission of Elsevier Science.

Figure 23.7 from Roth, I. (1949) Zur Entwicklungsgeschichte des Blattes, mit besonderer Berucksichigung von Stipular- und Ligularbildungen, *Planta*, 37, 299–336. © Springer-Verlag GmbH & Co.

Taxon index

Subject index

A genes 189, 208, 212–213, 220–226, 508, 514
Abaxial 4, 209, 214, 227, 249, 270–271, 275–86,
  290–291, 341, 343, 349, 399, 411, 413, 444,
  446, 448, 508
ABC Model 111, 127, 144, 179–181, 193–194,
  206–208, 212, 216, 220–222, 226, 229–230,
  508
ABERRANT LEAF AND FLOWER (ALF), *see*
  LEAFY
Abortion of organs 124–126, 136, 209, 222, 227,
  256, 260, 497
Accessory factors 1, 183
ACCTRAN 133, 333
Acropetal 130, 135, 406, 411, 414, 508
Acrotonic elongation 411
Actin gene family 331, 354, 482, 484
Actinomorphic flowers 1, 2, 3, 37, 39–40, 45, 128,
  131, 234, 23–28, 240, 249, 257, 268–273,
  290
Activator (Ac) 16–19, 23–24, 508
Acyanic flowers 164
Adaxial 4, 163, 166, 249, 270–291, 341, 343,
  349, 379, 395, 399–400, 406, 408, 411, 413,
  431, 435, 442–454, 497, 508
Adventitious 408, 508
AGAMOUS (AG) 80, 88, 101, 179–188,
  194–195, 212, 214, 220–229
Agamous-like genes 180–181, 184, 186–187,
  194–195, 208, 212–216, 220, 223–226, 229
AINTEGUMENTA (ANT) 406
Aleurone 16, 508
Allele 2, 10–11, 15–17, 22–29, 110–111,
  116–117, 141, 145, 147, 167, 236–237, 259,
  408, 421, 493
Allogamy 132, 142
Allopatry 113, 116, 147
Allopolyploidy 112, 114, 116, 236, 514
Alternation of generations 3, 28, 372
Amplified Fragment Length Polymorphism (AFLP)
  324–6, 508
Anagenesis 120
Anartiomorphic flowers 128
Anastomosing venation 342
Ancestral states and functions 52, 55, 58–59, 75,
  80, 86–87, 101, 103, 109, 112, 114, 116,
  130, 132, 134, 138, 173, 188–189, 190, 192,

  196–197, 247–249, 251, 253, 255, 257, 262,
  287, 290–299, 316, 320, 345, 352, 370,
  376–377, 392, 399, 432–433, 452, 494–502,
  508–9, 513–514
Androecium 3, 267, 273–274, 277–278, 280,
  283–284, 292, 511
Angiosperm origins, Darwin's abominable mystery
  85, 160, 173–174, 176–177, 181, 189
angustifolia (an) 413
Animals 6, 9–10, 28, 65–66, 75, 98, 109, 120,
  128, 135–136, 138, 144, 160–161, 170, 174,
  178–179, 183, 196, 198, 212, 248, 260, 287,
  315–317, 320, 322, 331, 354, 405, 462, 484
Anisophylly 65
ANITA clade 176
Anlage 406, 409, 412, 508
Antennae 129, 165
Anterior-posterior axis 178, 367, 484
Anther 73, 76, 80, 169, 209–211, 222, 226–227,
  270–272, 278, 280–287, 291, 513
Antheridium 391
Anthesis 62, 214, 270–271, 283–284, 301
Anthocyanin 16–17, 46, 53, 73, 162, 164–165
Anthophyte Theory 85, 93, 126, 175–176, 199,
  333, 508
APETALA1 (AP1) 88, 178, 180–182, 185, 189,
  195, 223, 225
APETALA2 (AP2) 179, 180–182, 189
APETALA3 (AP3) 88, 180–182, 184–185,
  190–192, 225, 497
Apocarpous 53, 176
Apomorphy (*see also* synapomorphy) 75, 100,
  508, 511, 513, 515
Aptian 86
Aquatic 142–143, 332, 336, 436, 439, 459, 479
Arber, A. 38, 342, 440, 446–447, 508
Archegonium 376
Areolae 470
Aril 53, 61
ARP (AS, RS, PHAN) genes 391–392
Ascidiate 176
ASYMMETRIC LEAVES1 (AS1) 391–392, 408,
  413, 419–422
Asymmetry
  diatoms 460, 471–2, 475, 478, 480, 483–484

Systematics Association Publications

1. Bibliography of key works for the identification of the British fauna and flora, 3rd edition (1967)[†]
Edited by G. J. Kerrich, R. D. Meikie and N. Tebble
2. Function and taxonomic importance (1959)[†]
Edited by A. J. Cain
3. The species concept in palaeontology (1956)[†]
Edited by P. C. Sylvester-Bradley
4. Taxonomy and geography (1962)[†]
Edited by D. Nichols
5. Speciation in the sea (1963)[†]
Edited by J. P. Harding and N. Tebble
6. Phenetic and phylogenetic classification (1964)[†]
Edited by V. H. Heywood and J. McNeill
7. Aspects of Tethyan biogeography (1967)[†]
Edited by C. G. Adams and D. V. Ager
8. The soil ecosystem (1969)[†]
Edited by H. Sheals
9. Organisms and continents through time (1973)[†]
Edited by N. F. Hughes
10. Cladistics: a practical course in systematics (1992)[*]
P. L. Forey, C. J. Humphries, I. J. Kitching, R. W. Scotland, D. J. Siebert and D. M. Williams
11. Cladistics: the theory and practice of parsimony analysis (2nd edition) (1998)[*]
I. J. Kitching, P. L. Forey, C. J. Humphries and D. M. Williams

[*]Published by Oxford University Press for the Systematics Association
[†]Published by the Association (out of print)

Systematics Association Special Volumes

1. The new systematics (1940)
Edited by J. S. Huxley (reprinted 1971)
2. Chemotaxonomy and serotaxonomy (1968)[*]
Edited by J. C. Hawkes
3. Data processing in biology and geology (1971)[*]
Edited by J. L. Cutbill
4. Scanning electron microscopy (1971)[*]
Edited by V. H. Heywood
5. Taxonomy and ecology (1973)[*]
Edited by V. H. Heywood
6. The changing flora and fauna of Britain (1974)[*]
Edited by D. L. Hawksworth
7. Biological identification with computers (1975)[*]
Edited by R. J. Pankhurst
8. Lichenology: progress and problems (1976)[*]
Edited by D. H. Brown, D. L. Hawksworth and R. H. Bailey
9. Key works to the fauna and flora of the British Isles and northwestern Europe, 4th edition (1978)[*]
Edited by G. J. Kerrich, D. L. Hawksworth and R. W. Sims

10. Modern approaches to the taxonomy of red and brown algae (1978)
Edited by D. E. G. Irvine and J. H. Price
11. Biology and systematics of colonial organisms (1979)*
Edited by C. Larwood and B. R. Rosen
12. The origin of major invertebrate groups (1979)*
Edited by M. R. House
13. Advances in bryozoology (1979)*
Edited by G. P. Larwood and M. B. Abbott
14. Bryophyte systematics (1979)*
Edited by G. C. S. Clarke and J. G. Duckett
15. The terrestrial environment and the origin of land vertebrates (1980)
Edited by A. L. Pachen
16. Chemosystematics: principles and practice (1980)*
Edited by F. A. Bisby, J. G. Vaughan and C. A. Wright
17. The shore environment: methods and ecosystems (2 volumes) (1980)*
Edited by J. H. Price, D. E. C. Irvine and W. F. Farnham
18. The Ammonoidea (1981)*
Edited by M. R. House and J. R. Senior
19. Biosystematics of social insects (1981)*
Edited by P. E. House and J.-L. Clement
20. Genome evolution (1982)*
Edited by G. A. Dover and R. B. Flavell
21. Problems of phylogenetic reconstruction (1982)
Edited by K. A. Joysey and A. E. Friday
22. Concepts in nematode systematics (1983)*
Edited by A. R. Stone, H. M. Platt and L. F. Khalil
23. Evolution, time and space: the emergence of the biosphere (1983)*
Edited by R. W. Sims, J. H. Price and P. E. S. Whalley
24. Protein polymorphism: adaptive and taxonomic significance (1983)*
Edited by G. S. Oxford and D. Rollinson
25. Current concepts in plant taxonomy (1983)*
Edited by V. H. Heywood and D. M. Moore
26. Databases in systematics (1984)*
Edited by R. Allkin and F. A. Bisby
27. Systematics of the green algae (1984)*
Edited by D. E. G. Irvine and D. M. John
28. The origins and relationships of lower invertebrates (1985)‡
Edited by S. Conway Morris, J. D. George, R. Gibson and H. M. Platt
29. Infraspecific classification of wild and cultivated plants (1986)‡
Edited by B. T. Styles
30. Biomineralization in lower plants and animals (1986)‡
Edited by B. S. C. Leadbeater and R. Riding
31. Systematic and taxonomic approaches in palaeobotany (1986)‡
Edited by R. A. Spicer and B. A. Thomas
32. Coevolution and systematics (1986)‡
Edited by A. R. Stone and D. L. Hawksworth
33. Key works to the fauna and flora of the British Isles and northwestern Europe, 5th edition (1988)‡
Edited by R. W. Sims, P. Freeman and D. L. Hawksworth
34. Extinction and survival in the fossil record (1988)‡
Edited by G. P. Larwood

35. The phylogeny and classification of the tetrapods (2 volumes) (1988)‡
Edited by M. J. Benton
36. Prospects in systematics (1988)‡
Edited by J. L. Hawksworth
37. Biosystematics of haematophagous insects (1988)‡
Edited by M. W. Service
38. The chromophyte algae: problems and perspective (1989)‡
Edited by J. C. Green, B. S. C. Leadbeater and W. L. Diver
39. Electrophoretic studies on agricultural pests (1989)‡
Edited by H. D. Loxdale and J. den Hollander
40. Evolution, systematics, and fossil history of the Hamamelidae (2 volumes) (1989)‡
Edited by P. R. Crane and S. Blackmore
41. Scanning electron microscopy in taxonomy and functional morphology (1990)‡
Edited by D. Claugher
42. Major evolutionary radiations (1990)‡
Edited by P. D. Taylor and G. P. Larwood
43. Tropical lichens: their systematics, conservation and ecology (1991)‡
Edited by G. J. Galloway
44. Pollen and spores: patterns and diversification (1991)‡
Edited by S. Blackmore and S. H. Barnes
45. The biology of free-living heterotrophic flagellates (1991)‡
Edited by D. J. Patterson and J. Larsen
46. Plant–animal interactions in the marine benthos (1992)‡
Edited by D. M. John, S. J. Hawkins and J. H. Price
47. The Ammonoidea: environment, ecology and evolutionary change (1993)‡
Edited by M. R. House
48. Designs for a global plant species information system (1993)‡
Edited by F. A. Bisby, G. F. Russell and R. J. Pankhurst
49. Plant galls: organisms, interactions, populations (1994)‡
Edited by M. A. J. Williams
50. Systematics and conservation evaluation (1994)‡
Edited by P. L. Forey, C. J. Humphries and R. I. Vane-Wright
51. The haptophyte algae (1994)‡
Edited by J. C. Green and B. S. C. Leadbeater
52. Models in phylogeny reconstruction (1994)‡
Edited by R. Scotland, D. I. Siebert and D. M. Williams
53. The ecology of agricultural pests: biochemical approaches (1996)**
Edited by W. O. C. Symondson and J. E. Liddell
54. Species: the units of diversity (1997)**
Edited by M. F. Claridge, H. A. Dawah and M. R. Wilson
55. Arthropod relationships (1998)**
Edited by R. A. Fortey and R. H. Thomas
56. Evolutionary relationships among Protozoa (1998)**
Edited by G. H. Coombs, K. Vickerman, M. A. Sleigh and A. Warren
57. Molecular systematics and plant evolution (1999)
Edited by P. M. Hollingsworth, R. M. Bateman and R. J. Gornall
58. Homology and systematics (2000)
Edited by R. Scotland and R. T. Pennington
59. The flagellates: unity, diversity and evolution (2000)
Edited by B. S. C. Leadbeater and J. C. Green

*Published by Academic Press for the Systematics Association
†Published by the Palaeontological Association in conjunction with Systematics Association
‡Published by the Oxford University Press for the Systematics Association
**Published by Chapman & Hall for the Systematics Association

*For Product Safety Concerns and Information please contact
our EU representative GPSR@taylorandfrancis.com Taylor & Francis
Verlag GmbH, Kaufingerstraße 24, 80331 München, Germany*

T - #0131 - 160425 - C4 - 244/170/30 - PB - 9780415257916 - Gloss Lamination